MODERN
BUSINESS
STATISTICS

Freund and Williams'

MODERN
BUSINESS
STATISTICS

REVISED BY

Benjamin Perles
Bentley College

Charles Sullivan
Bentley College

PRENTICE-HALL, INC.
Englewood Cliffs, New Jersey

13-589580-4

Library of Congress Catalog Card Number: 77-88574

Printed in the United States of America

Current Printing (last digit):
20 19 18 17 16 15 14 13 12 11

Prentice-Hall International, Inc., *London*
Prentice-Hall of Australia, Pty. Ltd., *Sydney*
Prentice-Hall of Canada, Ltd., *Toronto*
Prentice-Hall of India Private Limited, *New Delhi*
Prentice-Hall of Japan, Inc., *Tokyo*

to Barbara
and
Mary

Preface

To keep pace with the ever-increasing demand of business for college-trained personnel there has been, over the past few years, a continuous and general modernizing of approaches to the various areas of Business Administration. Old courses have been revised; new courses and entirely new departments have been added. As new problems have arisen in various fields of business and new methods of attacking them have appeared, the work of the colleges has been brought forward accordingly. The necessary adjustments have been relatively slow in the field of statistics, and until a few years ago many statistics courses were in almost every respect the same as those taught twenty or thirty years earlier. However, in the past few years there has been evidence—the appearance and acceptance of recently produced "modern" textbooks—that the college version of business statistics is about to catch up with the "statistics of business."

We would not like to be misunderstood: we do not suggest that the traditional, the so-called descriptive, methods no longer have a place in modern business statistics. One glance at the table of contents of this book should dispel any doubts as to our attitude in this respect. Descriptive statistics is being used in today's business, it will probably always be used, and we would be careless, indeed, if we were to underestimate its importance. However, our idea, and we are sure that of many of our colleagues, is that there has been a poor balance between the more traditional topics of purely descriptive statistics and the newer topics of statistical inference. It is well known today that many problems never before thought amenable to quantitative analysis (for example, predicting responses to advertising campaigns, locating oil wells, or finding customers) can be attacked by the methods of inductive statistics, and it is hard to see how the logical foundation underlying the statistical solutions to such problems can be ignored in an up-to-date treatment of business statistics.

It is the opinion of the authors that the concepts of modern statistics definitely are not beyond the understanding of any reasonably diligent college student. Indeed, this book has been keyed to the lowest possible level of mathematical rigor at which it is felt that modern statistics can effectively be taught. The mathematical training assumed of the reader does not go beyond a course in college algebra or equivalent preparation.

vii

The order and emphasis of the material covered in this book follows the current trend in the teaching of business statistics. Although the first five chapters deal primarily with descriptive methods, some of the basic ideas of statistical inference are introduced at a very early stage. Having learned to distinguish between samples and populations and being aware of the fact that they require different methods of analysis, the reader is prepared, for example, for the subsequent material on the standard deviation. The early distinction between σ and s in Chapter 4 was decided upon by the authors after a careful survey of professional statisticians, among them many teachers of statistics.

Since few instructors teach introductory statistics in exactly the same fashion, the authors would like to point out that there is room for considerable flexibility in the use of this text. For instance, those who may want to devote more time to graphical presentation can take up Sections 17.9 and 17.10, dealing with the charting of time series, and Appendix I, dealing with pictorial presentation, immediately after Section 2.3. Those who may want to devote more time to inductive methods can take up Chapter 11, an optional and somewhat more advanced chapter on tests of hypotheses, and Appendix II, a brief treatment of Quality Control, immediately after Chapter 10. Furthermore, the index number chapters or the time series chapters can be taken up at any time before the material on statistical inference, although the time series chapters should profitably be preceded by the chapter on linear regression.

Most of the examples used in the exercises and illustrations are based on actual experiments, surveys, and other kinds of studies, although in some instances the numbers were simplified and the sample sizes reduced, to avoid calculations which would serve only to distract the reader. The authors are greatly indebted to their many friends and colleagues who so generously provided problem material. The authors would also like to express their appreciation to their colleagues and students, whose helpful suggestions contributed greatly to the final version of the manuscript.

The authors are greatly indebted to the editorial staff of Prentice-Hall, Inc., for their courteous co-operation in the production of the book; to Miss Marianne Byrd, Mrs. Alice Kegan, Miss Gladys Brackney, and Miss Betty Prater for typing parts of various drafts of the manuscript; and above all to their wives for their cheerful encouragement in spite of the demands made on their husbands' time by the writing of this book.

Finally, the authors would like to express their appreciation and indebtedness to Professor R. A. Fisher and Messrs. Oliver and Boyd, Ltd., Edinburg, for permission to reprint parts of Tables III and IV from their book *Statistical Methods for Research Workers;* to Professor E. S. Pearson and the Biometrika trustees for permission to reproduce the material in Tables IV and V; to the American Society for Testing Materials for permission to reproduce the material in Table VII.

John E. Freund
Frank J. Williams

Revisors' Preface

As quantitative methods have become increasingly important in the various areas of business administration and in the social sciences, it has been necessary for college courses (and textbooks) in statistics to adapt to the resulting complex range of requirements. Indeed, in recent years, courses in business statistics have taken many different directions. Although one can defend specialization in selected areas or preoccupation with logical and mathematical details, the revisors feel very strongly about the need to preserve the approach of *Modern Business Statistics*, namely, its emphasis on the instruction in the use and appreciation of meaningful and well-established statistical techniques.

The future businessman should have knowledge of, and be skilled in, the employment of basic statistical tools and techniques. He should be aware of the problems involved in the collection of data, techniques of sampling and testing, and methods of analysis and interpretation. It is important for him to be acquainted with index numbers and their construction, to be able to read tables, charts, and graphs with great facility, and to know how to estimate and predict many different types of series. For these reasons, we feel that a basic text in business statistics should be especially strong in the above-mentioned business areas. *Modern Business Statistics* is just such a well-balanced text, and, in our enthusiasm, we have prevailed upon Messrs. Freund and Williams to grant us permission to revise this work.

The continuously changing nature of business data necessitated the updating of this widely used text. We have tried to preserve the book without disturbing its character and style; we have replaced old problems with more suitable, timely ones; and we have updated index number construction, base periods, technological illustrations, charts, graphs, and tables. The mathematical symbolism and formulas remain unchanged from the original version, except for a minor change in the notation for conditional probability. Other changes which might be of special interest to users of the original version of this book include the addition of a table of random numbers, a new table of the chi-square distribution, and a new table of confidence intervals for proportions. The order of topics and the scope of the text remain unchanged.

We are most grateful to Messrs. Freund and Williams for allowing us to revise this book. Sincere thanks are extended to Frederick K. Easter, Jr., editor, of Prentice-Hall, Inc., for his cooperation and encouragement in the preparation of this manuscript. The revisors are deeply indebted to their wives, Barbara and Mary, whose devoted interest made the completion of this undertaking possible.

Finally, the revisors acknowledge their indebtedness to the American Society for Testing Materials for granting permission to reproduce the material in Table VII; to Professor R. A. Fisher and Messrs. Oliver and Boyd, Ltd., Edinburg, for permission to reprint parts of Table IV from the book by R. A. Fisher, *Statistical Methods for Research Workers;* to E. S. Pearson and the Biometrika trustees for permission to reproduce the materials in Tables III, IVa, IVb, and V in this book.

Benjamin M. Perles
Charles M. Sullivan

Contents

Chapter 1. INTRODUCTION   1

    1.1 Introduction, 1
    1.2 Descriptive and Inductive Statistics, 3
    1.3 Sources of Data, 6
    1.4 The Direct Collection of Data, 7
    1.5 Published Data, 8
    1.6 Mathematical Prerequisites, 12
    1.7 Subscripts and Summations, 13

Chapter 2. FREQUENCY DISTRIBUTIONS   17

    2.1 Introduction, 17
    2.2 The Construction of Numerical Distributions, 19
    2.3 Graphical Presentations, 32
    2.4 The Construction of Categorical Distributions, 38
    2.5 Tabular Presentations, 40

Chapter 3. MEASURES OF LOCATION   43

    3.1 Introduction, 43
    3.2 The Arithmetic Mean, Ungrouped Data, 46
    3.3 The Arithmetic Mean, Grouped Data, 51
    3.4 The Median, Ungrouped Data, 57
    3.5 The Median, Grouped Data, 60
    3.6 The Mode, 63
    3.7 The Geometric and Harmonic Means, 66
    3.8 The Weighted Mean, 69
    3.9 Quartiles, Deciles, and Percentiles, 72
    3.10 Further Comparisons, 75

Chapter 4. MEASURES OF VARIATION   77

    4.1 Introduction, 77
    4.2 The Range, 80
    4.3 The Average Deviation, 81
    4.4 The Standard Deviation, Ungrouped Data, 84
    4.5 The Standard Deviation, Grouped Data, 91

Chapter 4. MEASURES OF VARIATION (Continued)

4.6 Further Measures of Variation, 95
4.7 Measures of Relative Variation, 95
4.8 Further Remarks About the Standard Deviation, 97

Chapter 5. FURTHER DESCRIPTIONS 100

5.1 Introduction, 100
5.2 Measures of Symmetry and Skewness, 104
5.3 Measures of Peakedness, 107

Chapter 6. PROBABILITY 110

6.1 The Meaning of Probability, 110
6.2 Some Rules of Probability, 113
6.3 Probabilities in Games of Chance, 120
6.4 Mathematical Expectation, 123

Chapter 7. THEORETICAL DISTRIBUTIONS 127

7.1 Introduction, 127
7.2 The Binomial Distribution, 131
7.3 The Mean and Standard Deviation of the Binomial Distribution, 138
7.4 Continuous Distributions, 142
7.5 The Normal Curve, 145
7.6 Some Applications, 152
7.7 The Binomial Distribution and the Normal Curve, 156
7.8 Fitting a Normal Curve to Observed Data, 160
7.9 Further Theoretical Distributions, 165

Chapter 8. SAMPLING DISTRIBUTIONS 169

8.1 Random Sampling, 169
8.2 Sampling Distributions, 172

Chapter 9. PROBLEMS OF ESTIMATION 186

9.1 Introduction, 186
9.2 The Estimation of Means (Large Samples), 188
9.3 The Estimation of Means (Small Samples), 195
9.4 The Estimation of Proportions, 199
9.5 Standard Errors and Probable Errors, 206
9.6 Sampling from Small Populations, 208

Chapter 10. TESTS OF HYPOTHESES   **212**

    10.1 Introduction, 212
    10.2 Type I and Type II Errors, 217
    10.3 Null Hypotheses and Significance Tests, 221
    10.4 Tests Concerning Proportions, 225
    10.5 Differences Between Proportions, 228
    10.6 Tests Concerning Means, 233
    10.7 Differences Between Means (Large Samples), 237
    10.8 Differences Between Means (Small Samples), 240

Chapter 11. FURTHER TESTS OF HYPOTHESES   **245**

    11.1 Tests Concerning k Proportions, 245
    11.2 The Analysis of an r by k Table, 253
    11.3 Tests of "Goodness of Fit," 257
    11.4 Tests Concerning k Means, 261
    11.5 Analysis of Variance, 268

Chapter 12. PROBLEMS OF SAMPLING   **270**

    12.1 Introduction, 270
    12.2 A Test of Randomness, 272
    12.3 Runs Above and Below the Median, 277
    12.4 Sample Designs, 279
    12.5 Double, Multiple, and Sequential Sampling, 283

Chapter 13. LINEAR REGRESSION   **286**

    13.1 Introduction, 286
    13.2 The Method of Least Squares, 288
    13.3 Linear Regression, 296
    13.4 Limits of Prediction, 299
    13.5 Multiple Linear Regression, 304

Chapter 14. CORRELATION   **307**

    14.1 The Coefficient of Correlation, 307
    14.2 The Interpretation of r, 314
    14.3 A Significance Test for r, 317
    14.4 The Calculation of r from Grouped Data, 320
    14.5 Rank Correlation, 327
    14.6 Multiple and Partial Correlation, 330
    14.7 The Correlation of Qualitative Data, 333

Chapter 15. INDEX NUMBERS: BASIC CONCEPTS   **338**

    15.1 Introduction, 338
    15.2 Purpose of the Index, 340

Chapter 15.   INDEX NUMBERS: BASIC CONCEPTS (Continued)

15.3   Availability and Comparability of Data, 341
15.4   Selection of the Items, 344
15.5   Choice of the Base Period, 345
15.6   Choice of the Weights, 347
15.7   Methods of Construction, 347

**Chapter 16.   INDEX NUMBERS: THEORY AND
APPLICATION**   350

16.1   Introduction, 350
16.2   Unweighted Index Numbers, 350
16.3   Weighted Index Numbers, 356
16.4   Weighted Averages of Price Relatives, 362
16.5   Chain Index Numbers, 365
16.6   Shifting the Base, 372
16.7   The Use of Index Numbers in Deflating, 373
16.8   Mathematical Properties of Index Numbers, 376
16.9   Current Problems, 378

Chapter 17.   TIMES SERIES ANALYSIS: BASIC CONCEPTS   381

17.1   Introduction, 381
17.2   The Behavior of Time Series, 384
17.3   Secular Trend, 384
17.4   Seasonal Variation, 387
17.5   Cyclical Variation, 388
17.6   Irregular Variation, 389
17.7   A Word About the Classical Approach, 390
17.8   The Preliminary Adjustment of Time Series, 391
17.9   Graphical Presentations of Time Series, 393
17.10 Logarithmic Line Charts, 395

Chapter 18.   TIME SERIES ANALYSIS: SECULAR TREND   400

18.1   Introduction, 400
18.2   Linear Trends: Semi-Averages, 402
18.3   Linear Trends: Least Squares, 404
18.4   Modified Trend Equations, 408
18.5   Parabolic Trends, 412
18.6   Exponential Trends, 416
18.7   Other Trend Curves, 419
18.8   The Smoothing of Time Series, 423
18.9   Forecasting, 426

Chapter 19. TIME SERIES ANALYSIS: SEASONAL AND CYCLICAL VARIATION 429

19.1 Seasonal Variation, 429
19.2 Some Preliminary Considerations, 431
19.3 The Method of Simple Averages, 432
19.4 The Ratio-to-Trend Method, 436
19.5 The Ratio-to-Moving Average Method, 440
19.6 Deseasonalized Data, 447
19.7 The Use of Seasonal Indexes in Forecasting, 449
19.8 Cyclical Variation, 453

Appendix I. PICTORIAL PRESENTATIONS 461

I.1 Introduction, 461
I.2 Bar Charts, 462
I.3 Pie Charts and Component Bar Charts, 466
I.4 Statistical Maps, 471

Appendix II. QUALITY CONTROL 474

II.1 Introduction, 474
II.2 The Control Chart, 476
II.3 $\bar{X}$ and R Charts, 478
II.4 $\bar{X}$ and σ Charts, 482
II.5 Control Charts for Attributes, 484
II.6 Acceptance Sampling, 489

Appendix III. CALCULATIONS WITH ROUNDED NUMBERS 492

III.1 Rounded Numbers, 492
III.2 Calculations with Rounded Numbers, 493

Appendix IV. THE USE OF LOGARITHM AND SQUARE ROOT TABLES 495

IV.1 The Use of Logarithm Tables, 495
IV.2 The Use of Square Root Tables, 497

STATISTICAL TABLES 499

ANSWERS TO ODD EXERCISES 521

INDEX 529

CHAPTER 1

Introduction

1.1 Introduction

In the last few decades, the growth of statistical ideas and statistical methods has made itself felt in almost every phase of human activity. In business, it has brought about drastic changes in production, in the efficient use of materials, in marketing, in various phases of business research, and in management. Here, statistical data and statistical techniques have become a vital factor in the decisions, analyses, and forecasts of the modern businessman.

There can be no doubt that it is impossible to understand the meaning and implications of most work done in business and economic research without having at least a speaking acquaintance with the subject of statistics. Clearly, numerical data derived from surveys, experiments, and other sources form the raw material on which analyses and forecasts are based, so it is essential to know how to squeeze usable information from such data. This, in fact, is the major objective of statistics.

Numerous textbooks have been written on business statistics, psychological statistics, educational statistics, agricultural statistics, medical statistics, and other specific areas of application. It is true, of course, that these diversified fields demand somewhat different and specialized techniques in particular problems; yet the fundamental principles that underlie all the various methods are identical regardless of the field of application. This will become evident to the reader once he realizes that *statistical methods in general are nothing but a refinement of everyday thinking*.

The approach we shall use in this elementary treatment of statistics is keynoted by the above statement; it is our goal to introduce the beginning student in business and economics to the ideas and the concepts which are fundamental to the understanding of modern statistics. Although the examples and exercises used in this text will deal primarily with subjects closely related to problems of business and economics, we shall repeatedly remind the reader that the various ideas and techniques are applicable also to other social sciences as well as to the natural sciences. It is hoped that the approach used in this text will not only provide the reader with a sound understanding of the scientific principles used in business or economic research and in subsequent planning and operations, but that it will also enable him to gain a better understanding of the scope and limitations of empirical knowledge in general.

As we have said before, the study of statistics may be directed toward applications in particular fields of inquiry. Furthermore, statistics may also be presented in varying degrees of mathematical refinement and in almost any balance between theory and application. Because it is, in our opinion, much more important to understand the *meaning and implications* of a few basic concepts than it is to memorize a large assortment of impressive sounding formulas, we shall have to sacrifice some of the mathematical detail that is sometimes covered in an elementary course in statistics. This is unfortunate in some respects, but it will avoid our getting lost in an excessive amount of detail which might easily obscure the most important issues. By stressing ideas over computational skills we hope to avoid the dangerous effect that often results from the indiscriminate application of statistical methods without a thorough understanding of the fundamental logical concepts that are involved.

Traditionally, business statistics has been looked upon as applying mainly to "front office" operations; hence, until recently, most elementary textbooks stressed the collection and sources of data and the presentation of numerical data in tables, charts, pictures, and maps. Although the importance of statistics in this phase of business cannot be denied, there has been a pronounced change in attitude toward the subject: business statistics is now viewed as providing quantitative bases for arriving at well-informed decisions with respect to *all* matters connected with the operation of a business. *In short, it is now recognized that, in its broadest sense, business statistics includes also the methods and inferences needed to provide an adequate flow of quality raw materials, to select and evaluate the performance of both machines and personnel, to design products and maintain their quality, and to evaluate new methods of production, advertising, and selling the goods and services of industry, both in the immediate present and in the near and distant future.* In view of all this, our goal is to introduce the reader to some of the methods of statistics, the traditional as well as the new, which are playing an ever increasing role in, as we said, *all matters connected with the operation of a business.*

1.2 Descriptive and Inductive Statistics

Everything dealing even remotely with the collection, analysis, interpretation, and presentation of numerical data may be classified as statistics. It includes even such diversified tasks as the computation of a ballplayer's batting average by a *team statistician;* the collection and presentation of data on births, marriages, and deaths as *vital statistics;* and the study of laws governing the behavior of atomic particles by a specialist in *statistical mechanics* or *quantum statistics.*

The word *statistics* itself can be given a variety of interpretations. For example, it is used in the plural to denote simply a collection of numerical data. Such statistics may be found in the *Economic Almanac*, the financial pages of newspapers, U.S. Bureau of the Census reports, the *Statistical Abstract of the United States*, the records of county clerks, or wherever numerical data are collected and recorded. The second meaning, also in the plural, is that of the totality of methods that are employed in the collection and analysis of numerical data. In this sense, statistics is a branch of applied mathematics, and it is this field of mathematics which we shall study in this book. In order to complete this linguistic study of the word *statistics*, we might also mention that the term *statistic*, in the singular, is used to denote a particular quantity such as an average, an index number, or a coefficient of correlation which one calculates on the basis of a given set of data.

The problems that we shall discuss in this book are essentially of two types, belonging either to the field of *descriptive statistics* or to that of *inductive statistics.* Although the term *descriptive statistics* is often used to denote merely the tabular or graphical presentation of data, we shall use it in a much wider sense. By descriptive statistics we shall understand any treatment of numerical data which does not involve generalizations. In contrast, we shall speak of *inductive statistics* the very moment that we make generalizations, predictions, or estimations.

To clarify this distinction with an example, let us consider a consumers' rating service which wants to compare 60 watt electric light bulbs produced by two different manufacturers. Taking *five* light bulbs of Brand A, they find that these bulbs burn out after 985, 863, 1024, 972, and 746 hours of continuous use. Repeating this test with *five* light bulbs of Brand B, they find that these bulbs burn out after 892, 1071, 993, 929, and 785 hours of continuous use. On the basis of this information we can say that the five Brand A light bulbs had an *average* lifetime of

$$\frac{985 + 863 + 1024 + 972 + 746}{5} = 918 \text{ hours}$$

while those of Brand B had an *average* lifetime of

$$\frac{892 + 1071 + 993 + 929 + 785}{5} = 934 \text{ hours}$$

What we have done so far belongs to the domain of descriptive statistics. We followed simple arithmetical rules in calculating the two averages which are, indeed, descriptive of the two sets of figures obtained in the test. However, if we concluded from this experiment that Brand B is superior to Brand A, that is, that the average lifetime of the light bulbs of Brand B is *in general* greater than the average lifetime of those of Brand A, our reasoning would have to go far beyond the information with which we were supplied and we would find ourselves in the domain of inductive statistics.

So long as we merely calculated the two averages, we did not add anything to the information with which we were supplied; we merely rearranged it in a different and possibly more useful form; this is characteristic of descriptive statistics. As soon as we generalized, we said more than we were given in the original data and this, in turn, is characteristic of inductive statistics.

In the given example it does not follow by any means that *in general* Brand B is superior to Brand A. As can be seen from the experiment, there are considerable differences even among the light bulbs manufactured by the same company. Hence, it is quite conceivable that Brand B showed up better in the test simply because we were "lucky" in picking Brand B light bulbs that happened to be of particularly good quality. It is equally conceivable that if we repeated the experiment, the result might easily be reversed if *by chance* we picked for our sample Brand B light bulbs of slightly inferior quality. Before reaching any conclusions we shall, therefore, have to investigate whether the difference of $934 - 918 = 16$ hours in the two averages might not be due to the variability (lack of uniformity) of the two products, together with chance factors involved in the selection of the particular samples used in the test.

If we decided on the basis of the above experiment that Brand B is in general superior to Brand A, we would be making a generalization that may or may not be correct and we would, consequently, be taking a risk. The careful evaluation, analysis, and control of the chances that must be taken when we make such a generalization is one of the main tasks of inductive statistics.

There are many questions that immediately come to one's mind when thinking about the above example. For instance, if the difference between the averages of the two samples had been *very large*, we might have been inclined to say that one brand is superior to the other without making a detailed statistical analysis. Similarly, if the difference had been *very small*, we might have been inclined to say that the products are of about equal quality. *But what if the difference were neither "very small" nor "very large"—where do we draw the line?* Also, how do we know that five light bulbs of each kind will provide us with enough information to reach any conclusion whatsoever? Would it not have been better, perhaps, to test 10 light bulbs of each kind, or possibly 20 or more? These are important questions, and they are the kind of questions we shall be able to answer with the methods of inductive statistics.

There is one other question that is extremely important whenever we wish to make a generalization like the one concerning the two brands of light bulbs, and that question is: *Was the experiment (or survey) conducted in such a way that a generalization is at all possible?* For instance, if the tests of the two kinds of light bulbs had been conducted at different voltages, the observed difference in the average lifetimes might easily have been caused by differences in voltage rather than by differences in quality. Although it would seem rather silly to conduct the experiment of our illustration by using different voltages, it is amazing how many surveys or experiments are conducted in such a way that no sensible statistical analysis and, of course, no generalizations are possible.

We have made this last point to emphasize that the statistical treatment of a problem does not merely consist of looking at a set of data, performing some calculations, and reaching a conclusion. The questions as to how the data were collected and how the whole experiment or survey was planned are of prime importance. Unless proper care is taken in the planning and design of an experiment (survey or other kind of investigation), we may not be able to reach any valid conclusion whatsoever. Generally speaking, poorly designed experiments cannot be salvaged by fancy mathematical or statistical techniques.

To summarize, we shall use the terms *descriptive statistics* and *inductive statistics* with reference to the kinds of problems we wish to solve—not with reference to a particular formula or statistic we may choose to employ. For example, we may calculate an average solely for the purpose of describing a set of data or we may use it to make generalizations or predictions.

1.3 Sources of Data

Because numerical data are the raw material of statistical investigations, one of the first steps in any statistical study must be the collection of suitable data. The only step that ordinarily precedes the collecting of data is the careful planning of the study and the precise formulation of its purpose, scope, and objectives.

Data needed for studies arising in the operation of a business organization will of necessity often have to come from the records of the organization itself. Data thus taken by a firm from its accounting records, payrolls, inventories, sales vouchers, and the like, for its own use in statistical investigations are called *internal data*. Although internal data are usually of considerable importance, in many situations they must be supplemented (or replaced) by numerical information from sources outside the firm. Data obtained from such sources as local, state, and national governments, private reporting organizations, trade associations, and trade publications are correspondingly referred to as *external data*.

Having made this distinction, we should not fail to note that some studies require both kinds of data. For instance, if a steel company wants to prepare a chart showing its stockholders how its accident record compares with that of the entire industry, data on the company's own accident rate will come from its internal records while the figures for the entire industry must be sought externally—perhaps, from a trade publication.

It is theoretically within the power of a firm to set up and maintain its own records in such a way as to provide it with whatever data are thought to be useful or necessary. Although this may not always be done and there may sometimes be regret for not having kept proper records, in principle, the collection of internal data generally does not pose any serious problems. The collection of data from external sources is another matter, however, and it is often quite troublesome. Sometimes data available from two or more sources do not agree, and sometimes data considered to be relevant to the solution of a problem are not available from any source, at least within limits imposed by practical considerations of cost and time.

External data are often classified as *primary*, meaning that they are published by the same organization by which they were collected, and *secondary*, meaning that they are published by an organization other than the one by which they were collected. For example, the Bureau of Labor Statistics collects the data needed for its *Consumer Price Index*, performs the necessary calculations, and publishes the index in its *Monthly Labor Review*. This is thus the primary source of these data. On the other hand, if we obtain the *Consumer Price Index* from the financial pages of a newspaper, we would refer to the newspaper as a secondary source. It is easy to see that a source can be primary for some data and secondary for others. The *Survey of Current Business*, a publication of the Department of Commerce, is the primary source for a great deal of data collected by this Department, but only a secondary source for data collected by the Department of Labor, the Federal Reserve Board, and various private organizations. Primary sources are generally preferred to secondary sources. There is not only the possibility of errors of transcription, but primary sources are often accompanied by better definitions or documentation.

1.4 The Direct Collection of Data

The internal records of a company and externally published data will often provide whatever information is needed in the solution of particular problems, but there are many situations in which the required information can be obtained only by conducting special surveys, experiments, and tests. The problem of how, where, and when to collect data that are needed to meet particular requirements is sometimes very difficult to solve, and it is, of course, a problem that is of vital concern to all agencies that specialize in the publication of primary data. In this section we shall briefly mention some of the major questions that arise in connection with the direct collection of data. Later on, for example, in Chapter 12, we shall study some of them in more detail.

Although it is sometimes feasible to collect *all* of the items, measurements, or observations, that are relevant to a given problem, this method is the exception rather than the rule. Even in simple problems involving relatively small numbers of individuals or items there are usually time and cost limitations which make it necessary to take only part, rather than all, of the data. In the language of statistics, we will often have to take *samples* from *populations* consisting of the totality of the items or observations with which we are concerned. Products are thus tested by checking, say, each tenth item coming off an assembly line; consumer preferences are investigated by interviewing samples of housewives in certain areas; and elections are predicted by taking sample polls of public opinion.

There are a great many important problems that arise in connection with the direct collection of data—problems such as whether or not to take a sample, and if so, what method to use in taking the sample, and how large a sample is needed to arrive at practically useful generalizations. All these questions will be studied later on in some detail.

In all statistical studies that use samples, great care must be exercised to ensure that the samples will lend themselves to valid generalizations; a key issue here is the question of *bias*. A sample is said to be *biased* if it is not representative of the population which it is supposed to represent. Every precaution must always be taken to avoid *inadvertent biases*, and it goes without saying that it is not very ethical to introduce *deliberate biases* to prove particular points. We shall have more to say about this in Chapters 8, 12, and 16.

In ordinary business applications, direct information is usually gathered by personal interviews, telephone calls, mail questionnaires, or any combination of the three. There is no general answer as to which of these methods is best, although mail questionnaires are often preferred because they are usually cheaper and easier to get into homes or concerns that are widely dispersed. Of course, what happens after a questionnaire has been distributed is another matter. For one reason or another, many people fail to return even well-constructed questionnaires unless the questions are devoted to matters in which they are particularly interested. It is easy to see how this may introduce a bias that would make it difficult to arrive at valid conclusions. Incidentally, the construction of questionnaires, whether they be used for personal interviews or sent through the mail, has become a science in itself and the subject of extensive study. Many a statistical investigation has been challenged on the grounds that questions were not properly asked. Clearly, "Why do you prefer Product A to Product B?" is apt to elicit a different reply than "Which do you prefer, Product A or Product B?"

Although mail questionnaires are often cheaper and, as we said earlier, easier to distribute, the advantage of personal interviews is that a tactful agent can often elicit answers relating to such personal matters as health, age, and income, which most persons would refuse to reveal in a mail questionnaire. Moreover, if we consider the cost per completed return, mail questionnaires do not always constitute the cheapest means of obtaining information.

Telephone calls are often used to get on-the-spot response, for example, to television viewing and other habits. Although this method of collecting information has been relatively fruitful in some areas, it is exposed to the criticism that it may lead to biased results because it reaches only telephone subscribers and, hence, not necessarily the entire population that may be relevant in a given situation.

1.5 Published Data

The federal government is easily the biggest collector and publisher of business data, and, consequently, the most important external source used by other organizations. Much of the data that the government collects and makes generally available is needed by the government itself in discharging its responsibilities, but other data, not needed specifically by the government, are collected and published in response to the private needs of groups of individuals or organizations that are large enough to justify the collection of these data at public expense. Various government agencies will provide special unpublished data to anyone requesting them, upon payment of a relatively slight charge.

Perhaps, the best way to get an over-all picture of what information collected by the federal government is available to a businessman is to look through the book by Hauser and Leonard mentioned in the Bibliography on page 16. Here we find a treatment of the government's activities in gathering statistics relating to wholesale, retail, and service trades; accounting, agriculture, minerals, and manufacturing; population, housing, construction, and labor; etc. Although this book stresses those series of data which serve as indicators of the general course of business activity and as guides to consumer and industrial markets, it also contains suggestions as to where more detailed information on other government statistics may be found.

With respect to the great mass of statistical material flowing from the federal government, let us call specific attention to the publications of the Departments of Commerce and Labor. Through its various offices and bureaus, the Department of Commerce periodically publishes basic information relating to both the domestic and international aspects of the economy of the country. Annually, the Bureau of the Census publishes the *Statistical Abstract of the United States*, a veritable storehouse of information gathered from various sources and relating to almost any aspect of the life of the nation. This bureau also collects and publishes the results of censuses of population, manufactures, distribution, housing, and agriculture; and it issues monthly trade reports giving data on inventories, sales, and the like, with reference to many lines of wholesale and retail business for the country as a whole and for selected cities. The Department of Commerce collects and publishes data on the trade of the United States and other countries.

One of the most important publications of the Department of Commerce is the *Survey of Current Business*, issued monthly by its office of Business Economics and supplemented by weekly information on some of the major series of data. Each year the January issue contains a review of the preceding year and the July issue contains detailed national income data. The statistical information that now appears regularly in the monthly issues of the *Survey of Current Business* is intended to bring up to date the data published in the 1967 edition of *Business Statistics*, a biennial supplement to the *Survey*, which presents statistical series, monthly, quarterly, or annually. The total number of series—that is, regularly published data on prices, production, sales, incomes, inventories, etc.—included in the *Survey* is in the neighborhood of 2,500. Most are on a monthly basis given for the past 13 months. Everything considered, the *Survey of Current Business* is without question one of the most valuable sources of business data.

The Department of Labor, too, is responsible for the collection, tabulation, and publication of some very important business data. Monthly indexes of consumer and wholesale prices are constructed by the Bureau of Labor Statistics and published in the Bureau's *Monthly Labor Review*. This publication also contains data concerning labor force and employment, labor turnover, earnings and hours, and work stoppages. The 1964–65 statistical supplement is the sixth and final supplement.

The material published by these two departments by no means exhausts the supply of statistical information coming from the federal government; much useful information bearing on various phases of business life is made available by other governmental units.

Although the federal government is the biggest single source of published business data, it is far from being the only one. In addition to the statistical work done by state and local governments and the United Nations, many private organizations collect and publish statistical data in an effort to fill particular needs. A good idea about the nature of data available from nongovernmental sources (as well as from governmental sources) may be obtained from the book by Coman referred to in the Bibliography on page 16. This book contains a valuable discussion of the entire problem of how to locate published business information quickly and efficiently. The word "information" in the title of this book is used broadly to include sources of qualitative as well as quantitative data.

The most important nongovernmental sources for statistical data are private statistical services, trade associations, trade publications, university research bureaus, commercial and financial periodicals, and specialized reporting agencies. University offices of business research often collect and publish data on employment, farm prices and marketings, construction contracts, department store sales, bank debits, and the like, for their respective states, and sometimes these figures are broken down on a city, county, or regional basis.

Of the private statistical services, Moody's Investors Service issues annually, in five volumes, financial data on industrial, transportation, and public utility corporations, as well as on banks and financial institutions, and governments and municipalities. The National Industrial Conference Board publishes an annual *Economic Almanac* containing data relating to activity in various fields, along with monthly publications giving data on wages, salaries, living costs, employment, etc. Other specialized reporting agencies provide information on such things as new car registrations and circulation of publications.

There are a number of periodicals that give data on commercial, financial, and general business conditions. Although they are published commercially, that is, for profit, to fill particular needs, many of them are highly respected, having long records of faithful and accurate reporting. Among such periodicals we find, for instance, the *Commercial and Financial Chronicle, Business Week, Dun's Review, Dun's Statistical Review, Barron's,* the *Wall Street Journal,* and the New York *Journal of Commerce and Commercial Digest.*

Statistical data are issued from time to time by various trade associations. For instance, the American Iron and Steel Institute, the American Petroleum Institute, and the Automobile Manufacturers Association issue annual statistical summaries of prices, production, sales, and the like. In addition, there is a large number of independent trade publications reporting on activities in particular fields. Almost all important industrial and commercial groups are covered by such publications, as is indicated by the names of *Textile World, Leather Manufacturer, Chemical Industry News, Automotive News, Chain Store Age, Engineering News Record, Power,* and many others. All these publications print important statistical data for their respective industries.

We would be careless, indeed, if we failed to add a word of caution about the indiscriminate use, or uncritical acceptance, of published data. To avoid serious mistakes, it is always important to check the precise definition of each term or explanation accompanying published data and usually this cannot be done by merely looking at the title. For instance, a set of data captioned "Employment" or "Total Employment" may refer to employment in all manufacturing industries, employment in manufacturing as well as nonmanufacturing industries, employment in the entire country, or perhaps it may exclude those who are self-employed. A careful search is often necessary to discover not only what the data are supposed to represent, but also what units the figures are expressed in, how these units are defined, and whether the definitions are consistent throughout so that comparisons between different periods can be made.

The availability of just such information as this is one of the most valuable features of data supplied by the federal government. For all data published by the government, it is possible to find a complete description of what data (rigorously defined) are contained, how, when, and where they were gathered, how they were processed, etc. Unfortunately, such information cannot always be obtained for data supplied by some of the other sources, and it is advisable, in that case, to proceed with utmost caution.

It is also important to keep in mind that published data are subject to the same pitfalls of *bias* as directly collected data. After all, somebody at one time had to collect them directly, and to what extent *interested* parties can be relied upon to furnish *disinterested* information can only be decided on an individual basis.

1.6 Mathematical Prerequisites

Many students face the study of statistics with mixed emotions. They know, or at least they are told, that on the one hand they cannot proceed to more advanced studies in their chosen field without having some understanding of statistical methods, while, on the other hand, they may distinctly remember difficulties experienced in their previous contacts with mathematics.

It is true, of course, that a limited amount of mathematics is a necessary prerequisite for any course in statistics on the college level. Indeed, a thorough study of the theoretical principles that underlie statistics would require a knowledge of mathematical subjects that are ordinarily taught only in advanced courses for students majoring in mathematics. Since this book is written for students with little background in mathematics, our aims and, therefore, also our prerequisities are considerably more modest.

Actually, the mathematics needed for this study of elementary statistics is amply covered in college algebra or any equivalent course in mathematics; even some knowledge of high school algebra would provide a sufficient foundation. Besides having a reasonable skill in the elementary arithmetic of addition, subtraction, multiplication, and division, it will help the reader to have some familiarity with the most common problems of high school or college algebra—for example, the representation of numbers with symbols such as x, y, and z, the solution of simple equations, substitutions, the use of the functional notation, logarithms, and tables. Logarithms are needed only in Sections 2.2, 3.7, 17.10, and 18.6; and in case the reader is not familiar with their use, a brief discussion of logarithm tables is included in Appendix IV. The subject of adding, subtracting, multiplying, and dividing rounded numbers is treated in Appendix III.

1.7 Subscripts and Summations

Since all of the formulas which we shall study in subsequent chapters will have to be applicable to *different* sets of data, we shall have to represent the figures (measurements or observations) to which these formulas are to be applied with some general symbols such as x, y, and z. Unless we introduce a slight modification, however, this kind of symbolism will lead to complications, because there are simply not enough letters in the alphabet. For example, if we wanted to represent the income of every wage earner in New York City by a different letter, we would easily use up the English, Greek, Russian, and Hebrew alphabets without accomodating even a small fraction of our data. This is why we shall follow the usual practice and use *subscripts*. Referring again to the example of Section 1.2, we could represent the "lifetimes" of the five light bulbs of Brand A by the symbols x_1, x_2, x_3, x_4, and x_5. Similarly, we could refer to the figures in the second sample (containing the light bulbs of Brand B) as y_1, y_2, y_3, y_4, and y_5, using different letters to distinguish between the two samples. If we want to discuss any one of these numbers in general, we shall refer to it as x_i or y_j where i and j are, so to speak, variable subscripts which, in this particular example, can take on the values 1, 2, 3, 4, or 5.

Instead of writing the subscripts as i and j we could just as well have used other letters such as k, l, m, . . . , and instead of x and y we could just as well have used other arbitrary letters or symbols. In general, it is customary to use different letters for different kinds of measurements and different subscripts for different individuals (different items). We might, thus, write x_{45} and x_{73} for the value of the *forty-fifth* and *seventy-third* stock on a certain list, and we might write x_{12}, y_{12}, and z_{12} for the income, savings, and assets of a given individual, namely, the 12th on our list.

In order to simplify formulas that will involve large sets of numerical data, let us now introduce the symbol Σ (capital Greek *sigma*, standing for S) which is merely a *mathematical shorthand notation*. By definition, we shall write

$$\sum_{i=1}^{n} x_i = x_1 + x_2 + x_3 + \cdots + x_n \qquad (1.7.1)$$

which reads: "the summation of x_i, i going from 1 to n." In other words, $\sum_{i=1}^{n} x_i$ stands for the *sum* of the x's having the subscripts 1, 2, . . . , and n. According to this definition we might write, for example,

$$\sum_{i=1}^{5} y_i^2 = y_1^2 + y_2^2 + y_3^2 + y_4^2 + y_5^2$$

or

$$\sum_{j=2}^{5} x_j f_j = x_2 f_2 + x_3 f_3 + x_4 f_4 + x_5 f_5$$

beginning in the last example with the subscript 2 and ending with the subscript 5.

Since summation signs will appear in many formulas later on, it will prove helpful to study some of their fundamental rules. These rules, three of which are given below, are relatively easy to understand and prove.

RULE A: *The summation of the sum (or difference) of two or more terms is equal to the sum (or difference) of their respective summations.*

Symbolically we can write this rule in the case of three terms as

$$\sum_{i=1}^{n} (x_i + y_i + z_i) = \sum_{i=1}^{n} x_i + \sum_{i=1}^{n} y_i + \sum_{i=1}^{n} z_i \qquad (1.7.2)$$

If we had wanted to use minus signs instead of plus signs, we could have done so on both sides of the equation. The proof of Rule A consists of showing that the two sides of the equation are identical when written in full without summation signs. It will be left as an exercise for the reader.

RULE B: $$\sum_{i=1}^{n} kx_i = k \cdot \sum_{i=1}^{n} x_i \qquad (1.7.3)$$

This rule states that the summation of a constant, k, times a variable, x_i, equals the constant times the summation of the variable. It is proved as follows:

$$\sum_{i=1}^{n} kx_i = kx_1 + kx_2 + kx_3 + \cdots + kx_n$$
$$= k(x_1 + x_2 + x_3 + \cdots + x_n)$$
$$= k \cdot \sum_{i=1}^{n} x_i$$

The third rule is

RULE C: $$\sum_{i=1}^{n} k = nk \qquad (1.7.4)$$

and it states that the summation of a constant, k, from 1 to n, equals the product of k and n. Writing as in the proof of Rule B

$$\sum_{i=1}^{n} kx_i = kx_1 + kx_2 + kx_3 + \cdots + kx_n$$

we can put all of the x_i equal to 1, getting

$$\sum_{i=1}^{n} k \cdot 1 = k \cdot 1 + k \cdot 1 + k \cdot 1 + \cdots + k \cdot 1$$

or

$$\sum_{i=1}^{n} k = nk$$

We could also have argued that since the constant k does not depend on the subscript i, we can immediately write

$$\sum_{i=1}^{n} k = k + k + k + \cdots + k = nk$$

EXERCISES

1. Prove RULE A when variable z is subtracted from the sum of variables x and y, and i varies from 2 to 5.

2. Write each of the following expressions as summations:
   (a) $x_1 + x_2 + x_3 + x_4 + \cdots + x_{15}$
   (b) $y_1 g_1 + y_2 g_2 + y_3 g_3 + y_4 g_4 + y_5 g_5$
   (c) $(x_1 - y_1) + (x_2 - y_2) + \cdots + (x_k - y_k)$
   (d) $(y_3)^2 z_3 + (y_4)^2 z_4 + (y_5)^2 z_5 + \cdots + (y_8)^2 z_8$

3. Write each of the following expressions without summation signs:
   (a) $\displaystyle\sum_{i=1}^{5} x_i f_i$ (d) $\displaystyle\sum_{j=2}^{5} (y_j + z_j)$
   (b) $\displaystyle\sum_{i=1}^{6} k x_i$ (e) $\displaystyle\sum_{j=1}^{6} (z_j - c)$
   (c) $\displaystyle\sum_{i=1}^{4} x_i^2$ (f) $\displaystyle\sum_{i=1}^{5} (x_i^2 y_i)$

4. Prove that $\displaystyle\sum_{i=1}^{m} (x_i - k) = \sum_{i=1}^{m} (x_i) - mk$

5. Given: $x_1 = 2$ $x_2 = -3$ $x_3 = 0$ $x_4 = 1$ $x_5 = 3$
   $f_1 = 3$ $f_2 = 7$ $f_3 = 15$ $f_4 = 8$ $f_5 = 2$
   $y_1 = 6$ $y_2 = -2$ $y_3 = 8$ $y_4 = 3$ $y_5 = 5$

   evaluate each of the following:
   (a) $\displaystyle\sum_{i=1}^{4} y_i$ (d) $\displaystyle\sum_{i=1}^{5} (y_i)^2$
   (b) $\displaystyle\sum_{i=2}^{5} x_i$ (e) $\displaystyle\sum_{i-1}^{5} (x_i + y_i)$
   (c) $\displaystyle\sum_{i-1}^{5} y_i f_i$ (f) $\displaystyle\sum_{i=1}^{5} (y_i^3 f_i)$

6. Prove that $\sum_{i=1}^{n} (x_i - k)^2 = \sum_{i=1}^{n} x_i^2 - 2k \sum_{i=1}^{n} x_i + nk^2$

7. Is the following a true statement? $[\sum_{i=1}^{n} x_i]^2 = \sum_{i=1}^{n} (x_i)^2$

Hint: Substitute $n = 2$ and check.

BIBLIOGRAPHY

The two reference books on sources of data which we mentioned in the text are

Coman, E. T., *Sources of Business Information*, rev. ed. Berkeley and Los Angeles: University of California Press, 1964.

Hauser, P. M., and Leonard, W. R., *Government Statistics for Business Use*, 2nd ed. New York: Wiley, 1956.

Further material on the collection of data, questionnaire construction, etc., may be found in

Croxton, F. E., Cowden, D. J., and Klein, S., *Applied General Statistics*, 3rd ed. Englewood Cliffs, N.J.: Prentice-Hall, Inc., 1967, Chap. 2.

Neter, J., and Wasserman, W., *Fundamental Statistics for Business and Economics*, 3rd ed. Boston: Allyn and Bacon, Inc., 1966, Chap 2.

Neiswanger, W. A., *Elementary Statistical Methods*, rev. ed. New York: Macmillan, 1956, Chap 3.

Stockton, John R., *Business Statistics*, 3rd ed. Cincinnati: South-Western Publishing Co., 1966, Chap 2.

Wasserman, Paul, et al., *Statistics Sources*, 2nd ed. Detroit: Gale Research, 1965.

and in many other introductory texts.

CHAPTER 2

Frequency Distributions

2.1 Introduction

Whereas the main objective of *inductive* statistics is to make generalizations and to go beyond the information with which we are supplied in a set of data, the main objective of *descriptive* statistics is to put the information contained in such data into a more usable form. In descriptive statistics we shall, thus, describe numerical data in the same way in which we might describe a person as being tall, weighing 164 pounds, and being 26 years old, or a business as being located in a certain area, dealing in certain kinds of goods, and being profitable. The descriptions which we have given in these two examples are, of course, far from complete, but for certain purposes they may very well be sufficient. As we shall see, the same will be true also for descriptions of numerical data.

Whenever we give a description we face the possibility of either describing *too much*, in which case "we might not see the forest for the trees," or describing *too little*, in which case the description may be worthless. Let us suppose, for instance, that we are given records of the weekly earnings of 1,843 stenographers employed in the Boston, Massachusetts metropolitan area in September 1967. Having this information, we could calculate their average weekly earnings as $87.50, describing, thus, the 1,843 individual earnings by means of a *single* number. There are some problems for which this single description would suffice. For example, it would be adequate if we wanted to compare the average earnings of these stenographers with the average earnings of some other occupation group. On the other hand, if we wanted to know what percentage of the stenographers had weekly earnings of $95.00 or more, this description would no longer be adequate. This emphasizes the point that *the type of description that is needed in any particular situation depends largely on what we intend to do with the description once it has been obtained.*

When dealing with large sets of data and facing the (not particularly pleasant) task of having to look at thousands, sometimes even millions, of numbers, we can often gain considerable information and get a good over-all picture of the situation by grouping our data into a number of classes (groups). Using records of the weekly earnings of the 1,843 stenographers mentioned above, we might, for example, construct the following table:

Earnings (in dollars)	Number of Stenographers
60.00– 64.99	4
65.00– 69.99	33
70.00– 74.99	208
75.00– 79.99	192
80.00– 84.99	285
85.00– 89.99	368
90.00– 94.99	404
95.00– 99.99	137
100.00–104.99	69
105.00–109.99	49
110.00–114.99	20
115.00–119.99	39
120.00 or more	35
Total	1,843

This kind of table is called a *frequency distribution;* it tells us how the individual wages are distributed among the different classes.

Although a frequency distribution presents data in a more *usable* form, it does omit some information. For example, by looking at the distribution we *cannot* tell what salaries were received by the 4 stenographers who belong to the first class. For all we know, they might all have received different salaries from $60 to $64.99. Also, the distribution does not tell us the lowest salary or the highest salary paid to the stenographers in this area. In order to recover any of this information we would have to go back to the original data.

Before going into some of the problems connected with the construction of frequency distributions, let us distinguish first between *numerical* and *categorical* frequency distributions. (We often omit the word "frequency" and speak simply of a distribution.) A grouping or distribution is said to be *numerical* if the data are grouped according to numerical size. The distribution given above is a numerical distribution, it shows how the wages of the stenographers are distributed by size. The following is another example of a numerical distribution; it shows how the residents of a certain community are distributed according to age:

Age (in years)	Number of Persons
0–9	43
10–19	36
20–29	128
. . .	. . .
etc.	etc.

A distribution is said to be *categorical* if the data are sorted into categories according to some qualitative description rather than numerical size. (Categorical distributions are also referred to as *qualitative* distributions and, similarly, numerical distributions are referred to as *quantitative* distributions.) We would obtain a categorical distribution if we classified college students into Freshmen, Sophomores, Juniors, Seniors, and Graduate Students, and recorded the number belonging to each group. If in the example of the stenographers we had been interested in the types of firms in which they were employed, we might have constructed the following categorical distribution:

Type of Industry	Number of Stenographers
Manufacturing	640
Public utilities	136
Wholesale trade	227
Retail trade	65
Finance	533
Services (hotels, TV stations, auto repair shops, etc.)	242

The following two sections deal with the construction of numerical distributions and their presentation in various alternate forms. Some problems dealing with the construction of categorical distributions are discussed briefly in Section 2.4. Later on, in Section 14.4, we shall treat the problem of grouping *paired* observations into *two-way frequency distributions*. By paired observations we mean, for example, the incomes and savings of a number of persons, the production costs and sales prices of a number of items, or the average prices charged for potatoes and beans in a number of stores.

2.2 The Construction of Numerical Distributions

The process of constructing a frequency distribution consists essentially of three steps. First, we choose the classes into which the data are to be grouped; then we sort our data by putting a check for each item into the appropriate class, and, finally, we count the number of checks in each class. (If the data are recorded on punch-cards, a procedure which is widely used for the handling of mass data, the sorting and counting can be done automatically in a single step.) Since the last two steps are purely mechanical, we shall concentrate on the first, namely, that of choosing appropriate classifications.

It is important to realize that the choice of the groupings (classes) is essentially arbitrary, and that it will depend largely on what we intend to do with our data once they have been grouped. Hence, it is difficult to give specific instructions which should invariably be followed in the construction of a frequency distribution.

The various decisions we have to make before we can begin to classify our *raw data* ("raw" in the sense that they have not yet been subjected to any kind of statistical treatment) may be summarized as follows: (a) we have to decide on the *number of classes* into which the data are to be grouped, and (b) we must decide on the *range of values* each class is to cover.

The first of these decisions is primarily a matter of judgment, because any decision to use 1, 3, 10, 15, or for that matter any arbitrary number of classifications, depends on what we intend to do with our data once they have been grouped. It also depends to a large extent on the nature of our data (for example, the total range of values covered) and, above all, on the actual number of items that we want to group. For instance, if we had as few as 10 numbers, we would accomplish very little by grouping them into 25 classes, most of which would be completely empty. Similarly, we might very well be giving away *too much* information if we grouped these 10 numbers into as few as 2 classes. (As a matter of fact, we would ordinarily not even think of grouping as few as 10 observations.)

In the case of the weekly earnings of the stenographers, we could have grouped our data into the following table:

Earnings (in dollars)	Number of Stenographers
60.00– 89.99	1090
90.00–119.99	718
120.00–149.99	35

But it must be apparent that by using such a coarse grouping we are giving away a considerable amount of information, and for many purposes this distribution may not be adequate. Although this is not meant to be a general rule and there are, of course, exceptions, *we seldom use fewer than 6 classes or more than 15.*

To consider an illustration, let us take the following data representing the size of orders (in dollars) received by a certain department of a large mail-order house:

21.36	5.45	19.84	29.34	10.85	34.82	19.71	20.84
10.37	22.50	32.50	18.49	22.49	17.50	12.25	11.50
33.55	19.87	20.63	6.12	12.72	24.15	36.90	23.81
18.25	26.70	24.25	31.12	7.83	11.95	17.35	33.82
26.43	12.73	8.89	19.50	17.84	26.42	22.50	5.57
24.97	37.81	27.16	23.35	25.15	34.75	13.84	23.05
14.67	24.81	15.95	27.48	21.50	16.44	24.61	10.00
27.49	17.75	31.84	18.75	26.80	21.75	28.40	22.46
24.76	15.10	23.11	30.26	16.30	18.64	9.36	17.89
17.45	28.50	13.52	21.50	14.59	29.30	15.12	29.65

Before making a definite decision on the number of classes, let us consider some other factors that are important in the construction of a frequency distribution.

It certainly would be embarrassing if after choosing a classification we were to find that it does not accommodate all of our data. In order to avoid this, we shall look for the smallest and largest values contained in the data, determining, thus, the range of values that will have to be covered. In our example the smallest value is $5.45, the largest value is $37.81, and we shall, therefore, have to cover, roughly, the interval from $5 to $38.

It is really not necessary that the classes we choose have all equal length; they need not cover equal ranges of values, but there are many reasons why this is desirable and *equal class intervals (classes of equal length) should be used whenever possible.* Aside from the fact that frequency tables with equal intervals are easier to read, there are other advantages which will become apparent as soon as we use grouped data for the calculation of further descriptions.

If in our example we used 7 classes, we could accommodate our data by choosing classes which cover a range of $5 each. We might, thus, be tempted to choose classes ranging from $5 to $10, from $10 to $15, from $15 to $20, etc. However, this classification does not tell us whether an order of $10 is to be put into the first class or into the second. Similarly, we would not know where to put orders of $15, $20, $25, and so forth, if such values actually occurred in our data. In order to eliminate an ambiguity of this type, let us state the general rule: *Overlapping class intervals must always be avoided.* By *overlapping intervals* we mean successive class intervals having one or more values in common.

Keeping this rule in mind, we might now choose the following classification: the first class ranges from $5.00 to $9.99, the second from $10.00 to $14.99, the third from $15.00 to $19.99, . . . , and the last from $35.00 to $39.99. Since the amounts which we want to group are given to the nearest cent, there will no longer be any ambiguity. Furthermore, the classes are of equal length and all of our data will be accommodated. Using this classification, we can now present our data, grouped, in the following table:

Size of Orders (in dollars)	Tally	Number of Orders
5.00– 9.99	LHT /	6
10.00–14.99	LHT LHT //	12
15.00–19.99	LHT LHT LHT ////	19
20.00–24.99	LHT LHT LHT LHT	20
25.00–29.99	LHT LHT ///	13
30.00–34.99	LHT ///	8
35.00–39.99	//	2

The tally shown in the middle of this table is helpful in the construction of a frequency distribution, but it is usually omitted in the final presentation. The numbers shown in the column on the right are referred to as the *class frequencies;* they give the number of cases that fall into each class.

Since the choice of a classification is largely arbitrary, there is no law which forbids us to use an alternative classification in which the first class goes, say, from $5.45 to $10.07, the second from $10.08 to $14.70, the third from $14.71 to $19.33, . . . , and the seventh from $33.23 to $37.85. Although this classification would also accommodate all of our data in classes which are of equal length and do not overlap, it has some very undesirable features. The resulting frequency table would not only be difficult to read, but the actual tally would be much more tedious. *Generally speaking, it is desirable to use equal class intervals that are easy to use; for example, intervals of 5, 10, 25, 100, . . . , units. It is also desirable that the values contained in each class be such that the tally is easy and the resulting table easy to read.*

If we plotted on a scale the class intervals which we originally chose in our example, we would find that there are small gaps between the classes. In Fig. 2.1 there is a small gap between $9.99 and $10.00,

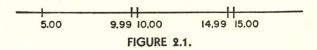

| 5.00 | 9.99 10.00 | 14.99 15.00 |

FIGURE 2.1.

another between \$14.99 and \$15.00, a third between \$19.99 and \$20.00, etc. Fortunately, this does not pose a problem in our example, because the data are such that they cannot fall into the gaps: they are given to the nearest cent. (If we had dealt with data given to *three* decimals, our table would not have accommodated a number such as 9.993 or 24.997 and we might, instead, have chosen classes going from 5.000 to 9.999, from 10.000 to 14.999, from 15.000 to 19.999, and so on.)

Since we shall meet some problems in which it will be desirable to eliminate these gaps (for example, in Sections 7.4, and 7.5, where we shall want to approximate the graph of a distribution with a smooth curve), we shall use the rather artificial device of "splitting the difference" and incorporating the halves of each gap into the two adjacent classes. Spreading our classification, thus, over a *continuous* scale, we shall say that the first class goes from 4.995 to 9.995, the second from 9.995 to 14.995, the third from 14.995 to 19.995, and the last from 34.995 to 39.995 (see Figure 2.2). (It should be noted that we adjusted the first and last classes as if there were additional classes to the left of the first and to the right of the last.)

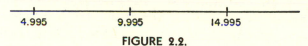

4.995   9.995   14.995

FIGURE 2.2.

The dividing lines between the successive classes as they are shown in Fig. 2.2 are called *class boundaries*. In our example the *lower class boundary* of the first class is 4.995 and its *upper class boundary* is 9.995. Similarly, the *lower class boundary* of the third class is 14.995 and the *upper class boundary* is 19.995.

The smallest and largest values that can fall into a given class are referred to as its *class limits*. In the table shown on page 22 we presented the classes by giving their class limits, the *lower class limit* of the first class being \$5.00, its *upper class limit* being \$9.99, the *lower class limit* of the fourth class being \$20.00, and the *upper class limit* of the sixth class being \$34.99.

To give another illustration of class boundaries and class limits, let us suppose that we want to make a study of the number of fatal automobile accidents occurring daily in the United States. Since we are dealing with whole numbers, the following classification may very well be adequate:

Number of Fatal Accidents
(class limits)

0–9
10–19
20–29

. . .

etc.

If we again divide the gaps evenly between adjacent classes, we will find that the boundary between the first two classes is 9.5, that between the next two classes is 19.5, and so forth. The relative artificiality of class boundaries is apparent from the fact that the lower and upper class boundaries of the first class are −0.5 and 9.5, respectively. Clearly, there can be no negative fractional part of a fatal accident, and it should be remembered that we use class boundaries and fill the gaps only because there are some problems in which it is desirable to use classifications which cover a continuous scale.

To complete our terminology, let us add that the length of a class interval, or simply the *class interval*, is, by definition, the difference between the upper and lower boundaries of a class. In the example of the automobile accidents, we shall, thus, say that the class interval is 10, while in the example of the mail-order house the class interval is 5. In distributions with *equal* class intervals, we can determine the length of the class interval also by calculating the difference between successive lower class limits (or successive upper class limits). Finally, the *class mark* is, by definition, the mid-point between the class boundaries (or class limits) of a class. In the accident example the class marks are 4.5, 14.5, 24.5, . . . , and in the case of the orders received by the mail-order house the class marks are 7.495, 12.495, 17.495, . . . , and 37.495.

When dealing with data containing a few values which are much greater (or much smaller) than the rest, we can often reduce the number of classes needed to accommodate our data by using so-called *open class intervals*. We used such an *open class* in the wage distribution on page 18 where the 68 highest earnings were classified under "$80.00 or more." Had we not used this device we might have needed four or five additional classes to accommodate a relatively small fraction of our data.

To consider another example, let us suppose that the first of the orders listed on page 21 was not $21.36 but $117.25. We would then have to group our data into a distribution covering the range from $5.45 to $117.25 and we might be tempted to construct the following table:

Size of Orders (in dollars)	Number of Orders
0.00– 19.99	37
20.00– 39.99	42
40.00– 59.99	0
60.00– 79.99	0
80.00– 99.99	0
100.00–119.99	1

It is apparent that this grouping tells us very little about the distribution of the vast majority of our data except, perhaps, that 79 of the 80 orders are for less than $40.00. For many practical purposes we will have discarded too much information by grouping the bulk of our data into as few as two classes.

In order to remedy this situation we could either use more classes, that is, a finer grouping, or we could use an *open class interval*. A class is said to be open if, instead of assigning definite values to its boundaries, we say " . . . or more," " . . . or less," "greater than . . . ," or "less than. . . . "

In the example of the mail-order house we *could* use more classes to accommodate the one large order of $117.25, but in order to retain sufficient information we would, probably, not want to use classes with intervals of more than $5.00. We would then have to take as many as 23 classes, the first going from $5.00 to $9.99 and the last from $115.00 to $119.99. Since 15 of these classes would be empty, this would, clearly, not be a very desirable solution. Instead, we could use an open class and present the orders in the following distribution:

Size of Orders (in dollars)	Number of Orders
5.00– 9.99	6
10.00–14.99	12
15.00–19.99	19
20.00–24.99	19
25.00–29.99	13
30.00–34.99	8
35.00–39.99	2
40.00 or more	1

This gives us a much clearer picture and also more information than the previous table, without making it necessary to use more than 8 classes.

In addition to their obvious advantages, open classes unfortunately have also some very definite disadvantages. Their best feature is that they can accommodate a wide range of values without requiring either a large number of classes or classes that hide too much of the relevant information. Their worst feature is that they do not tell us how much "or more" or "less than" a given value may fall. For example, the last distribution does not give us much information about the size of the largest order and, for all we know, it might have been $45.00 or $1,000.00. Classifications with open classes have the further disadvantage that they present us with difficulties in case we want to present them in graphical form (see page 32), and they make it impossible to calculate certain further descriptions (see page 63). Incidentally, a distribution may have an open class at either end or for that matter at both ends.

There are many situations in which it is not so much of interest to know how many cases fall into each class as it is to know how many of them fall above or below certain values. In our illustration, the management of the mail-order house might be interested in knowing how many of the orders were for *less than $10.00, less than $15.00, less than $20.00, etc.;* or how many of them were for *$5.00 or more, $10.00 or more, $15.00 or more, etc.* If this were the case we could easily convert our distribution into a *cumulative "less than" distribution* or a *cumulative "or more" distribution.*

To illustrate the meaning of these terms, let us first construct a cumulative "less than" distribution of the mail-order house data. Having already grouped these data into the distribution shown on page 22, this will not require much additional work. We can see by inspection that *none* of the orders were for less than $5.00 and that 6 of the orders were for less than $10.00. In order to determine the number of orders less than $15.00 we simply add the frequencies of the first two classes, obtaining $6 + 12 = 18$. Similarly, the number of orders less than $20.00 is $6 + 12 + 19 = 37$, the number of orders less than $25.00 is $6 + 12 + 19 + 20 = 57$, etc. The resulting *cumulative distribution* may then be presented as follows:

Size of Orders (in dollars)	Number of Orders
less than 5.00	0
less than 10.00	6
less than 15.00	18
less than 20.00	37
less than 25.00	57
less than 30.00	70
less than 35.00	78
less than 40.00	80

Had we been interested in showing how many of the orders were *above* or, better, *not below* given values, we could have cumulated (added) the frequencies beginning at the other end, and we would have obtained the following "or more" distribution:

Size of Orders (in dollars)	Number of Orders
5.00 or more	80
10.00 or more	74
15.00 or more	62
20.00 or more	43
25.00 or more	23
30.00 or more	10
35.00 or more	2
40.00 or more	0

Instead of saying "$5.00 or more," "$10.00 or more," . . . , we could also have said "more than $4.99," "more than $9.99," . . . , and in the previous distribution we could have said "$4.99 or less," "$9.99 or less," . . . , instead of "less than $5.00," "less than $10.00," . . . . It should be noted that if we had wanted to use "$5.00 or less," "$10.00 or less," . . . , or "more than $5.00," "more than $10.00," . . . , we could not have used the distribution on page 22: it would have been necessary to refer to the original raw data.

When dealing with large sets of data, we are often more interested in knowing the *percentage* of the total number of cases that fall into a given class than in knowing the actual class frequency. If this is the case, we can easily convert a given distribution into a *percentage distribution* by dividing each class frequency by the total number of cases and then multiplying by 100. (If we omitted the last step, and did not multiply by 100, we could call the resulting distribution a *distribution of proportions* instead of a percentage distribution.)

Since the total frequency (the total number of cases) in the "mail-order house" illustration was 80, we can easily convert the distribution of page 22 into a percentage distribution, by dividing each class frequency by 80 and then multiplying by 100. The resulting percentage distribution becomes:

Size of Orders (in dollars)	Percentage of Orders
5.00– 9.99	7.50
10.00–14.99	15.00
15.00–19.99	23.75
20.00–24.99	25.00
25.00–29.99	16.25
30.00–34.99	10.00
35.00–39.99	2.50

In identical fashion we can also convert cumulative frequency distributions into the corresponding *cumulative percentage distributions*. Taking, for instance, the "less than" distribution of page 27, we get, upon dividing each cumulative frequency by 80 and multiplying by 100, the following cumulative "less than" percentage distribution:

Size of Orders (in dollars)	Percentage of Orders
less than 5.00	0.00
less than 10.00	7.50
less than 15.00	22.50
less than 20.00	46.25
less than 25.00	71.25
less than 30.00	87.50
less than 35.00	97.50
less than 40.00	100.00

EXERCISES

1. A sporting goods store has grouped its sales data, given in dollars and cents, into a classification having the class limits $0.00–$9.99, $10.00–$19.99, $20.00–$29.99, $30.00–$39.99, $40.00–$49.99. Find (a) the corresponding class boundaries, (b) the class marks, and (c) the length of the class interval.

2. A lumber company has grouped the measurements of the diameters of a shipment of logs, given to the nearest tenth of an inch, into a table having the class boundaries 1.95–2.95, 2.95–3.95, 3.95–4.95, 4.95–5.95 inches. Find (a) the corresponding class limits, (b) the class marks, and (c) the class interval.

3. A set of measurements of the length of cash register tapes given to customers as receipts of purchase, measured to the nearest tenth of an inch, is grouped into a distribution having the class limits 4.0–4.9, 5.0–5.9, 6.0–6.9, 7.0–7.9, 8.0–8.9, 9.0–9.9 inches. Find (a) the corresponding class boundaries, (b) the class marks, and (c) the class interval.

4. The class marks of a distribution of prices of eggs per dozen, given to the nearest cent, are 42, 51, 60, 69 and 78 cents. Find the corresponding class limits.

5. If the apple crop (given in bushels) of 400 orchards in a particular county is such that the smallest is 103.4 and the largest is 1,080.5, construct a table with ten classes into which these quantities can be grouped.

6. The values (to the nearest dollar) of 300 packages shipped by a drug company varied from $104 to $968. Construct a table with nine classes into which these measurements might be grouped.

7. The following figures are prices (in dollars) asked for houses under $60,000 in the May 29, 1968, edition of *The Evening Tribune*, San Diego, California.

26,500	25,000	15,300	19,250	41,800	31,950
21,900	16,250	13,250	22,950	17,750	21,500
21,950	15,800	22,000	18,500	23,000	10,000
22,500	34,500	20,000	18,500	46,000	56,500
27,000	42,500	14,950	7,500	35,000	29,950
34,000	17,500	13,000	17,500	36,000	21,950
27,500	35,950	30,000	21,500	28,500	26,950
23,900	25,000	45,000	16,900	15,000	21,500
15,000	24,900	27,650	22,500	25,500	26,950
17,500	30,000	29,000	17,200	22,000	14,500
22,950	19,750	22,950	26,500	42,500	35,000
41,900	25,500	28,900	14,900	18,500	19,950
16,950	18,950	18,900	15,950	28,750	36,500
15,500	29,950	18,800	39,500	20,500	23,500

14,950	34,500	18,500	39,500	27,500	18,900
16,000	16,500	15,500	37,500	31,900	23,475
17,500	16,500	16,200	27,500	19,950	20,875
9,500	28,500	17,950	24,200	40,500	22,000
16,900	29,000	17,500	19,800	27,350	24,750

Group these prices into a frequency distribution using the classes $0–$9,999, $10,000–$19,999, $20,000–$29,999, ... Also construct a cumulative "less than" distribution of these prices.

8. When a machine in a factory is inoperative because of breakdown or other difficulties during working hours, the condition is called "downtime." An efficiency expert measures "downtimes" (in minutes during a certain period of time. The following are 100 consecutive downtimes (in minutes).

45	22	43	22	19	20	27	24	23	21
22	35	23	17	36	18	49	20	38	21
29	41	45	21	11	9	22	21	24	38
34	33	27	22	12	7	23	10	27	59
23	37	16	33	24	27	44	13	32	36
48	21	29	15	19	10	47	31	42	22
20	39	22	34	24	19	25	23	13	40
18	53	17	23	16	24	52	28	37	25
32	30	25	29	31	30	28	44	20	35
40	15	12	14	21	24	23	17	28	27

Group these "downtimes" into a frequency distribution having the classes 5–9, 10–14, 15–19, etc., minutes. Also construct the corresponding percentage distribution and cumulative "or more" percentage distribution.

9. According to the *Fortune Directory*, June 15, 1967, the following are the earnings of the 50 largest commercial banks in the United States as per cent of capital funds for the year 1966.

12.2	9.9	11.2	12.5	9.8
11.5	11.8	11.1	12.3	10.1
11.4	9.2	12.8	9.8	12.6
9.9	10.2	12.6	14.4	10.9
10.2	10.3	11.6	10.2	13.1
10.4	10.9	8.4	14.6	13.4
12.3	11.4	9.2	12.8	11.0
11.2	10.9	10.1	10.9	12.9
11.2	13.2	10.2	16.0	13.6
10.9	11.4	11.6	11.7	13.0

Group these percentages into a frequency distribution and construct the corresponding cumulative "less than" distribution.

10. The following are scores obtained in a secretarial aptitude test for a position with the United States Department of Labor:

```
102   89  207   86  101  127  198   36  187  169  173  130   89
135  121  149  117  142  100  138  115  124  164  135  142  123
152  106  130  123  123  158  171  188  172  154  184  105  128
131  122   94  103   59  129  107  154  115  138  107  154  158
106  150  136  169  112  103  108  112  144  113  128   70  167
135  199  140  219  131  126  105  150  114  138   95   78  160
113  101  158  135   86  113  156   98  171  116  144  144   79
148  125   87  143   92  166  115  145   96  195  129  118  122
129  115   83  183  112  146  143  118  179   86  162  142   65
122  113  110  128  118  126  114  203  113  123  180  135  150
132   72   69   31  118  134   67   99   56   99   80   94   98
148  115  120   59  138  182  158   98  136  140  156  150   86
105  204  132  191  145  164  110   76  163   98  106   98  132
103  113  169  172  160  151  127  139  101  122  141  160   83
102  183  105  106  115  202  123   98  150  211   82  179  138
123  172  135  106  154
```

Group these scores into a table having 10 classes and also construct a cumulative "less than" percentage distribution.

11. The weekly holdings of U.S. government securities (in millions of dollars) by member banks of the Federal Reserve System from May, 1967, through April, 1968, were: (Source: *Federal Reserve Bulletin,* May, 1968.)

45,597	46,833	47,220	48,970
45,929	46,931	47,865	49,344
45,543	46,716	47,868	49,210
45,530	46,504	47,837	48,443
45,726	46,249	48,396	48,724
45,955	46,693	48,902	49,077
45,596	46,207	48,853	49,278
45,654	46,055	48,708	49,840
45,940	46,452	48,937	49,621
46,809	46,976	49,298	49,843
47,158	49,563	49,183	50,153
46,471	49,802	48,758	50,219
46,715	47,098	49,105	49,927

Group these figures into a frequency distribution and construct a cumulative "less than" distribution.

12. The following are the number of physicians per 100,000 population in 110 selected large American cities in 1962: (Source: *Statistical Abstract of the United States 1967.*)

131	131	113	137	127	153	123	202	131	198	218	222	112
185	134	166	245	138	111	232	224	110	146	153	169	123
169	245	157	184	78	156	230	224	212	156	132	142	
141	166	190	154	137	129	152	256	171	188	161	128	
176	158	145	115	132	146	185	175	131	190	136	289	
165	116	198	130	108	95	211	126	204	154	185	119	
119	162	116	129	153	172	148	207	161	190	165	122	
194	129	176	127	192	144	169	178	140	296	149	144	
105	116	100	171	155	127	91	145	218	240	136	128	

Group these data into a table having the class limits 70–89, 90–109, 110–129, etc. Also construct a cumulative "or more" distribution.

13. In order to show how the choice of different classifications can alter the overall shape of a distribution, regroup the data of Exercise 12 (a) into a table having the class limits 78–117, 118–157, 158–197, etc., and (b) into a table having the class limits 70–109, 110–149, 150–189.

2.3 Graphical Presentations

Since many people have a strong aversion to anything dealing with numbers and tables, let us now see how frequency distributions may be represented in a more appealing form, that is, in a form that lends itself more readily to the human power of visualization.

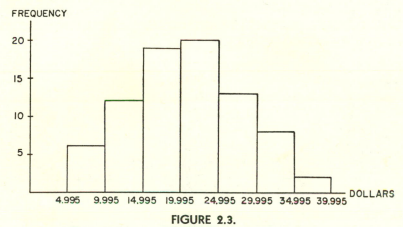

FIGURE 2.3.

The most widely used graphical presentation of a frequency distribution is the *histogram*. It is constructed by representing the measurements or observations constituting a set of data (in Figure 2.3 the order in dollars) on a *horizontal* scale and the class frequencies on a *vertical* scale.   The graph of the distribution is then constructed by drawing rectangles, the bases of which are supplied by the class intervals, and the heights of which are determined by the corresponding class frequencies.   It should be noted that on the horizontal scale

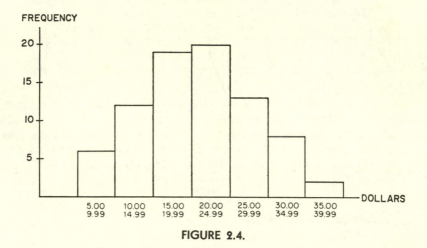

FIGURE 2.4.

we can either indicate the class boundaries, as in Figure 2.3, the class limits, as in Figure 2.4, or for that matter certain more or less arbitrary key values, as in Figure 2.5.   (If a histogram is to be used in a publication or in some other form of official presentation, it is usually easier to understand if we give the class limits even though the bases of the rectangles are determined by the class boundaries.)

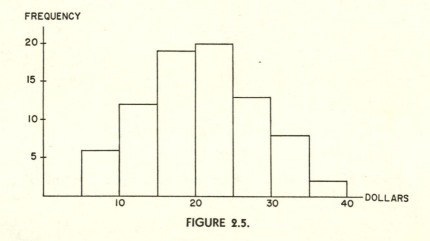

FIGURE 2.5.

There are a number of points that should be watched in the construction of a histogram. To begin with, this kind of figure cannot be used for distributions with open classes. Furthermore, a histogram can be very misleading if the distribution has *unequal class intervals* and we do not make a suitable adjustment. To illustrate this last point, let us regroup the mail-order house data by combining all orders from $20.00 to $29.99 into one class. The resulting frequency distribution will be

Values of Orders (in dollars)	Number of Orders
5.00– 9.99	6
10.00–14.99	12
15.00–19.99	19
20.00–29.99	33
30.00–34.99	8
35.00–39.99	2

and the corresponding histogram is shown in Figure 2.6, with the class frequencies again given by the heights of the rectangles.

If we look at Figure 2.6 we get the immediate impression that the vast majority of the orders are between $20.00 and $29.99. This is, of course, a mistake, since less than half of the data belong to this class. We make this mistake because we intuitively compare the *size*, that is, the areas, of the rectangles instead of their heights. In order to avoid getting such a misleading picture, it is necessary to make the

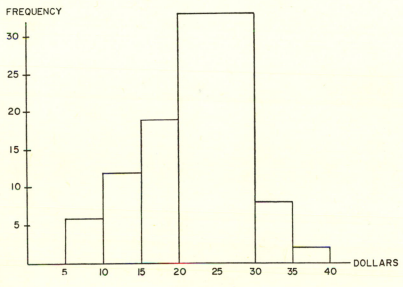

FIGURE 2.6.

following adjustment: *If one class interval is twice as wide as the others, we divide the height of its rectangle by two; similarly, if a class interval is three times as wide as the others, we divide the height of its rectangle by three, etc.* In following this practice we are actually representing the class frequencies by the *areas* of the rectangles instead of by their heights. In that case it is also better not to indicate the frequency scale but to give the class frequencies, as we have done in the adjusted histogram of Figure 2.7.

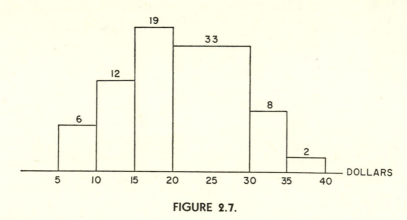

FIGURE 2.7.

The practice of representing class frequencies by means of areas is particularly desirable if we want to approximate histograms with continuous curves. For instance, if we wanted to approximate the histogram of Figure 2.4 with a smooth curve, we could say that the number of orders exceeding $29.99 is represented by the shaded area under the

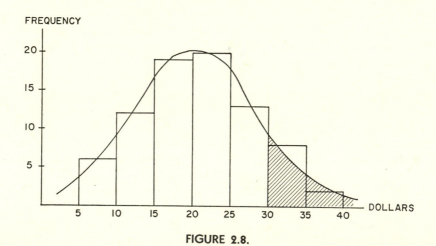

FIGURE 2.8.

curve of Fig. 2.8. Evidently, this area is approximately equal to that of the corresponding two rectangles of the histogram.

Frequency distributions may also be presented graphically as so-called *frequency polygons*. A frequency polygon (see Figure 2.9) is constructed by plotting the class frequencies of the various classes at their respective class marks, that is, the midpoints of the class intervals, and then connecting these points by means of straight lines. In

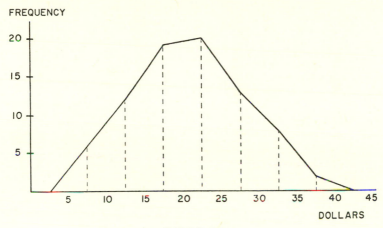

FIGURE 2.9.

order to "complete the picture" it is usually desirable to add one class at each end of the distribution. Since these extra classes have zero frequencies, both ends of the figure forming the frequency polygon will come down to the horizontal axis. It should be noted that frequency polygons present us with the same difficulties which we previously met in our study of histograms. They cannot be used for distributions with open classes, and a suitable adjustment, the same as in the case of the histogram, has to be made when there are unequal classes.

Had we applied the technique of constructing a frequency polygon to a cumulative distribution, we would have obtained a polygon that is generally known as an *ogive*. One difference between a frequency polygon and an ogive is that we plot cumulative frequencies instead of the ordinary class frequencies. Another distinction is that we do *not* plot the cumulative frequencies at the class marks. For example, it would only be reasonable to plot the cumulative frequencies corresponding to "$30.00 or less" or "less than $30.00" at $30.00, and it is important therefore to indicate whether the ogive represents a "less than" or an "or less" distribution. After all, *there is a difference between "$30.00 or less" and "less than $30.00."* (This difficulty could be avoided if the cumulative frequencies were plotted at the class boundaries instead of at the class limits.)

Figure 2.10 gives the ogive of the "less than" distribution of the mail-order house data which we have been discussing throughout this chapter, and we could, similarly, have drawn an ogive of the "or more" distribution. Incidentally, instead of indicating the class frequencies (or cumulative frequencies) on the vertical scale, we could also have used percentages. We would then have obtained the histograms, frequency polygons, and ogives of the corresponding percentage distributions.

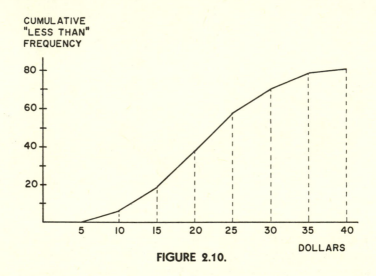

FIGURE 2.10.

Although the visual appeal of histograms and frequency polygons already exceeds that of a frequency table, there are methods of presentation which go one step further and present frequency distributions in a more dramatic way. We are referring here to the various kinds of pictorial presentations which will be treated briefly in Appendix I. (Some problems connected with the *presentation* of frequency tables in reports, publications, or advertisements will also be discussed in Section 2.5.)

EXERCISES

1. Construct a histogram, a frequency polygon, and an ogive of the distributions obtained either in Exercise 7, Exercise 8, Exercise 9, or Exercise 10 on pages 29 and 30.

2. Construct an ogive of the cumulative distribution obtained in Exercise 11 on page 30.

3. Construct a histogram, a frequency polygon, and an ogive of the distributions obtained in Exercise 12 on page 31.

4. Draw a histogram of the 1966 average monthly aid paid to families with dependent children, by states.

Average Monthly Payments	Number of States
$25– 49	1
50– 74	3
75– 99	3
100–124	13
125–149	11
150–174	9
175–199	8
200–249	2

Source: *Statistical Abstract of the United States 1967.*

5. The following is the distribution of the net paid circulation of the evening editions of 97 of the leading daily newspapers in the United States in 1967. (Figures based on averages over a 6 month period ending March 31, 1967.)

Net Paid Circulation (in thousands)	Number of Newspapers
0– 49	12
50– 99	16
100–149	30
150–199	7
200–249	11
250–299	6
300–399	7
400–499	4
500–599	0
600–699	3
700–799	1

Source: *Information Please Almanac 1968.*

Draw a histogram and a frequency polygon of this distribution.

6. Convert the distribution of Exercise 5 into a cumulative "less than" percentage distribution and draw the corresponding ogive.
7. Draw histograms of the two distributions of Exercise 13 on page 31.

2.4 The Construction of Categorical Distributions

Many of the comments we made in this chapter with regard to numerical distributions apply equally well to *categorical* (*or qualitative*) *distributions*. Again, we shall have to make decisions about the number of categories (classes) that are to be used, and we shall always have to make sure that all of our data are accommodated and that there will be no ambiguity or overlap.

Any decision about the number of classes to be used in a categorical distribution will have to be dictated partly by the nature of our data and partly by our objectives. For example, if we wanted to construct a table showing how industrial and commercial failures in the United States in 1966 are distributed among various industries or industrial groups, we could choose *five* categories and arrive at the following distribution:

Industrial Group	Number of Failures
Mining and manufacturing	1,852
Wholesale trade	1,255
Retail trade	6,076
Construction	2,510
Commercial service	1,368

For some purposes this distribution may very well be adequate, but if a finer classification were called for we could use, for example, the 39 categories listed by Dun and Bradstreet, Inc. Obviously, it would be absurd to decide in a problem like this that we would like to use 7, 8, or 12 classes without giving due consideration to the nature of our data.

In many studies we have to choose categories before the data are actually collected, and it is often very difficult to foresee all contingencies, to make sure that all of the data will be accommodated. For instance, if we wanted to determine the distribution of the makes of cars driven by the residents of a certain community, we might take a list of all standard makes and then in our survey put a check for each car in the appropriate class. This would not lead to complications unless we came upon some mechanically inclined individual who put a Packard engine on a Chrysler chassis and combined this with a body whose ancestry is partly Chevrolet and partly Ford. This is one reason it is often advisable to add a category captioned "others."

Another reason great care should be taken in the selection of categories is that it is easy to make the mistake of choosing overlapping classes. For example, if we tried to classify various food items sold in a supermarket and chose such categories as "meats and meat products," "frozen foods," etc., we might have some difficulty in deciding where to put "frozen hamburger steaks." Similarly, if we tried to classify the inhabitants of some country according to occupation, we would hardly know what to do with a *farm manager* if our table contained without qualifications the two categories "farmers" and "managers."   In this illustration it would probably be desirable to add a separate category for *farm managers*.

Although, when constructing a categorical distribution we do not have to worry about such mathematical details as class limits, class boundaries, class marks, and the like, we may find it much more difficult to find the categories in which the individual items belong.   If we work with a numerical distribution we have only to check the size of numbers, but in the case of a categorical distribution we often run into questions of *definition*.   For example, if we classified a number of industries as manufacturers of (a) *luggage*, (b) *handbags and small leather goods*, and (c) *gloves and miscellaneous leather goods*, we would hardly know how to classify these industries unless we were given clear-cut definitions as to what these categories mean.   It is for this reason that it is very often desirable to use the standard categories of the Bureau of the Census or other government agencies.   Some of the fields for which such standard categories are available are "Kind of business," "Occupation," "Industry," "Commodities," and "Crops."   These classifications and the corresponding definitions of the individual categories may be found in manuals prepared by various agencies of the federal government (for references see P. M. Hauser and W. R. Leonard, *Government Statistics for Business Use*, New York: John Wiley, 1956).

EXERCISES

1. Construct a table from a college or university catalogue showing how the faculty is distributed among the ranks of "Professor," "Associate Professor," "Assistant Professor," etc.
2. Using the vowels *a, e, i, o* and *u*, construct a distribution which shows the frequencies with which they occur in the first full paragraph on page 39 of this text.
3. Construct a distribution of listed *retail stores* from the yellow pages of your local telephone book.   Use the following categories:

    Florists
    Jewelers
    Women's Apparel
    Furniture
    Automobile Parts and Supplies
    Service Stations–Gasoline and Oil
    Lumber
    Restaurants
    Druggists

4. Use the latest issue of the *Monthly Labor Review* to construct a distribution showing how many of the food items listed showed an *increase, no change,* or a *decrease* in their retail price indexes for the most recent month listed as compared to the previous month.

2.5 Tabular Presentations

Having so far treated only problems dealing with the construction of numerical and categorical distributions, let us now mention some problems connected with their formal presentation in reports, advertisements, or other kinds of publications. If we wanted to publish the mail-order house distribution, which we have used as an illustration throughout this chapter, we might present it like this:

TABLE 1

Orders Received by Department K of the XYZ Mail-Order
House on April 5, 1968

VALUES OF ORDERS	NUMBER OF ORDERS
Total...................,	80
$ 5.00 to $ 9.99............	6
$10.00 to $14.99............	12
$15.00 to $19.99............	19
$20.00 to $24.99............	20
$25.00 to $29.99............	13
$30.00 to $34.99............	8
$35.00 to $39.99............	2

Source: Accounting Department of the XYZ Mail-order House.

Clarity is the most important feature of any tabular presentation of a frequency distribution, or, for that matter, the tabular presentation of any kind of statistical data. It may be accomplished by carefully selecting the type used for titles and captions, by using horizontal and vertical lines, proper spacing, leaders, and other devices which are more familiar to a printer than they are to a mathematician or a statistician.

The following are some of the most widely used rules for constructing tabular presentations:

1. *Number and Title:* It is usually desirable to give each table a number as well as a title. The number will provide an easy reference

and the title will make it possible to understand a table without having to refer to the accompanying text.

2. *Captions and Stubs:* Captions are the designations given to vertical columns, while stubs are those given to horizontal lines. Both should be brief, descriptive, and clearly defined. (If it is impossible to be brief and clear at the same time, it is usually better to make captions or stubs brief, supplying the clarification by means of footnotes.) Considerable care should be taken in choosing the type of captions and stubs to emphasize the most important features of the table.

3. *Rulings, Spacings, and Leaders:* Rulings and spacings serve more or less the same function, they serve to set apart the various components of a table. For example, in our illustration on page 40 we could have used a horizontal line instead of the extra white space in order to set the word "total" apart from the stubs of the frequency distribution. The modern tendency seems to be to use more white spaces and fewer rulings. Further emphasis may be given to different parts of a table by using light and heavy lines as well as double lines. The use of leaders (......) makes it easy to associate the correct number (frequency) with each class or category. A leader is particularly desirable if a stub occupies more than one line.

4. *Source reference:* The source of the data used in the construction of a table should always be indicated unless, of course, the data are original. It is customary to put a source reference either below a table, as we have done on page 40, or in parentheses immediately below the title.

5. *Footnotes:* Footnotes should be used for explanations and definitions needed for the sake of clarity that would otherwise clutter up the main body of the table. It is customary to put footnotes below the table but preceding the source reference.

6. *Arrangement of the Classes:* For numerical distributions the general rule is to arrange the classes in order of magnitude—that is, magnitude as shown in the class caption (stub) and not the magnitude of the class frequency—beginning with the smallest values at the top of the table. For categorical distributions there is no general rule, but, if possible, we can follow the order of listings suggested by government agencies, alphabetic order, or some obvious order suggested by the nature of our data.

There are numerous other points that might be mentioned about the construction of tabular presentations, and the reader will find most of them discussed in the *Bureau of the Census Manual of Tabular Presentation*. For example, we might also point out that, if necessary, *units of measurement* should be indicated in parentheses directly under the appropriate captions or, if too bulky, in footnotes below the table. Naturally, we do not *have* to obey the rules listed above (unless we happen to work for the federal government), but they are based on considerable experience and will usually lead to clear and professional looking tables.

BIBLIOGRAPHY

Further material on the construction of numerical and categorial frequency distributions may be found in

Croxton, F. E., and Cowden, D. J., *Practical Business Statistics*, 3rd ed. Englewood Cliffs, N.J.: Prentice-Hall, Inc., 1960, Chap. 8.

Lewis, E. E., *Methods of Statistical Analysis in Economics and Business*, 2nd ed. Boston: Houghton Mifflin Co., 1963, Chap. 2.

Spurr, W. A., and Bonini, C. P., *Statistical Analysis for Business Decisions*. Homewood, Ill.: R. D. Irwin, Inc., 1967, Chap. 4.

Waugh, A. E., *Elements of Statistical Method*. New York: McGraw-Hill, 1953, Chap. 3.

For a more detailed discussion of the problems connected with tabular presentatations see, for example,

Hanson, K. O., and Brabb, G. J., *Managerial Statistics*, 2nd ed. Englewood Cliffs, N.J.: Prentice-Hall, Inc., 1961, Chap. 10.

U.S. Department of Commerce, *Bureau of the Census Manual of Tabular Presentation*. Washington, D.C.: Government Printing Office, 1949.

For a discussion of what is to be avoided in the presentation of statistical data, consult

Huff, D. and Geiss, I., *How to Lie with Statistics*. New York: W. W. Norton & Co., Inc., 1954.

Reichmann, W. J., *Use and Abuse of Statistics*. London: Methuen & Co., Ltd., 1961.

CHAPTER 3

Measures of Location

3.1 Introduction

If we wanted to describe a house, a car, or a tree, our description would have to depend partly on the nature of the object itself and partly on the purpose we might have for giving the description. This argument applies also to the description of numerical data. The type of description we may choose or the statistical technique we may employ will also depend partly on the nature of the data themselves and partly on the purpose that we have in mind.

As we have pointed out in Chapter 1, the purpose of statistics is either to describe facts as they are contained in a set of data as such *or* to generalize and go beyond the information with which we are supplied. Correspondingly, we find ourselves either in the domain of descriptive statistics or in that of inductive statistics. In order to emphasize this distinction, let us now make the following definition: If a set of data consists of all conceivably (or hypothetically) possible observations of a certain phenomenon, we shall refer to it as a *population;* if it contains only part of these observations we shall refer to it as a *sample.*

If we are given complete records on the salaries paid to the executives of a certain railroad, these salaries may be looked upon as a *population.* Similarly, if we are given complete information about the prices of all items sold in a department store on a certain day, these prices constitute a *population.* In each case we are given all the facts about a certain phenomenon; in the first example it is the phenomenon of the salaries paid to executives by the particular railroad and in the second example it is the phenomenon of the prices of the items sold in the particular department store on the given day.

While looking at these two examples of populations, it may have occurred to the reader that *whether or not a given set of data is to be referred to as a population depends to some extent on how we look at it.* If we considered the population of the salaries paid to the executives of *all* railroads, then figures about the salaries paid to executives of a particular company would constitute merely a sample. Similarly, if we considered the population of the prices of all items sold in the given department store *throughout the year*, then the prices of the items sold on a particular day would also constitute merely a sample.

Whether or not a set of data is to be referred to as a population or a sample depends on what we intend to do with it. If we are not going to make generalizations about salaries paid by other railroads and if we are not going to make generalizations as to what these salaries might be in the future, we shall refer to the salary data of the particular railroad as a population. Similarly, we shall refer to the department store data as a population so long as we make no generalizations about the prices of items sold either on other days or in other stores.

To give one more example, let us suppose that the official election returns for a certain county show that 1,257 persons voted for Candidate X and that 483 voted for his opponent, Candidate Y. If Mr. X and Mr. Y were running for the office of county clerk, we would look upon the given information as constituting a *population*, for it contains all relevant facts. However, if these candidates were running for the U.S. Senate, the election returns of a single county would constitute merely a *sample*, the population being the election returns for the entire state.

In statistics, the word *sample* is used very much in its everyday connotation. If we interview 10 of a possible 500 employees of a certain company, we can consider the opinions which they express to be a sample of the opinions of *all* employees; if we measure the "lifetimes" of 5 light bulbs of a certain brand, we can consider these measurements to be a sample of the measurements we would have obtained if we had measured *all* similar light bulbs made by the same firm. The last illustration indicates why we often have to be satisfied with samples, being unable to get information about the complete population. (If we measured the "lifetimes" of *all* the light bulbs made by this firm, they would be in the unfortunate position of having none left to sell.)

In some books the reader will find the term "universe" used instead of "population." The very fact that these two terms are synonymous to a statistician makes it quite evident that neither is used in its colloquial sense. In statistics, both terms refer to sets of data consisting of all possible or, at least, *all hypothetically possible* observations of a certain phenomenon and *not* to human beings or animals.   We added the qualification "hypothetically possible" because we may wish to consider, for example, the results obtained in 10 flips of a coin as a sample from the population consisting of *all* possible flips of the coin. Clearly, this is a hypothetical population.

Although in the first part of this book we shall limit ourselves to problems of description, it is important even here to differentiate between sets of data that are populations and those that are samples. As we have pointed out earlier, the type of description we shall use in a given problem may very well depend on what we intend to do with our data, that is, it will depend on whether we want to generalize on the basis of a description of a sample or whether we merely want to describe a population.   In this chapter we shall sometimes use *different symbols*, depending on whether we are describing a population or a sample, and in Chapter 4 we shall even use *different formulas*, depending on whether we are describing a population or a sample.

In this and in the following two chapters we shall study *four* basic kinds of descriptions called *measures of location*, *measures of variation*, *measures of symmetry*, and *measures of peakedness*.   The first of these, which constitute the subject matter of this chapter, are also referred to as "measures of central values," "measures of central tendencies," and "measures of position."   They tell us, for example, the location of the "center" or the "middle" of a set of data, in other words, some sort of average, and, among other things, the location of a value which is such that it is exceeded only by 25 per cent of the data.

EXERCISES

1. Suppose we are given complete information about the number of policies sold by the Equitable Life Insurance Company during March, 1968, and the amount of each policy.   Give one example each of problems in which we would consider this set of data to be (a) a population and (b) a sample.
2. Suppose we are given the College Board scores of all freshmen entering a certain university in the fall of 1968.   Give one example each of problems in which we would consider this set of data to be (a) a population and (b) a sample.
3. Suppose we are given complete information about salaries paid to switchboard operators in Buffalo, New York, in September, 1968. Give one example each of problems in which we would consider this set of data to be (a) a population and (b) a sample.
4. Discuss whether the examples you used in Exercise 1, 2, or 3 to illustrate how the data can be looked upon as samples would really permit "reasonable" or "justifiable" generalizations.

3.2 The Arithmetic Mean, Ungrouped Data

There are many problems in which we would like to represent a set of numbers by means of a *single* number which is, so to speak, descriptive of the entire set of data. The most popular measure used for this purpose is what most laymen call an *average* and what statisticians call the *arithmetic mean*.† Given a set of n numbers, x_1, x_2, x_3, . . . , x_n, their arithmetic mean is defined as *their sum divided by n*. Symbolically, we can therefore write

$$\text{arithmetic mean} = \frac{\sum_{i=1}^{n} x_i}{n} \qquad (3.2.1)\star$$

To illustrate this formula, with which most readers must surely be familiar, let us suppose that we are given the following figures representing the annual number of work stoppages (in thousands) which occurred in the United States from 1962 to 1966:

$$3.6 \qquad 3.4 \qquad 3.7 \qquad 4.0 \qquad 4.4$$

On the basis of this information we can say that the "average" number of work stoppages, that is, the arithmetic mean, for the given period of time is

$$\frac{3.6 + 3.4 + 3.7 + 4.0 + 4.4}{5} = 3.8 \text{ work stoppages}$$

It is customary to abbreviate the term *arithmetic mean* and refer to this average simply as the *mean*. Since there also exists a *geometric mean* and a *harmonic mean* (see Section 3.7), it must be understood that whenever we speak of a *mean* we are actually referring to the arithmetic mean as defined by formula (3.2.1).

Since the mean is one of the most widely used statistical descriptions, it would seem desirable to give it a special symbol. In many books the reader will find the mean of the numbers x_1, x_2, . . . , x_n written as $\bar{x}$ and, similarly, the mean of y_1, y_2, y_3, . . . , y_n as $\bar{y}$, and the mean of z_1, z_2, . . . , z_n as $\bar{z}$. We shall deviate slightly from this practice and distinguish between the *mean of a sample*, for which we

---

† Statisticians do not like the term "average" because it has too loose a connotation. It has different meanings, for example, when we refer to a *batting average*, an *average income*, or an *average person*.

★ Formulas which are marked with a star are actually used for practical computations. This will make it easier for the reader to distinguish between formulas used for calculations and those given primarily for definitions or as part of derivations.

shall use the symbol $\bar{x}$ (or $\bar{y}$ or $\bar{z}$), and the *mean of a population,* for which we shall use the symbol μ (the Greek letter *mu*). The advantage of this symbolism will become apparent in later chapters where we shall use the mean of a sample as an estimate of the mean of the population from which the sample was drawn. Clearly, it would sound very confusing if we were to say that we use a mean to estimate a mean. Instead, we shall say that we use $\bar{x}$ to estimate μ. *The practice of using letters of the Roman alphabet for descriptions of samples and letters of the Greek alphabet for descriptions of populations has many advantages and is getting to be widely accepted.†*

To illustrate this usage of $\bar{x}$ and μ, let us refer again to the example given on page 4. Considering a sample of 5 light bulbs of a certain Brand A, we found that they had an "average lifetime" of $\bar{x} = 918$ hours. We took this sample because we were interested in the "average lifetime" of *all* 60 watt bulbs made by this firm: we were interested in μ, the mean of the "lifetimes" of *all* the 60 watt light bulbs made by this firm. Since we cannot determine μ exactly without burning out all of the bulbs, we have no alternative here but to generalize on the basis of a sample. We might, thus, *estimate* μ as being 918 hours. Whether or not such a generalization is justifiable is something we shall discuss in later chapters.

Some of the noteworthy properties of the arithmetic mean are that (1) most people understand what is meant by a *mean*, although they may not actually call it by that name; (2) it always exists, it can always be calculated for any kind of numerical data; (3) it is always unique, or, in other words, a set of numerical data has *one and only one* arithmetic mean; and (4) it takes into account each individual item. Whether or not this fourth property is really desirable is open to question, because a single extreme (very large or very small) item can affect the mean to such an extent that it is questionable whether it is really "representative" of our data.

Let us consider, for example, a party at which the ages of six guests are 19, 22, 20, 24, 20, and 75. If we said that the average age of these guests is

$$\frac{19 + 22 + 20 + 24 + 20 + 75}{6} = 30 \text{ years}$$

this might easily give a very misleading impression. The property of the mean that it is easily affected by one very large or very small item can have serious consequences in problems in which we use a *sample*

† In order to distinguish between descriptions of populations and samples, statisticians not only use different symbols, but they refer to the first as *parameters* and the second as *statistics.* We, thus, say that μ is a parameter and that $\bar{x}$ is a statistic.

mean for purposes of estimation. Let us suppose, for example, that in the test of the light bulbs which we mentioned in Chapter 1 and earlier in this section, the technician who conducted the tests made the careless mistake of recording the life times of the five light bulbs as

$$985, 863, 1024, 972, \text{ and } 246 \text{ hours}$$

instead of

$$985, 863, 1024, 972, \text{ and } 746 \text{ hours}$$

The resulting arithmetic mean is

$$\bar{x} = \frac{985 + 863 + 1024 + 972 + 246}{5} = 818 \text{ hours}$$

instead of the correct value of 918. One careless mistake can, thus, have a pronounced effect on $\bar{x}$ and it can lead to serious errors if we use this sample mean as an estimate of μ, the mean of the lifetimes of all similar light bulbs made by Firm A. (In Section 3.4 we shall discuss another kind of average, to be called the *median*, which has the advantage of being generally less easily affected by the kind of carelessness we have just discussed.)

Another desirable property of the mean is that if we are given the means of *several* sets of data, we can determine the over-all mean of the combined data without having to go back to the original data. Let us suppose, for example, that we are interested in studying the faculty salaries paid by a certain small university and that we are given the following table:

	Number	*Mean salary*
Professors.....................	16	$13,160
Associate professors............	13	10,040
Assistant professors............	22	8,210
Instructors....................	14	6,760

What we would like to know is the mean salary paid to these 65 college teachers. Of course, we could get this mean by referring to the treasurer's records and calculating the mean of the 65 salaries, but the purpose of this example is to show how the over-all mean may be obtained *without* having to go back to the raw data.

Generally speaking, if we have a set of n_1 numbers whose mean is $\bar{x}_1$, a set of n_2 numbers whose mean is $\bar{x}_2$, a set of n_3 numbers whose mean is $\bar{x}_3$, and a set of n_4 numbers whose mean is $\bar{x}_4$, the over-all mean of these $n_1 + n_2 + n_3 + n_4$ numbers is given by the formula

$$\bar{x} = \frac{n_1\bar{x}_1 + n_2\bar{x}_2 + n_3\bar{x}_3 + n_4\bar{x}_4}{n_1 + n_2 + n_3 + n_4} \tag{3.2.2}$$

Before we prove this, let us apply it first to the above example. Since $n_1 = 16$, $n_2 = 13$, $n_3 = 22$, $n_4 = 14$ and $\bar{x}_1 = 13,160$, $\bar{x}_2 = 10,040$, $\bar{x}_3 = 8,210$, and $\bar{x}_4 = 6,760$, we obtain by direct substitution

$$\bar{x} = \frac{16(13,160) + 13(10,040) + 22(8,210) + 14(6,760)}{16 + 13 + 22 + 14}$$

$$= \frac{616,340}{65} = \$9,482 \text{ (to the nearest dollar)}$$

To prove formula (3.2.2), let us write the total amount paid to the full professors as Σ_1, the total amount paid to the associate professors as Σ_2, the total amount paid to the assistant professors as Σ_3, and the total amount paid to the instructors as Σ_4. Since the over-all mean is by definition the total amount paid to all these teachers (the sum of all their salaries) divided by their number, we can write

$$\bar{x} = \frac{\Sigma_1 + \Sigma_2 + \Sigma_3 + \Sigma_4}{n_1 + n_2 + n_3 + n_4} \tag{3.2.3}$$

Furthermore, the means of the individual groups are

$$\bar{x}_1 = \frac{\Sigma_1}{n_1}, \qquad \bar{x}_2 = \frac{\Sigma_2}{n_2}, \qquad \bar{x}_3 = \frac{\Sigma_3}{n_3}, \qquad \bar{x}_4 = \frac{\Sigma_4}{n_4}$$

If we now multiply these four equations by n_1, n_2, n_3, and n_4, respectively, we get

$$n_1\bar{x}_1 = \Sigma_1, \qquad n_2\bar{x}_2 = \Sigma_2, \qquad n_3\bar{x}_3 = \Sigma_3, \qquad n_4\bar{x}_4 = \Sigma_4$$

and if we substitute these results into (3.2.3), we finally have

$$\bar{x} = \frac{n_1\bar{x}_1 + n_2\bar{x}_2 + n_3\bar{x}_3 + n_4\bar{x}_4}{n_1 + n_2 + n_3 + n_4}$$

which is the formula we set out to prove.

Had we considered an arbitrary number, say k, separate sets of data, we could similarly have shown that the over-all mean is

$$\bar{x} = \frac{\sum\limits_{i=1}^{k} n_i\bar{x}_i}{\sum\limits_{i=1}^{k} n_i} \tag{3.2.4}\star$$

where n_i stands for the number of items in the ith set of data and $\bar{x}_i$ for its mean. (It goes without saying that if we looked upon our data as *populations* we could use the same formula but write μ instead of $\bar{x}$ and μ_i instead of $\bar{x}_i$.)

Another important property of the mean, perhaps its most important one, will become apparent in later chapters where sample means will be used for purposes of estimation and decision. The mean will prove to be relatively *reliable*, it will usually not vary too much from sample to sample, at least not as much as some other kinds of statistical descriptions. We shall touch upon this point briefly in Section 3.4, where we shall compare the "reliability" of a mean with that of another measure of location.

EXERCISES

1. The 1966 net income as per cent of operating revenues of 10 leading U.S. railroads were 8.3, 9.8, 8.0, 12.2, 17.8, 15.9, 8.3, 6.7, 8.0, and 7.2 per cent. Find the mean of these 10 percentages, giving your answer to the nearest tenth of a per cent.
2. The following grades were received by 24 accounting students in a short quiz: 5, 6, 8, 9, 6, 4, 8, 6, 4, 9, 8, 9, 10, 3, 10, 7, 9, 5, 10, 6, 6, 7, 4, and 9. Find the mean of these grades.
3. The total monthly benefits paid under all Unemployment Insurance Programs from March, 1966, to February, 1967, were (in millions of dollars) 240.0, 161.4, 136.1, 123.4, 121.0, 152.0, 114.3, 100.4, 122.6, 166.4, 235.8, and 230.9. Find the mean of these monthly totals, giving your answer to the nearest tenth of a million.
4. An investment counselor recommended a portfolio of twenty stocks for appreciation. Find the mean of the current prices of these stocks which are 65, 58, 76, 46, 47, 67, 34, 64, 56, 46, 103, 85, 42, 56, 80, 111, 55, 74, 70, and 65 dollars per share.
5. The weights of 39 football players on the 1967 Boston Patriots team (rounded to the nearest 5 pounds) were:

270	190	250	205	180	240	255	250	190	220
220	200	255	255	250	240	190	245	195	260
250	185	245	220	270	255	240	210	210	260
230	180	195	180	185	220	220	250	265	

Source: Office of Boston Patriots.
Find the mean of these weights.

6. The consumer price index for food, as reported monthly in the *Monthly Labor Review* from January, 1966, to December, 1967, was:

111.4	113.5	115.6	114.7	113.9	115.9
113.1	113.8	115.6	114.2	115.1	115.7
113.9	114.3	114.8	114.2	116.0	115.6
114.0	115.8	114.8	113.7	116.6	116.2

Find the mean of this index for the given period of time, giving your answer to one decimal.

7. What is the average (mean) price of the houses referred to in Exercise 7 on page 28?
8. Find the mean of the 100 "downtimes" of Exercise 8 on page 29.
9. Find the mean of the earnings of the 50 largest commercial banks as per cent of capital funds for the year 1966, of Exercise 9 on page 29.

10. Find the mean score obtained by the 200 applicants for secretarial positions in Exercise 10 on page 30.

11. Calculate the average (mean) number of weekly holdings of U.S. Government Securities by member banks of the Federal Reserve on the basis of the figures given in Exercise 11 on page 30.

12. Calculate the mean number of physicians per 100,000 population for the figures given in Exercise 12 on page 31.

13. If one department store pays its 800 employees an average (mean) weekly wage of $68.57, a second department store pays its 1,000 employees an average weekly wage of $70.45, and a third department store pays its 1,200 employees an average weekly wage of $67.50, what is the overall mean of the wages received by these department store employees?

14. In 1965, 512,088,000 tons of bituminous coal were produced in the United States at an average (mean) value of $4.44 per ton. In the same year 14,866,000 tons of anthracite coal were produced at an average (mean) value of $8.21 per ton. Use formula (3.2.4) to find the average (mean) value of all coal produced in the United States.

15. The average (mean) weight of the four regular backs who started the 1967 season for the Patriots football team was 210 pounds while the mean weight of the seven regular linemen was 235 pounds. Find the overall average weight of this team. The above weights are rounded to the nearest 5 pounds.

3.3 The Arithmetic Mean, Grouped Data

If we want to describe a man's appearance, we can either look at him directly or we can base our description on a photograph. Similarly, we can describe a set of data by either looking directly at the raw data or by considering a distribution into which they have been grouped. Just as a direct description of a person is apt to be more accurate than one based on a photograph, we can also say a description based on raw data is apt to be more accurate than a description based on a frequency table. Although the difference between two such descriptions is often very slight, we shall distinguish in this and in subsequent chapters between descriptive measures obtained directly from raw data and the corresponding descriptive measures obtained from distributions.

Although the arithmetic needed for obtaining a mean does not go beyond addition and division, it would require a considerable amount of work if we had to find, for example, the mean of several thousand numbers without using punch-cards or other special equipment. Hence, when dealing with mass data it is often advisable to begin by grouping and then computing the mean (or other kinds of descriptions) from the resulting distribution.

Let us remember that whenever a set of data is grouped each measurement, so to speak, loses its identity. Instead of knowing the precise value of each item, we are left with a distribution which only tells us how many of the numbers fall into each class. Not knowing the actual value of each item, it might seem that it would no longer be possible to calculate their mean. This is true, unless we are willing to make some assumption concerning the distribution of the items *within* each class and to settle, thus, for an approximation. Let us assume, therefore, that *all measurements falling into a given class are located at its class mark, that is, at the midpoint of the class interval.* This assumption is not as unreasonable as it may seem. Although we cannot really expect all our numbers to be located at the class marks, some will fall above, some will fall below, and in calculating the mean the error introduced by our assumption will, thus, more or less "average out."

To illustrate how this assumption may be used to find the mean from grouped data, let us consider again the distribution of the mail-order house data, which was:

Size of Orders (in dollars)	Class mark	Frequency
5.00– 9.99	7.495	6
10.00–14.99	12.495	12
15.00–19.99	17.495	19
20.00–24.99	22.495	20
25.00–29.99	27.495	13
30.00–34.99	32.495	8
35.00–39.99	37.495	2
Total		80

According to the assumption we just made, these orders will be treated as if *six* had been for $7.495, *twelve* for $12.495, *nineteen* for $17.495, etc. To find the sum appearing in the numerator of the formula for the mean we shall, thus, have to add *six* times 7.495 to *twelve* times 12.495 to *nineteen* times 17.495 to *twenty* times 22.495, and so forth, and we get

$$\bar{x} = \frac{6(7.495) + 12(12.495) + 19(17.495) + 20(22.495) + 13(27.495) + 8(32.495) + 2(37.495)}{80}$$

$$= \$20.87$$

It is interesting to note that if we had calculated the mean of the 80 orders given on page 21 with Formula (3.2.1) we would have obtained $\bar{x} = \$21.05$. *The 18-cent difference between the two means must be attributed to the fact that some information is lost when data are grouped.*

To formalize what we have done in the calculation of the mean of the mail-order distribution, let us suppose that we are given a distribution whose *class marks* are x_1, x_2, x_3, . . . , x_k and whose *class frequencies* are f_1, f_2, f_3, . . . , f_k. Making the same assumption as before we shall assume that f_1 of the numbers are equal to x_1; f_2 of the numbers are equal to x_2; f_3 of them are equal to x_3, . . . ; and f_k of them are equal to x_k. We can, thus, write their mean as

$$\bar{x} = \frac{x_1 f_1 + x_2 f_2 + x_3 f_3 + \cdots + x_k f_k}{f_1 + f_2 + f_3 + \cdots + f_k}$$

where the numerator stands for the sum of all the numbers and the denominator stands for their total frequency. Using summation signs we can finally write

$$\bar{x} = \frac{\sum\limits_{i=1}^{k} x_i f_i}{\sum\limits_{i=1}^{k} f_i} \qquad (3.3.1)\star$$

for the mean of a distribution with k classes. (If we looked upon our data as a *population*, we would write μ instead of $\bar{x}$, but otherwise the formula would be the same.)

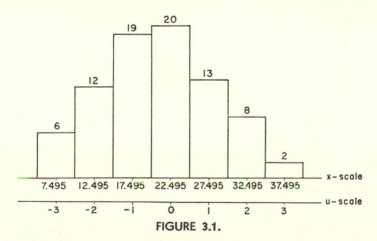

FIGURE 3.1.

The calculation of a mean of grouped data can still be quite tedious, particularly if the class marks x_i are large numbers or if they are given to many decimals. In order to reduce the necessary work, let us now demonstrate a standard trick, that of *changing the scale of measurement*. Let us consider Figure 3.1, which again represents the histogram of the mail-order distribution. The units of the horizontal scale, on which we have indicated the class marks, are dollars, and this scale is denoted as the x-scale. To simplify the numbers with which we have to work, let us imagine that we pick up the x-scale without changing the remainder of the diagram, and that we substitute in its place a new scale which, in Figure 3.1, is called the *u-scale*.   The new scale was chosen in such a way that *the class marks of the distribution coincide with the units of the new scale*.   The zero of the new scale was placed in line with the class mark of the fourth class and in the new scale the class marks of our distribution are at -3, -2, -1, 0, 1, 2, and 3.

The process of changing a scale (mathematicians refer to it as *performing a transformation* and some statisticians refer to it as *coding*) is not as far-fetched as it may seem.   It is used when we measure the length of an object in inches and feet or in centimeters and meters, and then change from one scale to the other, depending on the scale in which we want to perform our calculations and the scale in which we want to present our results. Another well-known example in which we sometimes change scales is that of the measurement of temperature.   Temperature is sometimes measured in the Fahrenheit scale, sometimes in the centigrade scale, and Figure 3.2 shows a thermometer having a centigrade scale on the left and a Fahrenheit scale on the right.   Whenever it is necessary to change temperatures from one scale to the other, the change may be made by the well-known formula

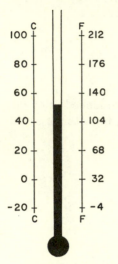

FIGURE 3.2.

$$F = \tfrac{9}{5}C + 32 \qquad (3.3.2)$$

What we are planning to do in the calculation of the mean from grouped data is very much the same.   In order to simplify our calculations, we shall first calculate the mean in the *u*-scale, and then find $\bar{x}$ by converting the result to the *x*-scale with a formula analogous to (3.3.2).   The reader can easily check for himself that in our example this formula must be

$$x = 5u + 22.495 \qquad (3.3.3)$$

Each measurement in the *u*-scale will have to be multiplied by 5 since the class interval is 5, and 5 units in the *x*-scale correspond to 1 unit in the *u*-scale (see Figure 3.1).   We also have to add 22.495 since this is the value of the *x*-scale which we chose as the zero point of the new scale.

More generally, if we introduce a new scale by assigning the values . . . , -4, -3, -2, -1, 0, 1, 2, 3, 4, . . . to the class marks of a distribution, the formula which will enable us to change from one scale to the other is

$$x = c \cdot u + x_0 \qquad (3.3.4)$$

where c stands for the length of the class interval (as measured in the original x-scale) and x_0 stands for the class mark which we choose as the zero of the new scale.

With formula (3.3.4) we will be able to change any measurement from the x-scale to the u-scale or vice versa, and, in particular, we will be able to write

$$\bar{x} = c \cdot \bar{u} + x_0 \qquad (3.3.5)$$

where $\bar{x}$ is the mean of the distribution in the original scale of measurement and $\bar{u}$ is its mean in the new scale.

Using formula (3.3.1), with x's replaced by u's, we have

$$\bar{u} = \frac{\sum\limits_{i=1}^{k} u_i f_i}{n} \qquad (3.3.6)$$

where the u_i are the class marks in the new scale and $n = \sum\limits_{i=1}^{k} f_i$ stands for the total frequency. Substituting this formula for $\bar{u}$ in (3.3.5), we obtain

$$\bar{x} = \frac{c \cdot \sum\limits_{i=1}^{k} u_i f_i}{n} + x_0 \qquad (3.3.7)\star$$

This formula is called *the short-cut formula for the calculation of the mean from grouped data.* (If we looked upon our data as a *population*, we would write μ instead of $\bar{x}$, but otherwise *nothing would be changed.*)

To illustrate the ease with which the mean of a distribution may be found by using (3.3.7), let us refer again to the mail-order distribution. Writing the original class marks and frequencies in the first and third columns, it will be convenient to use the following schema:

| Class Marks | | Frequency | |
x-scale	u-scale	f	uf
7.495	-3	6	-18
12.495	-2	12	-24
17.495	-1	19	-19
22.495	0	20	0
27.495	1	13	13
32.495	2	8	16
37.495	3	2	6
		Totals: 80	-26

Substituting the total frequency and the sum of the products $u_i f_i$ into short-cut formula (3.3.7), we immediately get

$$\bar{x} = \frac{5(-26)}{80} + 22.495$$
$$= -1.625 + 22.495 = \$20.87$$

It should be noted that this result is *identical* with the result obtained with formula (3.3.1) on page 52.

The short-cut method is so simple and it saves so much time and energy that it should always be used.　(About the only time the short-cut method does not provide considerable simplifications is when the class marks in the x-scale are already easy-to-use numbers such as 0, 1, 2, 3, etc.)　In order to reduce our work to a minimum, it is usually desirable to put the zero of the u-scale near the middle of the distribution, preferably at a class mark which has one of the largest frequencies.

The short-cut formula for the mean cannot be used for distributions with *unequal* class intervals, although there exists a modification which makes it applicable also in that case.　Of course, neither the short-cut formula nor formula (3.3.1) can be used for distributions with *open* classes at either end; in that case the mean cannot be found without going back to the raw data or making further assumptions.

EXERCISES

1. Compute the mean circulation of the 97 newspapers referred to in Exercise 5 on page 37 using formula (3.3.1).　Note that the class intervals are not all of equal width.
2. The following is a distribution of straight-time weekly earnings for Class A Bookkeeping-Machine Operators in the Boston, Massachusetts, metropolitan area, September, 1967.

Straight-time Weekly Earnings	Number of Workers
$ 75 and under $ 80	2
80 " " 85	26
85 " " 90	46
90 " " 95	62
95 " " 100	84
100 " " 105	50
105 " " 110	37
110 " " 115	6

Source: U.S. Dept. of Labor.

Find the mean of this distribution.

3. Use formula (3.3.7) to find the mean of whichever data you grouped among Exercises 7, 8, 9, or 10 on pages 29 and 30.
4. Find the mean of the distribution of Exercise 12 on page 31.
5. Find the means of both distributions constructed in Exercise 13 on page 31.
6. Find the average (mean) payment of aid to dependent children on the basis of the distribution in Exercise 4 on page 37. Use formula (3.3.1).

3.4 The Median, Ungrouped Data

A second measure of location that may be used to describe the "center," "middle," or "average" of a set of data is called the *median*. It is defined simply as *the value of the middle item (or the mean of the values of the two middle items) when the items are arrayed, that is, arranged in an increasing or decreasing order of magnitude*.

If we have an *odd* number of measurements there will always be a *middle item* whose value will, by definition, give us the median. For example, given the numbers

$$5 \qquad 8 \qquad 10 \qquad 12 \qquad 2$$

let us array them as follows:

$$2 \qquad 5 \qquad 8 \qquad 10 \qquad 12$$

We can then see by inspection that the median is 8. Given numbers

$$2 \qquad 4 \qquad 9 \qquad 9 \qquad 9 \qquad 9 \qquad 15$$

we can see by inspection that the median is 9. In the first case there are 5 numbers and the median is the 3rd largest; in the second case there are 7 numbers and the median is the 4th largest. Generally speaking, if n is *odd*, the location of the median of n numbers in an array is the $(n + 1)/2$ item. The median is the value of the number thus located. If there are 17 numbers, the median is the $(17 + 1)/2$th, or 9th, provided, of course, that these numbers are arranged according to size. Similarly, among 39 items the median is the $(39 + 1)/2$th, or 20th largest, and among 99 items the median is the $(99 + 1)/2$th, or 50th largest.

If several measurements are equal, it does not matter which one is written first, second, third, and so on. In the above example where we had 4 *nines*, it would be senseless to ask which particular *nine* is to be called the median. By definition, the median is the *value* of the middle item and in our example this value was 9.

If we have an *even* number of measurements, there will never be a middle item and according to our definition the median is given by the average (mean) of the values of the two middle items. For instance, given the numbers

$$3 \qquad 6 \qquad 8 \qquad 10 \qquad 12 \qquad 20$$

we find that the median is $(8 + 10)/2 = 9$. It is the mean of the 3rd and 4th largest of these 6 numbers. The median is, thus, halfway between the two central values of the *ordered* data, and, if it is interpreted correctly, the formula $(n + 1)/2$ will again give us the *position* of the median. For example, if we have 40 measurements, the median will be the $(40 + 1)/2$th, or 20.5th, largest item and we shall interpret this as meaning "halfway between the 20th and 21st."

In order to calculate the median of the mail-order house data, which we have been using throughout as an illustration, we shall first have to arrange the 80 orders (listed on page 21) according to size. Finding, thus, that the size of the 40th largest order is $21.36 and that of the 41st largest is $21.50, we obtain for the median

$$\frac{21.36 + 21.50}{2} = \$21.43$$

It should not surprise the reader that the median of these numbers does not coincide with their mean, which, as we pointed out on page 52, is $21.05. The median and the mean describe the center of a set of data *in a different way*, and the very fact that in this example they differ by as little as $21.43 - $21.05 = $0.38 will later, in Chapter 5, be interpreted as indicative of a further property of the given data, namely, their *symmetry*.

Beginning students often commit the error of mistaking the expression $(n + 1)/2$ for a formula that gives them the median. This is, of course, incorrect. The formula merely tells us how many of the ordered items we must count before we reach the particular measurement (or midpoint between measurements) whose value is the median.

Among the desirable properties of the median, we find that it always exists—that is, it may be found for any set of numerical data—it is always unique, and that it may be found with a minimum of mathematical calculations. Unfortunately, this last advantage is counterbalanced by the fact that we must first arrange our data according to size, and this can be a very tedious job. Another important property of the median which, incidentally, it does not share with the mean is that the median is *not* easily affected by extreme values. Referring again to the example in which the technician took an incorrect reading of the lifetime of a certain light bulb (see page 48), we find that his mistake will not affect the median. Our sample consisted of the numbers

985   863   1024   972   746

whose median is 972, and, as can easily be verified, its value will not change if we change the last reading from 746 to 246. (Had the technician misread 1024 as 124, the median would have been affected, but, generally speaking, we can say that the median is not so easily affected by extreme values as is the mean.)

Another advantage of the median which is not shared by the mean is that the median of grouped data (to be discussed in Section 3.5) may be found *even if the distribution is open at either end*. Furthermore, the median may be used to define the *middle* of a number of objects, properties, or qualities which do not permit a quantitative description.   It is possible, for instance, to rank a number of cakes according to their consistency or flavor and then choose the middle one as having "average" consistency or "average" flavor.

One of the undesirable properties of the median is that, if we have the medians of two or more sets of data, we *cannot* find the over-all median without going back to the raw data.   For example, the median of 15, 18, and 21 is 18, the median of 16, 40, and 43 is 40, and knowing only these two medians, we would never be able to obtain the result that the over-all median of the six numbers is 19.5.

So far we have not designated the median with a special symbol. This is due mainly to the fact that there is little agreement on this subject; most of the time the word "median" is spelled out in full, sometimes it is abbreviated to Med., Md, or M, and sometimes it is given as $\tilde{x}$.   We shall use the letter M for the median of a sample, but we shall not use a Greek letter for the median of a population, for the simple reason that when a sample median is used for purposes of estimation (or decision) it is usually used to estimated the population *mean* μ.   In many problems of estimation or decision we assume that the population mean and median *coincide*, and we can then use either M or $\tilde{x}$ as an estimate of μ.   When dealing with the median of a population, we shall follow common practice and refer to it by its full name.

We are now ready to illustrate what we meant on page 49 when we said that *the mean does not vary as much from sample to sample as some other kinds of descriptions*.   Let us suppose, for example, that a manufacturer of cake-mixes is experimenting with a new product and that he is interested in knowing the average height of a cake produced with this mix.   In other words, he wants to know μ, the true average height of all potential cakes that will be baked with this mix, and it is clear that he will have to be satisfied with an estimate based on a sample. Let us suppose furthermore that he sent 3 boxes of the new mix to each of three independent research laboratories, giving them specific instructions about pan-size, oven temperature, etc., and that the results are

Laboratory X: 1.7, 2.4, and 2.5 inches
Laboratory Y: 1.9, 1.9, and 2.5 inches
Laboratory Z: 1.8, 2.2, and 2.3 inches

If all three of these laboratories used the *mean* to estimate the true average height of *all* cakes baked with this mix, they would obtain 2.2, 2.1, and 2.1 inches, respectively. If, on the other hand, they used the *median*, their estimates would be 2.4, 1.9, and 2.2 inches. We can see from these results that the three means were relatively close to one another, varying from 2.1 to 2.2, while the medians were spread over the much wider interval from 1.9 to 2.4. It is in this sense that we say that the mean is more stable than the median, that it is subject to less chance variation.

Since this last illustration was artificial and somewhat exaggerated to prove our point, let us consider another example—one referring to a game of chance. Simultaneously rolling three dice and recording in each case the three numbers obtained as well as their median and mean, an actual experiment yielded the following results:

Rolls of Three Dice	Median	Mean
2, 6, 4	4	4
1, 5, 2	2	$2\frac{2}{3}$
3, 4, 4	4	$3\frac{2}{3}$
1, 6, 2	2	3
6, 2, 5	5	$4\frac{1}{3}$
5, 1, 3	3	3

If we study these results we find that there is, indeed, a greater variation between the medians than there is between the means. (Incidentally, μ of this example is 3.5. If we rolled a die a great number of times, 1, 2, 3, 4, 5, and 6 should each show up about one-sixth of the time, and *on the average* we should, thus, get $\dfrac{1 + 2 + 3 + 4 + 5 + 6}{6}$ = 3.5.) In case the reader is not convinced, he can easily conduct this experiment for himself. Although we cannot *guarantee* results, he will usually find that the medians are spread over a somewhat wider range of values than the means. This problem will be discussed further in Chapter 8.

3.5 The Median, Grouped Data

The definition of the median of grouped data is most easily understood with the aid of a diagram like that of Figure 3.3, in which we have again drawn the histogram of the mail-order distribution. *The median of a frequency distribution is the number which corresponds to the point of the horizontal scale through which a vertical line divides the total area of the rectangles of the histogram into two equal parts.* This means that the sum of the areas of the rectangles (or parts of rectangles) to the left of the dotted line in Figure 3.3 must equal the sum of the corresponding areas to its right. This definition of the median of grouped data agrees closely with that of the median of ungrouped data. As we pointed out earlier, the areas of the rectangles represent (are proportional to) the frequencies of the corresponding classes, and if the total area to the left of the median is equal to the total area to its right, the implication is that there are as many cases to the left of the median as there are to its right. Actually this is not *quite* correct, for it would have to depend on the distribution of the items *within* the class into which the median will fall. Our definition assumes that

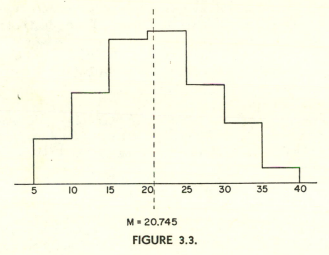

M = 20.745

FIGURE 3.3.

in this class the items are distributed evenly throughout the class interval.

In order to find the median of a set of grouped data consisting of n items we shall count $n/2$ items, starting from either end of the distribution. In contrast to finding the median of *ungrouped* data where we looked for the middle item (or items) and counted off $(n + 1)/2$, we are now looking for a number which divides the total area of the histogram into two equal parts, each part representing a total frequency of $n/2$.

If we now return to the mail-order distribution, we find that since there are 80 items we shall have to count 40 items starting at either end of the distribution. Beginning at the bottom, that is, with the smallest orders, it can be seen that 6 were for less than \$9.995, 18 were for less than \$14.995, 37 were for less than \$19.995, and 57 were for less than \$24.995 (see page 26). Needing 40 items on each side of the median, we find that there are too few items below \$19.995 and too many items below \$24.995. Hence, the median will have to lie somewhere in the interval from \$19.995 to \$24.995, and we shall have to count another 3 items in addition to the 37 items falling below \$19.995. Using the assumption made above, namely, that the 20 items falling into the next class are *evenly distributed* throughout the interval from \$19.995 to \$24.995, we obtain the median by adding $\frac{3}{20}$ of this class interval to its lower boundary of \$19.995. (When we said that the items were evenly distributed within this class, we meant that if we subdivided this class into any number of equal parts, their respective frequencies or fractional frequencies would all be the same.)

The median of a set of grouped data is therefore given by the lower boundary of the class into which the median must fall *plus* a fraction of its class interval, which depends on the number of items we still lack when reaching this lower boundary. The median of the mail-order distribution is thus

$$M = 19.995 + 5 \cdot \tfrac{3}{20} = \$20.745$$

or \$20.74 to the nearest cent.

Generally speaking, if L is the lower boundary of the class containing the median, f_M its frequency, c the class interval, and j the number of items we still lack when reaching L, then *the median is given by the formula*

$$M = L + c \cdot \frac{j}{f_M} \qquad (3.5.1)\star$$

Had we begun to count the necessary 40 items of our mail-order data from the *other end* of the distribution, we could not have used formula (3.5.1) unless we modified it by substituting U, the upper class boundary, for L, and changed the $+$ to a $-$. Since 23 of the orders exceed \$24.995 and 43 of them exceed \$19.995, we get

$$M = 24.995 - 5 \cdot \tfrac{17}{20} = \$20.745$$

which is identical with the result we obtained before.

In general, if the median of a distribution is computed in this fashion, we can write, corresponding to (3.5.1),

$$M = U - c \cdot \frac{j'}{f_M} \qquad (3.5.2)\star$$

where j' is the number of items we still lack when reaching U.

To give another illustration of the calculation of a median from grouped data, let us consider the distribution of the weekly earnings of 1,843 stenographers which we mentioned earlier on page 18:

Earnings (in dollars)	Number of Stenographers	Cumulative Frequency
60.00– 64.99	4	4
65.00– 69.99	33	37
70.00– 74.99	208	245
75.00– 79.99	192	437
80.00– 84.99	285	722
85.00– 89.99	368	1090
90.00– 94.99	404	1494
95.00– 99.99	137	etc.
100.00–104.99	69	
105.00–109.99	49	
110.00–114.99	20	
115.00–119.99	39	
120.00 or more	35	

Since $n = 1,843$, we shall have to count $1,843/2 = 921.5$ items from either end of the distribution. Using the cumulative frequencies, it can easily be seen that the median must fall into the sixth class, whose limits are \$85.00–89.99 and whose boundaries are \$84.995 and \$89.995. Since we lack $921.5 - 722 = 199.5$ items when reaching \$84.995, we get

$$M = 84.995 + 5 \cdot \frac{199.5}{368}$$

$$= 84.995 + 2.71$$

to the nearest cent. This figure represents the median of the weekly earnings of these stenographers, and it should be noted that we were able to find it in spite of the fact that the distribution has an open class at one end. We could not have found the *mean* of this distribution unless we had made some special assumption about the values of the 35 highest earnings.

3.6 The Mode

A third measure of "central" location is the *mode* and it is defined simply as *the value (or attribute) which occurs the most often, that is, with the highest frequency*. It applies to quantitative *and* qualitative data, as is illustrated by the following examples: if more banks pay 5 per cent interest on savings accounts than any other rate, we can say that 5 per cent is the *modal* rate; if more people die of heart disease than of any other cause, we can say that heart disease is the *modal* cause of death; and if more people live in 4-room apartments than in any other size apartment, we can say that 4 rooms is the *modal* size of an apartment.

One advantage of the mode is that it requires no calculations at all; we merely select the value (or category) which appears the most often.   Although the mode is in a sense *typical* of our data—*it represents the value or object which is the most common*—it also has some decided disadvantages.   If we deal with a set of measurements in which no two numbers are alike, it would be rather trivial to say that each one of them is a mode, that each one of them occurs with the highest frequency.   Instead, we shall say in a situation like this that the mode does not exist.   Furthermore, even if there is a mode, it need not be unique.   Given the numbers

63   65   65   65   70   72   74   79   79   79   82   82   85,

for example, we find that 65 and 79 both occur with the maximum frequency of *three* and we can, thus, say that this set of data has two modes or is "bimodal."

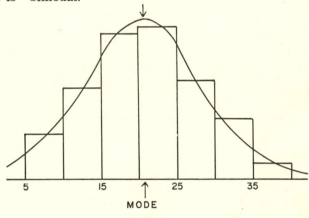

FIGURE 3.4.

Generally speaking, the mode's principal value lies in the fact that it can be used to describe qualitative data.   For example, if we wanted to compare consumers' preferences for different kinds of products, different kinds of packaging, or different kinds of advertising, we could compare the *modal preferences* expressed by different groups of people, where we could not have calculated a median or a mean.   Although relatively few people are familiar with the term "mode" itself, most people will have no difficulty in understanding what it means.

When dealing with quantitative data, the disadvantages of the mode outweigh its desirable features and it is actually seldom used. As we have said before, there are many problems in which the mode is either not unique or does not exist.   Moreover, the mode would be an extremely unstable measure of "central" location if it were used for purposes of estimation or decision (see page 59), and, finally, it does not lend itself to further manipulations.   For instance, we cannot take the modes of two sets of data and calculate the over-all mode of the combined data.

So far we have discussed only the mode of ungrouped data. When dealing with grouped data, it is customary to use the term *modal class* when referring to the class having the highest frequency. In our mail-order example the modal class is the one containing orders from $20.00 to $24.99.

It is also possible to *define* the mode of a distribution as *the class mark of the modal class* and if we followed this procedure we would say that the mode of our mail-order distribution is $22.495. Actually, there are more sophisticated ways of defining the mode of a distribution. A possible approach is to approximate the histogram of a distribution with a smooth curve (see Figure 3.4) and then define the mode as the value that corresponds to the highest point of this curve. We shall not go into this here, because there are few occasions where this method is used; suitable references may be found in the Bibliography on page 76.

EXERCISES

1. At a radar check point on U.S. Route 95, 29 cars were clocked at the following rates of speed (in miles per hour):

   55 51 55 48 45 57 71 46 50 52 61 54 55 58 49
   51 66 61 50 36 44 56 51 68 47 52 54 38 44

   Find the median of these speeds.

2. In a selection of 20 lots of 300 electronic components each, the following numbers of defective electronic components were found:

   4   2   4   4   4   5   9   2   3   3
   6   4   4   13   3   3   5   3   5   6

   Find the median of these numbers.

3. Find the median of the consumer price indexes given in Exercise 6 on page 50.

4. Find the median grade obtained by the 24 students of Exercise 2 on page 50.

5. Find the median of the 100 ungrouped "downtimes" of Exercise 8 on page 29.

6. Find the median number of physicians per 100,000 population in 110 selected large cities in the United States in 1962 on the basis of the ungrouped data of Exercise 12 on page 31.

7. Find the median payments of Aid to Dependent Children by state in 1966 on the basis of the distribution of Exercise 4 on page 37.

8. Find the median of the distribution of Exercise 2 on page 56.

9. Find the median of whichever data you grouped among Exercises 7, 8, 9, 10, 11, or 12 on pages 29 and 30.

10. Find the medians of both of the distributions obtained in Exercise 13 on page 31 for the number of physicians per 100,000 population.

11. In order to compare two brands of tires, a research organization tested 5 tires of each kind, measuring the mileage for which each tire gave adequate service. The results of this test were: The tires made by Firm *A* lasted 26,800, 22,300, 27,400, 24,000, and 23,500 miles, while those made by Firm *B* lasted 25,600, 23,400, 21,000, 26,000, and 25,000 miles. Comment on the claims made by *both* firms that "on the average" their tires showed up better in this test.

12. Find the mode of each of the following sets of numbers (if it exists):

    (a) 3, 7, 4, 5, 7, 5, 6, 6, 8, 6, 5, 6, 4
    (b) 92, 60, 108, 78, 114, 46, 62, 54, 70, 96, 44
    (c) 10, 14, 14, 17, 17, 15, 13, 14, 11, 17, 14, 17, 9

13. Find the modal speed (or speeds) for the data of Exercise 1 above.

14. Find the modal number of defective electronic components for the data of Exercise 2 above.

15. Find the modal classes and their class marks for the two distributions of Exercise 13 on page 31.

16. Find the mode of the letters of the alphabet used in the last paragraph of section 3.6 on page 65.

17. Thirty welders reported for work wearing shirts of the following colors: green, yellow, brown, blue, green, yellow, red, blue, brown, yellow, blue, black, blue, brown, red, blue, green, blue, yellow, red, blue, red, brown, blue, yellow, brown, blue, black, yellow and blue. Find the modal choice of color among this group of welders.

3.7 The Geometric and Harmonic Means

When first discussing the mean, we mentioned that its full name is the "arithmetic mean" in order to distinguish it from two other measures of location called the *geometric* and *harmonic means*. The first of these two other means is defined as follows: Given a set of numbers $x_1, x_2, x_3, \ldots, x_n$, the *geometric mean* is the nth root of their products, that is,

$$\text{geometric mean} = \sqrt[n]{x_1 \cdot x_2 \cdot x_3 \cdot x_4 \cdots \cdots x_n} \qquad (3.7.1)\star$$

Since the use of this formula evidently involves a considerable amount of work, it is only reasonable to ask why and when this special kind of "average" should be used. In actual practice, the geometric mean is used mainly in problems in which we are interested in averaging ratios, particularly index numbers, which will be discussed later in Chapters 15 and 16.

To illustrate the calculation of the geometric mean in a problem in which its use is appropriate, let us consider the following example: A certain store made profits of $5,000, $10,000, and $80,000 in 1966 1967 and 1968 respectively, and on the basis of this information, we are asked to determine the average rate of growth of this store's profits. Clearly, from 1966 to 1967 the store's profits *doubled*, and from 1967 to 1968 they were multiplied by 8. If we calculated the *mean* of these two numbers we would get $(2 + 8)/2 = 5$ and we could say that *on the average* the store's profits multiplied by 5 each year. This result can be very misleading; if we were told merely that the store made a profit of $5,000 in 1966 and that on the average its profits multiplied by 5 each year for the next two years, we might get the impressions that the store's profits in 1967 and 1968 were in the neighborhood of $25,000 and $125,000, respectively. Both of these figures are much too high.

If we calculate the geometric mean of these two rates of growth, we obtain

$$\text{geometric mean} = \sqrt[2]{2 \cdot 8} = \sqrt[2]{16} = 4$$

and we can say that the store's profits "on the average" quadrupled each year. If we applied this "average rate of growth" to the fact that the store made a profit of $5,000 in 1966 we might surmise that the store's profits in 1967 and 1968 were about $20,000 and $80,000, respectively. Whereas the 1967 figure is still too high, although not as much as before, the 1968 figure is now correct. This illustrates the type of problem in which the geometric mean provides a better "average" than the arithmetic mean.

Formula (3.7.1) is actually seldom used, because the necessary arithmetic is much too involved. Instead, we made use of the fact that $\log x \cdot y = \log x + \log y$, and $\log x^m = m \cdot \log x$, changing (3.7.1) into the *logarithmic* form of

$$\log G = \frac{\sum\limits_{i=1}^{n} \log x_i}{n} \qquad (3.7.2)\star$$

(Here G stands for the geometric mean.) With this formula we can find the logarithm of G by adding the logarithms of the x_i and dividing their sum by n. The antilogarithm of this result then gives the geometric mean. (The use of logarithm tables is explained in Appendix IV.)

To illustrate the use of (3.7.2), let us find the geometric mean of the numbers

$$101 \quad 99 \quad 108 \quad 96 \quad 103 \quad 79 \quad 85 \quad 100$$

which are the December, 1962, *Wholesale Price Indexes* of 8 selected major commodity groups.   Using Table VIII we find that

$$
\begin{aligned}
\log 101 &= 2.0043 \\
\log 99 &= 1.9956 \\
\log 108 &= 2.0334 \\
\log 96 &= 1.9823 \\
\log 103 &= 2.0128 \\
\log 79 &= 1.8976 \\
\log 85 &= 1.9294 \\
\log 100 &= 2.0000 \\
\hline
&= 15.8554
\end{aligned}
$$

and upon substitution into (3.7.2) we get

$$
\log G = \frac{15.8554}{8} = 1.9819
$$

Finally, referring again to Table VIII, we find that the geometric mean, the antilogarithm of 1.9819, is approximately equal to 96.

Some important aspects of the geometric mean are that it is used mainly to average *positive* numbers, since its value could otherwise be negative or imaginary, that like the arithmetic mean it takes into account each individual item, and that it is always uniquely defined. Furthermore, the geometric mean reduces to some extent the effect of a very large or very small number and this is one reason why it is sometimes preferred over the arithmetic mean.   Generally speaking, the geometric mean is nowadays rarely used.

The *harmonic mean* of n numbers x_1, x_2, x_3, . . . , x_n is defined as *n divided by the sum of the reciprocals of the x's*, that is

$$
\text{harmonic mean} = \frac{n}{\sum\limits_{i=1}^{n} \dfrac{1}{x_i}}
\tag{3.7.3}\star
$$

To illustrate the use of the harmonic mean, let us suppose that we have spent \$12 for eggs costing 40 cents a dozen and another \$12 for eggs costing 60 cents a dozen.   What we would like to know is these eggs' *average price* per dozen.   Perhaps, the obvious thing is to make the mistake of guessing that the average price is $(40 + 60)/2 = 50$ cents per dozen.   That this is incorrect may be seen from the following argument: For the first \$12 we get 30 dozen eggs at 40 cents a dozen and for the second \$12 we get 20 dozen eggs at 60 cents a dozen. Hence, we have bought a total of 50 dozen eggs for \$24 and the average price per dozen is $2,400/50 = 48$ cents. Had we tried to find the harmonic mean of 40 and 60 we would have obtained the correct answer

$$\frac{2}{\frac{1}{40} + \frac{1}{60}} = 48$$

as can easily be checked.

Since the harmonic mean is rarely used and even then only in very special kinds of problems, we shall not discuss it any further. (In case the reader is curious to learn more about such special averages as the geometric and harmonic means, he will find suitable references in the Bibliography on page 76.) We have given the numerical example mainly to illustrate that there *are* situations in which the harmonic mean provides a more appropriate "average" than the other measures of location treated in this chapter.

3.8 The Weighted Mean

There are many problems in which we cannot average quantities without paying some attention to their *relative importance* in the over-all situation we are trying to describe. For example, if there are two food stores in a given town selling butter at 79 cents a pound and 81 cents a pound, respectively, we cannot determine the over-all price paid for butter in this town unless we knew the number of pounds sold by each store. If most people buy their butter in the first store, the average price they pay per pound will be closer to 79 cents, and if most of them buy their butter in the second store, the average price will be closer to 81 cents. Hence, we cannot calculate a meaningful average unless we know the relative weight (the relative importance) carried by the numbers we want to average. Similarly, we might get a wrong picture if we calculated the *mean* of changes in the values of stocks without paying attention to the number of shares sold of each stock, and we could not average the number of accident fatalities reported by three different airlines per 1,000,000 passenger miles unless we knew the number of passenger miles flown by each airline.

To consider a particular example, let us suppose that a butcher sells three grades of beef for $0.69, $0.89, and $1.09 a pound and that we would like to know the average price he receives per pound of beef. Clearly, we cannot say that he receives an average of $(0.69 + 0.89 + 1.09)/3 = \$0.89$ per pound, unless by chance he happens to sell an equal number of pounds of each grade. If we are given the additional information that during a certain week he has sold 600 pounds of the cheapest grade, 300 pounds of the medium priced grade, and 100 pounds of the most expensive grade, we find that he received

$$600(0.69) + 300(0.89) + 100(1.09) = \$790.00$$

for $600 + 300 + 100 = 1000$ pounds of beef and that he, thus, received on the average \$0.79 per pound of beef. The average we have calculated here is called a *weighted mean*. We have averaged the three prices, giving due weight to the relative importance of each, namely, to the number of pounds sold of each grade.

In general, if we want to average a set of numbers $x_1, x_2, x_3, \ldots, x_n$ whose relative importance is expressed numerically by means of some numbers $w_1, w_2, w_3, \ldots, w_n$ called their *weights*, we shall use the *weighted mean*, which is defined by the formula

$$\bar{x}_w = \frac{\sum\limits_{i=1}^{n} x_i w_i}{\sum\limits_{i=1}^{n} w_i} \qquad (3.8.1)\star$$

To give another illustration, let us suppose that on a certain trip a motorist bought 12 gallons of gasoline at 32 cents per gallon, 18 gallons at 29 cents per gallon, and 50 gallons at 35 cents per gallon. In order to find the "average" price this motorist paid per gallon of gas, we can use formula (3.8.1), putting $x_1 = 32$, $x_2 = 29$, $x_3 = 35$ and $w_1 = 12$, $w_2 = 18$, $w_3 = 50$. Here the weights w_1, w_2, w_3, express the relative importance of the three prices of gasoline, and upon substitution we find that

$$\bar{x}_w = \frac{32(12) + 29(18) + 35(50)}{12 + 18 + 50}$$

$$= \frac{2656}{80} = 33.2 \text{ cents per gallon}$$

Although the choice of the weights did not present any problems in these two examples, their selection is not always quite so obvious. For example, if we wanted to construct a *cost-of-living* index, we would have to worry about the roles played by different commodities in the average person's budget; if we wanted to determine the average yield of wheat per acre in the United States on the basis of figures given for each state, we would have to use as weights the acreage of wheat cultivated in each state; and if we wanted to average the batting averages of several baseball players, we sould have to use as weight their respective "times at bat."

The weights that are most frequently used in the averaging of *prices* are the corresponding *quantities* consumed, sold, or produced. Incidentally, the formula which we derived earlier in this chapter for combining the means of several groups into an over-all mean is simply a special case of a weighted mean. In formula (3.2.4) we average the $\bar{x}_i$ by using as weights the number of items in each individual group.

EXERCISES

1. Without using logarithms, find (a) the geometric mean of the numbers 5 and 20, (b) the geometric mean of the numbers 4, 14, and 49, and (c) the geometric means of the numbers 1, 1, 3, and 27.

2. Use formula (3.7.2) to calculate the geometric mean of the numbers

    157   140   152   155   162   187   192   174   153   162

    standing for the 1967 indexes of farm real estate values in 10 different regions of the United States, 1957–1959 = 100. (Source: Agricultural Statistics.)

3. Use formula (3.7.2) to calculate the geometric mean of the numbers

    32.4   25.9   25.7   23.0   29.7   27.6   31.6   16.8   31.1

    representing the 1961 work injury frequency rates (per million man hours) in the heavy construction industry, by geographic regions of the U.S.

4. In a German vocabulary test a student answered 16 out of 100 questions correctly on the first try; 24 out of 100 on the second try a week later; and 81 out of 100 on the third try a week after that. (In the second try he did, thus, $\frac{24}{16} = 1.5$ times as well as on the first try.) Use formula (3.7.1) to find the "average improvement ratio" between the successive tries.

5. Find the harmonic mean of the numbers 6, 8, and 12.

6. If a motorist travels the first 20 miles of a trip at 30 miles per hour and the next 20 miles at 60 miles per hour, what is his average speed for these 40 miles? Would the harmonic mean give the correct answer?

7. If we spend $120 on books costing $2 each, $120 on books costing $3 each, $120 on books costing $4 each, and $120 on books costing $5 each, show that the harmonic mean of 2, 3, 4 and 5 will give the correct average price paid for these books.

8. Mr. Smith invests $500 at 4%, $1000 at 5% and $2000 at 6%. Use formula (3.8.1) to show that he will receive an average interest of 5.43 per cent on his investments.

9. A vendor at a football game sells 100 pennants at $1 each, 50 pennants at $2 each, and 20 pennants at $3 each. What is the vendor's average price received per pennant?

10. In 1960 there were 0.9 fatalities per 100,000,000 revenue passenger miles flown by scheduled air carriers in the United States, in 1964 there were 0.1, and in 1965 there were 0.4. Given that the total number of revenue miles flown by these carriers in 1960, 1964, and 1965 were approximately 30.6 hundred million, 44.1 hundred million, and 51.9 hundred million, respectively, find the weighted mean of the fatality rates for these three years.

11. The following table shows the *batting averages* and the *number of times at bat* of the 5 leading batters in the American League as of June 18, 1968: (Source: The *Boston Globe*.)

Player	Batting Average	Times at Bat
A	.337	202
B	.313	227
C	.297	155
D	.294	187
E	.289	194

Use formula (3.8.1) to find the combined batting average of these 5 players.

3.9 Quartiles, Deciles, and Percentiles

The various measures of location that we have discussed in this chapter might very well be called measures of *central* location. All but the mode, which need not be unique, provide us with single numbers, which are representative of the "center" or "middle" of our data. In this section we shall study methods of describing other kinds of locations. For example, we shall learn how to obtain values which are exceeded by 25 per cent, 10 per cent, or 3 per cent of our data, and values which exceed, say, 5 per cent, 25 per cent, or 40 per cent of our data. As there would be no sense in dividing small sets of data into 10 or even 100 equal parts, we shall discuss these measures only with reference to grouped data. (However, as can be seen from Exercise 5 on page 75, analogous measures *can* be defined for ungrouped data.)

The three *quartiles* of a distribution are defined as values which divide it into *four* parts containing equal numbers of observations such that one-fourth of the data fall below Q_1, the *first quartile*, half of the data fall below Q_2, the *second quartile*, and three-fourths of the data fall below Q_3, the *third quartile*. Making the same assumption as in the case of the median, which, incidentally, coincides with Q_2, we could say that *the quartiles divide the total area of the rectangles of a histogram into four equal parts*, as is illustrated in Figure 3.5. The necessary assumption is that *within* the classes containing the quartiles, the items are distributed evenly throughout the respective class invervals.

The steps needed for the calculation of the quartiles are almost identical with those needed to find the median. The only difference is that, starting at the bottom of the distribution, we now have to count $n/4$ cases to find Q_1 and $3n/4$ cases to find Q_3. If we started at the other end of the distribution we would correspondingly have to count $3n/4$ cases to find Q_1 and $n/4$ cases to find Q_3.

To illustrate the calculation of quartiles, let us again refer to the mail-order distribution, for which we have already found the median on page 62. Since there are 80 items, we must count $\frac{80}{4} = 20$ items starting at the bottom of the distribution to find Q_1 and $3 \cdot 80/4 = 60$ to find Q_3. As in the case of the median, we first have to find the classes into which the quartiles will fall; then we will have to add fractions of the class interval to account for the number of items we still lack when reaching these lower class boundaries. Since 18 of the orders fall below $14.995 (see page 26), we need 2 of the 19 items fall-

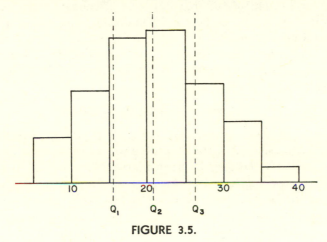

FIGURE 3.5.

ing into the next class in order to reach Q_1. Using a formula analogous to (3.5.1), we thus find that

$$Q_1 = 14.995 + 5 \cdot \tfrac{2}{19} = \$15.52$$

to the nearest cent. Similarly, since 57 of the orders fall below $24.995, we need 3 of the 13 items falling into the next class in order to reach Q_3. Again using a formula analogous to (3.5.1), we obtain

$$Q_3 = 24.995 + 5 \cdot \tfrac{3}{13} = \$26.15$$

to the nearest cent. Had we calculated Q_3 by counting 20 cases from the other end of the distribution, we would have obtained

$$Q_3 = 29.995 - 5 \cdot \tfrac{10}{13} = \$26.15$$

to the nearest cent, using in this case a formula analogous to (3.5.2).

If we use (3.5.1) or (3.5.2) for the determination of Q_1 or Q_3, it must be understood that L and U must now stand for the lower and upper boundaries of the class containing the respective quartile, and that f_M (we might now write f_Q) stands for the corresponding class frequency.

Deciles are values which, by definition, divide a distribution (or the area of the rectangles of its histogram) into 10 equal parts. We shall write D_1 for the *first decile* which exceeds the lowest 10 per cent of the data, D_2 for the *second decile* which exceeds the lowest 20 per cent of the data, and, in general, D_i for the *ith decile* which exceeds the lowest $i \cdot 10$ per cent of the data.

In order to find any one of the D_i (i can be either 1, 2, 3, . . . , or 9), we will have to count $i \cdot n/10$ of the total number of items starting at the bottom of the distribution. Although there is, actually, little value in calculating deciles when dealing with as few as 80 observations, we shall illustrate the calculation of deciles by referring again to the mail-order distribution. In order to find D_1, for example, we will have to count $1 \cdot 80/10 = 8$ of the items, and using a formula analogous to (3.5.1), we obtain

$$D_1 = 9.995 + 5 \cdot \tfrac{2}{12} = \$10.83$$

to the nearest cent. Six of the orders were for less than \$9.995 and we therefore needed 2 of the 12 orders which were in the next highest class. Similarly, D_6 may be determined either by counting $6 \cdot 80/10 = 48$ cases starting from the bottom of the distribution or by counting 32 cases starting at the other end. Using formulas analogous to (3.5.1) and (3.5.2), we thus obtain

$$D_6 = 19.995 + 5 \cdot \tfrac{11}{20} = \$22.74$$

or $$D_6 = 24.995 - 5 \cdot \tfrac{9}{20} = \$22.74$$

and the two answers are, of course, the same.

Percentiles are values which, by definition, divide a distribution (or the area of the rectangles of its histogram) into 100 equal parts. We shall write, for example, P_{15} for the *fifteenth* percentile which exceeds the lowest 15 per cent of the data, or P_{95} for the *ninety-fifth percentile* which is exceeded only by the highest 5 per cent. The calculation of percentiles is very similar to that of the median, quartiles, or deciles. To find P_{15} for our mail-order data, we will have to count 15 per cent of 80, namely, 12 cases starting at the bottom of the distribution. Using a formula analogous to (3.5.1), we get

$$P_{15} = 9.995 + 5 \cdot \tfrac{6}{12} = \$12.50$$

to the nearest cent. If we wanted to find P_{95} of the mail-order distribution, it would be less work to count 5 per cent of the cases starting at the *top* of the distribution, and we would thus get, with a formula analogous to (3.5.2),

$$P_{95} = 34.995 - 5 \cdot \tfrac{2}{8} = \$33.74$$

to the nearest cent.

EXERCISES

1. Find Q_1 and Q_3 for the distribution of newspaper circulations of Exercise 5 on page 37.
2. Find Q_1 and Q_3 of the distribution of average straight-time weekly earnings of Exercise 2 on page 56.
3. Calculate Q_1 and Q_3 for whichever data you grouped among Exercises 7, 8, 9, or 10 on pages 29 and 30.
4. Find Q_1 and Q_3 for the distribution of the number of physicians per 100,000 population obtained from the data of Exercise 12 on page 31.
5. It has been suggested that for ungrouped data the *first quartile* of n observations might be defined as the $(n + 3)/4$th largest. Check whether for $n = 9$ and $n = 13$ (unequal observations) this will yield a value which is exceeded by three times as many values as it exceeds.
6. Find D_1 and D_9 for the distribution used in Exercise 3 above.
7. Find D_2 and D_8 for the distribution used in Exercise 2 above.
8. Find D_3 and D_7 for the distribution of Exercise 4 above.
9. Find P_5 and P_{95} for the distribution of Exercise 3 above.
10. Find P_{15} and P_{85} for the distribution of Exercise 1 above.
11. Find P_{35} and P_{65} for the distribution of Exercise 4 above.

3.10 Further Comparisons

The various measures of location which we have studied in Sections 3.2 to 3.8 all describe the center of a set of data *in a specific way*. Whereas the median actually represents the *middle* of our data (it is exceeded by as many items as it exceeds), the mode represents the value which is the *most common* and, perhaps, the most typical, and the mean represents what might reasonably be called the *center of gravity* of our data. (Students of physics will note that a formula which looks exactly like (3.2.5) is used to find the center of gravity of a set of weights f_i that are arranged along a horizontal line at distances x_i from the origin.)

The question of what particular "average" should be used in a given problem is not easily answered. As we have seen, there are problems in which the nature of the data dictates the use of such special averages as the geometric or harmonic means. The nature of the data may also dictate the use of a weighted average to account for differences in the importance of various items. Although we have limited ourselves here to a discussion of the *weighted mean*, it should be noted that there are other weighted averages such as the *weighted geometric mean*, which will be mentioned briefly in Chapter 16. Finally, if we are dealing with qualitative data, we may have no choice but to use the mode, or in some isolated instances, perhaps, the median.

The *mean* is by far the most widely used measure of "central" location, and it will be used almost exclusively in the later chapters of this book, in which we shall study problems of estimation, decision, and prediction.

As there are many problems in which we can use the mean, the median, or some other measure of location, it is important that a choice be made *before* their values are actually calculated. As can be seen from Exercise 11 on page 66, it is possible to arrive at opposite results (or conclusions) depending on whether we use the median or the mean, and it would hardly be ethical to make this choice afterwards on the basis of the particular results we may want to prove. Later on, in Section 5.2, we shall see how differences between the mean, the median, and the mode, can actually be used to describe another feature, namely, the *symmetry* or *lack of symmetry*, of a distribution.

As we have repeatedly pointed out in Chapter 2, the grouping of data entails some loss of information, and we should therefore not be surprised to get slightly different values for the mean, median, or mode if we used different classifications of the *same* data. This grouping effect will not be too pronounced in case of the median or mean, but it can be considerable in case of the mode. (See, for example, Exercise 5 on page 57, Exercise 15 on page 66, and Exercise 10 on page 66.)

BIBLIOGRAPHY

For more elaborate ways of defining the mode of a distribution see

Croxton, F. E., Cowden, D. J., and Klein, S., *Applied General Statistics*, 3rd ed. Englewood Cliffs, N.J.: Prentice-Hall, Inc., 1967, Chap. 9.

Waugh, A. E., *Elements of Statistical Method*. New York: McGraw-Hill, 1952, p. 205. (This book also contains additional material on desirable and undesirable properties of the various measures of location.)

Further information about the geometric and harmonic means may be found in the two books mentioned above and, among others, also in

Neiswanger, W. A., *Elementary Statistical Methods*, rev. ed. New York: Macmillan, 1956, Chap. 9.

For additional references on measures of location see

Leabo, Dick A. and Smith, C. Frank., *Basic Statistics for Business and Economics*, Rev. ed. Homewood, Ill.: Richard D. Irwin, 1964.

Meyers, Cecil H., *Elementary Business and Economic Statistics*. Belmont, California: Wadsworth Publishing, 1966.

Neter, John, and Wasserman, William, *Fundamental Statistics for Business and Economics*, 3rd ed. Boston: Allyn & Bacon, 1966.

CHAPTER 4

Measures of Variation

4.1 Introduction

The mean, median, and most of the other "averages" discussed in the previous chapter provide us with single numbers which represent whole sets of data. Although the information contained in a measure of location may well provide an adequate description of a set of data in a limited number of problems, in general we will find it necessary to supplement it by describing additional features. The statistical measures to be studied in this chapter are called *measures of variation, spread,* or *dispersion.* The following are a few examples in which it is evident that our knowledge of the mean (or some other measure of location) must be supplemented with a description telling us something about the extent to which our data are dispersed—the extent to which they are spread out or bunched.

Let us suppose, for instance, that we want to compare two bonds, Bond X and Bond Y, and that over the last 6 years the closing price of Bond X for each year was such as to yield to maturity

6.0, 5.7, 5.6, 5.9, 6.1, and 5.5 per cent,

while, for the same period of time, the corresponding figures for Bond Y were

7.2, 7.7, 4.9, 3.1, 3.4, and 8.5 per cent.

As can easily be checked, the means of these two sets of percentages are both 5.8 and, hence, if we based our evaluation of the bonds entirely on their average yields, we would be led to believe that, at least for the given 6 years, they were equally good.

A more careful analysis reveals, however, that, whereas the closing price of Bond X was such that the yield to maturity was consistently close to 5.8 per cent (it varied from 5.5 to 6.1 per cent), the performance of Bond Y was much more erratic. In the fourth year its yield to maturity was as low as 3.1 per cent, and even though this was offset by 8.5 per cent in another year, it is clear that Bond Y provided a much *less consistent* investment than Bond X. In this example we have, thus, a problem in which an intelligent evaluation requires some measure of the fluctuations or variability of our data in addition to a measure of central location such as the mean.

Another illustration which demonstrates the need for measuring the variation of a set of data may be found in a problem dealing with the inspection of steel ordered for the construction of a certain bridge. Let us suppose that the construction of a bridge requires steel with a minimum compressive strength of 45,000 psi (pounds per square inch) and that, to play it safe, steel has been ordered with the specification that its average strength must be 60,000 psi. Let us suppose, furthermore, that after the first shipment is received an inspector tests 5 pieces of steel, obtaining the following measurements of compressive strength:

41,000,   44,250,   69,400,   70,000,   and   80,350 psi.

If the inspector reports merely that the mean of the 5 measurements is 61,000 psi and, thus, slightly above specification, we might be led to believe that the steel is satisfactory. However, if we look at the *variation* of the measurements, we find that they vary over a considerable range; 2 of the 5 pieces have a compressive strength below the minimum requirement, 45,000 psi, and a considerable portion of the shipment may very well be unsatisfactory. In a problem like this it is important, therefore, to consider variability in addition to an average. As a matter of fact, the original order for the steel should have contained some specification about permissible variability in addition to the specification about average strength.

The concept of variation or dispersion is of fundamental importance in the domain of inductive statistics, because it is here that we have to cope with questions of *chance variation*. To illustrate the meaning of this term, let us suppose that a balanced coin is flipped 100 times. Although we may *expect* to get 50 heads and 50 tails, we would certainly not be surprised if we got, say, 46 heads and 54 tails, 53 heads and 47 tails, or 41 heads and 59 tails. We might say that the occurrence of a few extra heads or a few extra tails could very well be ascribed to chance. In order to study this effect of chance, let us suppose that we repeatedly flip a balanced coin 100 times and that in 10 such "experiments" we observe

$$51, 54, 47, 56, 52, 45, 41, 58, 52, \text{ and } 49 \text{ heads.}$$

This gives us some idea about the magnitude of the fluctuations (variations) produced by chance in the number of heads obtained in 100 flips of a coin. A knowledge of this is important if we wanted to decide, for example, whether there is something wrong with a coin which in 100 tosses produced 30 heads and 70 tails. Judging by the experiment described above, it seems that most of the time we can expect to get anywhere from 40 to 60 heads (in our experiment the number of heads ranged from 41 to 58), and, hence, it would seem reasonable to conclude that there is something wrong with the coin which gave 30 heads and 70 tails. This argument has been presented on a rather intuitive basis in order to illustrate the need for describing or measuring variability in a problem of inductive statistics. The same problem will be discussed in detail in Chapters 7, 8, and 10.

To consider another example from inductive statistics in which the concept of variability plays an important role, let us return to the problem in which we tried to estimate μ, the true average height of *all* cakes that are baked with a certain new kind of mix. Taking the results of Laboratory X (see page 60), which obtained an average height of $\bar{x} = 2.2$ inches for a sample of 3 cakes, we might infer that $\mu = 2.2$ inches, or at least that μ is very close to 2.2. Clearly, this kind of reasoning would involve a considerable generalization since we would be using information about 3 cakes to estimate the true average height of, perhaps, millions of cakes which will be baked with this mix.

Whether or not such a generalization is reasonable or justifiable depends on many factors which will be discussed later in Chapter 9. For the moment we are interested only in *one* of these factors, that of the *variability of the population*. To explain the meaning of this term, let us consider the following two possibilities:

Case 1: The true average height is $\mu = 2.3$ inches, the cake mix is *very consistent*, and *all* cakes baked with this mix will have a height anywhere from 2.1 to 2.5 inches.

Case 2: The true average height is $\mu = 2.3$ inches, but the cake mix is *very inconsistent* and the height of the cakes will vary considerably, say, from 1.5 to 3.2 inches.

In the event of Case 1 we can be *sure* that for a sample of 3 cakes the average height $\bar{x}$ cannot possibly differ from μ by more than 0.2 inches. After all, the worst that can happen is that all three cakes

have a height of 2.1 inches or that all three have a height of 2.5.
(Regardless of how many observations there are in the sample, $\bar{x}$ cannot be less than 2.1 or greater than 2.5.)

In the event of Case 2 the situation is quite different.   Purely by
chance we might pick a sample in which all cakes have a height of
1.5 inches and $\bar{x}$ would then differ from μ by as much as 0.8 inches.
*This serves to illustrate that in order to evaluate the "goodness" of a
generalization or the "closeness" of an estimate, we will have to know
something about the variability of the population.*   If there is little variation, as in Case 1 where the heights of *all* cakes are close to 2.3, then
$\bar{x}$ is apt to be very close to μ; if there is a lot of variation, as in Case 2
where the heights of the cakes vary from 1.5 to 3.2, then $\bar{x}$ cannot be
expected to be as close to the true mean.

We have given these examples to demonstrate how the concept of
variability plays an important role in descriptive as well as inductive
statistics.   In the remainder of this chapter we shall study a number
of ways of describing (measuring) variability, spread, or dispersion.

4.2 The Range

The measure of variation which is easiest to explain and easiest to
obtain is the *range*.   The *range* of a set of numbers is defined very
simply as *the difference between the largest and the smallest*.   For
instance, the size of the largest order listed on page 21 is $37.81, the
size of the smallest is $5.45, and hence the range of the 80 orders is
$37.81 - 5.45 = \$32.36$.   Similarly, the range of the 5 measurements
of compressive strength on page 78 is $80,350 - 41,000 = 39,350$ psi,
and the range of the possible heights of the cakes in Case 2 on page 79
is $3.2 - 1.5 = 1.7$ inches.

Although the range is a measure of variation that is easily understood and does not require lengthy calculations, its usefulness is
limited considerably by its undesirable features.   Actually, it is used
primarily when we are interested in getting a quick, though perhaps
not very accurate, picture of the variability of a set of data.

The range is a poor measure of variability because it accounts
only for the two extreme values of our data.   It does not tell us anything about the dispersion of the remaining data except that they lie
on an interval whose length is given by the range.   If we investigate
the following 3 sets of data:

$$5, 17, 17, 17, 17, 17, 17, 17, 17, 17$$
$$5, \ 5, \ 5, \ 5, \ 5, 17, 17, 17, 17, 17$$
$$5, \ 6, \ 8, 10, 11, 14, 14, 15, 16, 17$$

it can easily be seen that in each case the range is equal to 12. Nevertheless, the dispersion of these sets of numbers is by no means the same. In the first, all but one of the numbers are 17's; in the second, half the numbers are 5's and half are 17's; and in the third, the numbers are spread all over the interval from 5 to 17.

If we were told that the range of the salaries paid by a certain manufacturer is $46,500, the highest salary being $50,000 and the lowest being $3,500, this would not really tell us very much about the dispersion of the salaries. For all we know, all but one of the employees might be getting $3,500. Of course, we could get a good picture of the variation of these salaries by looking at a distribution showing the number of employees making from $3,000 to $4,000, from $4,000 to $5,000, . . . , but let us remember that our goal is to describe this variation by means of a *single number*. The range provides us with such a single number, but, unfortunately, *it very often does not tell us enough about the dispersion of our data.*

In view of this shortcoming of the range (and there are other shortcomings that have not even been mentioned), we shall have to look for other ways of describing, or measuring, the rather loosely defined characteristic which we have been referring to as variation or dispersion.

To repeat, the main advantage of the range is that it can be obtained with a minimum of arithmetic. This is why the range is often used in *quality control* (see Appendix II), where it is important to keep a continuous check on the variability of raw materials, machines, or manufactured products without going through excessive calculations.

4.3 The Average Deviation

Since the dispersion of a set of numbers is *small* if the numbers are bunched very closely around the mean, and it is *large* if the numbers are spread over considerable distances away from the mean, we might define variation *in terms of the distances (deviations) by which the various numbers depart from the mean.* If we take a set of numbers x_1, x_2, . . . , x_n, which constitute a sample with the mean $\bar{x}$, we can write the amounts by which these numbers depart from the mean as $x_1 - \bar{x}$, $x_2 - \bar{x}$, . . . , $x_n - \bar{x}$. These quantities are called the *deviations from the mean,* and it would seem reasonable to use their average

$$\frac{\sum_{i=1}^{n} (x_i - \bar{x})}{n} \tag{4.3.1}$$

as a measure of variation.   Unfortunately, this average does not measure anything, for it can be shown that *it is always equal to 0.*

Using the rules about summations which we studied in Chapter 1, it can easily be seen that

$$\frac{\sum\limits_{i=1}^{n} (x_i - \bar{x})}{n} = \frac{\sum\limits_{i=1}^{n} x_i - \sum\limits_{i=1}^{n} \bar{x}}{n} = \frac{\sum\limits_{i=1}^{n} x_i}{n} - \bar{x}$$

and the last expression is 0, since by definition

$$\frac{\sum\limits_{i=1}^{n} x_i}{n} = \bar{x}$$

Although formula (4.3.1) does not give us a measure of variation, the original idea of expressing dispersion in terms of the deviations from the mean was not bad.   As it happened, we made the unfortunate choice of using the *mean* of the deviations which turned out to be 0 regardless of the values of the x_i.   We ran into difficulties because (unless the numbers are all equal) some of the deviations are *positive*, some are *negative*, and their sum is always equal to 0.   For example, the mean of 3, 12, 8, 17, and 5 is $\bar{x} = 9$, the deviations are -6, 3, -1, 8, -4, and $(-6) + 3 + (-1) + 8 + (-4) = 0$.

In view of our original goal of defining a measure of variation, it is apparent that we are interested in the *magnitude* of the deviations and not in their *signs*.   Hence, we could "ignore the minus signs" and define a measure of variation in terms of the *absolute values* of the deviations.   In mathematics, the absolute value of a *positive* number is the number itself, while the absolute value of a *negative* number is the number without its minus sign.   For example, the absolute value of 17 is 17, and the absolute value of -7 is 7.   It is customary to write the absolute value of x as $|x|$, and we have, for example,

$$|17| = 17 \qquad \text{and} \qquad |-7| = 7$$

Using the absolute values of the deviations from the mean, we can now define the following measure of variation, called the *average deviation* or the *mean deviation:*

$$\text{average deviation} = \frac{\sum\limits_{i=1}^{n} |x_i - \bar{x}|}{n} \qquad (4.3.2)^\star$$

It is similar to (4.3.1) except that the deviations from the mean are now replaced by their absolute values.

The steps needed to calculate an average deviation are (1) to find $\bar{x}$, (2) to determine the deviations $x_i - \bar{x}$ and their absolute values,

(3) to add the absolute values of the deviations, and (4) divide by n. As an illustration, let us find the average deviation of the following data representing the average relative humidity at 1:30 P.M. in Salt Lake City, Utah, for each month of the year:

$$71 \quad 64 \quad 53 \quad 43 \quad 37 \quad 32 \quad 28 \quad 28 \quad 31 \quad 42 \quad 59 \quad 70$$

First finding $\bar{x}$, we get

$$\bar{x} = \tfrac{558}{12} = 46.5$$

and then

x_i	$x_i - \bar{x}$	$\lvert x_i - \bar{x} \rvert$
71	24.5	24.5
64	17.5	17.5
53	6.5	6.5
43	− 3.5	3.5
37	− 9.5	9.5
32	−14.5	14.5
28	−18.5	18.5
28	−18.5	18.5
31	−15.5	15.5
42	− 4.5	4.5
59	12.5	12.5
70	23.5	23.5
558	0.0	169.0

$$\text{average deviation} = \tfrac{169}{12} = 14.1$$

It is really unnecessary to write down both the deviations and their absolute values, but it is a good check to see whether the sum of the deviations is equal to zero.

The average deviation, which we have calculated in this example, tells us that in Salt Lake City there are considerable fluctuations in relative humidity from month to month. Of course, we could have learned that much by simply looking at the data, but the average deviation tells us *more specifically* that on the average the monthly figures deviated by 14.1 from the annual average of 46.5.

For grouped data, the average deviation cannot be found unless we make some assumption about the distribution of the measurements *within* each class. Assuming, as we did for the mean of grouped data, that all measurements that lie in a class are located at its class mark, we get

$$\text{average deviation} = \frac{\displaystyle\sum_{i=1}^{k} \lvert x_i - \bar{x} \rvert \cdot f_i}{n} \qquad (4.3.3)^\star$$

where x_i stands for the class mark of the ith class, f_i for the corresponding class frequency, k for the number of classes, and n for the total number of measurements. Although formulas (4.3.2) and (4.3.3) defined average deviations of *samples*, analogous formulas may be used for average deviations of *populations* with the only distinction that μ is substituted for $\bar{x}$.

The calculation of an average deviation can be quite cumbersome, particularly if n, the number of measurements, is large and the mean has several decimals. If that is the case, it may be desirable to use the short-cut method referred to in the Bibliography on page 99. We shall not go into this here, because in actual practice the average deviation is rarely used. We have given it primarily as a stepping-stone for the definition of the *standard deviation*, the much more important measure of variation, which will be discussed in the next section.

Although the average deviation may have intuitive appeal as a measure of variation, the fact that it involves absolute values makes it very difficult to investigate how it "behaves" in problems of sampling, that is, how it is affected by chance. Indeed, the main drawback of the average deviation is that due to the absolute values it does not lend itself readily to further mathematical treatment.

EXERCISES

1. Find the range of the "downtimes" of Exercise 8 on page 29.
2. Find the range of the clerical aptitude scores of Exercise 10 on page 30.
3. Find the range of the weights of the football players of Exercise 5 on page 50.
4. Find the range of the number of physicians per 100,000 population for the data of Exercise 12 on page 31.
5. Find the average deviation of the 24 grades of Exercise 2 on page 50.
6. Calculate the average deviation of the 24 values of the Consumer Price Index given in Exercise 6 on page 50.
7. Determine the average deviation of the monthly figures under all Unemployment Insurance Programs of Exercise 3 on page 50. Round off the mean to the nearest tenth of a million.
8. Use formula (4.3.3) to find the average deviation of the mail-order house data whose distribution is given on page 52.
9. Use formula (4.3.3) to find the average deviation of whichever data you grouped among Exercises 7, 8, 9, or 10 on pages 29 and 30.

4.4 The Standard Deviation, Ungrouped Data

When we defined the average deviation, absolute values eliminated the signs of the deviations from the mean. Being interested in the size of the deviations and not in their signs, we can accomplish more or less the same by first squaring the deviations and then averaging their squares. (The square of a real number is never negative.) In order to compensate for the fact that we averaged the squared deviations, we finally take the square root of this average and use it as the definition of a new measure of variation. To express this symbolically, let us suppose that we are dealing with a set of numbers x_1, x_2, x_3, . . . , x_n, which constitute a *population*. (The reason for this qualification will become apparent in a moment.) Calling this new measure of variation the *standard deviation of the population* and representing it with the Greek letter σ (sigma), we can write

$$\sigma = \sqrt{\frac{\sum\limits_{i=1}^{n} (x_i - \mu)^2}{n}} \qquad (4.4.1)$$

This measure is also called the *root-mean-square deviation;* we are literally taking the square *root* of the *mean* of the *squared* deviations.

Having, thus, defined the standard deviation of a population, it would seem logical to substitute $\bar{x}$ for μ and write the standard deviation of a sample as

$$\sqrt{\frac{\sum\limits_{i=1}^{n} (x_i - \bar{x})^2}{n}} \qquad (4.4.2)$$

Although this definition is still used by many statisticians and may be found in many texts, the modern tendency is to make a slight modification, defining the *standard deviation of a sample* as

$$s = \sqrt{\frac{\sum\limits_{i=1}^{n} (x_i - \bar{x})^2}{n-1}} \qquad (4.4.3)$$

(Our use of the letters σ and s is consistent with the convention of using Greek letters for descriptions of populations and Roman letters for descriptions of samples.)

Before we attempt to justify the modification of dividing by $n - 1$ instead of n, let us point out that *there is no question here of one formula being right and the other being wrong.* We are trying to describe the variability of a set of data, and the choice of any particular description, that is, of any particular formula, is really arbitrary. *The only reason why we might prefer one formula over the other is that it provides a more useful description.* To see whether the standard deviation of a sample becomes a *more useful* measure of variation if we divide by $n - 1$ instead of n, we shall have to investigate first *for what purpose* this formula is to be used.

Generally speaking, whenever we measure the variability of a sample, we are directly or indirectly interested in getting an estimate of σ, the standard deviation of the population from which the sample was obtained. For example, on page 80 we said that one must know something about the variability of the population if one wants to judge how close a sample mean $\bar{x}$ can be expected to lie to the true mean μ. If, as in most problems of sampling, we have no information about the population except for the sample, we can learn about the variability of the population only by looking at the variability of the sample. Hence, we have no choice but to estimate σ on the basis of the information contained in the sample, and *it is in situations like this that we get a better estimate if we divide by $n - 1$ instead of n.* Although we shall not be able to prove this here, it can be shown that if we divided by n, our estimates of σ would on the average be too small (see Section 4.8).†

Division by $n - 1$ instead of n is particularly important when n is small. When n is large, say, 100 or more, the difference between formula (4.4.2) and (4.4.3) becomes negligible.

To illustrate the use of formulas (4.4.1) and (4.4.3), let us find s for the sample of the 5 Brand A light bulbs, which we used as an example on page 3. The measurements of the "lifetimes" of these light bulbs were

$$985, \ 863, \ 1024, \ 972, \ 746 \text{ hours}$$

and the calculation of s with formula (4.4.3) may be given in the following manner:

x_i	$x_i - \bar{x}$	$(x_i - \bar{x})^2$
985	67	4489
863	-55	3025
1024	106	11236
972	54	2916
746	-172	29584
4590	0	51250

$$\bar{x} = \tfrac{4590}{5} = 918$$

$$s = \sqrt{\tfrac{51250}{4}}$$

$$= \sqrt{12812.5}$$

Referring to Table IX, the use of which is explained in Appendix IV, we find that the standard deviation of the lifetimes of the 5 light bulbs is approximately 113 hours. By using this value as an *estimate* of σ, the standard deviation of the lifetimes of *all* Brand *A* 60 watt light bulbs, we will later on (in Chapter 9) be able to judge how close $\bar{x} = 918$ hours might be to μ, the true average lifetime of all these bulbs.

In the above example, the calculation of s was not very difficult, but it is easy to see that our work would have been much more involved if the mean had not been a whole number. In actual practice, formulas (4.4.1) and (4.4.3) are seldom used. In their place we use short-cut formulas which provide considerable simplifications *without being approximations*. To derive such a short-cut formula for s, let us start with s^2 which, incidentally, is called the *sample variance*, and write

$$s^2 = \frac{\sum\limits_{i=1}^{n} (x_i - \bar{x})^2}{n - 1} = \frac{\sum\limits_{i=1}^{n} (x_i^2 - 2x_i\bar{x} + \bar{x}^2)}{n - 1}$$

Applying Rules A and B of Chapter 1, this becomes

$$s^2 = \frac{\sum\limits_{i=1}^{n} x_i^2 - 2\bar{x} \sum\limits_{i=1}^{n} x_i + \sum\limits_{i=1}^{n} \bar{x}^2}{n - 1}$$

and after multiplying numerator and denominator by n, using Rule C of Chapter 1, and substituting $\sum\limits_{i=1}^{n} x_i/n$ for $\bar{x}$, we get

$$s^2 = \frac{n \cdot \sum\limits_{i=1}^{n} x_i^2 - (\sum\limits_{i=1}^{n} x_i)^2}{n(n - 1)}$$

The short-cut formula for the standard deviation of a sample can, thus, be written as

$$s = \sqrt{\frac{n \cdot \sum\limits_{i=1}^{n} x_i^2 - (\sum\limits_{i=1}^{n} x_i)^2}{n(n - 1)}} \qquad (4.4.4)\star$$

It is important to note that with this formula we can find s *without having to go through the process of actually finding the deviations from the mean*. We have only to add the x's, their squares, and then substitute the two sums into formula (4.4.4). Unless we make a mistake in arithmetic, formulas (4.4.3) and (4.4.4) will yield identical results; *the short-cut formula is not an approximation.*

Without going through a detailed derivation, we can write a similar short-cut formula for σ. Substituting n for the $n - 1$ appearing in the denominator of (4.4.4.), we get

$$\sigma = \sqrt{\frac{n \cdot \sum_{i=1}^{n} x_i^2 - (\sum_{i=1}^{n} x_i)^2}{n \cdot n}}$$

After taking n^2 out of the radical, *the short-cut formula for the population standard deviation becomes*

$$\sigma = \frac{1}{n} \sqrt{n \cdot \sum_{i=1}^{n} x_i^2 - (\sum_{i=1}^{n} x_i)^2} \qquad (4.4.5)\star$$

To illustrate the use of the short-cut formulas, let us again find s for the lifetimes of the 5 light bulbs. Since the measurements were 985, 863, 1024, 972, and 746 hours, we get

x_i	x_i^2
985	970225
863	744769
1024	1048576
972	944784
746	556516
4590	4264870

and

$$s = \sqrt{\frac{5(4264870) - (4590)^2}{5 \cdot 4}}$$

$$= \sqrt{12812.5}$$

$$= 113 \text{ hours (approximately)}$$

This agrees with the result obtained on page 87.

In this example the mean as well as the deviations were whole numbers, and it may have seemed simpler to use formula (4.4.3) than the short-cut technique. To demonstrate that formulas (4.4.4) and (4.4.5) do, in fact, provide considerable simplifications, let us find σ for a population consisting of the 12 numbers

15, 10, 12, 8, 7, 11, 20, 5, 14, 17, 16, 12

Using first formula (4.4.1) we get

x_i	$x_i - \mu$	$(x_i - \mu)^2$
15	2.75	7.5625
10	−2.25	5.0625
12	− .25	.0625
8	−4.25	18.0625
7	−5.25	27.5625
11	−1.25	1.5625
20	7.75	60.0625
5	−7.25	52.5625
14	1.75	3.0625
17	4.75	22.5625
16	−3.75	14.0625
12	− .25	.0625
147	0	212.2500

$$\mu = \tfrac{147}{12} = 12.25$$

$$\sigma = \sqrt{\frac{212.25}{12}}$$

$$= 4.2 \text{ (approx.)}$$

Using short-cut formula (4.4.5) we get

x_i	x_i^2
15	225
10	100
12	144
8	64
7	49
11	121
20	400
5	25
14	196
17	289
16	256
12	144
147	2013

$$\sigma = \tfrac{1}{12} \sqrt{12(2013) - (147)^2}$$

$$= \tfrac{1}{12} \sqrt{2547}$$

$$= 4.2 \text{ (approx.)}$$

In this example the mean was *not* a whole number, and the short-cut formula provided considerable simplifications. *Whereas formulas (4.4.1) and (4.4.3) give us a clearer picture of what is meant by a standard deviation, formulas (4.4.4) and (4.4.5) enable us to find s and σ with greater ease.*

In many examples the calculation of a standard deviation can be simplified further by *adding or subtracting an arbitrary number from each measurement.* It is not difficult to prove that if the same number is added to each measurement (or the same number is subtracted from each measurement), the value of the standard deviation is left unchanged. Even without giving a formal proof, it should be evident that the *variation or dispersion* of a set of data is not affected if the same constant is added to each number.

Had we used this trick in the last example, we might have subtracted 12 from each number, obtaining

$$3, \ -2, \ 0, \ -4, \ -5, \ -1, \ 8, \ -7, \ 2, \ 5, \ 4, \ 0$$

instead of the original measurements. The sum of the new numbers is 3, the sum of their squares is 213, and substitution into (4.4.5) yields

$$\sigma = \tfrac{1}{12} \sqrt{12(213) - (3)^2}$$
$$= \tfrac{1}{12} \sqrt{2547}$$

This is precisely what we had before.

Since the purpose of this trick is to reduce the size of the numbers with which we have to work, it is generally desirable to subtract a number that is close to the mean. In our example the mean was 12.25 and we subtracted 12.

In case the reader has some doubts as to when he should use σ and when he should use s, let us repeat what we said in Section 3.1: *Whether a set of numbers is to be looked upon as a population or a sample depends entirely on what we intend to do with it.* If the information contained in our data is to be used for generalizations, that is, predictions, estimations, or decisions *for which we do not have all the facts*, we are dealing with a sample and should use s. On the other hand, if our data contain *all possible* measurements of the phenomenon which we want to describe, we are dealing with a population and should use σ.

EXERCISES

1. Use formula (4.4.3) to find the standard deviation of the numbers 8, 12, 15, 5, 13, and 7, which constitute a sample of the number of absentees per day from the assembly line of a factory.
2. Find s for the following sample consisting of the weights of 8 people using (a) formula (4.4.3), and (b) formula (4.4.4). *Time yourself on the two parts to check how much time is saved by using the short-cut formula.*

   184, 163, 155, 178, 206, 142, 168, 215

3. Find s for the number of heads observed in the 10 experiments described on page 79. (Each experiment consisted of 100 flips of a coin.)
4. Find the standard deviation of the 5 measurements of compressive strength given on page 78.

5. The ages of the presidents of a group of six community colleges are as follows:

$$39, 40, 35, 51, 39, 48$$

Find σ using formula (4.4.1).

6. In their mid-term examination in economics 20 students obtained the following grades:

$$83, 70, 75, 92, 77, 45, 62, 65, 95, 50,$$
$$90, 67, 61, 66, 81, 68, 38, 54, 83, 78.$$

Find σ using formula (4.4.5) and the trick mentioned on page 90.

7. Find the standard deviation of the weights of the 39 players of the Boston Patriots' 1967 football squad using the data given in Exercise 5 on page 50.

8. Find σ for the 24 values of the consumer price index given in Exercise 6 on page 50.

4.5 The Standard Deviation, Grouped Data

To be able to calculate the standard deviation of grouped data, we shall again assume that all measurements contained in a class are located at its class mark. If we write x_i for the class mark of the ith class, f_i for its frequency, k for the number of classes, and n for the total number of items, formula (4.4.3) becomes

$$s = \sqrt{\frac{\sum_{i=1}^{k} (x_i - \bar{x})^2 \cdot f_i}{n - 1}} \tag{4.5.1}$$

(In this formula the squared deviations $(x_i - \bar{x})^2$ are added f_i times to account for the f_i values falling into the ith class.)

Formula (4.5.1) *defines* s for grouped data, but it is seldom used in actual practice. The calculation of s for grouped data can be simplified considerably by using the same change of scale which we introduced in Chapter 3 for calculating the mean of grouped data. Choosing a u-scale in which the class marks become . . . , -3, -2, -1, 0, 1, 2, 3, . . . , the *short-cut formula for calculating s from grouped data* can be written as

$$s = c \cdot \sqrt{\frac{n \cdot \sum_{i=1}^{k} u_i^2 f_i - (\sum_{i=1}^{k} u_i f_i)^2}{n(n - 1)}} \tag{4.5.2}^{\star}$$

Here c is the length of the class interval in the original scale and it is assumed that the classes are of equal length. A reference to the derivation of this short-cut formula is given on page 99.

Although this formula may look complicated, it actually makes the calculation of s very easy. Instead of having to work with the original class marks and the deviations from the mean, we have only to find the two sums

$$\sum_{i=1}^{k} u_i f_i \quad \text{and} \quad \sum_{i=1}^{k} u_i^2 f_i$$

and substitute them into (4.5.2).

When dealing with *populations*, formulas analogous to (4.5.1) and (4.5.2) may be obtained for σ by substituting n for $n-1$ in the denominators. After taking n^2 out of the radical, the *short-cut formula for calculating σ from grouped data* becomes

$$\sigma = \frac{c}{n} \sqrt{n \cdot \sum_{i=1}^{k} u_i^2 f_i - \left(\sum_{i=1}^{k} u_i f_i \right)^2} \qquad (4.5.3)\star$$

To illustrate the use of formulas (4.5.2) and (4.5.3), let us calculate s for the mail-order house data of Chapter 2. Introducing the same u-scale as on page 55, we get

x_i	f_i	u_i	u_i^2	$u_i f_i$	$u_i^2 f_i$
7.495	6	−3	9	−18	54
12.495	12	−2	4	−24	48
17.495	19	−1	1	−19	19
22.495	20	0	0	0	0
27.495	13	1	1	13	13
32.495	8	2	4	16	32
37.495	2	3	9	6	18
	80			−26	184

Substituting the totals into (4.5.2) gives

$$s = 5 \sqrt{\frac{80(184) - (-26)^2}{80 \cdot 79}}$$
$$= \$7.45.$$

The variation in the size of the orders is, thus, measured by the standard deviation of $7.45.

The standard deviation of grouped data is calculated on the assumption that all measurements belonging to a class are located at its class mark. The error which is introduced by this assumption, and which is appropriately called a *grouping error*, can be fairly large, particularly if the class interval is wide. A correction, called *Sheppard's correction*, which compensates for this error is mentioned in the Bibliography on page 99.

Having learned how to calculate standard deviations of grouped and ungrouped data, let us now show how the knowledge of a standard deviation can play an important role in the analysis of a commercial product. Referring again to the cake mix which we used as an example on page 59, let us suppose that in a sample of 20 cakes the heights had an average of $\bar{x} = 2.3$ inches and a standard deviation of $s = 0.2$ inches. Assuming that this is the only information available about the new mix, we may wish to infer that *the true mean and standard deviation of the heights of all potential cakes to be baked with the mix are $\mu = 2.3$ and $\sigma = 0.2$ inches.*

The true values of μ and σ are important to know, because there are many populations in which roughly

68 per cent of the values differ from the mean by less than one standard deviation,
95 per cent of the values differ from the mean by less than two standard deviations,
and more than 99 per cent of the values differ from the mean by less than three standard deviations.

Assuming that these percentages apply to our example and that 2.3 and 0.2 are good estimates of μ and σ, we can infer that 68 per cent of (all) the cakes will have heights from $2.3 - 0.2 = 2.1$ inches to $2.3 + 0.2 = 2.5$ inches; that 95 per cent will have heights from 1.9 to 2.7 inches; and that more than 99 per cent of the cakes will have heights from 1.7 to 2.9 inches. It is important to note that *these values are meant to apply to all the potential cakes that will be baked with the new mix* in spite of the fact that very few were actually baked. (Whether or not these values are very accurate is a question we shall not be able to investigate at this time.)

The knowledge of standard deviations is also important for the comparison of numbers belonging to *different* sets of data. Let us suppose, for example, that in a given month Brown, a salesman of industrial chemicals, made sales of $15,000, while Jones, a salesman of office supplies, made sales of $6,250. If that month's sales by salesmen of industrial chemicals and office supplies had, respectively, means of $13,000 and $5,000, and standard deviations of $2,000 and $500, we find that Brown's sales were

$$\frac{15,000 - 13,000}{2,000} = 1 \text{ standard deviation}$$

above the average of his group, while Jones' sales were

$$\frac{6{,}250 - 5{,}000}{500} = 2\tfrac{1}{2} \text{ standard deviations}$$

above the average of his group.   We now have two figures that can be compared, and we can say that Jones' performance was much better than Brown's. *Jones rated much higher among salesmen of office supplies than Brown rated among salesmen of industrial chemicals.*

What we have done here consisted of converting the two sales figures from dollars into so-called *standard units*.   If x is a measurement belonging to a set of data having the mean $\bar{x}$ (or μ) and the standard deviation s (or σ), then its value in *standard units* is

$$\frac{x - \bar{x}}{s} \quad \text{or} \quad \frac{x - \mu}{\sigma}$$

Standard units† thus tell us how many standard deviations an item is above or below the mean of the set of data to which it belongs.

EXERCISES

1. Find σ for the distribution of Exercise 2 on page 56.
2. Find the standard deviation of whichever data you grouped among those of Exercises 7, 8, 9, and 10 on pages 29 and 30.
3. Find s for the number of physicians per 100,000 population on the basis of the distribution constructed in Exercise 12 on page 31.
4. Find σ for the following distribution of 400 final examination grades in mathematics:

Grades	Frequency
20–29	3
30–39	20
40–49	40
50–59	72
60–69	177
70–79	62
80–89	22
90–99	4

5. Mr. Holmes received a grade of 90 in his final examination in Business Law and a grade of 84 in a final examination in Marketing.   Compare

---

† Standard units are also referred to as *standard scores* or *z-scores*.   The letter z is often used to represent

$$\frac{x - \mu}{\sigma} \quad \text{or} \quad \frac{x - \bar{x}}{s}$$

his performance in these two tests if the entire Business Law class' grades had a mean of 72 and a standard deviation of 10 while the Marketing class' grades had a mean of 64 and a standard deviation of 12.

6. Stock A has a normal (average) price of \$40 with a standard deviation of \$10 and it is currently selling for \$55. Stock B sells normally for \$20, has a standard deviation of \$4 and is currently selling for \$36. If an investor owns both kinds of stock and wants to dispose of one, which one should he sell in the absence of other information, and why?

4.6 Further Measures of Variation

In some problems it is convenient to use a measure of variation which is easier to find than the standard deviation, yet which is more informative than the range. Such a measure is the *interquartile range*, which gives the length of the interval containing the middle 50 per cent of the data. Its formula is

$$\text{Interquartile range} = Q_3 - Q_1 \qquad (4.6.1)\star$$

and once the quartiles are known it is easily found. An important feature of the interquartile range is that it accounts for the spread of the middle and, in that sense, most significant part of our data. Unlike the range, it is not affected by one very small or very large value.

In Section 3.9 we found that for the mail-order house data $Q_1 = \$15.52$ and $Q_3 = \$26.15$. Hence, the interquartile range of these data is \$26.15 − \$15.52 = \$10.63, and this gives us the length of the interval containing the middle 50 per cent of the orders.

Instead of the interquartile range, some persons use the *semi-interquartile* range, also called the quartile deviation. Its formula is

$$\text{Quartile deviation} = \frac{Q_3 - Q_1}{2} \qquad (4.6.2)\star$$

and it gives the average amount by which the two quartiles differ from the median.

4.7 Measures of Relative Variation

In science, the standard deviation of a set of measurements is often used as an indication of their accuracy. If we repeatedly measure a quantity—say, the height of a building, the weight of a bird, or the diameter of a ball bearing—we would hardly expect to get the same result each time. Hence, the amount of variation that we find in repeated measurements of *one and the same object* provides us with information about the accuracy of the measurements.

Let us suppose, for instance, that we have taken 5 measurements of the length of a certain object and that their standard deviation is 0.10 inches. *Will this enable us to decide whether our measurements are accurate?* · Clearly, such a decision can be reached only if we know the size of the object we are trying to measure. Had we measured the height of a tall building, then a standard deviation of 0.10 inches would imply that our measurements are *extremely accurate*. However, had we measured the diameter of a tiny ball bearing, this standard deviation would imply that our measurements are *highly inaccurate*.

This serves to illustrate why the standard deviation alone cannot serve as a measure of the accuracy of a set of measurements. What we need is a measure of *relative variation*, that is, a measure which expresses the magnitude of the variation relative to the size of the quantity that is being measured. The most widely used measure of relative variation is the *coefficient of variation*, V, which is defined as

$$V = \frac{s}{\bar{x}} \cdot 100 \qquad \text{or} \qquad V = \frac{\sigma}{\mu} \cdot 100 \qquad (4.7.1)\star$$

It can be seen from these formulas that the coefficient of variation gives the standard deviation as a *percentage* of the mean.

If in the above example the standard deviation of 0.10 inches refers to 5 measurements of the length of a room and if the mean of these 5 measurements is 240 inches, we get

$$V = \frac{0.1}{240} \cdot 100 \text{ per cent.}$$

This equals approximately *four hundredths of one per cent*, and it would seem that the measurements are extremely accurate.

By using the coefficient of variation it is possible to compare the dispersions of two or more sets of data *given in different units of measurement*. Instead of having to compare, say, the variability of *prices in dollars, ages in years*, and *weights in pounds*, we can compare the respective coefficients of variation, all of which are percentages. Of course, the coefficient of variation can also be used to compare the relative dispersion of several sets of data given in the same units of measurement. For instance, it can be used to investigate the relative variability of the prices or dividends of a number of stocks.

Alternate measures of relative variation may be defined in terms of statistical measures other than the standard deviation and the mean. Using the quartile deviation and the median we could write

$$\frac{(Q_3 - Q_1)/2}{M} \cdot 100$$

or substituting into this formula the mean of Q_1 and Q_3 for M, we get the more widely used *coefficient of quartile variation*

$$V_Q = \frac{Q_3 - Q_1}{Q_3 + Q_1} \cdot 100 \qquad (4.7.2)\star$$

This formula provides a convenient alternative to the coefficient of variation in problems in which we have calculated the quartiles but not the standard deviation.

EXERCISES

1. Use the result of Exercise 6 on page 91 to find the coefficient of variation of the grades obtained by the 20 students.
2. Use the results of Exercise 8 on page 91 and Exercise 6 on page 50 to find the coefficient of variation of the 24 values of the consumer price index.
3. Use the results of Exercise 3 on page 94 and Exercise 4 on page 57 to find the coefficient of variation for the number of physicians per 100,000 of population.
4. Find V_Q for the distribution of newspaper circulations using the values of Q_1 and Q_3 obtained in Exercise 1 on page 75.
5. Use the results of Exercise 4 on page 75 to find V_Q for the number of physicians per 100,000 of population.

4.8 Further Remarks About the Standard Deviation†

Unfortunately, there is no general agreement on the symbols and formulas used for standard deviations. In some books the same formula is used for standard deviations of samples and populations, in some books there are two different formulas for samples, and in some books there is a distinction between two kinds of populations with different formulas for their standard deviations.

In view of this lack of agreement on formulas and symbols, we would like to caution the reader to check the definition of each symbol whenever he uses other statistics books for reference or further study. Some of the formulas used in problems of estimation and in the testing of hypotheses will look different, depending on the symbols and formulas used for standard deviations. As we have said before, there is no question of one formula being right and the other being wrong, but no formula can be used correctly unless one knows the precise meaning of each symbol.

---

† This section may be omitted without loss of continuity.

On page 86 we stated without proof that if in the formula for s we divide by n instead of $n - 1$, we get an estimate of σ which will on the average be too small.   This argument may be supported by considering the identity

$$\frac{\sum\limits_{i=1}^{n} (x_i - \mu)^2}{n} = \frac{\sum\limits_{i=1}^{n} (x_i - \bar{x})^2}{n} + (\bar{x} - \mu)^2$$

which can be verified by applying the rules of summation given on page 14.   Here $\bar{x}$ is the mean of a sample of n observations x_i taken from a large population whose mean is μ.   This identity shows that, unless $\bar{x}$ happens to equal μ,

$$\frac{\sum\limits_{i=1}^{n} (x_i - \bar{x})^2}{n} \text{ will always be less than } \frac{\sum\limits_{i=1}^{n} (x_i - \mu)^2}{n}$$

since $(\bar{x} - \mu)^2$ cannot be negative.   Since σ is defined in terms of the deviations from μ, it would seem that a formula based on the deviations from the true mean should give a better estimate of σ than a formula based on the deviations from a sample mean.   Indeed, it can be shown that

$$\frac{\sum\limits_{i=1}^{n} (x_i - \mu)^2}{n}$$

provides an estimate which *on the average* equals σ^2.   Hence, according to what we said above,

$$\frac{\sum\limits_{i=1}^{n} (x_i - \bar{x})^2}{n}$$

will on the average be *less* than σ^2.   Of course, in actual practice we do not know μ and cannot use the deviations from the true mean, *but we can compensate for this by dividing by $n - 1$.*

The quantity $n - 1$ by which we divide in the formula for s is often referred to as the number of *degrees of freedom*.   The reason for this terminology may be explained as follows: On page 82 we showed that the sum of the deviations from the mean, $x_i - \bar{x}$, is *always equal* to 0.   Hence, if we know $n - 1$ of the deviations from the mean, the nth is automatically determined.   (For example, if we have 5 observations and 4 of the deviations from the mean are -4, 7, 5, and -6, then the 5th deviation from the mean *must* equal -2.)   Since the standard deviation measures variation in terms of the squared deviations from the mean, we can say that it is based on $n - 1$ *independent quantities;* since $n - 1$ of the deviations determine the nth, we say that we have $n - 1$ *degrees of freedom.*

BIBLIOGRAPHY

Short-cut formulas for the average deviation of grouped and ungrouped data are mentioned in

Waugh, A. E., *Elements of Statistical Method.* New York: McGraw-Hill, 1952, p. 132.

The derivation of a short-cut formula for the standard deviation of grouped data, that is, a formula analogous to (4.4.4) and (4.4.5), is given in

Freund, J. E., *Modern Elementary Statistics*, 3rd ed. Englewood Cliffs, N.J.: Prentice-Hall, Inc., 1967, pp. 63, 66.

Sheppard's correction, which compensates for the error introduced into the calculation of s or σ by the assumption that all measurements belonging to a class are located at its class mark, is discussed in

Mills, F. C., *Introduction to Statistics.* New York: Holt, Rinehart & Winston, Inc., 1956, p. 121.

Further problems dealing with the standard deviation and other measures of variation may be found in

Croxton, F. E., Cowden, D. J., and Klein, S., *Applied General Statistics*, 3rd ed. Englewood Cliffs, N.J.: Prentice-Hall, Inc., 1967, Chap. 10.
Neter, J., and Wasserman, W., *Fundamental Statistics for Business and Economics*, 3rd ed. Boston: Allyn and Bacon, Inc., 1966, Chap. 6.
Wallis, W. A., and Roberts, H. V., *Statistics: A New Approach.* Glencoe, Ill.: Free Press, 1956, Chap. 8.

and in many other introductory texts.

A discussion of why one divides by $n - 1$ rather than n in the formula for the sample variance may be found in many textbooks on mathematical statistics; for example, see

Freund, J. E., *Mathematical Statistics*, Englewood Cliffs, N.J.: Prentice-Hall, Inc., 1962.

CHAPTER 5

Further Descriptions

5.1 Introduction

In spite of the fact that frequency distributions can assume almost any shape or form, there are some standard types which fit most distributions that we meet in actual practice. Foremost among these is the *bell-shaped* distribution, which is illustrated by the histogram of Figure 5.1. Its name is self-explanatory. Many of the distributions we have met in previous chapters (for example, the mail-order house distribution, the distribution of the time lapses between

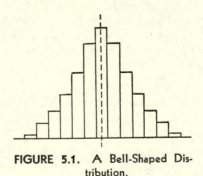

FIGURE 5.1. A Bell-Shaped Distribution.

eruptions of Old Faithful, and the distribution of the clerical aptitude scores) may reasonably be described as bell-shaped. Indeed, there exist theoretical reasons why, in many problems, we can actually *expect* to get bell-shaped distributions.

Although the distribution of the commercial bank earnings given in Exercise 9 on page 29 may also be described as bell-shaped (see Figure 5.2), this distribution provides the picture of a somewhat

100

"lopsided bell." In order to distinguish between the distributions of Figures 5.1 and 5.2, we shall refer to the first as *symmetrical* and the second as *skewed*.

A distribution is said to be symmetrical if we can fold its histogram along some dotted line (see Figure 5.1) and have the two halves more or less coincide. If a distribution has a "tail" at either end, such as the distribution of Figure 5.2, it is said to be skewed. Good examples of skewed distributions that we have already met are the distribution

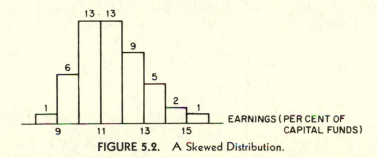

FIGURE 5.2. A Skewed Distribution.

of the amount of aid to dependent children per family (Exercise 4 on page 37) and the distribution of advertised prices of houses (Exercise 7 on page 28). The question of how to describe (measure) the degree to which a distribution is skewed will be discussed in Section 5.2.

Two other basic types of distributions which sometimes, though less frequently, arise in actual practice are *J-shaped* and *U-shaped* dis-

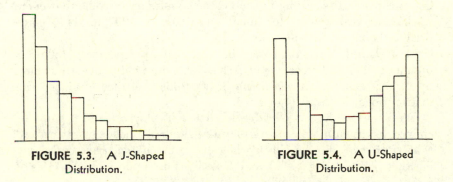

FIGURE 5.3. A J-Shaped Distribution.

FIGURE 5.4. A U-Shaped Distribution.

tributions. They are illustrated by the histograms of Figures 5.3 and 5.4 and their names literally describe the shapes of the distributions. An example of a *J*-shaped distribution is the distribution of the number of claims filed per year against holders of automobile liability policies. Most policy holders have no claims against them, some have one claim, fewer have two, etc. Another example of a *J*-shaped distribution is the distribution of the number of boats owned by individual families.

Most own no boats at all, some own one boat, fewer own two boats, and so forth.

In many texts, U-shaped distributions are illustrated with a distribution of the percentage cloudiness observed at a weather station. If there are many days on which the sky is completely clear, fewer days on which it is covered 10 per cent, fewer yet on which it is covered 20 per cent, . . . , and again more days on which it is covered 100 per cent, the resulting distribution of the percentage cloudiness will be U-shaped.

It is interesting to see how bell-shaped, J-shaped, and U-shaped distributions arise in games of chance. For instance, a symmetrical

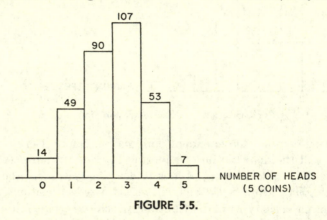

FIGURE 5.5.

bell-shaped distribution may be obtained by repeatedly flipping a number of coins and recording in each case the number of coins that come up heads. An actual experiment consisting of 320 flips of 5 coins produced the results shown in Fig. 5.5.

A *J-shaped distribution* may be obtained by repeatedly rolling a number of dice and recording in each case the number of dice that come up 6. An experiment consisting of 500 rolls of 4 dice produced the distributions shown in Figure 5.6.

To illustrate how a *U-shaped distribution* may arise in a game of chance, let us suppose that we play the game of "heads or tails," and that after each flip we check whether we are *ahead, even, or behind*. Writing W and L for "win" and "loss," let us suppose that a series of 6 flips produced the following result:

$$W \quad W \quad L \quad L \quad L \quad W$$

Clearly, after the first flip we are *one ahead*, after the second flip we are *two ahead*, after the third flip we are *one ahead*, after the fourth flip we are *even*, after the fifth flip we are *one behind*, and after the sixth flip we are again *even*. In this particular series of 6 flips we are, thus, ahead 3 times out of 6.

In an actual experiment consisting of 200 such series of 6 flips of a coin there were 67 series in which we were *never ahead,* 31 in which we were *ahead once,* 23 in which we were *ahead twice,* 13 in which we were *ahead three times,* 16 in which we were *ahead four times,* 18 in

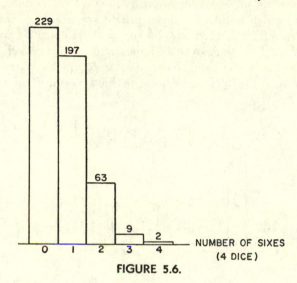

FIGURE 5.6.

which we were ahead *five times,* and 32 in which we were *always ahead.* These results are presented in the *U*-shaped distribution of Figure 5.7.

The result obtained in the previous paragraph illustrates the rather startling fact that *we are more likely to be ahead all the time or behind*

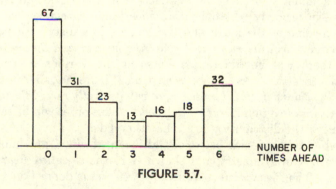

FIGURE 5.7.

all the time than ahead, say, half the time. (In case the reader does not trust our results, he is welcome to repeat the experiment for himself. See also Exercise 6 on page 130.)

What we have shown here really amounts to the fact that *once we get ahead* (in business or in a game of chance) *we are apt to stay ahead for a while* and *once we get behind we are apt to stay behind for some time.*

5.2 Measures of Symmetry and Skewness

Although there are many problems in which the mean and standard deviation provide sufficient information about a set of data, there are problems in which additional descriptions are needed. Indeed, two distributions can have identical means and standard deviations but differ considerably in their over-all appearance. The two distributions of Figure 5.8 have the same mean $\mu = 29.5$ and the same standard deviation $\sigma = \sqrt{145}$, yet the first is perfectly symmetrical while the

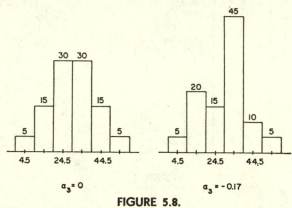

FIGURE 5.8.

second is somewhat lopsided or skewed. To distinguish between distributions like those of Figure 5.8, we can employ a statistical description of their symmetry or skewness.

The symmetry or lack of symmetry of a distribution may be expressed in terms of whatever difference there is between the mean and the median *or* the mean and the mode. If a distribution is perfectly symmetrical, *the mean will equal the median and the mode;* if it is not, there will be differences between the various measures of "central" location. As can be seen from Figure 5.9, *the mean of a distribution having a "tail" on the right will generally exceed the median, and the median will in turn exceed the mode.* This order will be reversed if the tail of the distribution is at the other side.

Possible discrepancies between the mean, median, and mode are used to measure symmetry and skewness in the so-called *Pearsonian formulas.* The *Pearsonian coefficient of skewness* is defined as

$$SK = \frac{\text{mean} - \text{mode}}{\text{standard deviation}} \qquad (5.2.1)\star$$

As in the case of the coefficient of variation where we divided by the mean, we are dividing by the standard deviation to make the formula independent of the scale of measurement. This is important because we are trying to describe the *shape* of a distribution and this should not be affected by a change in scale.

Formula (5.2.1) has the disadvantage that it involves the mode, which, as we saw in Section 3.6, may not exist, may not be unique, and which is difficult to define for grouped data. Let us, therefore, derive an alternate formula by making use of the fact that for moderately

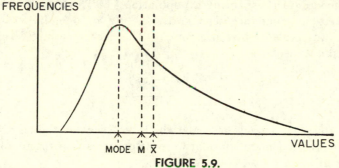

FIGURE 5.9.

skewed distributions the median will generally be about twice as far from the mode as from the mean (see Figure 5.9). In other words,

$$\text{mode} - \text{median} = 2(\text{median} - \text{mean})$$

or $$\text{mode} = 3(\text{median}) - 2(\text{mean}) \qquad (5.2.2)$$

Substituting this expression for the mode into (5.2.1), we obtain the alternate form of the *Pearsonian coefficient of skewness*, namely,

$$SK = \frac{3(\text{mean} - \text{median})}{\text{standard deviation}} \qquad (5.2.3)\star$$

Applying this formula to the two distributions of Figure 5.8, we find that for the first

$$SK = \frac{3(29.5 - 29.5)}{\sqrt{145}} = 0$$

and for the second

$$SK = \frac{3(29.5 - 31.72)}{\sqrt{145}} = -0.55$$

The Pearsonian coefficient of skewness will be *positive* if the mean exceeds the median and the mode, and it will be *negative* if the mean is less than the median and the mode. Hence, a distribution is said to be *positively skewed* if its tail is at the right (see Figure 5.9) and *negatively skewed* if its tail is at the left. Of course, for a perfectly symmetrical distribution the mean, median, and mode coincide and *SK* is equal to 0.

The most widely used measure of skewness is called α_3 (*alpha three*). It is defined as *the average of the cubed deviations from the mean divided by the cube of the standard deviation.* Symbolically, it is written as

$$\alpha_3 = \frac{\frac{1}{n} \sum\limits_{i=1}^{n} (x_i - \mu)^3}{\sigma^3} \tag{5.2.4}$$

for ungrouped data constituting a population.[†] (The average of the cubed deviations from the mean is divided by σ^3 to make α_3 independent of the scale of measurement.) The corresponding formula for grouped data is

$$\alpha_3 = \frac{\frac{1}{n} \sum\limits_{i=1}^{k} (x_i - \mu)^3 f_i}{\sigma^3} \tag{5.2.5}$$

and, with the same change of scale which we used for the short-cut formulas for the mean and standard deviation, it can be written as

$$\alpha_3 = \frac{c^3}{\sigma^3} \left[\frac{\sum\limits_{i=1}^{k} u_i^3 f_i}{n} - 3 \left(\frac{\sum\limits_{i=1}^{k} u_i^2 f_i}{n} \right) \left(\frac{\sum\limits_{i=1}^{k} u_i f_i}{n} \right) + 2 \left(\frac{\sum\limits_{i=1}^{k} u_i f_i}{n} \right)^3 \right] \tag{5.2.6}\star$$

Although this formula may look quite involved, it is not difficult to use. If we have already found $\sum\limits_{i=1}^{k} u_i f_i$ and $\sum\limits_{i=1}^{k} u_i^2 f_i$ in determining the standard deviation, we have only to calculate $\sum\limits_{i=1}^{k} u_i^3 f_i$ and then substitute these sums into (5.2.6). The symbol c stands, as before, for the length of the class interval.

The calculations needed to find α_3 for the second distribution of Figure 5.8 are shown in the following table:

Class	f_i	u_i	$u_i f_i$	$u_i^2 f_i$	$u_i^3 f_i$
0– 9	5	−2	−10	20	−40
10–19	20	−1	−20	20	−20
20–29	15	0	0	0	0
30–39	45	1	45	45	45
40–49	10	2	20	40	80
50–59	5	3	15	45	135
	100		50	170	200

$$\sigma = \tfrac{10}{100} \sqrt{100(170) - (50)^2}$$
$$= \sqrt{145} = 12 \text{ (approximately)}$$

and
$$\alpha_3 = \frac{10^3}{12^3} \left[\frac{200}{100} - 3 \left(\frac{170}{100} \right) \left(\frac{50}{100} \right) + 2 \left(\frac{50}{100} \right)^3 \right]$$
$$= -0.17.$$

[†] α_3 is seldom found unless n is very large and we shall use the same formula regardless of whether we are dealing with a sample or a population.

Had we calculated α_3 for the first distribution of Figure 5.8, the perfectly symmetrical distribution, we would have obtained $\alpha_3 = 0$.

5.3 Measures of Peakedness

There are some problems, though relatively few, in which it is of interest to describe also the *peakedness* or *kurtosis* of a distribution. The statistical measure which is commonly used for this purpose is called α_4 (*alpha four*), and it is defined as *the mean of the fourth powers of the deviations from the mean divided by the fourth power of the standard deviation*. Symbolically it is written as

$$\alpha_4 = \frac{\dfrac{1}{n} \cdot \sum_{i=1}^{n} (x_i - \mu)^4}{\sigma^4} \tag{5.3.1}$$

For grouped data we have analogous to (5.2.6) the *short-cut forumla*

$$\alpha_4 = \frac{c^4}{\sigma^4} \left[\frac{\sum\limits_{i=1}^{k} u_i^4 f_i}{n} - 4 \left(\frac{\sum\limits_{i=1}^{k} u_i^3 f_i}{n} \right) \left(\frac{\sum\limits_{i=1}^{k} u_i f_i}{n} \right) \right. $$
$$\left. + 6 \left(\frac{\sum\limits_{i=1}^{k} u_i^2 f_i}{n} \right) \left(\frac{\sum\limits_{i=1}^{k} u_i f_i}{n} \right)^2 - 3 \left(\frac{\sum\limits_{i=1}^{k} u_i f_i}{n} \right)^4 \right] \tag{5.3.2}\star$$

To calculate α_4 for the first distribution of Figure 5.8, we have to determine $\sum\limits_{i=1}^{k} u_i f_i$, $\sum\limits_{i=1}^{k} u_i^2 f_i$, $\sum\limits_{i=1}^{k} u_i^3 f_i$, and $\sum\limits_{i=1}^{k} u_i^4 f_i$, and then substitute these sums into (5.3.2). Using the same u-scale as on page 106, it can easily be checked that these sums are 50, 170, 230, and 770, respectively, and that $\alpha_4 = 2.633$.

If a distribution is *very peaked* and has relatively wide tails, it is referred to as *leptokurtic*, meaning "narrow humped." If a distribution is rather *flat in the middle* and has relatively thin tails, it is referred to as *platykurtic*, meaning "broad humped." In practice, we meet many distributions having the *bell shape* of the *normal distribution*, which will be discussed in Chapter 7. For these distributions α_4 will be approximately 3, and it has been the custom to call a distribution *leptokurtic* if its value of α_4 *exceeds 3* and *platykurtic* if its value of α_4 is *less then 3*. We can, thus, say that the distribution for which we obtained $\alpha_4 = 2.633$ is slightly platykurtic. The two distributions of Figure 5.10 have the same mean $\mu = 0$, the same standard deviation $\sigma = \sqrt{2}$, they are both symmetrical with $\alpha_3 = 0$, but the first is *leptokurtic* with $\alpha_4 = 3.2$, while the second is *platykurtic* with $\alpha_4 = 2.6$.

The descriptive measures which we have studied in the first part of this book were limited to descriptions of the most important features of frequency distributions or ungrouped data. Even though there are many other statistical descriptions that have not yet been discussed, we have come a long way in learning how to squeeze essen-

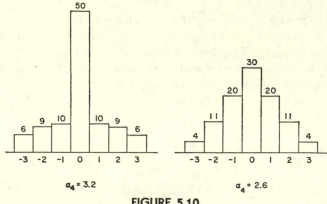

FIGURE 5.10.

tial information out of large sets of data by calculating a few well-chosen descriptions.

EXERCISES

1. Use formula (5.2.3) to measure the skewness of the mail-order house data. The mean, median, and standard deviation are given on pages 56, 62, and 92.
2. Use formula (5.2.3) to measure the skewness of whatever data were grouped among those of Exercises 7, 8, 9, or 10 on pages 29 and 30. The mean, median, and standard deviation were previously obtained in Exercise 3 on page 57, Exercise 9 on page 65, and Exercise 2 on page 94.
3. Use formula (5.2.3) to measure the skewness of the distribution of the number of physicians per 100,000 population obtained from the data which were grouped in Exercise 12 on page 31. The mean and standard deviation were previously determined in Exercise 4 on page 57 and Exercise 3 on page 94.
4. Verify that α_3 is equal to zero for the first distribution of Figure 5.8.
5. Find α_3 and α_4 for the mail-order house distribution of page 22, treating this distribution as a population.

BIBLIOGRAPHY

A more detailed treatment of the various measures of symmetry, skewness, and peakedness may be found in

Croxton, F. E., Cowden, D. J., and Klein, S., *Applied General Statistics*, 3rd ed.   Englewood Cliffs, N.J.: Prentice-Hall, Inc., 1967, Chap. 10.

Waugh, A. E., *Elements of Statistical Method*.   New York: McGraw-Hill, 1952, Chap. 8.

In view of the fact that the choice of statistical descriptions is essentially arbitrary, let us briefly refer to the question of *ethics*.   Data can be presented in a form in which they are *intentionally misleading* and particular measures can be chosen *to support preconceived notions or to "prove" things that we actually want to prove*.   Some of the "crimes" which come under the heading of "figures don't lie but liars sometimes figure" are discussed in

Huff, D., and Geis, I., *How to Lie with Statistics*.   New York: Norton and Co., 1954.

Wallis, W. A., and Roberts, H. V., *Statistics: A New Approach*.   Glencoe, Illinois: Free Press, 1956, Chap. 3.

CHAPTER **6**

Probability

6.1 The Meaning of Probability

The concept of probability plays an important role in many problems of everyday life, in business, in science, and particularly in statistics. Unfortunately, its study is complicated by the fact that the term "probability" itself is difficult to define; at least, there is no general agreement about the meaning of the term, and many people associate probability and chance with nebulous and mystic ideas.

So long as we limit ourselves to conversational language, it does not really matter whether we use such terms as "probably," "possibly," and "likely" without giving them a strict definition. However, we will find ourselves in difficulties if we use them in statistics (in science) without stating precisely what they are supposed to mean. We shall, therefore, have to explain what we mean when we say, for instance, "the probability that Candidate X will be elected is 0.75," "the probability that the shipment will arrive on time is 0.15," or "the probability that the price of a certain stock will go up is 0.80."

Although philosophical arguments about the various theories of probability make interesting reading (see Bibliography on page 126), we shall limit our discussion to the *objectivistic* point of view. After all, objectivity is the keynote of science and, in particular, it forms the basis for a scientific approach to problems of business and economics.

In statistics, probabilities are defined as *relative frequencies*, or to be more exact, as *limits of relative frequencies*. If we say "the probability that the shipment will arrive on time is 0.15," this means that in the long run 15 per cent of all similar shipments will arrive on time; if we say "the probability that it will rain in Blacksburg on the 4th of July is 0.26," this means that it rains there on the 4th of July about

110

26 per cent of the time; and if we say "the probability that a student who enters a given college will graduate is 0.40," this means that in the long run 40 per cent of the students who enter this college graduate. *The proportion of the time an event takes place is called its relative frequency, and the relative frequency with which it takes place in the long run is called its probability.*

If we say that the probability of getting *heads* with a balanced coin is $\frac{1}{2}$, this means that in the long run we will get 50 per cent heads and 50 per cent tails. It does not mean that we must necessarily get 5 heads and 5 tails in 10 flips of a coin or 10 heads and 10 tails in 20. No, we will not always get equal numbers of heads and tails, but *if we flip a balanced coin a large number of times we will usually get close to 50 per cent heads and 50 per cent tails.*

Since "in the long run" is not a very precise term, the probability of an event is actually defined *as the limit of the relative frequency with which it occurs.* If an event occurs x times out of n its relative frequency is x/n; the value which is approached by x/n when n becomes *infinite* is called the limit of the relative frequency. Since the concept of a limit belongs to calculus and is somewhat difficult to define, we shall be contented using the more intuitive yet easier to understand expression "in the long run."

The role played by probability in modern science is that of a substitute for certainty. There being few occasions in which we actually have *complete* information, that is, all the facts, we usually have no choice but to generalize from samples and hence, cannot be absolutely certain about our results. To illustrate how probability takes the place of certainty, let us suppose that Mr. White has recently opened a restaurant and that he asks an expert about the chances of its becoming a success. If the expert could tell him *for sure* that the restaurant will be a success, Mr. White would know precisely where he stands. Similarly, a negative reply would tell him exactly what to expect. Actually, the answer he will get from the expert, who is of course not a fortune teller, will merely tell him that his chances are fairly good, average, or pretty poor. Before we try to interpret such a reply in the light of what we have said about probabilities, let us ask the expert for a more precise evaluation of Mr. White's chances. Suppose he replies that the probability that Mr. White's restaurant will be a success is 0.90.—*Where do we stand now?—Would this entitle us to say that the expert was right if the restaurant becomes a success, or that he was wrong if it fails?* Let us check briefly what the expert means when he says that the probability is 0.90. According to the definition given above, a probability of 0.90 means that *in the long run something is going to happen 90 per cent of the time.* Hence, when the expert says that the probability is 0.90, he means that *among a large number of similar restaurants built in similar locations and under similar economic conditions 90 per cent will succeed and 10 per cent will fail.*

This illustrates the fact that in order to discuss the probability of a given event, we must actually refer to what will happen in the long run to a large number of similar events. There are many who view this with skepticism because they are unwilling to settle for anything but absolute certainty and absolute truth. Of course, it would be nice if we had a crystal ball and could make miraculous predictions, but if we want to be scientific and obtain knowledge from observations and experiments, we have to resign ourselves to the fact that all scientific predictions are similar in kind to the prediction of the expert mentioned above.

The foregoing may have given the impression that it is perfectly safe to make predictions in terms of probabilities since nobody can prove them right or wrong on the basis of single events. For instance, if a meteorologist claims with a probability of 0.80 that it will rain on a certain day, we cannot say whether he is right or wrong *regardless of what happens on the particular day to which he refers.* All he really says is that under similar conditions it will rain about 8 times out of 10, and this statement cannot be verified by considering one day.

This does not mean that since nobody can prove us wrong, we should go ahead and make wild predictions in terms of probabilities. It is as important in science as it is in business and every-day life to make correct predictions and correct decisions *as often as possible* and, hence, it is essential to know the correct values of probabilities. To give an example, if the probability of catching a cold at a football game is 0.30 if we do not wear a coat and 0.10 if we do, it would be smart *to play the odds* and wear a coat. Naturally, this will not protect us from catching colds, but *in the long run we will catch fewer.*

Since the next few chapters will be devoted to problems of inductive statistics, it will be of interest to see how probabilities are used to evaluate and judge the merits of the many generalizations that will be made. For instance, let us suppose that we are trying to predict an election and that we have polled a sample of the electorate. If on the basis of this information (and some fancy statistical techniques) we arrive at the conclusion that Candidate Y will receive anywhere from 57 to 59 per cent of the vote, *we cannot be absolutely certain that this is correct.* After all, we have made a generalization on the basis of incomplete information. Since we cannot be *certain*, we are open to the question as to "how sure" we really are that Candidate Y will receive from 57 to 59 per cent of the vote. Suppose we say that we are "95 per cent sure," or, in other words, that we assign our estimate a probability of 0.95.

To explain what this means, let us point out first that we are talking about a *specific* election. The probability of 0.95 is not meant to imply that Candidate *Y* would receive the stated per cent of the vote 95 per cent of the time *if he ran for office a great number of times*. No, if we assign to an estimate a probability of 0.95 this means that *we are using a method of estimation which in the long run will be successful 95 per cent of the time*. In other words, *the probability which we assign to an estimate is really a measure of the "goodness" of the method of estimation we employed*.

In the same way we shall express the "goodness" of decisions based on samples by giving the *success ratio* of the statistical techniques we employed. If we decide on the basis of certain grades (and certain statistical formulas) that Joe Brown is a better student than Harry Smith *and* if we say that we are 80 per cent sure that this decision is correct, *we are really saying that our decision was based on statistical techniques which will provide us with correct decisions 80 per cent of the time*. All probabilities that will be used to express "how sure" we are of estimates, predictions, and decisions will really express the "goodness" of the methods which we employ; in other words, they will stand for the proportion of the time that these methods will provide us with correct results.

6.2 Some Rules of Probability

In the study of probability there are essentially *three* kinds of problems. First, there is the question of what we mean when we say that a probability is, say, 0.80, 0.15, or 0.62. Then there is the problem of obtaining numerical probabilities, and, finally, there is the problem of using known probabilities to calculate others. Having resolved the first question by defining probabilities in terms of relative frequencies, the problem of obtaining numerical probabilities becomes a problem of *estimation*. For example, if we want to determine the probability that a shipment from a given firm will arrive on time, we refer to past experience and check what proportion of the time shipments from this firm have arrived on time. Assuming that what happened in the past is an indication of what will happen in the future, we thus get an estimate of the desired probability. In actual practice we estimate most probabilities by observing the relative frequency with which the appropriate events have occurred in the past.

Since the problem of estimation will be treated separately in Chapter 9, we shall investigate here only the third kind of problem, namely, that of calculating the probabilities of relatively complex events in terms of known (or assumed) probabilities of simpler kinds of events. This is often referred to as the *arithmetic or the calculus of probabilities*.

If we write the probability of the occurrence of event A as $P(A)$, the first fundamental rule of probability may be given as†

$$0 \leqslant P(A) \leqslant 1$$

This merely expresses the fact that a relative frequency (and the limit of a relative frequency) cannot be negative or exceed 1. After all, the proportion of the time that an event takes place cannot be negative and it cannot exceed 1, that is, an event cannot occur more than 100 per cent of the time.

Although the distinction is rather fine, $P(A) = 0$ does not mean that the occurrence of event A is beyond the realm of possibility and $P(A) = 1$ does not mean that the occurrence of event A is absolutely certain. It is customary to assign *zero probabilities* to events which in colloquial terms "would not happen in a million years" and *probabilities of one* to events whose occurrence is "practically certain." If a monkey is set loose on a typewriter, it is not impossible that he might (purely by chance) type the complete works of Shakespeare without a mistake. It is so unlikely, however, that we put the probability of this happening equal to 0. Similarly, we are *practically certain* that it will not snow in Los Angeles in the summer, although logically speaking it could, and we put this probability equal to 1. (For a further discussion of probabilities that can be put equal to 0 or 1, see the reference to E. Borel on page 126.)

A second rule of probability, which follows directly from the frequency definition, states that if the probability of the occurrence of event A is $P(A)$, the probability that it will *not* occur is

$$1 - P(A)$$

For instance, if the probability that it will rain on a certain day is 0.24, then the probability that it will *not* rain is $1 - 0.24 = 0.76$; if the probability that the value of a stock will go up is 0.63, then the

---

† There is little uniformity in the symbols used for probabilities. Instead of $P(A)$, some books use $Pr(A)$, $p(A)$, $pr(A)$, or simply p without any qualification about the event to which this probability is to refer.

In this and in subsequent sections we shall frequently use *inequality signs* to simplify our notation. In case the reader is not familiar with these symbols, let us briefly explain that a < b means "a is less than b" or "b is greater than a," while a > b means "a is greater than b" or "b is less than a." Similarly, a $\leqslant$ b means "a is less than or equal to b" and a $\geqslant$ b means "a is greater than or equal to b." Thus,

$$\$1.24 < \text{cost per item} < \$1.37$$

means that the cost per item is somewhere between $1.24 and $1.37, and

$$\text{Income} \geqslant \$5{,}000$$

means that the income is greater than or equal to $5,000.

probability that it will *not* go up is $1 - 0.63 = 0.37$; and if the probability that a certain train will arrive late is 0.60, then the probability that it will *not* be late is $1 - 0.60 = 0.40$. Clearly, if the train is late 60 per cent of the time it must be on time (or early) the remaining 40 per cent of the time.

The rule of probability which we shall mention next refers to special kinds of events, events that are *mutually exclusive*. *Two events are said to be mutually exclusive if the occurrence of either precludes the occurrence of the other*. If we toss a coin, heads and tails are mutually exclusive events; we can get one or the other but never both. Similarly, a person's being born in Chicago or in Detroit are mutually exclusive events and so are the election of a Republican or a Democratic President. In each of these examples the occurrence of one event renders the occurrence of the other impossible.

If two events A and B are mutually exclusive, the probability that A *or* B will occur is given by the following rule:

SPECIAL RULE OF ADDITION:
   If A and B are mutually exclusive, then
   $$P(A \text{ or } B) = P(A) + P(B) \qquad (6.2.1)\star$$

This formula tells us that if A and B are mutually exclusive, the probability of the occurrence of A or B is equal to the sum of the individual probabilities of A and B.

If, in a given subject, the probability that a student will receive a C is 0.58 and the probability that he will receive a B is 0.25, then the probability that he will receive either a C or a B is $0.58 + 0.25 = 0.83$. (In other words, if 58 per cent of the students get C's and 25 per cent get B's, then 83 per cent get either C's or B's.) Also, if the probability that there will be 2 persons in a tourist party stopped in the Smoky Mountains of Tennessee is 0.41 and the probability that there will be 3 is 0.15, then the probability that there will be either 2 or 3 is $0.41 + 0.15 = 0.56$. (Further examples illustrating the special rule of addition will be given in Section 6.3.)

Formula (6.2.1) can easily be generalized to apply to more than two mutually exclusive events. *A number of events are said to be mutually exclusive if the occurrence of any one precludes the occurrence of all of the others*. Given n mutually exclusive events $A_1, A_2, \ldots,$ and A_n, the probability that one of them will occur is

$$P(A_1 \text{ or } A_2 \text{ or } \cdots \text{ or } A_n) = P(A_1) + P(A_2) + \cdots + P(A_n)$$
$$(6.2.2)\star$$

If the probabilities that a person will some day die from heart disease, cancer, tuberculosis, or pneumonia are 0.53, 0.15, 0.01, and 0.03, respectively, then the probability that a person will die from one of these four causes of death is

$$0.53 + 0.15 + 0.01 + 0.03 = 0.72$$

Also, if the probability that a certain Mr. X will spend his summer vacation in Boston, Atlantic City, Yellowstone Park, Washington, D.C., or the Pocono Mountains is 0.12, 0.19, 0.27, 0.20, and 0.06, respectively, then the probability that he will spend his summer vacation in one of these 5 places is

$$0.12 + 0.19 + 0.27 + 0.20 + 0.06 = 0.84$$

If the probability that a man who enters a men's clothing store will buy a shirt is 0.04 and the probability that he will buy a tie is 0.05, we cannot use formula (6.2.1) to calculate the probability of his buying either a shirt or a tie. The two events are *not* mutually exclusive, he could very well buy a shirt *and* a tie, and the special rule of addition does not apply.

To formulate a general rule of addition for events which need not be mutually exclusive, we shall first have to explain what is meant by two events being *independent or dependent*. *Two events are said to be independent if the occurrence or non-occurrence of either in no way affects the occurrence of the other.* If A and B stand for getting *heads* in two successive flips of a coin, then A and B are independent; the second flip is in no way affected by what happened in the first. Similarly, if A stands for Mr. Jones' passing an examination in statistics and B stands for his having size 8D shoes, it is difficult to see how these events could possibly be dependent. On the other hand, if A stands for Mr. Brown's being intoxicated and B stands for his having an accident, these events are *not* independent.

If two events A and B are independent, the probability that A and B will *both* occur is given by the following rule:

SPECIAL RULE OF MULTIPLICATION:
If A and B are independent, then
$$P(A \text{ and } B) = P(A) \cdot P(B) \qquad (6.2.3)^*$$

This formula tells us that if A and B are independent, the probability that they will both occur is equal to the product of their individual probabilities.

To illustrate the use of this formula, let us suppose that the probability that it will rain in a certain town on the 4th of July is 0.30 and the probability that a Republican will be elected mayor of the town in the fall of the same year is 0.50. Since the two events are clearly independent, the probability that both of these things will happen is $(0.30)(0.50) = 0.15$. Also, if the probability that Henry Brown will get a promotion is 0.80 and the probability that his secretary will get married is 0.02, then the probability that both of these events will take place is $(0.80)(0.02) = 0.016$. (We are assuming that the secretary is not marrying Mr. Brown since the two events could otherwise easily be dependent.) Further applications of the special rule of multiplication will be given in Section 6.3.

To consider two events that are *not independent*, let A stand for a firm's spending a large amount of money on advertising and B for its showing an increase in sales. Of course, advertising does not guarantee higher sales, but the probability that the firm will show an increase in sales will be higher if A has taken place.

The probability that event B will take place provided that A has taken place is called the *conditional probability of B relative to A* and it will be written symbolically as $P(B|A)$. In the above example, $P(B|A)$ stands for the probability of an increase in sales provided that a considerable amount of money is spent on advertising. $P(A|B)$ stands for the probability that a considerable amount of money will be spent on advertising provided that there is an increase in sales. Similarly, if C and D stand, respectively, for a person's being a banker and making over \$10,000 a year, then $P(D|C)$ is the probability that a banker makes more than \$10,000 a year and $P(C|D)$ is the probability that someone who makes more than \$10,000 a year is a banker.

In terms of conditional probabilities we can now formulate a more general rule for the probability that two events A and B will *both* occur:

GENERAL RULE OF MULTIPLICATION:†

$$P(A \text{ and } B) = P(A) \cdot P(B|A) \qquad\qquad (6.2.4)\star$$
$$\text{or } P(A \text{ and } B) = P(B) \cdot P(A|B)$$

To illustrate the use of (6.2.4), let us suppose that A stands for a student's passing an examination in accounting while B stands for his passing an examination in business law. Then $P(B)$ is the probability of his passing in business law while $P(B|A)$ is the probability of his passing in business law provided that he passed in accounting. Since a student who is good in one of these subjects is apt to be good also in the other, the two events are *dependent* and $P(B|A)$ is greater than $P(B)$. If $P(A) = 0.80$, $P(B) = 0.75$, and $P(B|A) = 0.90$, we get, according to (6.2.4),

$$P(A \text{ and } B) = (0.80)(0.90)$$

$$= 0.72$$

for the probability that he will pass both exams.

---

† The two formulas are equivalent since we are merely interchanging A and B in the right-hand side of the equation and "A and B" is the same as "B and A." Incidentally, if A and B are *independent*, then $P(B|A) = P(B)$, $P(A|B) = P(A)$, and (6.2.4) reduces to (6.2.3).

To return to the problem of finding a rule of addition for events that need not be mutually exclusive, let us suppose that $P(A)$ is the probability that there will be sunshine (in a certain place at a certain time) and that $P(B)$ is the probability that there will be rain. If $P(A) = 0.70$ and $P(B) = 0.50$, formula (6.2.1) yields the *impossible* result

$$P(A \text{ or } B) = 0.70 + 0.50 = 1.20$$

By using the formula for mutually exclusive events *we have made the mistake of counting twice all days on which there was both sunshine and rain.* To compensate for this we can subtract the days (proportion of days) which were counted twice and use the following rule:

GENERAL RULE OF ADDITION:
$$P(A \text{ or } B) = P(A) + P(B) - P(A \text{ and } B) \qquad (6.2.5)\star$$

In this formula we actually subtract $P(A \text{ and } B)$, namely, the proportion of the cases that are counted twice in $P(A) + P(B)$. If in the previous example it is known that there is *both* sunshine and rain 40 per cent of the time we get

$$P(A \text{ or } B) = 0.70 + 0.50 - 0.40 = 0.80$$

instead of the incorrect value of 1.20 which we obtained with (6.2.1).

It should be noted that the "or" in $P(A \text{ or } B)$ is what is called the "inclusive or." $P(A \text{ or } B)$ stands for the probability that A, B, *or both occur* or, in other words, that at least one of the two events will occur. Also, if A and B are mutually exclusive, then $P(A \text{ and } B) = 0$, since A and B cannot both happen at the same time, and (6.2.5) reduces to (6.2.1).

If, in the example given on page 117, we are interested in the probability that the student will pass either in accounting or in business law, that is, in at least one of the two subjects, we obtain

$$P(A \text{ or } B) = 0.80 + 0.75 - 0.72 = 0.83$$

EXERCISES

1. If A stands for a man's having an MA degree and B stands for his making more than \$25,000 per year, write each of the following probabilities in symbolic form:
   (a) The probability that a man with an MA degree earns more than \$25,000 a year.
   (b) The probability that a man who earns more than \$25,000 a year has an MA degree.
   (c) The probability of having an MA degree and earning more than \$25,000 a year.

2. If A stands for a person's being married and B stands for his being a good poker player, what probabilities are expressed by

(a) $P(A|B)$ (c) $P(B|A)$

(b) $1 - P(A)$ (d) $P(A$ or $B)$

3. Which of the following events are mutually exclusive?
   (a) Having a surplus or a deficit on the balance sheet of a business corporation.
   (b) Living in Baltimore, Maryland, and attending school in Washington, D.C.
   (c) Being an alien and being President of the United States.
   (d) Being a lawyer and being a college professor.
   (e) Drawing a black card or a king out of an ordinary deck of playing cards.
   (f) Getting two tails in five tosses of a coin or getting three tails.
4. Which of the following events are independent?
   (a) Winning the door prize at two successive college dances.
   (b) Feeling depressed and winning on a lottery ticket.
   (c) Getting two black marbles in two drawings from a bowl if the first marble is not replaced before the second is drawn.
   (d) Having large feet and having a large income.
   (e) Earning a large salary and paying a high income tax.
5. A secretary can bring her own lunch, eat in the company cafeteria, or eat in the restaurant across the street. If the probability that she brings her own lunch is 0.11, and the probability that she eats in the company cafeteria is 0.66, what is the probability that on a certain day she does not eat in the restaurant across the street? (Assume that there are no other alternatives.)
6. If the probability of drawing an ace from a deck of cards is $\frac{1}{13}$, the probability of drawing a red queen is $\frac{1}{26}$, and the probability of drawing the jack of diamonds is $\frac{1}{52}$, what is the probability of drawing either an ace, a red queen, or the jack of diamonds?
7. If the probability of rolling a *seven* with a pair of dice is $\frac{1}{6}$, what is the probability of rolling 2 *sevens* in a row?
8. Suppose that the probability of getting a busy signal when calling a friend is 0.03. What is the probability of getting busy signals in two calls to this friend on two different days?
9. If the probability that a married man will vote in a certain primary election is 0.40 and the probability that a woman will vote provided that her husband votes is 0.80, what is the probability that a husband and wife both vote in this election?
10. If the probability that it rains on an April day is 0.45, and the probability that it rains on an April day if it rained the day before is 0.90, what is the probability that it will rain on two successive April days?
11. In a certain lottery, the probability of drawing a number which is divisible by 2 is $\frac{1}{2}$, the probability of drawing a number divisible by 3 is $\frac{1}{3}$, and the probability of drawing a number divisible by 6 is $\frac{1}{6}$.

What is the probability of drawing a number which is divisible either by 2 or by 3?

12. Among the members of a certain club, 10 per cent are bankers and 20 per cent have incomes of over $10,000 a year. If it is known that 80 per cent of the bankers have incomes of over $10,000 a year, what percentage of the club members with incomes of over $10,000 are bankers? What percentage of the club members are either bankers or have an income of over $10,000 a year? (*Hint:* Use formulas (6.2.4) and (6.2.5).)

6.3 Probabilities in Games of Chance

Although we are not directly concerned in this book with games of chance, many of the rules of probability are easier to explain and easier to understand with reference to such games. If cards are thoroughly shuffled, dice are properly shaken before they are rolled, numbered slips are thoroughly mixed, and coins are properly flipped, it is generally assumed that the different results that can arise have equal probabilities. In that case many probabilities may be calculated with the following rule:

SPECIAL RULE FOR EQUIPROBABLE EVENTS:†

If N mutually exclusive events have equal probabilities, one of them must occur, and S of them are labeled "successes," then the probability of a "success" is S/N.

This rule can easily be derived from those given in Section 6.2 (see Bibliography on page 126).

To illustrate the special rule for equiprobable events, let us find the probability of drawing a king out of a well-shuffled deck of cards. Since there are 4 kings and 52 cards, $S = 4$, $N = 52$, and if each card has an equal chance of being drawn the probability of getting a king is $S/N = \frac{4}{52}$. Similarly, there are 13 spades, 2 red queens, 1 ace of clubs, and the probabilities of getting a spade, a red queen, or the ace of clubs are $\frac{13}{52}$, $\frac{2}{52}$, and $\frac{1}{52}$, respectively.

To give an example in which formula (6.2.1) applies, let us find the probability of getting a 5 or a 6 with one die. Since a die has

---

† In some books this rule is used as a *definition* of probability. This is objectionable because we would be using the very term we are trying to define. To define probability in terms of *equiprobable* or *equally likely* events is circular reasoning and, hence, fallacious.

six faces, two of which will yield a "success," we have $N = 6$, $S = 2$, and the desired probability is $S/N = \frac{2}{6}$. We could also have argued that the probability of getting a 5 is $\frac{1}{6}$, the probability of getting a 6 is $\frac{1}{6}$, and since the two events are mutually exclusive the probability of getting a 5 or a 6 is $\frac{1}{6} + \frac{1}{6} = \frac{2}{6}$.

To consider a slightly more complicated example, let us look for the probability of obtaining a 7 or an 11 with a pair of dice. In this case $N = 36$; there are 36 possible outcomes of the form *1 and 6, 6 and 1, 2 and 5, 5 and 2, 3 and 4, 2 and 2, 1 and 4,* . . . . Since 7 may be obtained by rolling either *6 and 1, 1 and 6, 2 and 5, 5 and 2, 3 and 4,* or *4 and 3*, $S = 6$ and the probability of getting a 7 is $S/N = \frac{6}{36}$. Similarly, 11 may be obtained by rolling either *5 and 6* or *6 and 5* and the probability of getting an 11 is $S/N = \frac{2}{36}$. According to formula (6.2.1), the probability of rolling a 7 or an 11 is then $\frac{6}{36} + \frac{2}{36} = \frac{8}{36}$. We could also have argued that there are altogether 8 ways of getting a 7 or an 11, $S = 8$, and the probability of a "success" is $S/N = \frac{8}{36}$.

To illustrate formula (6.2.3), let us find the probability of getting *heads* in two successive flips of a coin. For each flip there are two equiprobable outcomes, heads and tails, and the probability for each is $\frac{1}{2}$. Since two successive flips of a coin are presumably independent, we find that according to (6.2.3) the desired probability is $\frac{1}{2} \cdot \frac{1}{2} = \frac{1}{4}$. We could also have argued that there are 4 equiprobable outcomes *heads and heads, heads and tails, tails and heads, tails and tails*, and that therefore the probability of getting 2 heads is $\frac{1}{4}$, the probability of getting 1 head and 1 tail is $\frac{2}{4}$, and the probability of getting 2 tails is $\frac{1}{4}$.

Let us now find the probability of drawing two aces in succession from a standard deck of cards, assuming that the first card is not replaced before the second is drawn. Since there are 4 aces among the 52 cards, the probability of obtaining an ace in the first draw is $\frac{4}{52}$. After that there are only 3 aces among the remaining 51 cards and the probability of getting an ace in the second draw is $\frac{3}{51}$. According to (6.2.4) the probability of getting aces in both draws is $\frac{4}{52} \cdot \frac{3}{51} = \frac{1}{221}$. In this problem the two events are *not* independent since the probability of getting an ace in the second draw depends on the outcome of the first. If the first draw is an ace, there are 3 aces left and the probability of getting an ace in the second draw is $\frac{3}{51}$; if the first draw is *not* an ace, there are 4 aces left and the probability of getting an ace in the second draw is $\frac{4}{51}$.

Finally, to illustrate (6.2.5), the general rule of addition, let us ask for the probability of getting either a red card or a queen in one draw from an ordinary deck of cards. Since there are 26 red cards, 4 queens, and 2 red queens, the probabilities of getting a red card, a queen, or a red queen are $\frac{26}{52}$, $\frac{4}{52}$, and $\frac{2}{52}$, respectively. Hence, according to (6.2.5) the probability of obtaining either a red card or a queen is

$$\frac{26}{52} + \frac{4}{52} - \frac{2}{52} = \frac{28}{52}$$

To obtain this result we could also have argued that there are 26 red cards, 2 black queens, and hence 28 cards which will yield a "success." Substituting $S = 28$ and $N = 52$ into the special formula for equiprobable events, we find immediately that the desired probability is $\frac{28}{52}$. In this last example the two events (getting a red card or a queen) are *not* mutually exclusive since there are two red queens.

EXERCISES

1. What is the probability of drawing either a jack or a queen out of a standard deck of 52 playing cards?
2. What is the probability of rolling a 2 or a 4 with one die?
3. Find the probability of drawing a 5, 6, 7, 8, 9, or 10 from an ordinary deck of playing cards.
4. What is the probability of rolling a 5 with a pair of dice?
5. A club with 12 members elects its president and vice-president by drawing lots. What is the probability that Mr. Jones, a member of this club, will be elected either president or vice-president?
6. If the numbers from 1 to 100 are written on slips of paper that are thoroughly mixed in a goldfish bowl, what is the probability of drawing a number which is divisible by 13?
7. An urn contains 8 red marbles and 10 white ones. What is the probability of drawing a red marble out of this urn?
8. What is the probability of getting either a red jack or a black king in one draw from a standard deck of cards?
9. What is the probability of not rolling 2 *sixes* in succession with one die?
10. A shipment of 20 television tubes contains 4 defectives. If each tube has an equal chance of being selected, what is the probability of selecting two tubes which are both defective?
11. What is the probability of getting 4 heads in 4 successive flips of a coin? Enumerate all possible cases and use the special formula for equiprobable events. Does this example suggest how formula (6.2.3) might be generalized to apply to 4 independent events?
12. Find the probability of drawing a card which is either a heart or smaller than 8 from a standard deck of 52 playing cards.
13. What is the probability of getting two black cards in two successive draws from an ordinary deck of cards if (a) the first card is replaced before the second is drawn and (b) if it is not replaced?
14. What is the probability of getting an ace and a king (not necessarily in that order) in two draws from an ordinary deck of cards if the first card is not replaced before the second is drawn?

15. What would be the answer to Exercise 14 if the first card were replaced before the second is drawn?

6.4 Mathematical Expectation

If in the game of "heads or tails" we received $1.00 for each head and nothing for each tail, we would *in the long run* win $1.00 about 50 per cent of the time. On the average we would thus get $0.50 per toss and this is referred to as our *mathematical expectation*. More generally, *if the probability of winning an amount A is $P(A)$, our mathematical expectation is $A \cdot P(A)$, namely, the product of the amount we stand to receive and the probability of obtaining it.* For example, if the grand prize in a lottery is $1,000 and we are given 1 of the 5000 tickets that are being sold, the probability of our holding the winning ticket is $P(A) = \frac{1}{5000}$ and our mathematical expectation is $A \cdot P(A) = 1000 \cdot \frac{1}{5000} = \$0.20$. Obviously, we cannot win $0.20 in this lottery, but if we bought tickets in many such lotteries we would (in the long run) win about *once* in 5000 tries and receive, thus, an average of $0.20 per ticket.

The above game of heads or tails was not very realistic, because it was arranged so that we could not possibly lose. To make it equitable, let us suppose that we are given $1.00 for each head and that we lose $1.00 for each tail. In the long run we will win $1.00 about half the time, lose $1.00 about half the time, and *on the average* we will break even. In order to make the concept of a mathematical expectation apply also to this example, let us now make the following definition: *If the probabilities of winning amounts A_1, A_2, . . . , or A_n are $P(A_1)$, $P(A_2)$, . . . , and $P(A_n)$, our mathematical expectation is*†

$$\sum_{i=1}^{n} A_i \cdot P(A_i) = A_1 \cdot P(A_1) + A_2 \cdot P(A_2) + \cdots + A_n \cdot P(A_n)$$

$$(6.4.1)\star$$

It should be noted that the amounts A_i referred to in this definition need not be positive. If we stand to *win* $1.00, the corresponding A is $+1$ and if we stand to *lose* $1.00, the corresponding A is -1. In the last example we can, thus, say that the probability of winning a dollar is $P(A_1) = \frac{1}{2}$, the probability of losing a dollar is $P(A_2) = \frac{1}{2}$, and the mathematical expectation is $(1.00)(\frac{1}{2}) + (-1.00)(\frac{1}{2}) = 0$. *A mathematical expectation of zero defines what we mean by a fair or equitable game.*

---

† It is assumed that the possibilities of winning these amounts are mutually exclusive.

To give a second illustration of formula (6.4.1), let us suppose that someone suggests the following game: We are to give him $1.00 for each roll of a die and he will pay us back $3.00 if he rolls a 6, $1.50 if he rolls a 5, $0.75 if he rolls a 4, and nothing if he rolls a 1, 2, or 3. We will, thus, *win* $2.00 if he rolls a 6, $0.50 if he rolls a 5, and *lose* $0.25 if he rolls a 4, and $1.00 if he rolls a 1, 2, or 3. Writing these four amounts as A_1, A_2, A_3, and A_4 we have

$$A_1 = 2.00 \qquad P(A_1) = \tfrac{1}{6}$$
$$A_2 = 0.50 \qquad P(A_2) = \tfrac{1}{6}$$
$$A_3 = -0.25 \qquad P(A_3) = \tfrac{1}{6}$$
$$A_4 = -1.00 \qquad P(A_4) = \tfrac{1}{2}$$

and the mathematical expectation is

$$(2.00)(\tfrac{1}{6}) + (0.50)(\tfrac{1}{6}) + (-0.25)(\tfrac{1}{6}) + (-1.00)(\tfrac{1}{2}) = -0.125$$

This means that *on the average* we stand to lose 12.5 cents each time that we play this game. Evidently, it is not an equitable game.

It has been suggested that a person's behavior might be called *rational* if he always chooses the alternative which has the highest mathematical expectation. Although this may seem to be a reasonable criterion, it involves a number of difficulties. There are many practical problems in which we do not know the probabilities associated with the various eventualities and cannot even get reasonably good estimates. Also, it is often difficult to ascribe suitable values to the A_i.

Let us return for a moment to the example of the lottery which we discussed on page 123 and let us suppose that we are asked to pay $0.25 per ticket. We then stand to lose $0.25 with a probability of 4999/5000, we stand to win $999.75 ($1000 minus the $0.25 we paid for the ticket) with a probability of 1/5000 and our mathematical expectation is

$$(-0.25)\,\frac{4999}{5000} + (999.75)\,\frac{1}{5000} = -0.05$$

This quantity is negative: on the average we stand to lose $0.05 per ticket, and according to the above criterion it would be *irrational* to buy a ticket. And yet, we have overlooked the fact that the holder of a ticket may get some pleasure merely from having the ticket and this may make it worthwhile to him even if he loses. Such a secondary gain would be extremely difficult to express in monetary terms.

To give a second example, let us suppose that a manufacturer has to decide whether to leave a certain factory where it is or to move it to another location. Even though one alternative may have a higher mathematical expectation than the other, purely in terms of financial gains, it does not necessarily follow that it would actually be the wiser move.   Such factors as good will of customers and creditors, the happiness of employees, and other more or less intangible factors are very difficult to evaluate in terms of dollars and cents.   We might conclude by saying that *mathematical expectations can be very useful guides for making decisions but there are usually other factors that have to be taken into account.*

EXERCISES

1. A lottery sells 200,000 tickets at $1.00 each and plans to give a grand prize of $20,000. What is the mathematical expectation for each ticket?

2. Suppose that someone will give us $1.00 each time that we roll a *six* with one die.   How much should we pay him when we roll a 1, 2, 3, 4, or 5 to make this game equitable?

3. Our opponent, in a game of chance, will give us $10.00 if we draw a deuce and $16.00 if we draw an ace from an ordinary deck of 52 playing cards.   If we draw neither a deuce nor an ace we are to pay him $2.34. What is our mathematical expectation?

4. A builder is bidding on a construction job which promises a profit of $36,000 with a probability of $\frac{4}{5}$ or a loss (from rework, late performance penalties, etc.) of $12,000 with a probability of $\frac{1}{5}$.   What is the builder's expected profit if he gets the job?

5. The probability that Mr. Jones will sell his house at a loss of $1,000 is $\frac{3}{18}$, the probability that he will break even is $\frac{5}{18}$, the probability that he will sell it at a profit of $1,000 is $\frac{7}{18}$, and the probability that he will sell it at a profit of $2,000 is $\frac{3}{18}$.   What is his expected profit?

6.† A company which operates a chain of lunchrooms plans to install a new unit in either of two locations.   The company figures that the probability of a unit's being successful in location X is $\frac{3}{4}$ and that, if it is successful, it will make an annual profit of $4,000.   If it is not successful, however, the company will lose $1,000 a year.   The probability of a unit's succeeding in location Y is only $\frac{1}{2}$, but if it does succeed the annual profit will be $6,000.   If it does not succeed in location Y, the annual loss will be $1,200.   Where should the company locate the new unit so as to maximize its expectation?   Suppose that the probability of success in location Y were $\frac{5}{8}$, would that affect the company's decision?   Where should the company locate the new unit if there were no information available about the probability of success

---

† Exercises 4, 5, and 6 provide a glimpse at the very interesting subject of *Decision Theory*.   If the reader is interested in the type of questions asked in these exercises, he should find it stimulating to read I. D. J. Bross, *Design for Decision*, New York: Macmillan, 1953.

in either location and its management consisted of confirmed optimists? What if it consisted of confirmed pessimists?

BIBLIOGRAPHY

An interesting discussion of various philosophers' views on objective probabilities, subjective probabilities, and logical probabilities may be found in

Nagel, E., *Principles of the Theory of Probability*. Chicago: University of Chicago Press, 1939.   (Published as part of the *International Encyclopedia of Unified Science*.)

The question of what probabilities are to be considered as negligible or practically certain from the human perspective, terrestrial perspective, cosmic perspective, and super-cosmic perspective is discussed in

Borel, E., *Elements of the Theory of Probability* (trans. by John E. Freund). Englewood Cliffs, N.J.: Prentice-Hall, Inc. 1965.

A derivation of the special rule for equiprobable events is given in

Freund, J. E., *A Modern Introduction to Mathematics*. Englewood Cliffs, N.J.: Prentice-Hall, Inc., 1956, p. 464.

Further problems dealing with probabilities in games of chance may be found in most textbooks of college algebra. An interesting treatment of this subject is given in

Levinson, H. C., *Chance, Luck and Statistics*. New York: Dover Publications, Inc., 1963.
Weaver, W., *Lady Luck—The Theory of Probability*. Garden City, N.Y.: Doubleday & Co., Inc., 1963.

More detailed, though still elementary, treatments of the theory of probability may be found in

Goldberg, S., *Probability—An Introduction*. Englewood Cliffs, N.J.: Prentice-Hall, Inc., 1960.
Mosteller, F., Rourke, R. E. K., and Thomas, G. B., *Probability with Statistical Applications*. Reading, Mass.: Addison–Wesley, 1961.

CHAPTER 7

Theoretical Distributions

7.1 Introduction

In the first part of this book we studied distributions which were obtained by grouping observed data. Now we shall study distributions which can be *expected* on the basis of previous experience or theoretical considerations. To illustrate the importance of expected distributions we have only to consider the proprietor of a shoe store, who must know something about the distribution of the sizes of his potential customers' feet; the executive of a publishing house, who must know something about the distribution of the public's literary tastes; or the manager of a restaurant, who must know something about the distribution of people's likes and dislikes for various foods. Unless he knows reasonably well what to expect, the proprietor of the shoe store may be stuck with shoes which he cannot sell, the executive of the publishing house may put out a book which nobody wants to read, and the manager of the restaurant may find himself with a refrigerator full of lamb chops and not enough steaks.

In contrast to these three examples in which the expected distributions have to be based on past experience, there are many situations in which an expected distribution can be based on *theoretical considerations*. To give an example of a theoretical distribution, let us consider the game of "heads or tails," and let us suppose that in order to test whether a coin is balanced we flip it 100 times, obtaining 53 heads and 47 tails. Before we can decide whether this coin is balanced, we must know what to *expect* from a balanced coin. Since the probabilities of getting heads and tails with a balanced coin are both equal to $\frac{1}{2}$, it would seem reasonable to say that we *expect* as many heads as tails or, in other words, the following distribution:

	Frequency
Heads	50
Tails	50

It is important to note that we have been using the word "expect" in the sense of a mathematical expectation, that is, *in the sense of an average.* The fact that the probabilities for heads and tails are both $\frac{1}{2}$ does not imply that we must necessarily always get 50 heads and 50 tails. Sometimes we will get 49 and 51, sometimes 53 and 47, sometimes 44 and 56, but if we repeat this game a great many times we will *on the average* get close to 50 heads and 50 tails.

In order to convert the distributions of the observed and expected number of heads and tails into *numerical distributions* we shall refer to *tails* as "0 heads" and *heads* as "1 head." We, thus, count the number of heads observed in each flip of the coin and we can present the result of our experiment and the corresponding expectations as the following *numerical distributions:*

Experiment		*Corresponding Expectations*	
Number of Heads	*Observed Frequency*	*Number of Heads*	*Expected Frequency*
0	47	0	50
1	53	1	50

(Later, in Chapter 11, we shall learn how to decide on the basis of such distributions whether discrepancies between observed and expected frequencies may reasonably be attributed to chance.)

To give a slightly more complicated example of a theoretical distribution, let us find the frequencies with which we can expect to get 0, 1, and 2 heads in, say, 120 flips of *two* balanced coins. As we pointed out on page 121, there are four equally likely possibilities, namely,

$$TT \quad TH \quad HT \quad HH$$

where H stands for heads and T for tails. Hence, the probability for 0 heads is $\frac{1}{4}$, the probability for 1 head is $\frac{1}{4} + \frac{1}{4} = \frac{1}{2}$, and the probability for 2 heads is $\frac{1}{4}$. We thus expect to get 0 heads $\frac{1}{4}$ of the time, 1 head $\frac{1}{2}$ the time, 2 heads $\frac{1}{4}$ of the time and the expected distribution for 120 flips of a balanced coin is

Number of Heads	*Expected Frequency*
0	30
1	60
2	30

If *three* coins are flipped simultaneously, there are 8 possible out-comes which (if the coins are balanced) may be considered as equi-probable.   They are

$$TTT \quad TTH \quad THT \quad HTT$$
$$HHT \quad HTH \quad THH \quad HHH$$

Since we can thus expect to get 0 heads $\frac{1}{8}$ of the time, 1 head $\frac{3}{8}$ of the time, 2 heads $\frac{3}{8}$ of the time, and 3 heads $\frac{1}{8}$ of the time, the *expected* distribution for 160 tosses of three balanced coins is

Number of Heads	Expected Frequency
0	20
1	60
2	60
3	20

The frequencies with which we can expect to get 0 heads, 1 head, 2 heads, and so on, depend, of course, on the total number of flips.   In the first example we considered 100 flips of *one* coin, in the second we considered 120 flips of *two* coins, and in the third we considered 160 flips of *three* coins.   In order to make the theoretical distributions which we have discussed apply to more general problems, we could present the *probabilities* of getting the different numbers of heads instead of the *expected frequencies*.   The table for *one* coin could be rewritten as

Number of Heads	Probability
0	$\frac{1}{2}$
1	$\frac{1}{2}$

and this is appropriately called a *probability distribution*.   Similarly, the probability distribution for *two* balanced coins is

Number of Heads	Probability
0	$\frac{1}{4}$
1	$\frac{2}{4}$
2	$\frac{1}{4}$

and that for *three* balanced coins is

Number of Heads	Probability
0	$\frac{1}{8}$
1	$\frac{3}{8}$
2	$\frac{3}{8}$
3	$\frac{1}{8}$

If we know the probability distribution, we can always find the expected frequencies for *any* number of flips by multiplying the probabilities by the total number of flips.

The last two probability distributions are presented graphically by the histograms of Figure 7.1. It should be noted that by drawing the rectangles so that there are no gaps, we are again "spreading" the classes over a continuous scale. Although the number of heads

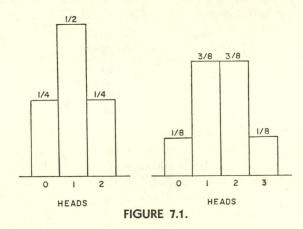

FIGURE 7.1.

must be 0, 1, 2, . . . , and cannot be $\frac{1}{2}$ or $\frac{3}{4}$, the bases of the rectangles go from $-\frac{1}{2}$ to $\frac{1}{2}$, from $\frac{1}{2}$ to $1\frac{1}{2}$, from $1\frac{1}{2}$ to $2\frac{1}{2}$, and so on.

EXERCISES

1. Flip *four* coins 240 times and construct a distribution showing the frequencies with which 0, 1, 2, 3, and 4 heads occurred. Also find the corresponding expected frequencies by enumerating the 16 equiprobable cases $TTTT$, $TTHT$, $HHTT$, ...

2. Use the result of Exercise 1 to draw histograms of the observed and expected distributions.

3. Flip *five* coins 320 times and construct a distribution showing the frequencies with which 0, 1, 2, 3, 4, and 5 heads occurred. Also find the corresponding probability distribution and expected distribution for 320 flips by enumerating the 32 equiprobable cases $HHTTT$, $TTHTT$, $THTHT$, . . .

4. Draw a histogram of the probability distribution for the number of heads obtained with 5 balanced coins.

5. Roll one die 180 times and construct a table showing the *relative frequencies* with which the different faces appeared. Also draw a histogram of this distribution and superimpose on it the corresponding probability distribution.

6. Suppose that a balanced coin is flipped 4 times and that after each flip we check whether we are ahead, even, or behind as in the example

on page 102. Enumerate the 16 equiprobable outcomes to find the probabilities of being ahead 0, 1, 2, 3, or 4 times out of 4. Also draw a histogram of this U-shaped probability distribution.

7.2 The Binomial Distribution

The probability distributions which we studied in the previous section were easy to obtain by enumerating in each case all equiprobable outcomes. To consider a slightly more difficult example, let us now find the probability distribution for getting 0, 1, and 2 *sixes* in two successive rolls of a balanced die (*or in 1 roll of a pair of dice*). Letting S and N stand for "six" and "not a six," the four possible outcomes are

$$SS \quad SN \quad NS \quad NN$$

analogous to the HH, HT, TH, and TT for the tosses of two coins. However, whereas HH, HT, TH, and TT were *equiprobable*, SS, SN, NS, and NN are not. Since the probability of *rolling a six* with one die is $\frac{1}{6}$, the probability of *not rolling a six* is $\frac{5}{6}$, and since two successive rolls of a die are *independent*, we can use formula (6.2.3), the special rule of multiplication, and write

the probability of getting SS is $\frac{1}{6} \cdot \frac{1}{6} = \frac{1}{36}$

the probability of getting SN is $\frac{1}{6} \cdot \frac{5}{6} = \frac{5}{36}$

the probability of getting NS is $\frac{5}{6} \cdot \frac{1}{6} = \frac{5}{36}$

the probability of getting NN is $\frac{5}{6} \cdot \frac{5}{6} = \frac{25}{36}$

Total $\qquad \frac{36}{36} = 1$

It follows that the probability for 0 *sixes* is $\frac{25}{36}$, the probability for 2 *sixes* is $\frac{1}{36}$, and since SN and NS, which both yield 1 *six*, are mutually exclusive, the probability for 1 *six* is $\frac{5}{36} + \frac{5}{36} = \frac{10}{36}$. The probability distribution of the number of sixes obtained in two successive rolls of a die can, therefore, be written as

Number of Sixes	Probability
0	$\frac{25}{36}$
1	$\frac{10}{36}$
2	$\frac{1}{36}$
Total	$\frac{36}{36} = 1$

If we were interested in determining the probability distribution of the number of sixes obtained in 3 rolls of a die (or in 1 roll of 3 dice), we would have to consider the 8 possible outcomes

$$NNN \quad NNS \quad NSN \quad SNN$$
$$NSS \quad SNS \quad SSN \quad SSS$$

Since these arrangements of the letters S and N are *not* equiprobable, we shall have to calculate, as before, the probability associated with each possible outcome. Using a generalization of (6.2.3) to the effect that if three events A, B, and C are independent, the probability that all three will occur is

$$P(A \text{ and } B \text{ and } C) = P(A) \cdot P(B) \cdot P(C)$$

we find that

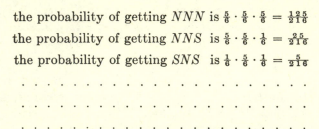

the probability of getting NNN is $\frac{5}{6} \cdot \frac{5}{6} \cdot \frac{5}{6} = \frac{125}{216}$

the probability of getting NNS is $\frac{5}{6} \cdot \frac{5}{6} \cdot \frac{1}{6} = \frac{25}{216}$

the probability of getting SNS is $\frac{1}{6} \cdot \frac{5}{6} \cdot \frac{1}{6} = \frac{5}{216}$

. . . .

. . . .

. . . .

After adding the probabilities of the different arrangements yielding 0, 1, 2, and 3 sixes, respectively, we find that the desired probability distribution is

Number of Sixes	Probability
0	$\frac{125}{216}$
1	$\frac{75}{216}$
2	$\frac{15}{216}$
3	$\frac{1}{216}$

The histograms of the two J-shaped distributions of the number of sixes obtained in 2 and 3 rolls of a balanced die are shown in Figure 7.2.

The process of enumerating all possible outcomes and calculating the probability associated with each would be rather tedious if we were interested, for example, in determining the probability distribution of the number of sixes obtained in, say, 10 or 20 rolls of a die. For 10 rolls we would have to consider 1,024 different arrangements of the letters S and N and for 20 rolls we would have to consider over a million. Although there may be few occasions in which the probabilities connected with 20 rolls of a die are of vital concern, there are many practical problems in which we must know probabilities that are of a similar type. For example, a quality control engineer may want to know the probability of getting 4 defective tubes in a sample of 20 if it is assumed that the sample is taken from a very large lot in which 1 per cent of the tubes are defective; or an election forecaster may want to know the probability of getting a sample in which 12 persons out of 20 are for Candidate X if it is assumed that the probability of any one voter being for Candidate X is 0.70. In both of these examples we are interested in the probability that an event will take place a certain number of times out of 20. (Of course, there is nothing special about the number 20, which we have used merely as an example. The quality control engineer might similarly be interested in the probability of getting 6 defective tubes in a sample of 100, and the forecaster might be interested in the probability of getting 350 votes for Candidate X in a sample of 500.)

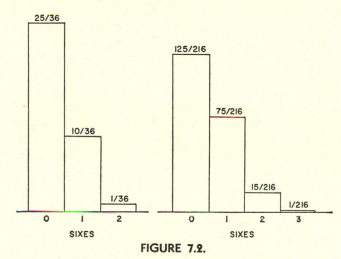

FIGURE 7.2.

To study this problem in a more general fashion, let us again refer to a game of chance and let us look for the probability of getting 2 *sixes* in 5 rolls of a balanced die. If we enumerate the 32 possible arrangements of 5 letters S and N, we find that 10 of them, namely,

$$SSNNN \quad SNSNN \quad SNNSN \quad SNNNS \quad NSSNN$$
$$NSNSN \quad NSNNS \quad NNSSN \quad NNSNS \quad NNNSS$$

contain 2 letters S and 3 letters N. Making use of a further extension of formula (6.2.3), the special rule of multiplication, to the effect that if $A_1, A_2, \ldots, A_k$ are *independent events*, the probability that they will *all* occur is

$$P(A_1 \text{ and } A_2 \text{ and } A_3 \ldots \text{ and } A_k) = P(A_1) \cdot P(A_2) \cdot \ldots \cdot P(A_k)$$

we find that each of the 10 arrangements in which there are two S's and three N's has the probability $(\frac{1}{6})^2 \cdot (\frac{5}{6})^3$. In each case there are two factors $\frac{1}{6}$ corresponding to the two letters S and *three* factors $\frac{5}{6}$ corresponding to the three N's. Since the 10 arrangements are, furthermore, *mutually exclusive*, we find that the probability of getting exactly 2 *sixes* in 5 rolls of a balanced die is 10 *times* $\left(\frac{1}{6}\right)^2 \cdot \left(\frac{5}{6}\right)^3$ or $\frac{1250}{7776}$.

Let us now investigate the more general problem of finding the probability that *a given event will take place exactly x times in n "trials," in other words, x times out of n, if the probability that it will take place in any one trial is equal to some number p.* Referring to the occurrence of the event as a "success" and its non-occurrence as a "failure," we will thus be interested in the probability of "x successes in n trials."[†] In order to determine this probability we could refer to a success as S, a failure as F, and enumerate as before the various arrangements of x letters S and $n - x$ letters F. Then we could find the probabilities associated with these arrangements and their sum would give us the desired probability.

Since the probability of a "success" is p and the probability of a "failure" is $1 - p$, the probability of getting x successes and $n - x$ failures (x letters S and $n - x$ letters F) *in a given order* is $p^x(1 - p)^{n-x}$. In this expression there are x factors p for the x letters S, and $n - x$ factors $1 - p$ for the $n - x$ letters F. Since the probability is the same regardless of the order, we can obtain the desired probability of getting "x successes in n trials" by multiplying $p^x(1 - p)^{n-x}$ by the total number of distinct ways in which x letters S and $n - x$ letters F can be arranged. Representing the number of such arrangements with the symbol $\binom{n}{x}$, we can finally write the probability of getting "x successes in n trials" as

$$P(x;n) = \binom{n}{x} \cdot p^x(1 - p)^{n-x} \qquad (7.2.1)\star$$

Although this formula may look complicated, let us point out that in its derivation we merely duplicated the steps which we used earlier to find the probability of getting 2 *sixes* in 5 rolls of a die.

Formula (7.2.1) represents a special kind of probability distribu-

---

[†] Although we are using the term "success," this does not imply that the event to which we are referring must necessarily be advantageous or desirable. If an importer wants to know the probability that 2 of 50 shipments will get lost at sea, he is interested in the probability of "2 successes in 50 trials"; if a doctor wants to know the probability that 3 of 20 pneumonia patients will die, he is interested in the probability of "3 successes in 20 trials"; and if an economist wants to know the probability that 4 of 100 stores will go bankrupt, he is interested in the probability of "4 successes in 100 trials." This usage of "success" and "failure" is a carry-over from the days when probabilities were considered only with reference to games of chance.

tion which is called the *binomial distribution*.† It is a probability *distribution* because it gives the probabilities of 0, 1, 2, 3, . . . , and n successes if we substitute $x = 0, 1, 2, 3, . . .$, and n. The quantities $\binom{n}{x}$, which incidentally are *not* fractions, are called *binomial coefficients* (see footnote). They may be obtained by enumerating all possible arrangements, as we did in the case of the dice, or they may be calculated with the formula

$$\binom{n}{x} = \frac{n \cdot (n - 1) \cdot (n - 2) \cdot \ldots \cdot (n - x + 1)}{x \cdot (x - 1) \cdot (x - 2) \cdot \ldots \cdot 2 \cdot 1} = \frac{n!}{x!(n - x)!}$$

(7.2.2)

where $n!$ (*n-factorial*) equals $n \cdot (n - 1) \cdot (n - 2) \cdot \ldots \cdot 2 \cdot 1$ and by definition $0! = 1$. To simplify our calculations, the coefficients $\binom{n}{x}$ are given in Table VI on page 507 for certain values of x and n.

To illustrate the use of (7.2.1), let us find the probability of getting 4 heads and 6 tails in 10 flips of a coin. Substituting $n = 10$, $x = 4$, and $p = \frac{1}{2}$, we obtain

$$P(4;10) = \binom{10}{4} (\tfrac{1}{2})^4 (1 - \tfrac{1}{2})^6$$

Since $\binom{10}{4} = 210$ according to Table VI, we find that the probability for 4 heads and 6 tails is

$$P(4;10) = 210 \cdot \frac{1}{16} \cdot \frac{1}{64} = \frac{105}{512}$$

or approximately 0.2. Similarly, the probability of getting exactly 1 *six* in 5 rolls of a die is

$$P(1;5) = \binom{5}{1} \cdot (\tfrac{1}{6})^1 \cdot (1 - \tfrac{1}{6})^4$$

$$= 5 \cdot \frac{1}{6} \cdot \frac{625}{1296}$$

$$= \frac{3125}{7776}$$

---

† The term *binomial* derives from the fact that the expressions yielded by (7.2.1) for $x = 0, 1, 2, . . .$, and n are the corresponding terms of the binomial expansion of $[(1 - p) + p]^n$. As can easily be checked in any College Algebra text,

$$[(1 - p) + p]^n = \binom{n}{0} \cdot (1 - p)^n + \binom{n}{1} \cdot p \cdot (1 - p)^{n-1}$$

$$+ \binom{n}{2} \cdot p^2 \cdot (1 - p)^{n-2} + \ldots + \binom{n}{n} \cdot p^n$$

where $\binom{n}{0}$, $\binom{n}{1}$, $\binom{n}{2}$, etc., represent the binomial coefficients.

or approximately 0.4. In this example we substituted $n = 5$, $x = 1$, $p = \frac{1}{6}$, and $\binom{5}{1} = 5$.

To give an example in which we compute the entire probability distribution, let us suppose that the probability of a person answering a mail questionnaire is 0.20 and let us find the probabilities of getting 0, 1, 2, 3, 4, or 5 responses to a questionnaire which is mailed to 5 individuals. Here $n = 5$, $p = 0.2$, and according to Table VI the binomial coefficients are 1, 5, 10, 10, 5, 1. Substituting these values

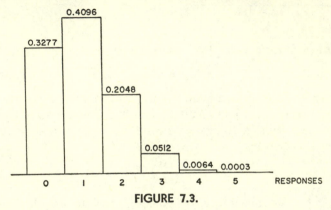

FIGURE 7.3.

together with $x = 0$, 1, 2, 3, 4, and 5 into (7.2.1) and rounding the answers to four decimals, we obtain

$$
\begin{aligned}
P(0;5) &= 1 \cdot (0.2)^0 (0.8)^5 = 0.3277 \\
P(1;5) &= 5 \cdot (0.2)^1 (0.8)^4 = 0.4096 \\
P(2;5) &= 10 \cdot (0.2)^2 (0.8)^3 = 0.2048 \\
P(3;5) &= 10 \cdot (0.2)^3 (0.8)^2 = 0.0512 \\
P(4;5) &= 5 \cdot (0.2)^4 (0.8)^1 = 0.0064 \\
P(5;5) &= 1 \cdot (0.2)^5 (0.8)^0 = 0.0003
\end{aligned}
$$

This distribution, which is also presented graphically in Figure 7.3, shows that *almost one third of the time we can expect to get no response at all, 41 per cent of the time we will get 1 response, 20 per cent of the time we will get 2, 5 per cent of the time we will get 3, and less than 1 per cent of the time we will get 4 or 5.*

The binomial distribution has many important applications which will be discussed later in Chapters 9 and 10. Although formula (7.2.1) could always give us the necessary probabilities, it is actually seldom used. It would be much too tedious, for example, to use it for finding the probability of getting 5 defective ball bearings in a sample of 400 if the probability that any one ball bearing is defective is 0.01. Similarly, it would require an enormous amount of arithmetic to use this formula for finding the probability of getting anywhere from 46 to 54 heads in 100 tosses of a balanced coin. In the last example we would have to calculate $P(x;n)$ for $n = 100$, $p = \frac{1}{2}$, and $x = 46, 47, \ldots$, and 54. Instead of using formula (7.2.1), we can either refer to the special tables which are mentioned on page 168 or employ the approximations discussed in Sections 7.7 and 7.9.

Before we proceed to further problems dealing with the binomial distribution, let us add a word of caution about its use. *Formula (7.2.1) applies only if the probability of a "success" remains constant from trial to trial and if the trials are, furthermore, independent.* The formula cannot be used, therefore, to find the probability that it will rain, say 25 out of 100 consecutive days. Not only does the probability of rain vary for a period extending over more than 3 months, but the "trials" are not independent. As we have pointed out in an earlier example, the probability that it will rain on a given day will depend on whether it did or did not rain on the previous day.

EXERCISES

1. Find the probability of getting 4 heads and 8 tails in 12 flips of a balanced coin. Use formula (7.2.1).
2. Find the probability of rolling exactly 3 *sixes* in 4 rolls of a balanced die.
3. A bowl contains 15 red marbles, 15 blue marbles, and 30 white ones. What is the probability of getting a red marble in 3 out of 6 draws if each marble is replaced before the next one is drawn? Could formula (7.2.1) be used if the marbles were not replaced?
4. Find the probability of getting at least 7 heads in 10 flips of a balanced coin. Use formula (7.2.1), with $x = 7, 8, 9$, and 10, and add the results in accordance with formula (6.2.2).
5. In a certain hospital, 50 per cent of all children born are girls. What is the probability that among 8 children born on a given day, there are three girls and five boys?
6. If the probability that any one plane is shot down in a raid is 0.05, find the probability that 2 out of 4 planes will be shot down.
7. There are 10 questions and 5 answers to each question in a multiple choice test. If each question is answered by rolling a die until 1, 2, 3, 4, or 5 appears and the appropriate alternative is marked, what is the probability of getting exactly 3 correct answers?
8. Find the binomial distribution of the number of tails obtained in 6 flips of a balanced coin. Also draw a histogram of this distribution.
9. If the probability that an insurance salesman will make a sale on his first visit to a new customer is 0.30, find the probabilities that among four salesmen selling insurance 0, 1, 2, 3, or 4 will make a sale on a first visit to a new customer. Also draw a histogram of this binomial distribution.

10. 50 black marbles and 10 white ones are contained in a bowl. Find the binomial distribution for the number of white ones obtained in 6 draws if it is assumed that each marble is replaced before the next one is drawn.
11. Calculate the binomial distribution for $n = 6$ and $p = \frac{1}{5}$.
12. Calculate the binomial distribution for $n = 4$ and $p = 0.05$.

7.3 The Mean and Standard Deviation of the Binomial Distribution

In Chapter 3 we defined the mean of a distribution (the mean of grouped data constituting a population) as

$$\mu = \frac{\sum\limits_{i=1}^{k} x_i f_i}{n}$$

assuming that all values falling into the ith class are equal to the class mark x_i or, in other words, that the x_i occur with frequencies f_i. If we now divide the n which appears in the denominator into the f_i, the formula for the mean can be rewritten as

$$\mu = \sum_{i=1}^{k} x_i \cdot \frac{f_i}{n} \tag{7.3.1}$$

It should be noted that in this formula each x_i is multiplied by the proportion f_i/n, namely, by the *relative frequency* with which it occurs.

In order to define the mean of a *probability distribution* we shall substitute for the relative frequencies f_i/n the probabilities $P(x_i)$, namely, the probabilities of getting x_i and write

$$\mu = \sum_{i=1}^{k} x_i \cdot P(x_i) \tag{7.3.2}$$

To justify this definition, let us remind the reader that *we originally defined probabilities as limits of relative frequencies*.

In the case of the *binomial distribution*, the x_i stand for the number of successes 0, 1, 2, 3, . . . , and n, while $P(x_i)$ is the probability of getting "x_i successes in n trials," as given by formula (7.2.1). If we now substitute the probabilities of (7.2.1) into (7.3.2) and perform certain simplifications, it can be shown that the formula for the *mean of a binomial distribution* reduces to

$$\mu = n \cdot p \tag{7.3.3}\star$$

The result expressed by this formula should really have been expected. If a balanced coin is flipped 100 times, we would expect to get *on the average* 50 heads, and formula (7.3.3) yields, indeed, $\mu = np$

$= 100 \cdot \frac{1}{2} = 50$. Similarly, we would expect to get *on the average* 20 sixes in 120 rolls of a die and for $n = 120$ and $p = \frac{1}{6}$ formula (7.3.3) yields $\mu = np = 120 \cdot \frac{1}{6} = 20$. Although the derivation of formula (7.3.3) is not very difficult, we shall not give it here. If the reader is interested in seeing a proof, he will find a suitable reference on page 168.

Instead of proving the formula for the mean of a binomial distribution let us merely verify it by considering a special case and calculating μ in two ways: (a) with formula (7.3.2) which defined the mean of a probability distribution and (b) with the formula $\mu = np$. Using the probability distribution of the number of sixes obtained in 3 rolls of a die which we derived on page 132, we can write

x_i	$P(x_i)$	$x_i \cdot P(x_i)$
0	$\frac{125}{216}$	0
1	$\frac{75}{216}$	$\frac{75}{216}$
2	$\frac{15}{216}$	$\frac{30}{216}$
3	$\frac{1}{216}$	$\frac{3}{216}$
		$\frac{108}{216}$

Since the mean of a probability distribution is by definition the *sum* of the products $x_i \cdot P(x_i)$, we obtain

$$\mu = \sum_{i=1}^{k} x_i \cdot P(x_i) = \frac{108}{216} = \frac{1}{2}$$

for the binomial distribution having $n = 3$ and $p = \frac{1}{6}$. If we substitute the same values of n and p into formula (7.3.3) we get

$$\mu = np = 3 \cdot \frac{1}{6} = \frac{1}{2}$$

and this shows that values obtained with the two formulas are identical. Incidentally, this result means that *on the average* we can expect to get 0.5 *sixes* in 3 rolls of a die. Sometimes we will get 0, sometimes 1, sometimes 2, sometimes 3, and *on the average* 0.5.

In Chapter 4 we defined the standard deviation of a distribution (the standard deviation of grouped data constituting a population) as

$$\sigma = \sqrt{\frac{\sum_{i=1}^{k} (x_i - \mu)^2 \cdot f_i}{n}}$$

assuming again that all values falling into the ith class are equal to the class mark x_i. Dividing n into the f_i, as we did on page 138, this formula becomes

$$\sigma = \sqrt{\sum_{i=1}^{k} (x_i - \mu)^2 \cdot \frac{f_i}{n}}$$

If we now substitute for the relative frequencies f_i/n the probabilities of getting x_i, we can define the *standard deviation of a probability distribution* as

$$\sigma = \sqrt{\sum_{i=1}^{k} (x_i - \mu)^2 \cdot P(x_i)} \qquad (7.3.4)$$

To illustrate the calculation of the standard deviation of a probability distribution, let us again use the distribution of the number of sixes obtained in 3 rolls of a die. Having already shown above that $\mu = \frac{1}{2}$, we can write

x_i	$x_i - \mu$	$(x_i - \mu)^2$	$P(x_i)$	$(x_i - \mu)^2 \cdot P(x_i)$
0	$-\frac{1}{2}$	$\frac{1}{4}$	$\frac{125}{216}$	$\frac{125}{864}$
1	$\frac{1}{2}$	$\frac{1}{4}$	$\frac{75}{216}$	$\frac{75}{864}$
2	$\frac{3}{2}$	$\frac{9}{4}$	$\frac{15}{216}$	$\frac{135}{864}$
3	$\frac{5}{2}$	$\frac{25}{4}$	$\frac{1}{216}$	$\frac{25}{864}$
				$\frac{360}{864}$

Substitution into (7.3.4) yields the result that

$$\sigma = \sqrt{\tfrac{360}{864}} = \sqrt{\tfrac{5}{12}}$$

which is approximately 0.645.

It should be apparent from this example that the calculation of σ with formula (7.3.4) can entail a good deal of work. This is true particularly if n is *large* and p is such that the calculation of the probabilities and the squared deviations is rather involved. *In actual practice formula (7.3.4) is never used to find the standard deviation of a binomial distribution.* Instead we use the formula

$$\sigma = \sqrt{np(1 - p)} \qquad (7.3.5)\star$$

which may be derived by substituting the binomial probabilities of (7.2.1) for the $P(x_i)$ and performing certain simplifications. A reference to the derivation of (7.3.5) is given in the Bibliography on page 168.

Let us verify this simple formula for the standard deviation of a binomial distribution with reference to the distribution of the number of sixes obtained in 3 rolls of a die. Substituting $n = 3$ and $p = \frac{1}{6}$ into (7.3.5), we obtain

$$\sigma = \sqrt{3 \cdot \tfrac{1}{6} \cdot \tfrac{5}{6}} = \sqrt{\tfrac{5}{12}}$$

and it can be seen that this is identical with the result which we obtained with formula (7.3.4).

In Chapter 4 we introduced the standard deviation as a *measure of variation*, and $\sigma = \sqrt{np(1 - p)}$ tells us, indeed, *how much variation there is due to chance in the number of successes we might obtain in n trials.* To give the reader some idea about a possible application of this formula for σ, let us repeat what we said on page 93: for certain distributions, 95 per cent of the cases differ from the mean by less than *two* standard deviations. If we now consider the distribution of the number of heads obtained in 300 flips of a coin we find that

$$\mu = 300 \cdot \tfrac{1}{2} = 150$$

and

$$\sigma = \sqrt{300 \cdot \tfrac{1}{2} \cdot \tfrac{1}{2}} = 8.66 \text{ (approximately)}$$

Assuming that the above argument applies, we can say that if a balanced coin is flipped 300 times, *95 per cent of the time* we will get anywhere from $150 - 2(8.66)$ to $150 + 2(8.66)$ or, roughly, from 133 to 167 heads. (For a further discussion of this, see Sections 7.6 and 7.7.)

The reader may have noted that we described binomial distributions (and other probability distributions) by using the Greek letters μ and σ, which we reserved originally for *populations*. Indeed, statisticians often refer to binomial distributions as *binomial populations*. If a balanced coin is flipped, say, 100 times, the number of heads observed constitutes a *sample* of the results we might get if we repeated this experiment over and over again. The *hypothetical population* from which we are, thus, sampling consists of *all* possible "experiments" consisting of 100 flips of a balanced coin, and it is characterized by the corresponding binomial distribution.

EXERCISES

1. Use formula (7.3.2), formula (7.3.4), and the probabilities given on page 129 to find μ and σ for the distribution of the number of heads obtained when tossing three balanced coins. Check your results against formulas (7.3.3) and (7.3.5).

2. Use formula (7.3.2), formula (7.3.4), and the probabilities given on page 131 to find μ and σ for the distribution of the number of sixes rolled with a pair of balanced dice. Check your results against formulas (7.3.3) and (7.3.5).

3. Use formulas (7.3.3) and (7.3.5) to find the mean and standard deviation of binomial distributions having

   (a) $n = 100$ and $p = \tfrac{1}{2}$
   (b) $n = 16$ and $p = \tfrac{1}{4}$
   (c) $n = 400$ and $p = \tfrac{1}{10}$
   (d) $n = 225$ and $p = \tfrac{1}{5}$

4. Use formulas (7.3.3) and (7.3.5) to find the mean and standard deviation of the binomial distribution of Exercise 8 on page 137.

5. Use formulas (7.3.3) and (7.3.5) to find the mean and standard deviation of the binomial distribution of Exercise 10 on page 138.

7.4 Continuous Distributions

It is often tempting to approximate the zigzag appearance of a histogram, be it that of an observed distribution or a probability distribution, by means of a smooth curve. It was partly for this reason that we introduced class boundaries in Chapter 2, eliminating thus the gaps between adjacent classes and spreading the distribution over a continuous scale. We also used this device earlier in this chapter for the histograms shown in Figures 7.1, 7.2, and 7.3 because the number of heads observed in n flips of a coin, the number of sixes obtained in n rolls of a die, or more generally the number of "successes" observed in n trials are *discrete* variables. *A variable is said to be discrete if it assumes only a finite number of values or as many values as there are whole numbers.* The number of heads which we obtain in, say, 10 flips of a coin is a discrete variable because it cannot assume values other than 0, 1, 2, 3, . . . , and 10. In this case there is a finite number of values—11 to be exact. If we flipped a coin until the first head appears, the number of the particular flip on which this occurs is also a discrete variable. However, in this case the variable can assume the values 1, 2, 3, . . . , that is, as many values as there are whole numbers.

In contrast to discrete variables, we shall say that a *variable is continuous if it can assume all values of a continuous scale.* Such quantities as length, time, and temperature are measured on continuous scales and their measurements may be referred to as continuous variables. (Nevertheless, measurements of these quantities are usually rounded off, which again makes them discrete. Even though time could be measured on a continuous scale, it is usually rounded off to the nearest second, tenth of a second, or hundredth of a second, and even though a person's height could be measured on a continuous scale, it is usually rounded off to the nearest inch, tenth of an inch, or sixteenth of an inch.)

When we first discussed histograms in Chapter 2, we pointed out that the frequencies, percentages, (and we might now add probabilities) which are associated with the various classes are represented by the *areas* of the rectangles. For example, the areas of the rectangles of Figure 7.4 represent the probabilities of getting 0, 1, 2, . . . , and 10 heads in 10 flips of a balanced coin, or, better, they are proportional to these probabilities. If we now look at Figure 7.5, which is an enlargement of a portion of Figure 7.4, it is apparent that the area of rectangle $ABCD$ is approximately equal to the shaded area under the continuous curve. Since the area of rectangle $ABCD$ represents (is proportional to) the probability of getting 3 heads in 10 flips of a balanced coin, we can say that this probability is also represented by

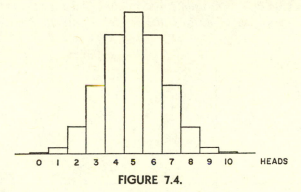

FIGURE 7.4.

the shaded area under the continuous curve which approximates the histogram. *More generally, if a histogram is approximated by means of a smooth curve, the frequency, percentage, or probability, of any given class is represented by (proportional to) the corresponding area under the curve.*

If we approximated the distribution of 1965 money income of families in the United States with a smooth curve, we could determine the proportion of incomes falling into any given class by looking at the corresponding area under the curve. By comparing the shaded area of Figure 7.6 with the total area under the curve we find that in 1965 roughly 25 per cent of the families had incomes of $10,000 or more. It can, similarly, be seen from Figure 7.6 that about 25 per cent of the families had incomes of $4,000 or less and approximately 50 per cent had incomes of $7,000 or more. We obtained these percentages by (mentally) dividing the corresponding areas under the curve by the total area under the curve, which, after all, represents 100 per cent of the families.

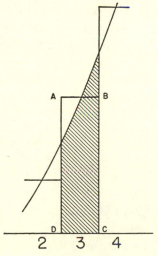

FIGURE 7.5.

Had we drawn the curve of Figure 7.6 so that the total area under the curve is equal to 1, the proportion of the families belonging to any (income) class would have been given directly by the corresponding area under the curve.   Indeed, we shall refer to a curve as a *distribution curve* or a *continuous distribution*† if the area under the curve between two numbers *a* and *b* (see Figure 7.7) is *equal* to the proportion of the cases falling between *a* and *b*.   In other words, *we shall refer to a curve as a continuous distribution if the area under the curve between a and b equals the probability of getting a value between a and b.*

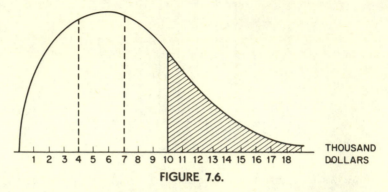

FIGURE 7.6.

For example, if the curve of Figure 7.8 is a distribution curve which approximates the probability distribution of the number of heads obtained in 100 flips of a balanced coin, the probability of getting anywhere from 46 to 54 heads is given by the shaded area under the curve, and in this example it is approximately equal to 0.63.   It should be noted that we have shaded the area between 45.5 and 54.5 and *not*

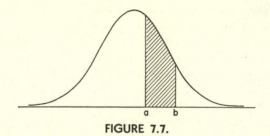

FIGURE 7.7.

the area between 46 and 54.   As we pointed out earlier, the number of heads obtained in *n* flips of a coin is a discrete variable and its probability distribution cannot be approximated with a smooth curve unless we let the intervals from $-\frac{1}{2}$ to $\frac{1}{2}$, from $\frac{1}{2}$ to $1\frac{1}{2}$, from $1\frac{1}{2}$ to $2\frac{1}{2}$, . . . , represent 0 heads, 1 head, 2 heads, ...   If we thus spread the discrete variable over a continuous scale, "46 to 54 heads" is represented by the interval from 45.5 to 54.5.

---

† It is also customary to use the terms *density function* or *probability density function.*

Continuous distributions play a very important role in statistical theory. They provide close approximations for many distributions but, what is even more important, they provide the basis for most of the theory used in problems of estimation, prediction, and in the testing of hypotheses.

Since continuous distributions can always be looked upon as close approximations to histograms, *we can define the mean and standard deviation of a continuous distribution in the following way:* If a continuous distribution is approximated with a sequence of histograms having narrower and narrower classes, the means of the distributions represented by the histograms will approach the mean of the continu-

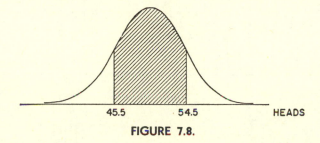

FIGURE 7.8.

ous distribution and the standard deviations will approach the standard deviation of the continuous distribution. *Intuitively speaking, the mean and the standard deviation of a continuous distribution measure the identical features as the mean and the standard deviation of an ordinary distribution, namely, its center and its spread.* More rigorous definitions of the mean and standard deviation of a continuous distribution are referred to on page 168. We shall not present them here because they cannot be given without calculus.

7.5 The Normal Curve

There is one theoretical distribution, the *normal curve*, which is in many respects the cornerstone of modern statistical theory. Its mathematical treatment dates back to the eighteenth century when scientists observed an astonishing degree of regularity in errors of measurements, that is, in repeated measurements of one and the same quantity. They found that the patterns (distributions) which they observed were closely approximated by a continuous distribution which they referred to as the "normal curve of errors" and attributed to the laws of chance. The mathematical properties of this continuous distribution and its theoretical basis were first investigated by Pierre Laplace (1749–1827), Abraham de Moivre (1667–1745), and Carl

Gauss (1777–1855). In honor of the last, normal curves are also referred to as *Gaussian distributions*.

There are a number of ways in which the normal curve may be introduced. Although, historically speaking, the normal curve was originally related to "laws of error," it is easier to look upon this distribution as a continuous distribution which provides a very close approximation to a binomial distribution when n, the number of trials, is large. Figure 7.9 contains the histograms of binomial distributions

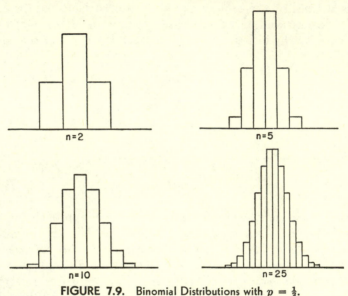

FIGURE 7.9. Binomial Distributions with $p = \frac{1}{2}$.

having $p = \frac{1}{2}$ and $n = 2, 5, 10$, and 25, respectively, and it can be seen that with n increasing these distributions approach a *symmetrical bell-shaped pattern*, namely, that of the normal curve as shown in Figure 7.10.†

The normal curve is a bell-shaped curve which extends indefinitely in both directions. Although this may not be apparent from Figure 7.10, the curve comes closer and closer to the horizontal axis without

---

† Mathematically speaking, it is better to say that if we converted x, the number of successes obtained in n trials, into *standard units* (see page 94), *for large n* the distribution of

$$z = \frac{x - \mu}{\sigma} = \frac{x - np}{\sqrt{np(1 - p)}}$$

approaches the normal curve. This formulation is preferable because the mean and standard deviation of the binomial distribution will both become infinite when n becomes infinite. A rigorous proof that for large n the binomial distribution approaches the normal curve is referred to on page 168.

ever reaching it, no matter how far we might go in either direction. Fortunately, it is seldom necessary to extend the tails very far, because the area under the curve becomes negligible if we go more than 4 or 5 standard deviations away from the mean.

An important property of a normal curve is that it is completely determined if we know its mean and standard deviation. The equa-

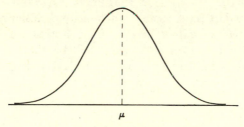

FIGURE 7.10.

tion of a normal curve whose mean and standard deviation are μ and σ is

$$ y = \frac{1}{\sqrt{2\pi}\,\sigma} e^{-\frac{1}{2}\left(\frac{x-\mu}{\sigma}\right)^2} \tag{7.5.1} $$

where π, the ratio of the circumference of a circle to its diameter, is approximately 3.14159 and e, the basis of Naperian logarithms, is approximately 2.71828. If we know μ and σ, we can substitute any

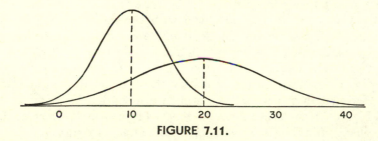

FIGURE 7.11.

value of x into (7.5.1) and calculate y. This will give us the *height* of the curve corresponding to any given value of x.

Formula (7.5.1) is seldom used in applications, because we are usually interested in *areas under the normal curve* instead of its height. As we pointed out earlier, *the proportion of the cases falling between two numbers or the probability of getting a value between two numbers is given by the corresponding area under the curve.* Since areas under normal curves are needed for many of the statistical techniques which we shall meet in the next few chapters, they have been tabulated and are shown in Table I on page 501.

Before we can discuss the use of this table, we shall have to explain what is meant by a normal curve in its *standard form*. As we pointed out earlier, the equation of the normal curve depends on μ and σ and we will, thus, get *different curves* if we substitute different values for μ and σ. For instance, Figure 7.11 shows the superimposed graphs of two normal curves, one having $\mu = 10$, $\sigma = 5$ and the other having $\mu = 20$ and $\sigma = 10$. This means that unless we found a more desirable alternative, we would have to construct *separate* tables of normal curve areas for each pair of values of μ and σ. Fortunately, this "impossible" task will not be necessary. We will be able to determine normal curve areas regardless of μ and σ, by tabulating only the areas

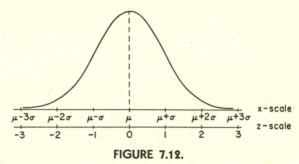

FIGURE 7.12.

under the normal curve having $\mu = 0$ and $\sigma = 1$. *Such a normal curve with zero mean and unit standard deviation is referred to as a standard normal distribution.*

If we are given a normal curve with the mean μ and standard deviation σ, we can always convert it into a standard normal distribution by performing the change of scale indicated in Figure 7.12. Whereas in the original scale (the x-scale) the mean and standard deviation are μ and σ, in the new scale (the z-scale) they are 0 and 1. The fact that in the z-scale the standard deviation is 1 can be seen from Figure 7.12. As can easily be checked, the formula which enables us to change from the x-scale to the z-scale and vice versa is

$$z = \frac{x - \mu}{\sigma} \qquad (7.5.2)\star$$

and this is precisely the formula which we used in Chapter 4 (page 94) to convert measurements into *standard units* or *standard scores*. Hence, when we change a normal distribution into the standard form, we are merely transforming the scale of measurement into standard units.

If in a given problem we want to determine an area under a normal curve whose mean and standard deviation are *not* 0 and 1, we have only to change the x's to z's and then use Table I on page 501. This table contains the normal curve area shaded in Figure 7.13 for values of z from 0.00 to 3.09.   In other words, *the entries in Table I equal the areas under the normal curve between the mean ($z = 0$) and the given values of z.*   For example, the entry corresponding to $z = 1.24$ is

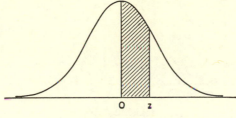

FIGURE 7.13.

0.3925, and this measures the shaded area of Figure 7.14, namely, the area between $z = 0$ and $z = 1.24$.

Table I does not contain entries corresponding to *negative* values of z.   Since the normal curve is symmetrical, we can find the area

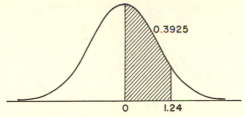

FIGURE 7.14.

between, say, $z = -1.50$ and $z = 0$ by looking up the area corresponding to $z = 1.50$.   We will, thus, find that the desired area, the shaded area of Figure 7.15, is 0.4332.

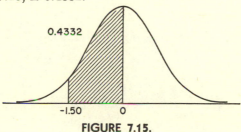

FIGURE 7.15.

If we are interested in determining a normal curve area *to the right of a positive value of z,* we have only to subtract the tabular value from 0.5000.   Since the normal curve is symmetrical, the area to the right of the mean is 0.5000 and the area to the right of a positive value of z is 0.5000 *minus* the tabular value given for z.   For instance, to find the area to the right of $z = 0.35$, we subtract 0.1368 (the entry given in Table I for $z = 0.35$) from 0.5000, getting $0.5000 - 0.1368 = 0.3632$ (see Figure 7.16).

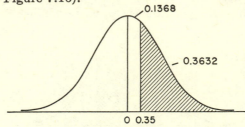

FIGURE 7.16.

If we want to find an area *to the left of a positive value of z*, we merely *add* 0.5000 to the tabular value given for z. For instance, the area to the left of $z = 2.15$, the shaded area of Figure 7.17, is $0.4842 + 0.5000 = 0.9842$.

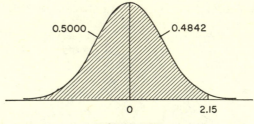

FIGURE 7.17.

In most problems dealing with normal curve areas it is best to make a drawing showing the area which we need. It will then be apparent whether we have to add 0.5000, subtract the tabular value from 0.5000, or perform some other calculation. For instance, to find

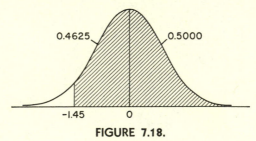

FIGURE 7.18.

the area to the right of $z = -1.45$ we note from Figure 7.18 that we must add the tabular value of $z = 1.45$ to 0.5000 and we get $0.4265 + 0.5000 = 0.9265$.

There are also problems in which we will need normal curve areas lying *between* two given values of z. If both z's are on the *same* side of the mean, if they are both positive or both negative, the area between them is given by the *difference* of their tabular values. For instance, the normal curve area between $z = 0.73$ and $z = 1.64$ is $0.4495 - 0.2673 = 0.1822$ (see Figure 7.19). If the two z's are on

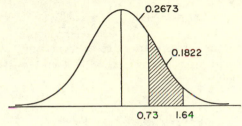

0.2673

0.1822

0.73   1.64

FIGURE 7.19.

opposite sides of the mean—in other words, if one is positive and the other negative—the area between them is given by the *sum* of their tabular values. For instance, the normal curve area between $z = -0.50$ and $z = 0.75$ is $0.1915 + 0.2734 = 0.4649$ (see Figure 7.20).

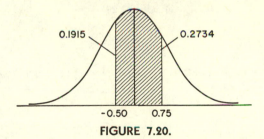

0.1915   0.2734

-0.50   0.75

FIGURE 7.20.

In some problems we are given areas under the normal curve and are then asked to find the corresponding values of z. For instance, we may want to find a z which is such that the area to its right is 0.1000. As it is apparent from Figure 7.21 that this z will have to

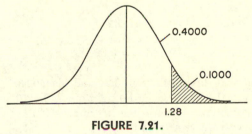

0.4000

0.1000

1.28

FIGURE 7.21.

correspond to an entry of 0.4000, we find that the closest value in Table I is $z = 1.28$.

To give an example of a problem in which we first have to convert into standard units, let us suppose that a normal curve has $\mu = 24$,

$\sigma = 12$, and that we want to find the area between $x_1 = 17.4$ and $x_2 = 58.8$ (see Figure 7.22). Writing

$$z_1 = \frac{17.4 - 24}{12} = -0.55$$

$$z_2 = \frac{58.8 - 24}{12} = 2.90$$

we find that the areas corresponding to these z's are 0.2088 and 0.4981 and that the desired area between $x_1 = 17.4$ and $x_2 = 58.8$ is 0.2088 + 0.4981 = 0.7069.

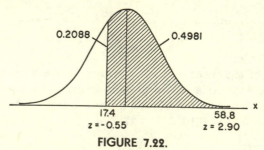

FIGURE 7.22.

EXERCISES

1. If the mean and standard deviation of a set of measurements are 100.0 pounds and 2.0 pounds, respectively, change each of the following measurements into standard units:

(a) 102.0 pounds   (c) 97.0 pounds
(b) 105.0 pounds   (d) 95.5 pounds

2. Find the normal curve area which lies
(a) to the right of $z = 1.34$
(b) to the left of $z = 2.08$
(c) to the right of $z = -0.67$
(d) to the left of $z = -1.25$
(e) between $z = 1.13$ and $z = 1.42$
(f) between $z = -0.53$ and $z = -2.14$
(g) between $z = -1.19$ and $z = 3.09$

3. Find z if
(a) the normal curve area between 0 and z is 0.0040
(b) the normal curve area to the right of z is 0.1170
(c) the normal curve area to the left of z is 0.9772
(d) the normal curve area to the right of z is 0.9938
(e) the normal curve area between $-z$ and z is 0.9902

7.6 Some Applications

When we discussed applications of the standard deviation in Chapter 4, we mentioned that there are many distributions in which certain fixed percentages of the cases fall within *one* standard deviation of the mean, *two* standard deviations of the mean, and so forth. The kind of distributions we were referring to were normal curves, or, better, distributions of observed data which can be approximated closely with normal curves. *With the use of Table I we will now be able to verify the percentages given on page* 93.

If a measurement is one standard deviation above or below the mean, its z value is $+1$ or -1 (see Figure 7.12) and according to Table I the normal curve area between $z = -1$ and $z = 1$ is $0.3413 + 0.3413 = 0.6826$. *This means that if a distribution can be approximated closely with a normal curve, about 68 per cent of the cases fall within one standard deviation from the mean.*

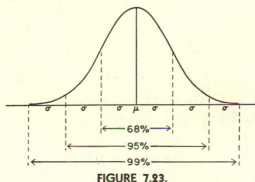

FIGURE 7.23.

The same argument can also be used to show that the normal curve area between $z = -2$ and $z = 2$ is $2 \cdot 0.4772 = 0.9544$ and that the normal curve area between $z = -3$ and $z = 3$ is $2 \cdot 0.4987 = 0.9974$. In other words, *if a distribution can be approximated closely with a normal curve, about 95 per cent of the cases fall within two standard deviations from the mean and more than 99 per cent of the cases fall within three standard deviations from the mean.* The interval which covers one standard deviation on either side of the mean is often referred to as the *one-sigma range*. Similarly, we use the terms *two-sigma range* and *three-sigma range* when referring to the corresponding intervals shown in Figure 7.23.

In the three examples which follow, we shall assume that we are dealing with data whose distribution can be approximated closely with a normal curve. Obviously, there would otherwise be no justification for using normal curve theory and normal curve areas.

Example 1. In a large shipment of ball bearings the diameters have a mean of 0.397 inches and a standard deviation of 0.005 inches. Assuming that the distribution of these diameters can be approximated closely with a normal curve, what percentage of the ball bearings have a diameter of 0.400 inches or more?

Treating these diameters as if they were measured on a continuous scale, our answer will be given by the normal curve area to the right of $x = 0.400$ (see Figure 7.24).† Since

$$z = \frac{0.400 - 0.397}{0.005} = 0.60$$

we find that the corresponding tabular value is 0.2257 and that the desired area to the right of $z = 0.6$ is $0.5000 - 0.2257 = 0.2743$.

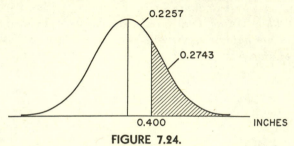

FIGURE 7.24.

Hence, slightly more than 27 per cent of the ball bearings have diameters of 0.400 inches or more.

> *Example 2.* The grades obtained by a large number of students in a final examination in marketing had a mean of 68 and a standard deviation of 8.2. Assuming that these grades are approximately normally distributed, below which grade will we find the lowest 10 per cent of the class?

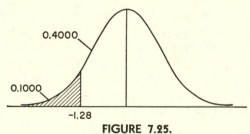

FIGURE 7.25.

This problem differs from the previous example in that we are now given a percentage (a normal curve area) instead of a value of x or z. It is apparent from Figure 7.25 that we shall first have to find the z which corresponds to an area of $0.5000 - 0.1000 = 0.4000$ and then convert it into a numerical grade. Since the z value whose area is closest to 0.4000 is 1.28, substitution into (7.5.2) yields

$$-1.28 = \frac{x - 68}{8.2}$$

---

† If we were given that the measurements are rounded to 3 decimals, we would have to look for the normal curve area to the right of 0.3995.

Solving this equation for x, we get $x = 68 - 10.496 = 57.504$, and we can say *the lowest 10 per cent of the class had grades of 57 or less.*

Example 3. The number of telephone calls made in a certain community (daily) between 1 P.M. and 2 P.M. have a mean of 248 and a standard deviation of 26. What percentage of the time will there be anywhere from 229 to 280 phone calls made in this community between 1 P.M. and 2 P.M.? It will be assumed that the distribution can be approximated closely with a normal curve.

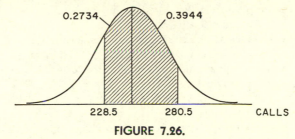

FIGURE 7.26.

Since we are dealing with a *discrete* variable, we shall have to look for the normal curve area between 228.5 and 280.5 (see Figure 7.26). Substituting $x_1 = 228.5$ and $x_2 = 280.5$ into formula (7.5.2), we get

$$z_1 = \frac{228.5 - 248}{26} = -0.75$$

$$z_2 = \frac{280.5 - 248}{26} = 1.25$$

and the corresponding normal curve areas are 0.2734 and 0.3944. Hence, the desired area is $0.2734 + 0.3944 = 0.6678$ and we can say that *roughly 67 per cent of the time there will be anywhere from 229 to 280 calls made in this community between 1 P.M. and 2 P.M.* We can also say that $0.5000 - 0.3944 = 0.1056$ or *10.56 per cent of the time there will be more than 280 calls* and $0.5000 - 0.2734 = 0.2266$ or *22.66 per cent of the time there will be fewer than 229.*

EXERCISES

1. The cedar fence posts produced by a certain lumber company have a mean length of 72 inches and a standard deviation of 1 inch. Assuming that the lengths of the posts can be measured to any desired degree of accuracy and that the distribution of the lengths of these posts can be approximated closely with a normal curve, find
   (a) the percentage of posts which have a length of 74.5 inches or more
   (b) the percentage of posts having lengths from 70.0 to 72.5 inches

    (c) the length below which we will find the shortest 20 per cent of the posts.

2. A certain type of automobile tire has a mean life of 30,000 miles and a standard deviation of 1,000 miles. Assuming that the distribution of the lifetimes, *which are measured to the nearest mile*, can be approximated closely with a normal curve, find
    (a) the percentage of the tires which have a lifetime of more than 32,000 miles
    (b) the value above which we will find the best 25 per cent of the tires
    (c) the proportion of the tires that have a lifetime from 28,500 miles to 31,500 miles, inclusive.

3. The attendance at an athletic stadium is normally distributed with a mean of 20,000 and standard deviation of 2,000 persons. Find
    (a) the value below which we will find the lowest 70 per cent of these attendance figures
    (b) the percentage of attendance figures which fall between 18,500 and 21,000 persons
    (c) the percentage of attendance figures that differ from the mean by 3,000 persons or more.

4. A set of final examination grades in statistics is approximately normally distributed with a mean of 75 and a standard deviation of 5. If the highest 15 per cent of the students are to get A's and the lowest 10 per cent F's, which grade is the lowest A and which grade is the highest F?

7.7 The Binomial Distribution and the Normal Curve

In Section 7.5 we pointed out that when p equals $\frac{1}{2}$ and n is large, the binomial distribution can be approximated very closely with a normal curve. In fact, normal curve areas can be used to evaluate binomial probabilities even when n is relatively small and p differs from $\frac{1}{2}$. To illustrate this *normal curve approximation of the binomial distribution*, let us first consider the probability of getting 5 heads in 12 tosses of a balanced coin. Substituting $n = 12$, $x = 5$, and $p = \frac{1}{2}$ into formula (7.2.1), we get

$$P(5;12) = \binom{12}{5} \cdot \left(\frac{1}{2}\right)^5 \left(\frac{1}{2}\right)^7 = 792 \cdot \frac{1}{32} \cdot \frac{1}{128} = \frac{792}{4096}$$

or approximately 0.1934. To determine the normal curve approximation of this binomial probability, we shall have to find the area between 4.5 and 5.5 (see Figure 7.27). Since

$$\mu = np = 12(\tfrac{1}{2}) = 6$$

and $\sigma = \sqrt{np(1 - p)} = \sqrt{12(\tfrac{1}{2})(\tfrac{1}{2})} = 1.732$

we find that

$$z_1 = \frac{4.5 - 6}{1.732} = -0.87$$

and

$$z_2 = \frac{5.5 - 6}{1.732} = -0.29$$

The corresponding areas are 0.3078 and 0.1141 and the desired probability is given by their difference, namely, $0.3078 - 0.1141 = 0.1937$.

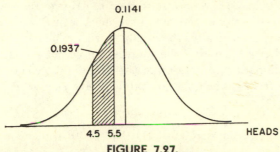

FIGURE 7.27.

Clearly, the difference between the values obtained with formula (7.2.1) and the normal curve approximation is negligible.

To give an example in which p is not equal to $\frac{1}{2}$, let us find the probability of getting "6 successes in 16 trials" when the probability of a success is $\frac{1}{5}$. To determine the normal curve approximation of

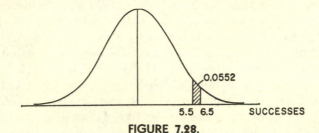

FIGURE 7.28.

this probability, we shall have to find the area between 5.5 and 6.5, namely, the shaded area of Figure 7.28. Since

$$\mu = np = 16(\tfrac{1}{5}) = 3.2$$

and

$$\sigma = \sqrt{16(\tfrac{1}{5})(\tfrac{4}{5})} = 1.6$$

we find that

$$z_1 = \frac{5.5 - 3.2}{1.6} = 1.44$$

and

$$z_2 = \frac{6.5 - 3.2}{1.6} = 2.06$$

The corresponding entries in Table I are 0.4251 and 0.4803 and the desired probability is 0.4803 − 0.4251 = 0.0552. *This result agrees very closely with the value of 0.0550 given in the National Bureau of Standards Table of Binomial Probabilities* (see Bibliography on page 168).

These two examples should help to convince the reader that binomial probabilities can be approximated very closely with corresponding areas under normal curves. Generally speaking, the approximation is very good unless p is *very close* to 0 or 1. However, even when p is fairly close to 0 or 1, the normal curve method will yield good results provided that, numerically, the z-values with which we have to work are not too large. To give one more example, let us use the normal curve approximation to determine the probability of getting "2 successes in 49 trials" when the probability of a success is 0.01. Since

$$\mu = 49(0.01) = 0.49$$

and $$\sigma = \sqrt{49(0.01)(0.99)} = 0.70$$

the necessary z values are

$$z_1 = \frac{1.5 - 0.49}{0.70} = 1.44$$

and $$z_2 = \frac{2.5 - 0.49}{0.70} = 2.87$$

The corresponding areas are 0.4251 and 0.4979, and the desired probability is approximately 0.0728. *This agrees very closely with the value of 0.0733 given in the National Bureau of Standards Table.*

The normal curve approximation of the binomial distribution is of tremendous value in problems in which the use of formula (7.2.1) would involve a prohibitive amount of work. Suppose, for example, that the probability that a person replies to a mail questionnaire is 0.20 and that we want to know the probability of getting *at least* 15 replies to 100 questionnaires. (In other words, we want to find the probability of getting "at least 15 successes in 100 trials" when the probability of a success is $\frac{1}{5}$.) If we tried to solve this problem by using the formula of the binomial distribution we would have to find the *sum* of the probabilities corresponding to $x = 15, 16, 17, \ldots$, and 100, or those corresponding to $x = 0, 1, 2, \ldots$, and 14. Clearly, this would involve an enormous amount of work. By using the normal curve approximation we have only to find the shaded area of Figure 7.29, namely, the area to the right of 14.5. Since

$$\mu = 100(\tfrac{1}{5}) = 20$$

and $$\sigma = \sqrt{100(\tfrac{1}{5})(\tfrac{4}{5})} = 4$$

we find that the z value corresponding to 14.5 is

$$z = \frac{14.5 - 20}{4} = -1.38$$

and that the desired probability is $0.4147 + 0.5000 = 0.9147$. *This means that we can expect to get at least 15 replies to 100 questionnaires*

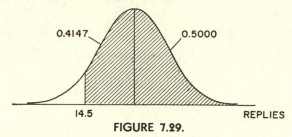

FIGURE 7.29.

about 91 per cent of the time. (This answer is, of course, contingent upon the value $p = \frac{1}{5}$ being the correct probability of getting a reply.)

To give another example in which it is impractical to use formula (7.2.1), let us determine the probability of getting anywhere from 25 to 40 sixes in 180 rolls of a balanced die. By using the normal curve approximation we have only to find the shaded area of Figure 7.30, namely, the area between 24.5 and 40.5. (Note that in all of these examples we have made the necessary adjustments to account for the

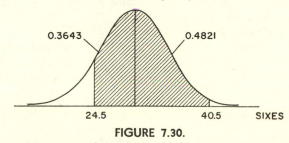

FIGURE 7.30.

fact that the number of "successes" is a discrete variable while the normal curve is continuous.) Since the mean and standard deviation of a binomial distribution having $n = 180$ and $p = \frac{1}{6}$ are

$$\mu = 180(\tfrac{1}{6}) = 30$$

and

$$\sigma = \sqrt{180(\tfrac{1}{6})(\tfrac{5}{6})} = 5$$

we find that the z values corresponding to 24.5 and 40.5 are

$$z_1 = \frac{24.5 - 30}{5} = -1.10$$

$$z_2 = \frac{40.5 - 30}{5} = 2.10$$

and that the desired probability of getting anywhere from 25 to 40 sixes in 180 rolls of a balanced die is $0.3643 + 0.4821 = 0.8464$.

EXERCISES

1. Find the probability of getting 6 heads in 10 tosses of a balanced coin by using

   (a) formula (7.2.1)   (b) the normal curve approximation

2. Find the probability of getting 2 sixes in 8 rolls of a balanced die by using

   (a) formula (7.2.1)   (b) the normal curve approximation

3. If a manufacturer knows that on the average 2 per cent of his product is slightly defective, what is the probability that in a lot of 100 pieces 3 will be slightly defective? Use the normal curve method.

4. If panelists on a television quiz show are stumped on the average in 1 out of 4 questions, what is the probability that they will miss exactly 4 out of 20 questions? Use the normal curve approximation.

5. The probability that Candidate A will be favored by the voters in an impending election is 0.55. What is the probability of getting a sample of 100 voters in which fewer than 50 favor Candidate A? Use the normal curve method.

6. If the favorite color of 65 per cent of all people is red, find the probability that in a sample of 1,000 people more than 680 will prefer red. Use the normal curve approximation.

7. Use the normal curve approximation to find the probability of getting anywhere from 180 to 220 heads in 400 flips of a balanced coin.

8. Use the normal curve approximation to find the probability of getting anywhere from 70 to 85 successes in 200 trials if the probability of a success is 0.40.

7.8 Fitting a Normal Curve to Observed Data

There are a number of ways in which we can test whether an observed distribution fits the pattern of a normal curve. A rather crude yet simple method involves the use of a special kind of graph paper called *probability graph paper* or *normal curve paper*.† If the cumulative "less than" percentage distribution of a set of data *which fits closely to a normal curve* is plotted on this kind of paper, the points will lie on a straight line (or reasonably close to a straight line).

To illustrate this technique, let us consider the following distribution of the weights of 300 army recruits:

---

† Probability graph paper may be obtained in most college bookstores and from dealers in art supplies. Instructions for "do it yourself" probability paper may be found on page 176 of A. E. Waugh, *Elements of Statistical Method*, 3rd ed. (New York: McGraw-Hill, 1952).

Weights (in pounds)	Number of Recruits (Frequency)	Percent of Recruits
150–158	9	3
159–167	24	8
168–176	51	17
177–185	66	22
186–194	72	24
195–203	48	16
204–212	21	7
213–221	6	2
222–230	3	1
	300	100

Converting this distribution first into a cumulative percentage distribution (see page 28), we get

Weights (in pounds)	Cumulative Percentage
less than 149.5	0
less than 158.5	3
less than 167.5	11
less than 176.5	28
less than 185.5	50
less than 194.5	74
less than 203.5	90
less than 212.5	97
less than 221.5	99
less than 230.5	100

Before we actually plot this cumulative distribution, let us briefly investigate the scales of normal curve paper. As can be seen from Figure 7.31, the cumulative percentage scale is already marked off in the rather unusual pattern which makes the paper suitable for our particular purpose. The other scale consists of equal subdivisions which are not numbered and which, in our problem, will be used to indicate the class boundaries of 149.5, 158.5, and so forth.

If we now plot a point corresponding to the cumulative percentage of each class boundary, we obtain the points shown in Figure 7.31. It would seem reasonable to say on the basis of this figure that the points lie *very close* to a straight line and, hence, that the *original distribution can be approximated fairly closely with a normal curve*. It should be noted that we did not plot points corresponding to the first and last class boundaries. As was pointed out on page 146, we will never reach 0 or 100 per cent of the area under a normal curve, no matter how far we go in either direction.

A disadvantage of the technique we have just discussed is that we must decide *subjectively* whether the points fall "reasonably close to a straight line." It is surprising to see how close the points can be to a straight line even though a distribution is fairly skewed.

A more scientific way of testing whether an observed distribution fits the pattern of a normal curve consists of the following *two* steps: first we calculate the proportions (and frequencies) which we could expect to find in the different classes if we had a normal distribution

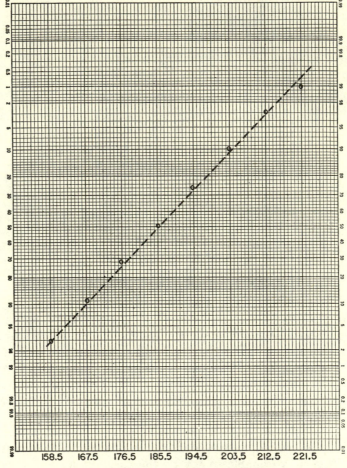

FIGURE 7.31 Probability Graph Paper.

with the same mean and standard deviation as our data, and then we compare these *expected normal curve frequencies* with those of the original distribution. For the time being, we shall only demonstrate the calculation of expected normal curve frequencies; the criterion used to compare the two sets of frequencies will be given later in Section 11.3.

To illustrate the determination of expected normal curve frequencies, let us refer again to the weight distribution of page 161. As can easily be checked, the mean of this distribution is 184.3 and its standard deviation is 14.54.   Before tackling the problem as a whole, let us illustrate the general approach by calculating the expected normal curve frequency of one class, say, the class which goes from 204 to 212.   The method which we shall use is really the same as the one which we used in Example 3 on page 155; we are again interested in finding the normal curve area between two given numbers.   Since the class boundaries are 203.5 and 212.5, the z values in which we are interested are

$$z_1 = \frac{203.5 - 184.3}{14.54} = 1.32$$

and

$$z_2 = \frac{212.5 - 184.3}{14.54} = 1.94$$

The corresponding normal curve areas (see Table I) are 0.4066 and 0.4738, and the area between 203.5 and 212.5 is $0.4738 - 0.4066 = 0.0672$.   *This means that 6.72 per cent of the weights could be expected to fall into the given class if the distribution were really close to a normal curve.*   Since 6.72 per cent of 300 (the total number of weights) is 20.16, we can say that the *expected normal curve frequency* of the given class is approximately 20.2.   It is interesting to note how close this value is to the actual frequency of 21.

If we want to apply this technique of finding expected normal curve frequencies to the entire distribution, we can arrange the necessary calculations as follows:

Class Limits (1)	Class Boundaries (2)	z (3)	Normal Curve Areas (Table I) (4)	Difference of Successive Areas in Col. 4 (5)	Normal Curve Frequencies (6)	Observed Frequencies (7)
	149.5	−2.39	0.4916			
150–158				0.0300	9.0	9
	158.5	−1.77	0.4616			
159–167				0.0846	25.4	24
	167.5	−1.16	0.3770			
168–176				0.1716	51.5	51
	176.5	−0.54	0.2054			
177–185				0.2373	71.2	66
	185.5	0.08	0.0319			
186–194				0.2261	67.8	72
	194.5	0.70	0.2580			
195–203				0.1486	44.6	48
	203.5	1.32	0.4066			
204–212				0.0672	20.2	21
	212.5	1.94	0.4738			
213–221				0.0210	6.3	6
	221.5	2.56	0.4948			
222–230				0.0045	1.4	3
	230.5	3.18	0.4993			

Whereas column (1) contains the class limits, column (2) contains the class boundaries, and column (3) contains the corresponding z's.   Column (4) contains the normal curve areas (as given in Table I), and column (5) contains the *differences* of successive areas of column (4), except for the fourth entry, which is the *sum* of 0.2054 and 0.0319. (The reason for this exception is that this is the only area between z values which differ in sign.) Finally, we multiply each area (proportion or probability) of column (5) by 300 and we get the *expected normal curve frequencies* which are shown in column (6).

Once we have calculated the expected normal curve frequencies, we can compare them with the observed frequencies, which were originally given on page 161 and which are copied again in column (7). Since we will not be able to put this comparison on a precise basis until we reach Section 11.3, we shall be satisfied here with presenting

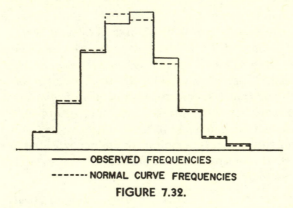

——— OBSERVED FREQUENCIES

------- NORMAL CURVE FREQUENCIES

FIGURE 7.32.

merely the superimposed histograms of the distributions of columns (6) and (7). It would seem reasonable to say on the basis of Figure 7.32 that there is a close agreement between the two sets of frequencies and that, therefore, *the original distribution fits closely to the pattern of a normal curve.*

EXERCISES

1. Use probability graph paper to check whether the mail-order distribution given on page 22 fits reasonably well to a normal curve.
2. Plot the cumulative percentage distribution of whichever data were grouped among Exercises 7, 8, 9, and 10 on page 29 on probability graph paper and decide whether the distribution fits reasonably well to a normal curve.
3. Plot the cumulative percentage distribution of the number of physicians per 100,000 population (see Exercise 12 on page 31) on probability paper and check whether this distribution fits reasonably well to a normal curve.
4. Find the expected normal curve frequencies of the mail-order distribution given on page 22, whose mean and standard deviations are

equal to \$20.87 and \$7.45, respectively. Also plot superimposed histograms of the observed distribution and the corresponding normal curve frequencies.

5. Find the expected normal curve frequencies of whichever data were grouped among Exercises 7, 8, 9, and, 10 on page 29, using the mean and standard deviation which were found in Exercise 3 on page 57 and Exercise 2 on page 94. Also, plot superimposed histograms of the observed frequencies and the corresponding normal curve frequencies.

6. Find the expected normal curve frequencies of the distribution of physicians per 100,000 population whose mean and standard deviation were obtained in Exercise 4 on page 57 and Exercise 3 on page 94. Also, plot superimposed histograms showing the observed frequencies and the corresponding expected normal curve frequencies.

7.9 Further Theoretical Distributions

The fact that most of this chapter was devoted to the binomial and normal distributions may have given the erroneous impression that they are the only two distributions that matter in statistical theory. Although it is true that the binomial distribution and the normal distribution play important roles in many applications, their indiscriminate use can lead to ridiculous and misleading results. As was pointed out earlier, the binomial distribution must be replaced with another probability distribution if the trials are not independent or if the probability of an individual success does not remain constant from trial to trial. Similarly, there are problems in which we have good reasons to expect continuous distributions other than the normal curve. As we shall see in Chapters 9, 10, and 11, other continuous distributions, among them the t-distribution, the chi-square distribution, and the F-distribution, play very important roles in problems of estimation, prediction, and the testing of hypotheses.

When we originally discussed the normal curve approximation of the binomial distribution, we pointed out that the approximation should not be used if p is *very close* to 0 or *very close* to 1. *If p, the probability of an individual success, is very small and n is large*, the binomial distribution can be approximated very closely with a probability distribution which is called the *Poisson distribution*.† The formula of this distribution is

$$P(x) = \frac{\mu^x \cdot e^{-\mu}}{x!}$$
(7.9.1)★

---

† If the probability of a "success" is close to 1, the probability of a "failure" is close to 0. Hence, if we interchange "success" and "failure" we can convert a problem in which the probability of a success is close to 1 into a problem in which the probability of a success is close to 0.

where $x! = x \cdot (x - 1) \cdot (x - 2) \cdot \ldots \cdot 2 \cdot 1$ (see page 135) and e is the irrational number which appeared also in the formula of the normal curve (see page 147). As before, x stands for the number of successes, $P(x)$ for the probability of getting x successes, and μ is the mean. A rigorous proof of the fact that formula (7.2.1), the formula of the binomial distribution, approaches formula (7.9.1) when p becomes very small and n becomes very large (while $\mu = np$ is fixed) is referred to on page 168.

To illustrate the use of the Poisson distribution, let us suppose that a piano is advertised for sale in a newspaper having 100,000 readers. Assuming that the probability that any one reader will respond to the advertisement is $p = 1/50,000$, let us find the probabilities of getting 0, 1, 2, 3, 4, . . . , responses to the ad.

Essentially, this is a problem of finding the probabilities of getting "x successes in 100,000 trials" when the probability of an individual success is $p = 1/50,000$. Since the use of formula (7.2.1) is ruled out for practical reasons and p is too small to use the normal curve approximation, we shall use formula (7.9.1), the formula of the Poisson distribution. In order to use this formula we must first calculate μ, which according to (7.3.3) is

$$\mu = np = 100,000 \cdot \frac{1}{50,000} = 2$$

(*On the average* we can, therefore, expect 2 responses to such an ad.) Substituting $\mu = 2$ and $x = 0, 1, 2, \ldots$, into (7.9.1), we obtain the following probabilities rounded to four decimals:

Number of Responses	Probability
0	0.1353
1	0.2707
2	0.2707
3	0.1804
4	0.0902
5	0.0361
6	0.0120
7	0.0034
8	0.0009
9	0.0002

Logically speaking, it is possible to get more than 9 responses to the ad, but the probability of this happening is very small (about 0.000046). The histogram of this Poisson distribution is shown in Figure 7.33.

The Poisson distribution has many important applications. It is used, for example, in the field of casualty insurance, where the probability that any given house will be destroyed by fire is very small while n, the number of insured houses, is very large. Similarly, the probability that a person will get killed in an automobile accident is very small while n, the number of people who ride in cars, is very large. The Poisson distribution is also used in problems dealing with the inspection of manufactured products when the probability that any one piece is defective is very small and the lots are very large. Since the use of formula (7.9.1) involves a good deal of arithmetic, it is usually

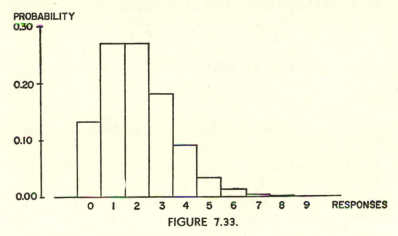

FIGURE 7.33.

easier to refer to a special table of Poisson probabilities (such as the table mentioned on page 168).

EXERCISES

1. In the manufacture of a certain kind of button, a factory produces *on the average* one defective button among every 100 buttons. If the buttons are packed in boxes of 300, what percentage of these boxes would you expect to have no defective buttons? (Use $e^{-3} = 0.0498$; also, by definition $0! = 1$.)

2. Suppose that *on the average* 1 car in 1,000 has a flat tire while traveling on the New Jersey Turnpike. If on a certain day 10,000 cars travel on this turnpike, what is the probability that exactly 8 cars will have flat tires? (Use $e^{-10} = 0.000045$.)

3. A life insurance salesman sells on the average 3 life insurance policies per week. Use formula (7.9.1) with $\mu = 3$ to calculate the probability that in a given week he will sell only one life insurance policy. (Use $e^{-3} = 0.0498$).

4. Assume that it is known from past experience that in a certain plant there are on the average 5 industrial accidents per month. Use

formula (7.9.1) to find the probability that in a given month there will be fewer than 5. (*Hint:* Add the probabilities for $x = 0, 1, 2, 3$, and 4; also use $e^{-5} = 0.0067$.)

BIBLIOGRAPHY

Binomial probabilities for $n = 2$ to $n = 49$ may be found in

Tables of the Binomial Probability Distribution, National Bureau of Standards Applied Mathematics Series No. 6. Washington, D.C.: U.S. Government Printing Office, 1950,

and for $n = 50$ to $n = 100$ in

Romig, H. G., *50–100 Binomial Tables.* New York: Wiley and Sons, 1953.

The derivation of the formulas for the mean and standard deviation of the binomial distribution may be found in

Freund, J. E., *Mathematical Statistics.* Englewood Cliffs, N.J.: Prentice-Hall, Inc., 1962.

and the definition of the mean and standard deviation of a *continuous* distribution is given in

Wilks, S. S., *Elementary Statistical Analysis.* Princeton, N.J.: Princeton University Press, 1948, pp. 116–17.

Proofs showing that the binomial distribution approaches a normal curve when p is fixed and n goes to infinity may be found in many textbooks of mathematical statistics. A proof showing that the binomial distribution approaches the Poisson distribution when p becomes very small, n becomes very large, and $u = np$ is fixed, may be found in the book by Wilks mentioned above (pp. 133–34) and, among others, in

Hoel, P. G., *Introduction to Mathematical Statistics,* 3rd ed. New York: Wiley and Sons, 1962, p. 68.

The most widely used table of Poisson probabilities is

Molina, E. E., *Poisson's Exponential Binomial Limit.* New York: D. Van Nostrand, 1942.

CHAPTER 8

Sampling Distributions

8.1 Random Sampling

In Chapter 3 we stated that a *population* consists of all conceivably (or hypothetically) possible observations relating to a given phenomenon and that a *sample* is simply part of a population. Since the fundamental objective of inductive statistics is to make inferences about populations on the basis of samples, let us now see under what conditions this may or may not be possible. Suppose that we want to study wholesale prices of farm products and that the only information we have at our disposal pertains to the wholesale prices of eggs. It would hardly seem reasonable to suppose that we can use information about eggs alone to arrive at sweeping generalizations about wholesale prices of farm products in general. Similarly, we can hardly expect to arrive at reasonable generalizations about personal incomes in general if we are supplied only with information about the incomes of doctors *or* use information about traffic on the Pennsylvania Turnpike on the 4th of July to arrive at general conclusions about traffic on this super-highway throughout the year. These examples are perhaps somewhat extreme, but they serve to illustrate that we shall have to be very careful whenever we want to generalize on the basis of a sample.

The whole problem of when and under what conditions samples permit "reasonable" generalizations is not easily answered. In the theory we shall develop in the next three chapters we will always assume that we are dealing with *random samples*—special kinds of samples which will be defined below. More general problems of sampling will be treated later in Chapter 12.

If we sample from a *finite* population, we shall say that *a sample*

169

is random if every item in the population has an equal chance of being included in the sample.† This definition implies that the selection of the sample should be left to *chance*, and it is, indeed, common practice to base the selection of random samples on some form of gambling device.

If we wanted to study the merchandizing practices of 516 retail food stores operating in a certain community, we could select a random sample of, say, 10 of these stores by writing the name of each store on a slip of paper, mixing the slips thoroughly, and then drawing 10 slips. Since the construction of suitable gambling devices becomes rather tedious when the number of items is large, we can leave this part of the job to others by selecting our sample with the use of a table of *random numbers.* The sample page shown on page 171 comes from Table X, Random Numbers, which was constructed with the use of some gambling device supposedly giving each entry an equal probability of being 0, 1, 2, 3, . . . , or 9.

To illustrate how Table X, Random Numbers (or *random digits*, as they are also called) are used in the selection of a sample, let us refer again to the example of 516 retail food stores from which we want to select a random sample of size 10. (A sample is said to be of *size n* if it contains *n* measurements or observations.) Numbering the stores from 1 to 516 or, better, numbering them 001, 002, 003, . . . , 515, 516, we arbitrarily pick from a page of the table of random numbers three consecutive columns and a row from which to start. Suppose that we pick columns 6, 7, and 8 of the random numbers shown on page 171 and that we start with row 11. Reading down the columns (marked with the answers), it can easily be seen that our sample will contain the stores having the numbers

476   436   063   325   081   473   370   305   178   184

In the selection of this sample we ignored numbers exceeding 516. If, by chance, the same number appears more than once, we would ignore it after it appeared for the first time. If we had wanted to be "even more random" in the previous example, we could have left the selection of the page, columns, or rows in the table of random numbers to chance by using a gambling device or, perhaps, another page of random numbers.

---

† A population is said to be *finite* if it contains a finite number of elements. For example, the population consisting of the salaries paid to employees of a certain firm is finite and so is the population consisting of the closing prices listed by the New York Stock Exchange on a certain day. In contrast to finite populations, we shall say that a population is *infinite* if there is no limit to the number of elements which it contains. For instance, the population consisting of *all possible* flips of a coin is an infinite population.

STATISTICAL TABLES

TABLE X

Random Numbers†

Col. 8
Col. 7
Col. 6

04433	80674	24520	18222	10610	05794	37515
60298	47829	72648	37414	75755	04717	29899
67884	59651	67533	68123	17730	95862	08034
89512	32155	51906	61662	64130	16688	37275
32653	01895	12506	88535	36553	23757	34209
95913	15405	13772	76638	48423	25018	99041
55864	21694	13122	44115	01601	50541	00147
35334	49810	91601	40617	72876	33967	73830
57729	32196	76487	11622	96297	24160	09903
86648	13697	63677	70119	94739	25875	38829
Row 11 → 30574	47609	07967	32422	76791	39725	53711
Row 12 → 81307	43694	83580	79974	45929	85113	72268
Row 13 → 02410	54905	79007	54939	21410	86980	91772
18969	75274	52233	62319	08598	09066	95288
87863	82384	66860	62297	80198	19347	73234
68397	71708	15438	62311	72844	60203	46412
28529	54447	58729	10854	99058	18260	38765
44285	06372	15867	70418	57012	72122	36634
86299	83430	33571	23309	57040	29285	67870
84842	68668	90894	61658	15001	94055	36308
56970	83609	52098	04184	54967	72938	56834
83125	71257	60490	44369	66130	72936	69848
55503	52423	02464	26141	68779	66388	75242
47019	76273	33203	29608	54553	25971	69573
84828	32592	79526	29554	84580	37859	28504
68921	08141	79227	05748	51276	57143	31926
36458	96045	30424	98420	72925	40729	22337
95752	59445	36847	87729	81679	59126	59437
26768	47323	58454	56958	20575	76746	49878
42613	37056	43636	58085	06766	60227	96414
95457	30566	65482	25596	02678	54592	63607
95276	17894	63564	95958	39750	64379	46059
66954	52324	64776	92345	95110	59448	77249
17457	18481	14113	62462	02798	54977	48349
03704	36872	83214	59337	01695	60666	97410
21538	86497	33210	60337	27976	70661	08250
57178	67619	98310	70348	11317	71623	55510
31048	97558	94953	55866	96283	46620	52087
69799	55380	16498	80733	96422	58078	99643
90595	61867	59231	17772	67831	33317	00520
33570	04981	98939	78784	09977	29398	93896
15340	93460	57477	13898	48431	72936	78160
64079	42483	36512	56186	99098	48850	72527
63491	05546	67118	62063	74958	20946	28147
92003	63868	41034	28260	79708	00770	88643
52360	46658	66511	04172	73085	11795	52594
74622	12142	68355	65635	21828	39539	18988
04157	50079	61343	64315	70836	82857	35335
86003	60070	66241	32836	27573	11479	94114
41268	80187	20351	09636	84668	42486	71303

† Based on parts of *Table of 105,000 Random Decimal Digits*, Interstate Commerce Commission, Bureau of Transport Economics and Statistics, Washington, D.C.

Having defined random samples and having demonstrated how random samples may be obtained with the use of random numbers, we might add that this is very often easier said than done. If we were asked to estimate the average diameter of 500,000 ball bearings on the basis of a random sample of 50, it would hardly be practical to number these ball bearings 000001, 000002, 000003, and so on, and proceed as in our previous example. In a situation like this we often have little choice but to proceed according to the dictionary definition of the word randomness and proceed "haphazardly without definite aim or purpose," hoping that this will give us a sample to which we can apply statistical theory that is otherwise reserved for random samples. This question will be discussed further in Chapter 12.

8.2 Sampling Distributions

Let us now illustrate the most fundamental concept of inductive statistics, that of a *sampling distribution*, with the concrete (though fictitious) example of *estimating* the average height of the adult male residents of a certain community. Let us suppose that if we actually made an exhaustive survey, that is, if we actually measured the height of each adult male resident of the community, we would obtain the following distribution:

Height (in inches)	Frequency
55–57	7
58–60	116
61–63	761
64–66	2379
67–69	3472
70–72	2379
73–75	761
76–78	116
79–81	7
	9998

whose mean is $\mu = 68$ and whose standard deviation is $\sigma = 3.44$.

What we have done here is referred to in statistics as *exhaustive sampling*. We actually obtained the entire population, the heights of all adult male residents of the given community, and we then found its mean *without having to make any generalizations*. In actual practice, statisticians seldom resort to exhaustive sampling. *It is not only at times impossible or unfeasible, but it is usually impractical and unnecessary.* It is *impossible* if we sample from an infinite population like that consisting of all possible flips of a coin and it is *unfeasible* if the sampling is destructive. A sampling procedure is said to be *destructive* if it necessitates damaging or destroying the products or objects that are being measured. If we wanted to determine the average lifetime of all the TV tubes made by a certain firm, we could hardly test them all; if we did, there would be none left to sell. (Sampling is also said to be *destructive* if it changes the nature of a product so that it can no longer be used for the purpose for which it was intended.)

Even if we do not damage or destroy the objects with which we are concerned, there is seldom any need to resort to exhaustive sampling. In the above example, exhaustive sampling would be *impractical* because the cost of locating, visiting, and then measuring the height of each adult male resident would be prohibitive. However, even if we were willing to spend the money, this would be a waste because we can usually get sufficiently accurate information on the basis of relatively small random samples. For instance, a random sample consisting of the opinions expressed by 500 voters may well be adequate to predict an election, so it would be a waste of time, money, and effort, to take a sample of 1,000, 5,000, 10,000 or make an exhaustive survey. Considerations of this type are very important because we shall see later on that *gains in accuracy are usually not proportional to increases in the size of samples*.

Let us now return to the problem of estimating the average height of the adult male residents of the given community and *let us forget about the complete (exhaustive) survey which yielded the distribution shown on page* 172. Instead, let us take a random sample of size 5 and use its mean as an estimate of μ. (If this were not a fictitious example, we would probably take a sample of size 25, 50, or more; we shall limit ourselves to 5 observations, to simplify calculations.) Let us finally assume that we obtain the following 5 measurements rounded to the nearest inch:

$$66 \quad 61 \quad 66 \quad 70 \quad 68$$

Since the mean of this sample is

$$\bar{x} = \frac{66 + 61 + 66 + 70 + 68}{5} = 66.2$$

we shall estimate that μ, the mean of the entire population, is 66.2. In other words, we make the generalization that the average height of *all* the adult male residents of the given community is 66.2 inches.

Having made a statement about the average height of several thousand persons after having measured as few as 5, we might well be asked any one or all of the following questions:

1. Can we really expect a sample mean to be "reasonably" close to the mean of the population from which the sample was obtained?

2. How sure are we that our sample mean will not differ from μ by more than 1 inch, 2 inches, or, say, 3.1 inches?

3. If we were to repeat this experiment and take several random samples of 5 measurements each, how closely could we expect their means to be clustered around μ?

These questions are very important and they should be answered, but before we can do so, we must conduct a further experiment, specifically, the experiment which was suggested in question 3.

It is clear that we cannot reasonably expect every sample mean to coincide with the mean of the population. As a matter of fact, if we peek on page 172, we find that $\mu = 68$ and that our estimate of 66.2 is, thus, off by 1.8 inches. If the sample mean had been 67.4, our estimate of μ would have been closer and if the sample mean had been 61.8, it would have been worse. In order to see how sample means fluctuate from sample to sample, let us now take 50 *separate* random samples of 5 measurements each from the given population which, as the reader will recall, consists of the heights of the adult male residents of a certain community. The following are 50 such random samples.

Sample 1: 68, 66, 67, 71, 72
Sample 2: 72, 69, 67, 64, 64
Sample 3: 73, 70, 64, 69, 66
Sample 4: 69, 68, 73, 71, 69
Sample 5: 66, 61, 66, 70, 68
Sample 6: 68, 67, 66, 65, 66
Sample 7: 68, 73, 66, 65, 65
Sample 8: 60, 68, 72, 69, 78
Sample 9: 65, 70, 70, 63, 73
Sample 10: 66, 73, 66, 65, 65
Sample 11: 74, 67, 71, 70, 68
Sample 12: 65, 70, 69, 68, 74
Sample 13: 67, 64, 70, 69, 63
Sample 14: 77, 70, 70, 69, 70
Sample 15: 67, 69, 73, 69, 62
Sample 16: 61, 67, 67, 67, 67
Sample 17: 70, 64, 66, 67, 64
Sample 18: 68, 69, 71, 66, 74
Sample 19: 67, 68, 72, 75, 63
Sample 20: 71, 70, 67, 65, 70
Sample 21: 73, 66, 68, 66, 69
Sample 22: 66, 69, 67, 65, 69
Sample 23: 59, 69, 66, 70, 70
Sample 24: 67, 65, 64, 61, 66
Sample 25: 64, 70, 69, 72, 68

Sample 26: 68, 68, 67, 59, 68
Sample 27: 68, 74, 63, 70, 64
Sample 28: 65, 70, 72, 71, 67
Sample 29: 71, 68, 65, 68, 69
Sample 30: 68, 65, 66, 70, 68
Sample 31: 70, 56, 68, 75, 71
Sample 32: 64, 72, 68, 70, 68
Sample 33: 63, 70, 74, 73, 80
Sample 34: 68, 70, 69, 67, 73
Sample 35: 68, 68, 70, 66, 76
Sample 36: 71, 70, 68, 62, 68
Sample 37: 64, 69, 66, 66, 63
Sample 38: 71, 74, 65, 62, 72
Sample 39: 68, 70, 66, 67, 69
Sample 40: 72, 68, 71, 65, 67
Sample 41: 62, 72, 67, 67, 71
Sample 42: 65, 68, 63, 65, 76
Sample 43: 66, 73, 70, 70, 73
Sample 44: 66, 70, 69, 71, 64
Sample 45: 64, 64, 64, 70, 64
Sample 46: 73, 70, 65, 69, 67
Sample 47: 70, 66, 72, 73, 69
Sample 48: 67, 69, 64, 66, 68
Sample 49: 67, 66, 69, 69, 67
Sample 50: 68, 67, 70, 67, 69

Having obtained all these samples, our next step is to calculate their means and study their distribution. The 50 means are

68.8	67.2	68.4	70.0	66.2	66.4	65.4	69.4	68.2	67.0
70.0	69.2	66.6	71.2	68.0	65.8	66.2	69.6	69.0	68.6
68.4	67.2	66.8	64.6	68.6	66.0	67.8	69.0	68.2	67.4
68.0	68.4	72.0	69.4	69.6	67.8	65.6	68.8	68.0	68.6
67.8	67.4	70.4	68.0	65.2	68.8	70.0	66.8	67.6	68.2

and, as can easily be checked with formulas (3.2.1) and (4.4.4), the mean of these 50 $\bar{x}$'s is 68.03 and their standard deviation is 1.55. Symbolically, we shall write

$$\bar{x}_{\bar{x}} = 68.03 \quad \text{and} \quad s_{\bar{x}} = 1.55$$

using the subscript $\bar{x}$ to indicate that we are referring to the mean and standard deviation of the 50 $\bar{x}$'s.

In order to get an over-all picture of the 50 means, let us group them into the following distribution:

$\bar{x}$	Frequency
64.5–65.4	3
65.5–66.4	6
66.5–67.4	8
67.5–68.4	14
68.5–69.4	11
69.5–70.4	6
70.5–71.4	1
71.5–72.4	1

This distribution is called an *experimental sampling distribution*. It consists of the means obtained in an *experiment* in which we took repeated samples from the same population.

Although the term "experimental sampling distribution" is used here for the first time, we already met such distributions in Chapter 7. If 4 balanced coins are tossed 160 times (see Exercise 1 on page 130), the resulting distribution showing the frequencies with which 0, 1, 2, 3, and 4 heads occurred is an *experimental sampling distribution*. If a balanced die is rolled 120 times (see Exercise 5 on page 130), the resulting distribution showing the proportion of the time that the different faces appeared is an *experimental sampling distribution*. In the first example each sample consists of tossing 4 balanced coins and in the second example each sample consists of a roll of a balanced die.

Our purpose in making the reader construct these experimental sampling distributions was to provide experimental support for the corresponding *probability distributions*. Such a comparison can also be made in the sampling experiment which we have been discussing in this section. Corresponding to the experimental sampling distribution of the 50 means shown above, we might now ask *what is the probability of obtaining a mean which lies between 67.5 and 68.4, what is the probability of obtaining a mean which lies between 68.5 and 69.4, what is the probability of obtaining a mean which lies between 66.5 and 67.4, and so forth.*

A distribution which shows the probabilities of obtaining different values of $\bar{x}$ is called a *theoretical sampling distribution of $\bar{x}$.* Since this concept may not be too easy to grasp, let us illustrate it with a relatively simple example. Let us suppose that we take 6 slips of paper on which we write the numbers 1, 3, 5, 7, 9, and 11. If we now draw a sample of 2 of these slips, we will get either *1 and 3, 1 and 5, 1 and 7, 1 and 9, 1 and 11, 3 and 5, 3 and 7, 3 and 9, 3 and 11, 5 and 7, 5 and 9, 5 and 11, 7 and 9, 7 and 11,* or *9 and 11.* Considering these 15 possibilities to be *equally likely,* we can construct the theoretical sampling distribution for the mean of a sample of size 2 from this population by calculating the mean of each of the 15 possible samples and assigning it a probability of $\frac{1}{15}$. This will give us the following *theoretical sampling distribution:*

$\bar{x}$	Probability
2	$\frac{1}{15}$
3	$\frac{1}{15}$
4	$\frac{2}{15}$
5	$\frac{2}{15}$
6	$\frac{3}{15}$
7	$\frac{2}{15}$
8	$\frac{2}{15}$
9	$\frac{1}{15}$
10	$\frac{1}{15}$

This distribution tells us that if we draw two of the slips, the probability that the *mean* of the two numbers will be 4 is $\frac{2}{15}$, the probability that the *mean* of the two numbers will be 6 is $\frac{3}{15}$, the probability that the *mean* of the two numbers will be 9 is $\frac{1}{15}$, and so on.

Returning now to the problem dealing with the heights of the 9,998 persons mentioned on page 172, we could duplicate what we have done in the above example by enumerating all possible samples of size 5, calculating their means, and constructing the theoretical sampling distribution of $\bar{x}$ by assigning each possible sample an equal chance of being selected. This would be an enormous task, but it would give us the *probabilities* of getting various values of $\bar{x}$ in a random sample of size 5.

Fortunately, it will not be necessary to perform the steps outlined above. Instead we shall refer to two very important theorems about theoretical sampling distributions of $\bar{x}$. The first of these gives us general formulas for the mean and standard deviation of a theoretical sampling distribution of $\bar{x}$.†

THEOREM 8.1: *If random samples of size n are taken from a population which has the mean μ and the standard deviation σ, the theoretical sampling distribution of $\bar{x}$ has the mean μ and the standard deviation $\sigma/\sqrt{n}$.*

One aspect of this theorem which should not be surprising is that the mean of the theoretical sampling distribution of $\bar{x}$ *equals* the mean of the population. In our illustration the mean of the 50 $\bar{x}$'s was 68.03, while the mean of the population was 68 (see pages 175 and 172). The reader may wish to verify that in the example dealing with the 6 slips of paper, the mean of the population as well as the mean of the theoretical sampling distribution is equal to 6.

A very interesting feature of Theorem 8.1 is that the standard deviation of the theoretical sampling distribution of $\bar{x}$ is given by the formula $\sigma/\sqrt{n}$. Following the customary symbolism and terminology, we shall write this standard deviation as $\sigma_{\bar{x}}$ and call it the *standard error of the mean*. We thus have

$$\sigma_{\bar{x}} = \frac{\sigma}{\sqrt{n}} \qquad (8.2.1)\star$$

where σ is the standard deviation of the population and n the sample size.

The standard error of the mean plays an important role in inductive statistics, because it measures the *variation* of the theoretical sampling distribution of $\bar{x}$. In other words, *it tells us how much sample means can be expected to vary from sample to sample.*

It should be interesting to check how this theoretical value compares with the value which we actually obtained for the standard deviation of the 50 $\bar{x}$'s. Since the population had a standard deviation of 3.44 (see page 172) and n was 5, formula (8.2.1) gives

$$\sigma_{\bar{x}} = \frac{3.44}{\sqrt{5}} = 1.54$$

This is extremely close to the value of $s_{\bar{x}} = 1.55$ which we obtained in our experiment.

It is apparent from the formula for $\sigma_{\bar{x}}$, in which σ is divided by the square root of n, that the standard error of the mean *decreases* when

---

† This theorem applies only for samples taken from *very large* populations so that a sample of size n constitutes but a minute fraction of the population. If each sample constitutes an appreciable portion (more than 5 per cent) of the population, the theorem will have to be modified as explained in Section 9.6.

we increase the size of the sample. This means that when n is large, *a sample mean will be a more reliable estimate of* μ, that is, *when n becomes large and we have more information, a sample mean can be expected to be closer to the mean of the population.*

To illustrate how the variability of a sampling distribution decreases with increasing n, let us convert the 50 samples listed on page 174 into 25 samples of size 10 by combining Samples 1 and 2, Samples 3 and 4, Samples 5 and 6, and so forth. If we calculate the means of these 25 samples we obtain

68.0	69.2	66.3	67.4	67.4
69.6	68.9	66.9	67.9	68.8
67.8	65.7	67.3	68.4	67.8
68.2	70.7	68.7	67.2	68.3
67.6	69.2	67.0	68.4	67.9

and these values may be grouped into the following distribution:

$\bar{x}$	Frequency
65.5–66.4	2
66.5–67.4	5
67.5–68.4	11
68.5–69.4	5
69.5–70.4	1
70.5–71.4	1

As can easily be checked with formulas (3.2.1) and (4.4.4), the mean of the 25 new $\bar{x}$'s is 68.03 and their standard deviation is

$$s_{\bar{x}} = 1.07$$

This agrees very closely with the theoretical values which we should expect according to Theorem 8.1. Substituting $\sigma = 3.44$ and $n = 10$ into formula (8.2.1), we obtain

$$\sigma_{\bar{x}} = \frac{3.44}{\sqrt{10}} = 1.09$$

Clearly, there is less variation between the 25 means based on samples of size 10 than there is between the 50 means based on samples of size 5; the standard deviations of the corresponding sampling distributions are 1.07 and 1.55, respectively. Had we begun our experiment with samples of, say, 25 observations, the standard deviation of our experimental sampling distribution should have been close to $3.44/\sqrt{25} = 0.69$ (see Exercise 3 on page 183).

Although it may be important to know the mean and standard deviation of a theoretical sampling distribution of $\bar{x}$, knowledge of these two values alone will not enable us to calculate the probabilities of obtaining various values of $\bar{x}$.   To be able to calculate such probabilities, we shall have to refer to a second theorem—the *Central Limit Theorem.*

> THEOREM 8.2 (Central Limit Theorem): *If n is large, the theoretical sampling distribution of $\bar{x}$ can be approximated very closely with a normal curve.*

A noteworthy aspect of this theorem is that it contains no specification about the population.   (To be precise, we might add that the population must have a *finite* standard deviation σ.)   *This means that when n is large, the theoretical sampling distribution of $\bar{x}$ will be close to a normal curve regardless of the shape of the distribution of the population.*†

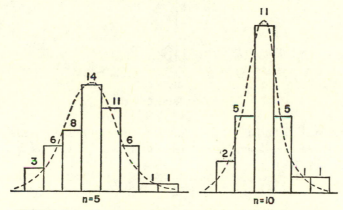

FIGURE 8.1.   Experimental Sampling Distributions of the Means.

Since the n which we used in our example was relatively small (it was equal to 5), let us add that Theorem 8.2 applies also when n is small, *provided that the distribution of the population can be approximated closely with a normal curve.*   As can easily be seen, the distribution of the 9,998 heights given on page 172 is very close to a normal curve, and this (as well as a good deal of luck) explains why our experimental sampling distributions (see Figure 8.1) can be approximated closely with normal curves.   The diagrams shown in Figure 8.1 present the *experimental sampling distributions* as well as the normal curves which approximate the corresponding *theoretical sampling distributions.*

---

† It is difficult to make any precise statement as to how large n has to be before we can approximate the theoretical sampling distribution of $\bar{x}$ with a normal curve.   Unless the distribution of the population has a very unusual shape, the approximation will generally be good even if n is relatively small.

Let us now illustrate how Theorems 8.1 and 8.2 can be used in our problem to calculate the probabilities of getting various values of $\bar{x}$. For instance, let us find the probability of getting a random sample of size 5 (from the population which consists of the 9,998 heights) for which $\bar{x}$ lies between 68.5 and 69.5.   According to Theorem 8.2, we can say that the theoretical sampling distribution of such sample means can be approximated closely with a normal curve, and according to Theorem 8.1, we can say that the mean and standard deviations of this normal curve are $\mu = 68$ and $\sigma_{\bar{x}} = 1.54$, respectively (see page 177).   *Hence, the desired probability is given by the shaded area of Figure 8.2.*   To find this area we have only to calculate the z values corre-

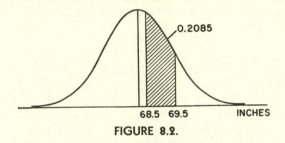

FIGURE 8.2.

sponding to 68.5 and 69.5, look up the corresponding normal curve areas in Table I, and then take their difference.   We, thus, get

$$z_1 = \frac{68.5 - 68}{1.54} = 0.32$$

$$z_2 = \frac{69.5 - 68}{1.54} = 0.97$$

and the desired probability is $0.3340 - 0.1255 = 0.2085$ or approximately 0.21.   Using the identical technique we can calculate the probability of getting an $\bar{x}$ which lies between 67.5 and 69.4, an $\bar{x}$ which lies between 60.5 and 67.3, an $\bar{x}$ which lies between 69.5 and 72.1, and the like.   In other words, we can calculate all of the probabilities which relate to the *theoretical sampling distribution of $\bar{x}$* for random samples of size 5 from the given population.

The main objective of the preceding discussion was to introduce the concepts of experimental and theoretical sampling distributions. These concepts are of fundamental importance in statistics and they are by no means limited to the study of sample means.   For instance, we could have found the *medians* of the 50 samples given on page 174 and studied their distribution in order to get some idea as to how much medians vary from sample to sample.   This would have given us the following values:

68	67	69	69	66	66	65	69	70	66
70	69	67	70	69	67	66	69	68	70
68	67	69	65	69	68	68	70	68	68
70	68	73	69	68	68	66	71	68	68
67	65	70	69	64	69	70	67	67	68

If we now group these 50 sample medians, using the same class intervals as for the distribution of the 50 means, we obtain the following distribution:

Median	Frequency
63.5–64.4	1
64.5–65.4	3
65.5–66.4	5
66.5–67.4	7
67.5–68.4	13
68.5–69.4	11
69.5–70.4	8
70.5–71.4	1
71.5–72.4	0
72.5–73.4	1

Comparing the experimental sampling distribution of the 50 medians with that of the 50 means (or the histograms of Figures 8.3

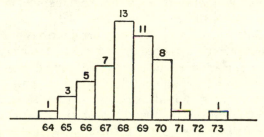

FIGURE 8.3. Experimental Sampling Distribution of the Medians.

and 8.1) it is apparent that the first has a *greater variation* than the second. Whereas $s_{\bar{x}}$ equalled 1.55, it can easily be shown that the standard deviation of the 50 medians is $s_M = 1.73$. This agrees with what we said on page 59, where we pointed out that the mean is *more reliable* than the median or, in other words, that *sample means do not fluctuate as much from sample to sample as the corresponding medians.*

Analogous to Theorems 8.1 and 8.2, there also exists a theorem pertaining to the theoretical sampling distribution of the median. As we shall formulate it here, it applies only to samples from large populations whose distributions can be approximated closely with normal curves.

THEOREM 8.3: *If random samples of size n are taken from a large population which has the mean μ, the standard deviation σ, and which can be approximated closely with a normal curve, the theoretical sampling distribution of the median has the mean μ and the standard deviation* $1.25 \cdot \dfrac{\sigma}{\sqrt{n}}$.† *If n is large, the theoretical sampling distribution of the median can be approximated closely with a normal curve.*

We can thus write the *standard error of the median*, namely, the standard deviation of the *theoretical* sampling distribution of the median, as

$$\sigma_M = 1.25 \cdot \frac{\sigma}{\sqrt{n}} \qquad (8.2.2)\star$$

Since the standard deviation of the population from which we took the 50 samples was 3.44, we should have *expected* the standard deviation of our experimental sampling distribution of the 50 medians to be

$$\sigma_M = 1.25 \cdot \frac{3.44}{\sqrt{5}} = 1.92$$

This is somewhat more than the value of $s_M = 1.73$ which we obtained. However, if we had continued the experiment and taken more than 50 samples, the standard deviation of the medians should before long have come close to the theoretical value of 1.92.

We could continue our study of sampling distributions by calculating the standard deviation, the range, the mode, the geometric mean, and the average deviation of each of the 50 samples and group them into the corresponding experimental sampling distributions. Then we could calculate the standard deviation of each sampling distribution, to measure how much the respective statistic varies from sample to sample. (A *statistic* is a quantity which is calculated from a sample.) We could also check how close each of these standard deviations is to the corresponding *standard error*, that is, to the corresponding standard deviation which we should *expect* according to appropriate theory.

Our entire discussion of sampling distributions was motivated by the three questions asked on page 173. Although we have not yet explained in so many words *whether we can really expect a sample mean to be close to the mean of the population* and *how sure we are that the error, which we make when using x̄ as an estimate of μ, does not exceed, say, 1, or 2,* we now have the necessary tools to give suitable answers. Reserving this task for the following chapter, let us conclude our discussion of sampling distributions with the following observation: *In*

---

† The figure given is actually $\sqrt{\dfrac{\pi}{2}}$, or approximately 1.25.

actual applied problems we do not have 50 samples or 100 samples, we have only 1. Hence, we shall have to use what we have learned in this section to justify generalizations based on 1 sample mean, *or* 1 sample median, *or* 1 sample standard deviation, ... In practical situations we cannot repeat experiments 50 times to see how much a mean or some other statistic varies from sample to sample; we did so in this chapter only to give a concrete illustration of a sampling distribution.

EXERCISES

1. Select the name of a plumber from the yellow pages of the local telephone directory by means of a table of random numbers.

2. Assume that a table of random numbers is used first to pick a page in the local phone directory and then to select a name on that page. Indicate why such a procedure will not yield a strictly random selection.

3. If we take the 50 samples given on page 174 and combine Samples 1 to 5, Samples 6 to 10, and so on, we obtain 10 samples of size 25 whose means are

$$68.12, \quad 67.28, \quad 69.00, \quad 67.84, \quad 67.12,$$
$$67.68, \quad 69.48, \quad 67.76, \quad 67.76, \quad 68.28$$

Use formula (4.4.4) to find the standard deviation of these 10 means and compare it with the value which we should *expect* according to formula (8.2.1).

4. Take 100 milk-bottle tops, metal-rim tags, or other small symmetrical objects and record on them the numbers from 1 to 11 with the following frequencies:

Number to Be Written on Tag	Frequency
1	1
2	3
3	7
4	12
5	17
6	20
7	17
8	12
9	7
10	3
11	1

This provides a population whose mean and standard deviation are $\mu = 6$ and $\sigma = 2$, and which can be approximated closely with a normal curve. Take 50 samples of size 4, replacing each tag before the next one is drawn, and calculate the mean as well as the median of each of these 50 samples. Then use formula (4.4.4) to calculate the standard deviations of the 50 means and the 50 medians and compare them with the values which we should *expect* according to Theorems 8.1 and 8.3. (*Hint:* The quantities to be substituted into the formulas for the standard errors are $\sigma = 2$ and $n = 4$.)

5. Repeat the experiment of Exercise 4 with the use of random numbers (those given on page 171) by using the following scheme:

Number to Be Recorded	Random Numbers
1	00
2	01 to 03
3	04 to 10
4	11 to 22
5	23 to 39
6	40 to 59
7	60 to 76
8	77 to 88
9	89 to 95
10	96 to 98
11	99

For instance, if we obtained the random numbers 23, 74, 41, and 81, we would record the numbers 5, 7, 6, and 8; if we obtained the random numbers 42, 97, 09, and 53, we would record the numbers 6, 10, 3, and 6.

6. What happens to the standard error of the mean if we increase the size of our sample ninefold?

7. What happens to the standard error of the mean if we change our sample size from 16 to 400?

8. Random numbers can also be used to simulate an experiment consisting of the tosses of a certain number of coins. Letting the digits 0, 2, 4, 6, and 8 stand for *heads* and the digits 1, 3, 5, 7, and 9 stand for *tails*, use the random numbers given on page 171 to simulate an experiment consisting of 160 tosses of 4 balanced coins. Draw superimposed histograms of the experimental sampling distribution and the expected sampling distribution which was obtained earlier in Exercise 1 on page 130.

9. Verify that a *median* based on a sample of size 100 and a *mean* based on a sample of size 64 are about *equally reliable* estimates of μ, the mean of the population from which these samples are obtained.

10. A random sample of size 81 is to be taken from a population whose mean and standard deviations are $\mu = 79.8$ and $\sigma = 5.4$. Use Theorems 8.1 and 8.2 to find the probability that the mean of this sample will lie between 80.7 and 82.1. (Treat the variable under consideration as a continuous variable; in other words, find the area under a normal curve between 80.7 and 82.1.)

11. A random sample of size 75 is to be taken from a population whose mean and standard deviation are $\mu = 112$ and $\sigma = 10$. Use Theo-

rems 8.1 and 8.2 to find the probability that the mean of this sample will lie between 110.5 and 113.5. (Treat the variable under consideration as a continuous variable, that is, find normal curve area between 110.5 and 113.5.)

12. With reference to the sample discussed in Exercise 11, find the probability that its *median* will lie between 110.5 and 113.5.

BIBLIOGRAPHY

The following are some of the most widely used tables of random numbers:

Interstate Commerce Commission, Bureau of Transport Economics and Statistics, *Table of 105,000 Random Decimal Digits*. Washington, D.C.: Government Printing Office, 1949.

Kendall, M. G., and Smith, B. B., *Tables of Random Numbers Tracts for Computers No. XXIV*. Cambridge: Cambridge University Press, 1939.

Owen, D. B., *Handbook of Statistical Tables*. Reading, Mass.: Addison-Wesley, 1962.

The Rand Corporation, *A Million Random Digits with 100,000 Normal Deviates*. Glencoe, Ill.: Free Press, 1955.

Tippett, L. H. C., *Random Sampling Numbers, Tracts for Computers No. XV*. London: Cambridge University Press, 1927.

A proof of Theorems 8.1 and 8.2 may be found in most textbooks on mathematical statistics, for instance, in

Hoel, P. G., *Introduction to Mathematical Statistics*, 3rd ed. New York: Wiley and Sons, 1962, Chap. 6.

A proof of Theorem 8.3 is given in

Kendall, M. G., and Stuart, A., *The Advanced Theory of Statistics*, 2nd ed., *Vol. I*. New York: Hafner Publishing Co., 1963, p. 237.

CHAPTER 9

Problems of Estimation

9.1 Introduction

Many problems of statistics deal with the estimation of unknown quantities such as means, standard deviations, and percentages. To mention a few examples, a chamber of commerce official may wish to *estimate* the average amount of money spent by a tourist during a week's stay in a resort; a television producer may wish to *estimate* what proportion of the viewing audience is dialed to his show; a manufacturer of tires may wish to *estimate* how much variability there is in his product; an efficiency expert may wish to *estimate* the average time a housewife takes to iron a shirt; and a quality control engineer may wish to *estimate* what proportion of a large shipment of electronic equipment has imperfections.

To give a concrete example, let us consider a time and motion study in which we are interested in estimating the average time that it takes a secretary to execute a certain order form which goes from one department to another. Let us suppose that an experiment in which a random sample of 40 secretaries were timed in the performance of this task yielded the following results (in seconds):

46.2	61.9	52.5	57.3	51.8	38.0	53.7	56.1	65.4	48.5
51.6	43.0	47.8	60.5	71.1	62.3	56.6	52.5	43.9	52.0
58.1	66.5	33.9	42.7	46.4	53.8	61.2	55.3	48.5	42.9
40.7	52.4	46.6	55.6	58.3	50.4	63.8	35.0	49.5	53.2

The mean of this sample is $\bar{x} = 52.2$ seconds and, in the absence of other information, we shall use it as an estimate of μ, the actual average time that it takes a secretary to execute the given form.

We are now in the same position in which we were on page 173 when we estimated the average height of a large group of persons by using the mean of a random sample of size 5. In order to avoid questions like those asked on page 173, namely, questions about the possible accuracy of our estimate, let us now see how we might formulate our results so that such questions are automatically answered. To consider a few possibilities, let us take a look at the following alternatives:

Alternative 1: The average time that it takes a secretary to execute the given form is estimated as 52.2 seconds. This figure is the mean of a random sample of 40 observations.

Alternative 2: The average time that it takes a secretary to execute the given form is estimated as 52.2 seconds. This figure is the mean of a random sample of 40 observations whose standard deviation is 8.52 seconds.

Alternative 3: The average time that it takes a secretary to execute the given form is estimated as 52.2 seconds. We are 95 per cent sure (we can assert with a probability of 0.95) that the error of this estimate is less than 2.64 seconds.

Alternative 4: We are 95 per cent sure (we can assert with a probability of 0.95) that the interval from 49.56 to 54.84 seconds contains the actual average time that it takes a secretary to execute the given form.

Inasmuch as it mentions the size of the sample, Alternative 1 presents an improvement over merely stating that the estimate is 52.2. Although it is true that we are apt to have more confidence in an estimate which is based on 40 observations than in an estimate which is based on 5 or 10, knowledge of the sample size alone will not enable us to answer questions about the accuracy of an estimate or a method of estimation.

Alternative 2 gives all the information needed to answer the first two questions raised on page 173, but it has the disadvantage of requiring a knowledge of statistics to translate the given values of n and s into an evaluation of the accuracy of our estimate.

Alternative 3 gives the estimate together with an appraisal of its accuracy. It presents the generalization in terms that are easy to understand and in a form that is easy to use for further calculations. The only question that might arise is what we mean by "being 95 per cent sure." (Of course, we did not explain how we arrived at the figure 2.64; this will be taken care of later on.)

So far we have estimated the average time that it takes a secretary to execute the given form as 52.2 seconds. In the language of statistics, this figure is referred to as a *point estimate;* it consists of a single number or, in other words, a single point of the scale. The disadvantage of a point estimate is that we would have to be very optimistic to suppose that such an estimate actually *equals* the quantity which we want to estimate. As a matter of fact, we saw in Chapter 8 that most of our sample means were *not* exactly equal to μ and this is why a point estimate must always be supplemented with a statement about its accuracy—a statement about the possible size of our error.

Since we cannot expect each sample mean to equal the population mean which we want to estimate, it would seem reasonable to give ourselves some leeway by estimating μ to lie on some *interval.* Whereas we are *practically certain* in our example that the mean of the population is not exactly 52.2, we can assert with a suitable probability that it is contained in an interval, say, the interval from 49.56 to 54.84. Such an interval is referred to as an *interval estimate.*

In Alternative 4 we gave an interval estimate for the average time that it takes a secretary to execute the given form. What remains to be explained is how such an interval is obtained and what we mean by "being 95 per cent sure" that the interval covers the true mean. *After all, it either does or does not contain the quantity we are trying to estimate.*

The foregoing discussion is by no means limited to the estimation of means. We can, similarly, use point estimates and interval estimates to estimate a population standard deviation, a proportion, or in general any parameter (description) of a population. For instance, if we wanted to estimate the percentage of car owners who will buy new cars within a year's time, we could base our estimate on a random sample in which, say, 100 car owners said that they would buy new cars while 400 said that they would not. We could then give the *point estimate* that 20 per cent of all car owners will buy new cars within a year's time, or we could give the *interval estimate* that, say, anywhere from 16 to 24 per cent of all car owners will buy new cars within a year's time.

9.2 The Estimation of Means (Large Samples)

Let us now demonstrate how we calculated the maximum error of 2.64 mentioned in Alternative 3 and the interval from 49.56 to 54.84 mentioned in Alternative 4 (see page 187). According to Theorem 8.1 we know that *if $\bar{x}$ is the mean of a random sample of size n, the theoretical sampling distribution of $\bar{x}$ has the mean μ and the standard deviation $\sigma_{\bar{x}} = \sigma/\sqrt{n}$, where μ and σ are the mean and standard deviation of the population from which the sample was obtained.*† According to Theorem

---

† As was pointed out on page 177, the formula $\sigma_{\bar{x}} = \sigma/\sqrt{n}$ applies only for samples taken from very large populations, so the sample constitutes but a small

8.2 we know that *if n is large, the theoretical sampling distribution of x̄ can be approximated closely with a normal curve.*

Using Theorem 8.2, let us suppose that the normal curve of Figure 9.1 represents a theoretical sampling distribution of x̄ and that the indicated z is such that 95 per cent of the area under the curve lies between $-z$ and z. As can easily be checked in Table I on page 501, this z value is 1.96. Since the theoretical sampling distribution of x̄ provides us with the probabilities of obtaining x̄'s that lie in any given

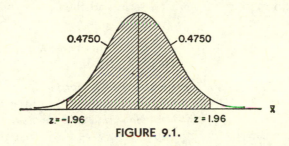

0.4750   0.4750

z = -1.96   z = 1.96

FIGURE 9.1.

interval, we can now say that if a sample mean is converted into *standard units*, the probability that its z value will lie between -1.96 and 1.96 is 0.95.

As was explained on page 94, a measurement is converted into standard units by subtracting the mean of the distribution to which it belongs and then dividing by its standard deviation. Since we are talking about the theoretical sampling distribution of x̄, whose mean and standard deviation are μ and $\sigma/\sqrt{n}$, the formula for z becomes

$$z = \frac{\bar{x} - \mu}{\sigma/\sqrt{n}} \qquad (9.2.1)$$

In view of what we said above, we can now assert with a probability of 0.95 that if x̄ is the mean of a random sample of size n (and n is large), $z = \dfrac{\bar{x} - \mu}{\sigma/\sqrt{n}}$ will lie between -1.96 and 1.96. *In other words*

fraction of the population. *The methods which we shall discuss in this section apply, therefore, only when n constitutes but a small portion of the population.* When n constitutes an appreciable portion of the population, say 5 per cent or more, these methods will have to be modified as indicated in Section 9.6.

*we can assert with a probability of 0.95 that the following inequality will
hold:*

$$-1.96 < \frac{\bar{x} - \mu}{\sigma/\sqrt{n}} < 1.96 \qquad (9.2.2)$$

Multiplying each term by $\sigma/\sqrt{n}$, (9.2.2) becomes

$$- 1.96 \frac{\sigma}{\sqrt{n}} < \bar{x} - \mu < 1.96 \frac{\sigma}{\sqrt{n}} \qquad (9.2.3)$$

and after subtracting $\bar{x}$ from each term and then multiplying by -1,
we finally get

$$\bar{x} - 1.96 \frac{\sigma}{\sqrt{n}} < \mu < \bar{x} + 1.96 \frac{\sigma}{\sqrt{n}} \qquad (9.2.4)$$

*We can now claim with a probability of 0.95 that the interval from $\bar{x} -$
$(1.96\sigma/\sqrt{n})$ to $\bar{x} + (1.96\sigma/\sqrt{n})$ contains μ.* The interval given in
(9.2.4) is called a *95 per cent confidence interval.* We are 95 per cent
confident (we can assert with a probability of 0.95) that this interval
contains the mean of the population.

The result which we have obtained has the unfortunate feature that
σ must be known in order to calculate the confidence interval given in
(9.2.4). Since this is not the case in most practical problems in which
we want to estimate a mean, we shall replace σ with an *estimate,*
namely, with the sample standard deviation s. Since sample standard
deviations cannot be expected to provide close estimates of σ unless
n is reasonably large, we shall make this substitution only for *large
samples.* It is customary to refer to a sample as *"large"* if n is 30 or
more and as *"small"* if n is less than 30.

Substituting s for σ, we can now write the *95 per cent confidence
interval for* μ as

$$\bar{x} - 1.96 \frac{s}{\sqrt{n}} < \mu < \bar{x} + 1.96 \frac{s}{\sqrt{n}} \qquad (9.2.5)\star$$

To supplement our vocabulary we might add that $\bar{x} - (1.96s/\sqrt{n})$
is referred to as the *lower confidence limit,* $\bar{x} + (1.96\ s/\sqrt{n})$ as the
upper confidence limit, and 0.95 as the *confidence coefficient.*

Returning now to the time and motion example described on page
186, we find that $n = 40$, $\bar{x} = 52.2$, and (as can easily be checked)
$s = 8.52$. Substituting these values into (9.2.5), we get

$$52.2 - 1.96 \frac{8.52}{\sqrt{40}} < \mu < 52.2 + 1.96 \frac{8.52}{\sqrt{40}}$$

or $49.56 < \mu < 54.84$

and we can assert with a probability of 0.95 that the interval from 49.56 to 54.84 seconds contains the true average time that it takes a secretary to execute the given form. *This is precisely the interval which we gave in Alternative 4.*

The fact that the confidence interval of (9.2.5) can be asserted with a probability of 0.95 is to be interpreted as follows: *In a given problem, the quantity which we want to estimate either does or does not lie in the interval which we calculate according to (9.2.5). However, if we calculate 95 per cent confidence intervals in many different problems, our intervals will in the long run "do their job" 95 per cent of the time. In other words, in the long run 95 per cent of the confidence intervals calculated according to (9.2.5) will contain the means we are trying to estimate.*

If someone were to ask us whether we are *certain* that the interval from 49.56 to 54.84 seconds contains the true average time that it takes a secretary to execute the given form, our answer would, of course, have to be "No." However, in view of the fact that we are using a method which, in the long run, works 95 per cent of the time, we should be willing to give pretty good odds that it does. As a matter of fact, *fair* odds would be 19 to 1 (95 to 5) that the interval does contain the true mean.

In the preceding discussion we employed a confidence coefficient of 0.95. Since there are situations in which many persons would be reluctant to use methods which, in the long run, work only 95 per cent of the time, let us now modify the confidence interval of (9.2.5) so that it applies also to other degrees of confidence.

The reader will recall that the figure 1.96 was obtained by looking for a z value which is such that 95 per cent of the area under the normal curve lies between $-z$ and z. If we want to change the confidence coefficient in (9.2.5) to 0.98 (or 0.99), we only have to substitute for 1.96 a z value which is such that 98 per cent (or 99 per cent) of the area under the normal curve lies between $-z$ and z. As can easily be checked in Table I, 98 per cent of the area lies between $z = -2.33$ and 2.33, while 99 per cent of the area lies between $z = -2.58$ and $z = 2.58$. Substituting these values for 1.96 into (9.2.5), we can write the corresponding *98 per cent confidence interval for μ* as

$$\bar{x} - 2.33\,\frac{s}{\sqrt{n}} < \mu < \bar{x} + 2.33\,\frac{s}{\sqrt{n}} \qquad (9.2.6)\star$$

and the corresponding 99 per cent confidence interval for μ as

$$\bar{x} - 2.58\,\frac{s}{\sqrt{n}} < \mu < \bar{x} + 2.58\,\frac{s}{\sqrt{n}} \qquad (9.2.7)\star$$

Had we wanted to calculate a 98 per cent confidence interval in the time and motion example, substitution into (9.2.6) would have yielded

$$52.2 - 2.33 \frac{8.52}{\sqrt{40}} < \mu < 52.2 + 2.33 \frac{8.52}{\sqrt{40}}$$

or $$49.06 < \mu < 55.34$$

We could, thus, have asserted with a probability of 0.98 that the interval from 49.06 to 55.34 seconds contains the true average time that it takes a secretary to execute the given form. *This illustrates the fact that "the surer we want to be, the less we have to be sure of." If we increase the degree of certainty, the confidence interval becomes wider and tells us correspondingly less about the quantity which we want to estimate.*

If $\bar{x}$ is used as a *point estimate* of μ, our *error* is given by their difference, namely, $\bar{x} - \mu$. Since this difference is the middle term of (9.2.3), we can rewrite this inequality as

$$-1.96 \frac{\sigma}{\sqrt{n}} < \text{error} < 1.96 \frac{\sigma}{\sqrt{n}} \qquad (9.2.8)$$

This means that if $\bar{x}$ is used as an estimate of μ, we can assert with a probability of 0.95 that our error lies between $-1.96\sigma/\sqrt{n}$ and $1.96\sigma/\sqrt{n}$, or, in other words, that we are "off" by less than $1.96\sigma/\sqrt{n}$.

Substituting again s for σ and limiting ourselves to *large* samples, we can formulate this result as follows:

THEOREM 9.1: *If $\bar{x}$, the mean of a random sample of size $n \geqslant 30$, is used as an estimate of μ, we can assert with a probability of 0.95 that the size of our error is less than $1.96s/\sqrt{n}$.*

(It is understood, of course, that the μ referred to in this theorem is the mean of the population from which the sample is obtained.)

Returning again to the time and motion example of page 186, let us now substitute $n = 40$ and $s = 8.52$ into the formula of Theorem 9.1.   Getting

$$1.96 \frac{s}{\sqrt{n}} = 1.96 \frac{8.52}{\sqrt{40}} = 2.64$$

we can now claim with a probability of 0.95 that if we estimate the average time that it takes a secretary to execute the given form as 52.2 seconds, the error of this estimate is less than 2.64 seconds. *This is precisely what we said in Alternative 3 on page* 187.

Using the same argument as on page 191, we can change the probability in Theorem 9.1 to 0.98 or 0.99 by substituting 2.33 or 2.58 for

1.96.   In the time and motion example we can thus assert with a probability of 0.98 that our estimate of 52.2 seconds is "off" by less than

$$2.33 \frac{s}{\sqrt{n}} = 2.33 \frac{8.52}{\sqrt{40}} = 3.14 \text{ seconds}$$

or with a probability of 0.99 that it is "off" by less than

$$2.58 \frac{s}{\sqrt{n}} = 2.58 \frac{8.52}{\sqrt{40}} = 3.48 \text{ seconds}$$

Since in a given problem the difference between $\bar{x}$ and μ must be either *less than* $1.96s/\sqrt{n}$ or else it must be $1.96s/\sqrt{n}$ *or more*, let us explain briefly what we mean when we say that we are 95 per cent sure that our error is less than a certain quantity.   *If in various problems in which $\bar{x}$'s are used as estimates of μ's we claim that our error is less than $1.96s/\sqrt{n}$, we will in the long run be right 95 per cent of the time.* Sometimes our error will exceed the calculated value of $1.96s/\sqrt{n}$, sometimes it will be less, but *in the long run* our error will be *less than* $1.96s/\sqrt{n}$ ninety-five per cent of the time.

The method which we discussed can also be used to determine the sample size needed to attain a desired degree of accuracy.   To give an illustration, let us suppose that we want to estimate the current average annual income of college graduates who graduated 10 years ago.   Let us suppose, furthermore, that we want to be able to assert with a probability of 0.95 that our estimate will be within $100 of the correct value.   The question is, *how large a sample do we need to attain this degree of accuracy?*

According to what we said on page 192, we can assert with a probability of 0.95 that if $\bar{x}$ is used as an estimate of μ, our error will be less than $1.96\sigma/\sqrt{n}$.   Since this quantity is supposed to equal $100, we can write

$$1.96 \frac{\sigma}{\sqrt{n}} = 100 \tag{9.2.9}$$

Unfortunately, (9.2.9) involves σ, the population standard deviation which is usually unknown.   In order to use it, we shall thus have to make some assumption about the possible value of σ or, perhaps, base an estimate of σ on previous studies of a similar nature.   Assuming that in our example σ can be expected to be in the neighborhood of $1000, substitution into (9.2.9) yields

$$1.96 \frac{1000}{\sqrt{n}} = 100$$

Solving for n, we get 384.2, and this means that *a random sample of size 385 will suffice to give us the desired degree of accuracy.*

To formulate this problem in a more general fashion, let us suppose that we want to use the mean of a random sample as an estimate of the mean of a population *and* that we want to be able to assert with a probability of 0.95 that our error will be less than some quantity E. Proceeding as in the numerical illustration, we can write

$$1.96 \frac{\sigma}{\sqrt{n}} = E$$

and upon solving for n this becomes

$$n = \left[\frac{1.96 \cdot \sigma}{E} \right]^2 \qquad\qquad (9.2.10)\star$$

As we pointed out earlier, this formula cannot be used unless we have some idea about the possible size of σ.

If in the above we wanted to be 98 or 99 per cent sure that our error is less than E, we would only have to modify (9.2.10) by substituting 2.33 or 2.58 for 1.96. Had we wanted to be 99 per cent sure that our error is less than \$100 in the example dealing with the incomes of college graduates, we would have had

$$n = \left[\frac{2.58 \cdot 1000}{100} \right]^2$$

Solving for n, we find that a sample of size 666 would have sufficed to attain this higher degree of confidence.

EXERCISES

1. A random sample of 81 tractor-trailer trucks traveling on the Massachusetts Turnpike between Boston, Massachusetts, and Worcester, Massachusetts, has an average gross weight of 48,000 pounds with a standard deviation of 3,600 pounds.
   (a) Find a 95 per cent confidence interval for the average gross weight of all tractor-trailer trucks traveling this route.
   (b) What can we assert with a probability of 0.95 about the possible size of our error if we estimate the average gross weight of all these trucks as 48,000 pounds?
2. A random sample of 36 tires produced by a certain firm lasted an average of 25,000 miles with a standard deviation of 2,000 miles.
   (a) Construct a 98 per cent confidence interval for the average lifetime of all tires manufactured by the given firm.
   (b) What can we assert with a probability of 0.95 about the possible size of our error if we estimate the average lifetime of all these tires as 25,000 miles?

3. A random sample of 49 delinquent accounts at a certain telephone company has an average size of $28.00 with a standard deviation of $3.20.
   (a) Construct a 99 per cent confidence interval for the average size of all delinquent accounts at the telephone company.
   (b) What can we assert with a probability of 0.98 about the possible size of our error if we estimate this average as $28.00?

4. The I.Q.'s of a random sample of 225 college students enrolled at a certain college have a mean of 110 and a standard deviation of 12.5.
   (a) Find a 95 per cent confidence interval for the average I.Q. of all the students in this college.
   (b) What can we assert with a probability of 0.99 about the possible size of our error if we estimate this average I.Q. as 110?

5. In a study conducted to determine the average size of the outstanding accounts of a wholesale plumbing dealer, a random sample of 36 accounts had an average size of $847 with a standard deviation of $53. If we estimate the average size of *all* the accounts from which this sample was obtained as $847, with what probability can we assert that our error is less than $14? (*Hint:* Find a z value which is such that $z \cdot 53/\sqrt{36} = 14$ and then find the normal curve area between $-z$ and z.)

6. With what probability can we assert in Exercise 2 that our estimate of the average lifetime of these tires will be within 350 miles of the correct value? (*Hint:* Proceed as in Exercise 5.)

7. If we wanted to determine the average clerical aptitude of a large group of people (as measured by a certain test), how large a random sample would we need to be able to assert with a probability of 0.95 that our sample mean will be within 2.5 points of the true mean? Assume that it is known from previous studies that $\sigma = 15$.

8. What sample size would be needed in Exercise 7 if we wanted to be 99 per cent sure that our error will be less than 2.5?

9. An efficiency expert wants to determine the average time that it takes a worker to assemble the parts of an electric lamp. How large a sample will he need to be able to assert with a probability of .95 that his sample mean will differ from the true mean by less than 30 seconds? Assume that it is known from other studies that measurements of this kind can be expected to have a standard deviation of $\sigma = 100$ seconds.

10. What sample size would the efficiency expert of Exercise 9 have to take if he wanted to be 99 per cent sure that his sample mean will be off by less than 3 seconds?

9.3 The Estimation of Means (Small Samples)

In Section 9.2 we repeatedly stressed the fact that the methods which we introduced are to be used only for *large* samples, that is, when n is 30 or more. To develop corresponding theory for *small samples*, let us now consider the sampling distribution of the statistic

$$t = \frac{\bar{x} - \mu}{s} \cdot \sqrt{n} \qquad (9.3.1)\star$$

where $\bar{x}$ and s are the mean and standard deviation of a random sample of size n from a population which has the mean μ, the standard deviation σ, and which can be approximated closely with a normal curve.

If we wanted to, we could calculate $\bar{x}$ and s for each of the 50 samples listed on page 174 and, using the fact that $\mu = 68$, determine the corresponding values of t. These 50 t's could then be grouped into an *experimental sampling distribution*. The problem of finding the corresponding *theoretical sampling distribution* was first investigated by W. S. Gosset. In a paper published in 1908 he derived the equation

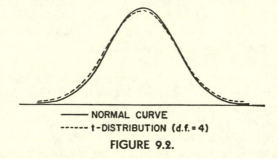

——— NORMAL CURVE
------ t-DISTRIBUTION (d.f. = 4)

FIGURE 9.2.

of this theoretical distribution which is nowadays referred to as the *Student-t distribution*. At the time, Gosset was employed by a well-known Irish brewery which did not permit the publication of research done by its staff. He chose the pen name "Student," and hence the name "Student-t distribution."

The Student-t distribution, an illustration of which is shown in Figure 9.2, is a *symmetrical* distribution and, as in the case of the normal curve, it will suffice to tabulate its area for positive values of t. Unfortunately, there is the complication that the equation of the Student-t distribution depends on the size of the sample or, better, on a quantity referred to as "*the number of degrees of freedom*." As was explained on page 98, the quantity $n - 1$ by which we divide in the formula for s is referred to as the *number of degrees of freedom* since $n - 1$ of the deviations $x_i - \bar{x}$ automatically determine the nth.†

---

† At this stage, it is difficult to explain why one should want to assign a special name to $n - 1$, which, after all, is only the *sample size minus 1*. As we shall see in Section 10.7, there are other applications of the Student-t distribution in which the number of degrees of freedom is defined in a slightly different way.

Since the formula of the Student-t distribution and, hence, the areas under the curve depend on the number of degrees of freedom, it would be impractical to tabulate this theoretical distribution as we tabulated the normal curve. Instead of giving a complete table for each value of $n - 1$, we shall list only the values of t above which we find 10, 5, $2\frac{1}{2}$, 1, and $\frac{1}{2}$ per cent of the area under the curve. Symbolically, we shall write $t_{.025}$ for the value of t which is exceeded by 2.5 per cent of the area under the curve, $t_{.01}$ for the value of t which is exceeded by 1 per cent of the area under the curve, and so forth. Table II, on page 502, contains the values of $t_{.10}$, $t_{.05}$, $t_{.025}$, $t_{.01}$, and $t_{.005}$ for $d.f.$ (the number of degrees of freedom) going from 1 to 29.

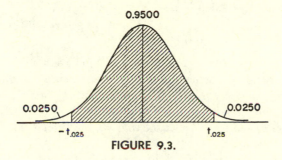

FIGURE 9.3.

Since 95 per cent of the area under the Student-t distribution lies between $-t_{.025}$ and $t_{.025}$ (see Figure 9.3), we can assert with a probability of 0.95 that any t, calculated according to (9.3.1) on the basis of a random sample, satisfies the inequality

$$-t_{.025} < t < t_{.025} \tag{9.3.2}$$

Substituting into this inequality $t = \dfrac{\bar{x} - \mu}{s} \cdot \sqrt{n}$, we get

$$-t_{.025} < \frac{\bar{x} - \mu}{s} \sqrt{n} < t_{.025}$$

which can be rewritten as

$$\bar{x} - t_{.025} \frac{s}{\sqrt{n}} < \mu < \bar{x} + t_{.025} \frac{s}{\sqrt{n}} \tag{9.3.3}\star$$

We now have a 95 per cent confidence interval for μ to be used when n is less than 30, that is, for small samples. The appropriate value of $t_{.025}$ is to be looked up in Table II with d.f. (the number of degrees of freedom) equal to $n - 1$.

Comparing (9.3.3) with (9.2.5), the corresponding 95 per cent confidence interval for *large* samples, we find that the only difference is that $t_{.025}$ takes the place of 1.96. As a matter of fact, we can similarly construct small sample formulas for 98 and 99 per cent confidence intervals by replacing 2.33 and 2.58 with $t_{.01}$ and $t_{.005}$, respectively.

To illustrate the use of (9.3.3), let us take the experiment referred to on page 59, in which we wanted to estimate the average height of all cakes baked with a new kind of mix. Taking the results obtained by Laboratory X, whose cakes had heights of 1.70, 2.40, and 2.50 inches, it can easily be shown that in this sample $\bar{x} = 2.20$ and $s = 0.44$ inches. Since $n = 3$, the number of degrees of freedom is $3 - 1 = 2$, and according to Table II, the appropriate value of $t_{.025}$ is 4.303. Substituting all these values into (9.3.3) we get

$$2.20 - 4.303 \frac{0.44}{\sqrt{3}} < \mu < 2.20 + 4.303 \frac{0.44}{\sqrt{3}}$$

or
$$1.11 < \mu < 3.29$$

We can thus assert with a probability of 0.95 that the interval from 1.11 to 3.29 inches contains the true average height of all cakes that are baked with the new mix.† *This illustrates the fact that although we can make logically correct inferences on the basis of very small samples, our results are apt to be rather vague, that is, our confidence intervals are apt to be rather wide.*

If the mean of a *small* random sample is used as a point estimate of μ, we can refer to the Student-t distribution to appraise the possible size of our error. The only change needed in Theorem 9.1 is that 1.96 must be replaced with $t_{.025}$. (If we wanted to use probabilities of 0.98 or 0.99, we would similarly substitute $t_{.01}$ or $t_{.005}$ for 2.33 or 2.58.)

To consider an illustration, let us take the sample referred to on page 78, which consisted of the following measurements of compressive strength:

$$41{,}000 \quad 44{,}250 \quad 69{,}400 \quad 70{,}000 \quad 80{,}350 \text{ psi.}$$

The mean and standard deviation of this sample are $\bar{x} = 61{,}000$ and $s = 17{,}350$, and since for $5 - 1 = 4$ degrees of freedom $t_{.025} = 2.776$, we find that

$$t_{.025} \frac{s}{\sqrt{n}} = 2.776 \frac{17{,}350}{\sqrt{5}} = 21{,}540 \text{ psi.}$$

If we estimate the average strength of *all* the steel from which this sample was obtained as being 61,000 psi, we can assert with a probability of 0.95 that our error is less than 21,540 psi.

---

† We are tacitly assuming that the sample is random and that, furthermore, the population consisting of the heights of all cakes baked with the new mix can be approximated closely with a normal curve.

EXERCISES

1. Construct a 95 per cent confidence interval for the mean of the population from which each of the following random samples was obtained:
   (a) $n = 25$, $\bar{x} = 14.5$ lbs., $s = 2.5$ oz.
   (b) $n = 11$, $\bar{x} = 250.3$ gals., $s = 32.1$ gals.
   (c) $n = 5$, $\bar{x} = \$1,500.00$, $s = \$124.30$.

2. Construct a 98 per cent confidence interval for μ for each of the samples of Exercise 1.

3. Construct a 99 per cent confidence interval for μ for each of the samples of Exercise 1.

4. A random sample of 10 tires is taken from a large shipment of super-Brand X tires. The sample indicated that the tires wore out on the average at 42,500 miles and had a standard deviation of 2,850 miles. Construct a 95 per cent confidence interval for the average life of the entire shipment of tires.

5. What can we say with a probability of 0.98 about the possible size of our error, if in Exercise 4 we estimate the average lifetime of all tires in the shipment as being 42,500 miles?

6. A random sample of 20 delegates spent an average of $247.69 (with a standard deviation of $21.73) while attending a national political convention. Find a 98 per cent confidence interval for the average amount spent by a delegate attending this convention.

7. What can we assert with a probability of 0.99 about the possible size of our error, if in Exercise 6 we estimate the average amount spent by a delegate as being $247.69?

8. One phase of a housing study conducted by a business bureau in a large city is concerned particularly with the rents charged for two-room unfurnished apartments in a certain section of the city. A random sample of 16 such units showed an average monthly rental of $80.00 with a standard deviation of $12.00. Find a 95 per cent confidence interval for the true average rental charged for two-room unfurnished apartments in this section of the city.

9. If, in Exercise 8, the bureau estimates the average rental as $80.00, what can they say with a probability of 0.95 about the possible size of the error of this estimate?

9.4 The Estimation of Proportions

Many problems in business and economics deal with the estimation of percentages, proportions, and probabilities. To mention a few examples, we may wish to estimate what *percentage* of American housewives prefer Soap A to Soap B, what *proportion* of vacationists travel by air, or we may wish to estimate the *probability* that a certain kind of business venture will succeed. In principle, these problems are the same since a percentage is merely a proportion multiplied by 100 and a probability is a proportion "in the long run."

The information that is usually available for the estimation of a proportion is the relative frequency with which an appropriate event has occurred. If an event occurs x *times out of* n, the relative frequency of its occurrence is x/n and we can use this *sample proportion* as an estimate of p, the true proportion which we want to estimate. For example, if in a random sample of 500 housewives, 320 prefer Soap A while 180 prefer Soap B, $x/n = 320/500 = 0.64$ and we might claim that $p = 0.64$ or that 64 per cent of *all* housewives prefer Soap A over Soap B.

If p, the proportion which we want to estimate, is not close to 0 or 1 *and* if the population is large, the theoretical sampling distribution

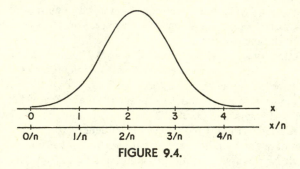

FIGURE 9.4.

of x, the number of successes in n trials, can be approximated closely with a normal curve whose mean and standard deviation are

$$\mu = np \qquad \text{and} \qquad \sigma = \sqrt{np(1-p)} \qquad (9.4.1)$$

This argument is based on the fact that if the population is large, the probability of a success is for all practical purposes equal to p in each trial and the probability of getting x successes in n trials is, thus, given by the binomial distribution.†

If we consider the normal curve of Figure 9.4, which is supposed to represent a sampling distribution of x, the *number of successes* in n trials, it is easy to see that we can convert it into a sampling distribution of the *proportion of successes* by performing the indicated change of scale. (After all, the proportion of successes is simply the number of successes *divided by n*.)

---

† Strictly speaking, the binomial distribution applies only if the population is *infinite* or if we sample *with replacement* so that the probability of a "success" remains the same. If the population is small, we cannot assume that the probability of a success remains constant from trial to trial, it changes the moment one or more items are removed, and it will be necessary to make the modification discussed in Section 9.6.

If we now divide formulas (9.4.1) by n, we find that in the new scale, the x/n scale, the formulas for the mean and standard deviation become

$$\mu = p \quad \text{and} \quad \sigma = \sqrt{\frac{p(1-p)}{n}} \qquad (9.4.2)\star$$

Hence, if p is not too close to 0 or 1 and the population from which we are sampling is large, the theoretical sampling distribution of x/n can be approximated closely with a normal curve having the mean and standard deviation given in (9.4.2). Using the symbolism and terminology of Section 8.2, we shall refer to the standard deviation of this sampling distribution as the *standard error of a proportion* and write it as

$$\sigma_{x/n} = \sqrt{\frac{p(1-p)}{n}} \qquad (9.4.3)\star$$

In order to avoid confusion, let us repeat that $\sqrt{p(1-p)/n}$ measures how much the *proportion of successes* can be expected to vary from sample to sample while $\sqrt{np(1-p)}$ measures how much the *number of successes* can be expected to vary from sample to sample.

Duplicating the argument presented on page 189, we can now claim that if a sample proportion x/n is converted into *standard units*, the probability that its z-value will lie between -1.96 and 1.96 is 0.95. Such a z value is obtained with the formula

$$z = \frac{\frac{x}{n} - p}{\sqrt{\frac{p(1-p)}{n}}} \qquad (9.4.4)\star$$

and we can thus assert with a probability of 0.95 that the following inequality holds:

$$-1.96 < \frac{\frac{x}{n} - p}{\sqrt{\frac{p(1-p)}{n}}} < 1.96 \qquad (9.4.5)$$

Multiplying each term by $\sqrt{p(1-p)/n}$, (9.4.5) can be rewritten as

$$-1.96\sqrt{\frac{p(1-p)}{n}} < \frac{x}{n} - p < 1.96\sqrt{\frac{p(1-p)}{n}} \qquad (9.4.6)$$

If we manipulated the inequality given in (9.4.6) so that the middle term is p and the other two terms consist of expressions that can be calculated without knowledge of p, we would obtain a *95 per cent confidence interval for p.*   Since this would involve a good deal of work, we shall leave it to others, using instead tables constructed specially for this purpose.   Table V on page 506 provides 95 per cent confidence limits of p for random samples of size 8, 10, 12, 16, 20, 24, 30, 40, 60, 100, 200, 400, and 1,000.   For other values of n we will literally have to read between the lines.   (Similar tables for 98 and 99 per cent confidence limits are referred to in the Bibliography of page 211.

It may be seen in Table V that the sample proportions x/n with values from 0.00 to 0.50 are located on the bottom, with the corresponding p values on the left side of the chart.   Sample proportions x/n with values from 0.50 to 1.00 are located on the top, with corresponding p values on the right side of the chart.

To illustrate the use of Table V for a proportion which lies between 0.00 and 0.50, let us suppose that in a random sample of 200 cigarette smokers, 48 state that they prefer Brand A to all other brands of cigarettes, while 152 express a preference for some other brand.   The sample proportion of smokers preferring Brand A is $48/200 = 0.24$, and we mark this value on the *bottom horizontal scale* (the x/n scale) of Table V (see Figure 9.5).   We then go up vertically from this point

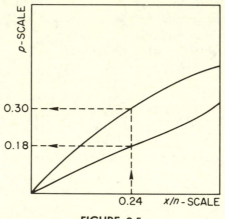

FIGURE 9.5.

until we reach the two curves that are labeled 200.   The two values on the *left vertical scale* that correspond to the points at which we cut these curves, finally, give us the desired confidence limits for p.   In our example these values are approximately 0.18 and 0.30.

To illustrate the use of Table V for a proportion which lies between 0.50 and 1.00, let us suppose that in a random sample of 400 women, 280 preferred Brand X perfume, while 90 preferred Brand Y.   The sample proportion of women preferring Brand X is $280/400 = 0.70$, and we mark this value on the *top horizontal scale* (the x/n scale) of

Table V (see Figure 9.6).   We then go down vertically from this point until we reach the two curves that are labelled 400.   The two values of the *right vertical scale* that correspond to the points at which we cut these curves give us the desired confidence limits for p.   In our example these values are approximately 0.65 and 0.75.

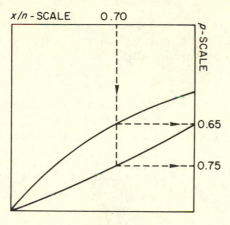

FIGURE 9.6.

In Section 9.2 we *approximated* the standard error of the mean by writing $s/\sqrt{n}$ instead of $\sigma/\sqrt{n}$.   Doing the same thing here, we shall substitute x/n for p and *approximate* (9.4.3), the formula for the standard error of x/n, by writing

$$\sqrt{\frac{\frac{x}{n}\left(1 - \frac{x}{n}\right)}{n}} \quad \text{instead of} \quad \sqrt{\frac{p(1 - p)}{n}}$$

Substituting this approximation into (9.4.6), we get

$$-1.96\sqrt{\frac{\frac{x}{n}\left(1 - \frac{x}{n}\right)}{n}} < \frac{n}{x} - p < 1.96\sqrt{\frac{\frac{x}{n}\left(1 - \frac{x}{n}\right)}{n}} \qquad (9.4.7)$$

$$\text{or} \quad \frac{x}{n} - 1.96\sqrt{\frac{\frac{x}{n}\left(1 - \frac{x}{n}\right)}{n}} < p < \frac{x}{n} + 1.96\sqrt{\frac{\frac{x}{n}\left(1 - \frac{x}{n}\right)}{n}} \qquad (9.4.8)\star$$

and we now have an approximate 95 per cent confidence interval for p. (If we wanted to change the degree of confidence to 98 per cent or 99 per cent, we would only have to substitute 2.33 or 2.58 for 1.96.)

Had we used (9.4.8) in the illustration of 200 cigarette smokers given above, we would have obtained the following confidence interval for the true proportions of smokers preferring Brand A to all other brands of cigarettes:

$$0.24 - 1.96 \sqrt{\frac{(0.24)(0.76)}{200}} < p < 0.24 + 1.96 \sqrt{\frac{(0.24)(0.76)}{200}}$$

$$0.181 < p < 0.299$$

It is interesting to note that this is almost identical with the result obtained previously with the use of Table V.   We leave it as an exercise for the reader to use formula (9.4.8) to obtain the confidence interval for the true proportion of women preferring Brand X perfume to Brand Y.

If a sample proportion is used as a *point estimate* of p, our error is $(x/n) - p$.   Writing "error" instead of $(x/n) - p$ for the middle term of (9.4.7), we arrive at the following theorem:

THEOREM 9.2: *If x/n, the sample proportion observed in a large random sample, is used as an estimate of p, we can assert with a probability of 0.95 that the size of our error is less than*

$$1.96 \sqrt{\frac{\dfrac{x}{n}\left(1 - \dfrac{x}{n}\right)}{n}}$$

To illustrate the use of Theorem 9.2, let us suppose that a random sample of 600 eligible voters included 330 votes for Candidate X and 270 votes for Candidate Y.   If on the basis of this sample we claim that Candidate X will receive $\frac{330}{600} = 0.55$ (or 55 per cent) of the total vote, we will have to add that we can only be 95 per cent sure that the error of this estimate is less than

$$1.96 \sqrt{\frac{(0.55)(0.45)}{600}} = 0.04 \text{ (or 4 per cent)}.$$

According to (9.4.6) we can say with a probability of 0.95 that if x/n is used as an estimate of p, the error will be less than $1.96 \sqrt{p(1 - p)/n}$. Consequently, if we want to be 95 per cent sure that our error is less than some quantity E, we can write

$$1.96 \sqrt{\frac{p(1 - p)}{n}} = E$$

and, solving for n, this becomes

$$n = p(1 - p) \left[\frac{1.96}{E}\right]^2 \qquad (9.4.9)\star$$

Unfortunately, this formula cannot be used directly to determine the sample size needed to attain a desired degree of accuracy; it involves

p, the unknown quantity which we want to estimate.   However, since p must lie between 0 and 1, it can be shown that $p(1 - p)$ is *at most* equal to $\frac{1}{4}$ and instead of (9.4.9) we can write

$$n \leqslant \frac{1}{4}\left[\frac{1.96}{E}\right]^2 \qquad (9.4.10)\star$$

The fact that $p(1 - p)$ cannot exceed $\frac{1}{4}$ may be verified either by completing the square or with elementary differential calculus.   *It is important to note that in this formula E is to be written as a proportion and not as a percentage.*

To illustrate the use of (9.4.10), let us suppose that we want to estimate the proportion of teenagers (in a certain city) who go to the movies at least once a week and that we want to be 95 per cent sure that our estimate will not be off by more than 0.02.   Substituting $E = 0.02$ into (9.4.10), we obtain

$$n \leqslant \frac{1}{4}\left[\frac{1.96}{0.02}\right]^2 = 2401$$

and if we base our estimate on a random sample of size 2401 we can be *at least* 95 per cent sure that our sample proportion will not differ from the true proportion by more than 0.02.   We are "at least" 95 per cent sure because the sample size of 2401 may be larger than necessary to assure the desired degree of accuracy.

In the last example we assumed that we had no idea about the possible value of p and we, therefore, used (9.4.10) instead of (9.4.9). Had we known from past experience (or other sources of information) that the proportion which we want to estimate is in the neighborhood of 0.80, we could have substituted this value into (9.4.9), getting

$$n = (0.80)(0.20)\left[\frac{1.96}{0.02}\right]^2 = 1537$$

This demonstrates that information about the possible value of p may enable us to reduce the required size of the sample.

EXERCISES

1. In a random sample of 250 housewives in a certain city, 90 stated that they own dishwashing machines while 160 stated that they do not.
   (a) Use Table V to construct a 95 per cent confidence interval for the actual proportion of housewives who own dishwashing machines.
   (b) If, on the basis of this sample, we estimated the proportion of housewives owning dishwashing machines as being 0.36, what could we say with a probability of 0.95 about the possible size of our error?

2. A random sample of 200 sheets of plywood (taken from a very large shipment) contains 36 with imperfections.
   (a) Use Table V to construct a 95 per cent confidence interval for the actual proportion of imperfect sheets of plywood in this shipment.
   (b) Use Theorem 9.2 to appraise the possible size of our error if we estimated this proportion as being 0.18. Use a probability of 0.98.
   (c) Use (9.4.8), with 2.58 substituted for 1.96 to construct a 99 per cent confidence interval for the actual proportion of imperfect sheets of plywood in this shipment.

3. In a random sample of 400 families owning color television sets in Cicero, Illinois, it was revealed that 160 sets were then tuned to Network C, broadcasting in black and white.
   (a) Use Table V to construct a 95 per cent confidence interval for the actual proportion of families with color television sets in Cicero, Illinois, who at the given time were tuned to the black and white broadcast of Network C.
   (b) Use (9.4.8), with 2.33 substituted for 1.96, to construct a 98 per cent confidence interval for the desired proportion.

4. In a random sample of 1,000 eligible voters, 580 were for a new issue of municipal bonds while 420 were against it.
   (a) Use Table V to construct a 95 per cent confidence interval for the actual proportion of eligible voters who are for the new issue of bonds.
   (b) If we estimated the proportion of voters who are for the new issue of bonds as being 0.58, what could we say with a probability of 0.95 about the size of the error of this estimate?
   (c) Use (9.4.8) to construct a 95 per cent confidence interval and compare it with the one obtained in (a).

5. In a random sample of 3,200 secretaries, 2,529 expressed a preference for electric typewriters while 671 expressed a preference for manual typewriters. Use (9.4.8) to construct a 95 per cent confidence interval for the proportion of secretaries preferring electric typewriters over manual typewriters.

6. In a random sample of 10 professional athletes, 3 stated that they drank Brand X Coffee, while the other 7 preferred other brands or did not drink coffee. Use Table V to construct a 95 per cent confidence interval for the true proportion of athletes who drink Brand X Coffee. *Note that the sample proportion of 0.30 does not lie in the middle of the confidence interval. This is explained by the fact that the binomial distribution is not symmetrical unless $p = \frac{1}{2}$. If p is not equal to $\frac{1}{2}$, the probability of underestimating p by some quantity is no equal to the probability of overestimating p by the same quantity.*

7. Suppose that we want to estimate the proportion of coeds who get married within 3 years after graduating from college. How large a random sample will we need to be able to assert with a probability of at least 0.95 that our error will be less than 0.05?

8. What is the smallest random sample a public opinion poll will have to take in order to be able to assert with a probability of at least 0.95 that their sample proportion will be "off" by less than 0.03?

9. If we wanted to estimate the proportion of men who prefer French cuffs to buttoned wristbands, how large a sample would we need to be able to assert with a probability of at least 0.98 that our sample proportion will differ from the true proportion by less than 0.04? (*Hint:* Use (9.4.10) with 2.33 substituted for 1.96.)

10. We know from past experience that the proportion of wrist watches shipped by a certain firm which can be expected to be defective is in the neighborhood of 0.04. We want to estimate what proportion of a new large shipment from that firm is defective. Use (9.4.9) to determine the sample size needed to enable us to assert with a probability of 0.95 that our sample proportion will be within 0.005 of the actual proportion.

11. How large a sample would we have to take in Exercise 8 if we knew that the proportion which we want to estimate is in the neighborhood of 0.60?

9.5 Standard Errors and Probable Errors

Although we have discussed here only the construction of confidence intervals for μ and p and the evaluation of the accuracy of point estimates of μ and p, a similar approach can be used in the estimation of other population parameters. By studying the theoretical sampling distribution of appropriate statistics, statisticians have developed formulas giving confidence intervals for population standard deviations, medians, quartiles, coefficients of variation, and the like. In principle, the concepts and methods are always the same, and the main difficulty lies in the fact that some of these theoretical sampling distributions are of a rather complicated nature. *Fortunately, when n is large many of the sampling distributions which we need in actual practice can be approximated closely with normal curves.*

The last sentence implies that if S is some statistic calculated on the basis of a *large* random sample, we can often write a 95 per cent confidence interval for the population parameter which S is supposed to estimate as

$$S - 1.96\sigma_S < population\ parameter < S + 1.96\sigma_S \quad (9.5.1)\star$$

Here σ_S is the standard deviation of the theoretical sampling distribution of S or, in other words, *the standard error of S*. If S happened to be a *sample mean*, σ_S would equal $\sigma_{\bar{x}}$ and (9.5.1) would reduce to (9.2.5); if S happened to be a *sample proportion*, σ_S would equal $\sigma_{x/n}$ and (9.5.1) would reduce to (9.4.6).   Formulas for the standard errors of various other statistics are referred to in the Bibliography on page 211.

When we first introduced standard deviations in Chapter 4, we showed how important it can be to have some knowledge about the standard deviation of a population.   If s, the standard deviation of a large random sample, is used as an estimate of σ, we can write (9.5.1) as

$$s - 1.96 \frac{\sigma}{\sqrt{2n}} < \sigma < s + 1.96 \frac{\sigma}{\sqrt{2n}} \qquad (9.5.2)$$

since, for large samples, the standard error of s may be approximated with the formula

$$\sigma_s = \frac{\sigma}{\sqrt{2n}} \qquad (9.5.3)\star$$

Rewriting (9.5.2) as

$$\frac{s}{1 + \dfrac{1.96}{\sqrt{2n}}} < \sigma < \frac{s}{1 - \dfrac{1.96}{\sqrt{2n}}} \qquad (9.5.4)\star$$

we now have a (large sample) 95 per cent confidence interval for σ.   (Corresponding 98 and 99 per cent confidence intervals may be obtained by replacing 1.96 with 2.33 and 2.58, respectively.)

Referring again to the time and motion example used in Section 9.1, let us substitute $n = 40$ and $s = 8.52$ into (9.5.4).   Getting

$$\frac{8.52}{1 + \dfrac{1.96}{\sqrt{80}}} < \sigma < \frac{8.52}{1 - \dfrac{1.96}{\sqrt{80}}}$$

or $\qquad\qquad 6.98 < \sigma < 10.92$

we can thus assert with a probability of 0.95 that the interval from 6.98 to 10.92 seconds contains σ, the standard deviation of the times secretaries take to execute the given form.

In Section 9.2 we arbitrarily chose 0.95, 0.98, and 0.99 as the probabilities in terms of which we appraised the possible size of our error. Had we wanted to use a probability of 0.50, we would only have had to substitute 0.6745 for 1.96.   (As can be checked in a table of normal curve areas more detailed than Table I, 50 per cent of the area under the normal curve lies between $z = -0.6745$ and $z = 0.6745$.   Our table shows only that this z must lie between 0.67 and 0.68.)   We could then have said that there is a *fifty-fifty chance* that $\bar{x}$ differs from μ by less than

$$0.6745 \cdot \sigma_{\bar{x}} \qquad \text{or} \qquad 0.6745 \frac{\sigma}{\sqrt{n}}$$

and this quantity is referred to as the *probable error of the mean.*

More generally, *the probable error of a statistic S is defined as 0.6745 times its standard error* or, symbolically, as

$$P.E._S = 0.6745 \cdot \sigma_S$$

We thus have, for example,

$$P.E._{\bar{x}} = 0.6745 \frac{\sigma}{\sqrt{n}}$$

$$P.E._{\frac{x}{n}} = 0.6745 \sqrt{\frac{p(1-p)}{n}}$$

$$P.E._s = 0.6745 \frac{\sigma}{\sqrt{2n}}$$

In order to use these formulas we usually substitute s for σ and x/n for p. Nowadays, probable errors are rarely used save for applications in military technology, where they are employed in connection with problems of gunnery and bombardment.

EXERCISES

1. Use the sample of Exercise 1 on page 194 to construct a 95 per cent confidence interval for σ, the standard deviation of the gross weights of all tractor-trailer trucks traveling on the Massachusetts Turnpike between Boston, Massachusetts and Worcester, Massachusetts.
2. Use the data of Exercise 4 on page 195 to construct a 95 per cent confidence interval for σ, the standard deviation of the I.Q.'s of all the college students enrolled at a certain college.

9.6 Sampling from Small Populations

The methods we have discussed in this chapter were based on a number of assumptions. Foremost, we always assumed that our samples were random and that we sampled from *very large* populations. To illustrate the need for the latter assumption, let us suppose that we want to determine the proportion of doctors living in a certain town who smoke cigars. Let us suppose, furthermore, that there are 10 doctors in this town and that among the 9 that can be reached, 3 smoke cigars and 6 do not.

If we used our so-called standard technique, we would find that for $n = 9$ and $x/n = \frac{3}{9}$ Table V yields a 95 per cent confidence interval going from 0.07 to 0.69. Aside from the fact that our sample was not random, it can easily be seen that what we have done here cannot be correct. Clearly, if the 10th doctor smokes cigars, $p = \frac{4}{10} = 0.40$, and if he does not, $p = \frac{3}{10} = 0.30$. Since p must be either 0.30 or

0.40, it would seem rather nonsensical to say that we are 95 per cent sure that p lies between 0.07 and 0.69.

The fact that a sample constitutes an appreciable portion of a population may be accounted for by making a suitable modification in the formula for the standard error of the appropriate statistic. If N stands for the size of the population and n, as always, for the size of the sample, the formulas for $\sigma_{\bar{x}}$ and $\sigma_{x/n}$ may be written as

$$\sigma_{\bar{x}} = \frac{\sigma}{\sqrt{n}} \cdot \sqrt{\frac{N - n}{N - 1}} \qquad (9.6.1)\star$$

and

$$\sigma_{x/n} = \sqrt{\frac{p(1 - p)}{n}} \cdot \sqrt{\frac{N - n}{N - 1}} \qquad (9.6.2)\star$$

It should be noted that these formulas are obtained by multiplying the previous formulas for $\sigma_{\bar{x}}$ and $\sigma_{x/n}$ by the factor

$$\sqrt{\frac{N - n}{N - 1}} \qquad (9.6.3)$$

To demonstrate that this refinement may be omitted when n is small compared to N, let us evaluate the factor $\sqrt{(N - n)/N - 1)}$ for $n = 100$ and $N = 10,000$. In that case

$$\sqrt{\frac{N - n}{N - 1}} = \sqrt{\frac{10,000 - 100}{10,000 - 1}} = 0.995$$

and it is apparent that we might just as well put this factor equal to 1. It would seem that there is no need in this case to make an adjustment for the size of the population. Of course, the adjustment factor should be used if a sample constitutes an appreciable portion of a population, say arbitrarily, 5 per cent or more. If we wanted to estimate the average size of the savings of 400 families on the basis of a random sample of 200, the adjustment factor would be

$$\sqrt{\frac{N - n}{N - 1}} = \sqrt{\frac{400 - 200}{400 - 1}} = 0.708$$

and this would make a difference of almost 30 per cent in the width of the resulting confidence interval or in the appraisal of our error.

It is important to note that the adjustment factor given in (9.6.3) is never greater than 1. This means that it is always "safe" to use the methods which we studied earlier in this chapter, namely, the methods in which we assumed that we are sampling from a very large population. By ignoring the factor

$$\sqrt{\frac{N - n}{N - 1}}$$

we would only be making our confidence intervals wider than necessary and the appraisal of our error larger than necessary. In the example of the cigar smoking doctors it would have been correct, though trivial, to say that we are *at least* 95 per cent sure that p lies between 0.07 and 0.69.

EXERCISES

1. Supposing that there are 800 college students in the college referred to in exercise 4 on page 195, use the modification discussed in this section and the data given in that exercise to construct a 95 per cent confidence interval for the average I.Q. of these 800 college students.
2. A sample of 100 plastic rods selected from a shipment of 2,000 plastic rods has an average diameter of 0.354 inches with a standard deviation of 0.048 inches. Find a 98 per cent confidence interval for the average diameter of these 2,000 plastic rods.
3. Supposing that there are 1,000 sheets of plywood in the shipment referred to in Exercise 2 on page 206, construct a 99 per cent confidence interval for the proportion of imperfect sheets of plywood in this shipment.
4. Supposing that there are 8,000 eligible voters in the town referred to in Exercise 4 on page 206, construct a 99 per cent confidence interval for the actual proportion of eligible voters who favor the new bond issue.

BIBLIOGRAPHY

An informal discussion of the problem of estimation and confidence intervals, given under the heading of "How to be Precise though Vague," may be found in

Moroney, M. J., *Facts from Figures*. London: Penguin Books, 1956, Chap. 14.

For other introductory treatments of the methods discussed in this chapter, see

Neter, J., and Wasserman, W., *Fundamental Statistics for Business and Economics*, 3rd ed. Boston: Allyn and Bacon, Inc., 1966, Chap. 9.
Wilks, S. S., *Elementary Statistical Analysis*. Princeton: Princeton University Press, 1948, Chap. 10.

Formulas for the *standard errors* of medians, quartiles, coefficients of variation, and other statistics may be found in

Waugh, A. E., *Elements of Statistical Method*. New York: McGraw-Hill, 1952, Chap. 9.

Tables for 98 and 99 per cent confidence intervals for p (analogous to Table V) are given in

Pearson, E. S. and Hartley, H. O., *Biometrika Tables for Statisticians*, Vol. I, 3rd ed., Cambridge: University Press, 1966.

CHAPTER 10

Tests of Hypotheses

10.1 Introduction

One of the most frequently used terms in modern statistics is the word "decision," and this is due to the fact that *statistics plays an ever increasing role in the construction and analysis of criteria on which decisions are based.* If we must decide, for instance, whether to invest in real estate or in government bonds, whether to advertise in a newspaper or by television, whether to buy new machinery or repair the old, or whether to open a new branch in Charleston or in Atlanta, we always face the possibility that the alternative we choose will turn out to be the least profitable one. No matter how we decide problems arising in the day-by-day operation of a business, in industrial or economic planning, in science, or in everyday life, we must always face the risk of making a wrong choice. *It is the task of statistics to evaluate such risks and, if possible, to provide criteria which minimize the chances of making wrong decisions.* Excluded from our discussion will be decisions that are merely a matter of taste. There would be no question in that case whether a decision is right or wrong, true or false.

To give an example which is typical inasmuch as it illustrates the basic concepts underlying statistical decision criteria, let us consider a problem faced by a company which sends out thousands of mailings a year in an attempt to sell house dresses, shirts, raincoats, ties, and other apparel, by mail. In the past, their direct mail appeal has yielded a response of 10 per cent and the management of the company is interested in the possibility of changing to a more expensive appeal involving swatches of materials, prizes, and other inducements. Naturally, it is hoped that this more expensive appeal will yield a higher response.

212

Having no information about the potential of the more expensive appeal, it is decided to send out a trial mailing to 1000 persons. *The question is how to interpret the results of this mailing.* Variability of responses from one mailing to the next is well known to the officers of the company and they all agree that anything less than the 100 responses expected with the old appeal will not constitute evidence that the new one is better. To settle the question, an advertising executive suggests that the company decide to *retain the old appeal if the number of responses is less than 120 and to switch to the new appeal if there are 120 or more.*

The above provides a clear-cut criterion for deciding whether to retain the old appeal or to switch to the new one. The only trouble is that this criterion is by no means infallible. It could easily happen,

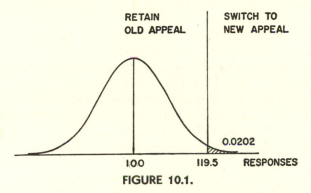

FIGURE 10.1.

purely by chance, that there will be 120 or more responses to the new appeal in spite of the fact that *on the average* the new appeal will yield the same percentage response as the old one. If this happened, the company would waste thousands of dollars on the more expensive appeal without getting anything in return. Before adopting the above criterion, it would seem wise for the management of the company to investigate what the chances are that the criterion will, thus, lead to the wrong decision.

In the language of probability, they will have to determine *the probability of getting "120 or more successes in 1,000 trials" if the probability of an individual success is 0.10.* (If the new appeal is exactly as effective as the old one, the probability that any person will respond is still 0.10.) We shall ignore the possibility that the new appeal might be worse than the old one. As the reader will be able to verify later on, if the new appeal is worse, the probability that the old appeal will be retained anyhow is very high (see Exercise 4 on page 217).

Using the normal curve approximation to the binomial distribution, the desired probability may be found by determining the shaded area of Figure 10.1, namely, the area under the normal curve to the right of 119.5. Since the formulas for the mean and standard deviation of the binomial distribution are

$$\mu = np \quad \text{and} \quad \sigma = \sqrt{np(1 - p)}$$

[see (7.3.3) and (7.3.5)], substituting $n = 1,000$ and $p = 0.10$ yields

$$\mu = 1000(0.10) = 100$$

and

$$\sigma = \sqrt{1000(0.10)(0.90)} = 9.5$$

The z value of 119.5, the dividing line of the criterion, is thus

$$z = \frac{119.5 - 100}{9.5} = 2.05$$

and the corresponding entry in Table I is 0.4798. *Hence, the probability that the criterion will lead to the adoption of the more expensive appeal even though it is only as good as the old one is 0.5000 − 0.4798 = 0.0202 or, roughly, 0.02.*

Whether or not this probability presents an acceptable risk is a matter of executive decision. If the risk is considered to be *too high*, the criterion could be modified so that the switch to the new appeal is made only if the number of responses exceeds, say, 124 (see Exercise 1 on page 216). If it is considered to be *too low*, the criterion could be modified so that the switch to the new appeal is made if the number of responses exceeds, say, 114 (see Exercise 2 on page 217).

The risk of switching to the more expensive appeal when actually it is not better than the old one is only one of the risks which the company must face. There is the possibility that the company will *not* switch to the new appeal when it is really better. Let us suppose, for instance, that the switch to the new appeal will be financially worth while if it increases the percentage of responses to 14 per cent. If this were the case, the company would lose money if, by chance, the sample yielded fewer than 120 responses and the old appeal is retained.

The probability of this happening is *the probability of getting "fewer than 120 successes in 1,000 trials" if the probability of an individual success is 0.14.* Using again the normal curve approximation to the binomial distribution, the desired probability may be found by evaluating the shaded area of Figure 10.2. Substituting $n = 1,000$ and $p = 0.14$ into the formulas for the mean and standard deviation of the binomial distribution, we get

$$\mu = 1000(0.14) = 140$$

and

$$\sigma = \sqrt{1000(0.14)(0.86)} = 11.0$$

The z value of the dividing line of the criterion is then

$$z = \frac{119.5 - 140}{11.0} = -1.86$$

and the corresponding entry in Table I is 0.4686. *Hence, the probability that the criterion will lead to the rejection of the new appeal even though it increases the average percentage of responses to 14 per cent is 0.5000 − 0.4686 = 0.0314 or, roughly, 0.03.*

If it is felt that this probability entails *too high* a risk, the criterion could be changed so that the old appeal is retained only if the number of responses is, say, below 115. Unfortunately, this change would automatically *increase* the probability of switching to the new appeal

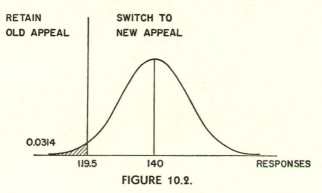

FIGURE 10.2.

when actually it presents no improvement over the old one. It will be left to the reader to show in Exercise 2 on page 217 that while the probability of mistakenly retaining the old appeal is thus reduced from 0.03 to 0.01, the probability of mistakenly switching to the new one is increased from 0.02 to, roughly, 0.06. This illustrates the fact that by *reducing* the probability of one kind of error we *increase* the probability of another.

In the above, we arbitrarily assumed that the new appeal is worth while only if it increases the percentage of responses to 14 per cent or more. For the sake of argument, let us now suppose that the new appeal is worth while even if it increases the average percentage of responses only to 12 per cent. In that case, the probability of making the error of rejecting the new appeal even though it increases the percentage of responses to 12 per cent is given by the shaded area of Figure 10.3.

Substituting $n = 1,000$ and $p = 0.12$ into the formulas for the mean and standard deviation of the binomial distribution, we now have

$$\mu = 1000(0.12) = 120$$

$$\sigma = \sqrt{1000(0.12)(0.88)} = 10.3,$$

the z value of the dividing line of the criterion becomes

$$z = \frac{119.5 - 120}{10.3} = -0.05,$$

and the corresponding entry in Table I is 0.0199. *Hence, the probability that the original criterion will lead to the rejection of the new appeal even though it increases the average percentage of responses to 12 per cent is 0.5000 − 0.0199 = 0.4801 or, roughly, 0.48.*

Clearly, a decision criterion can hardly be described as satisfactory if there is almost a *fifty-fifty* chance of making an error of this kind.

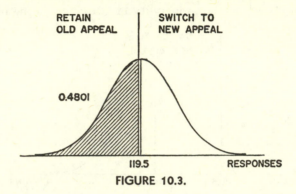

FIGURE 10.3.

In order to improve this situation without increasing the probability connected with the other kind of error, we have little choice but to increase the size of the sample (see Exercise 3 on page 217). It stands to reason that, if we want to decide between two values that are very close together (in our example, $p = 0.10$ and $p = 0.12$), we will need a very large sample to keep all the probabilities of making wrong decisions reasonably small.

The purpose of the discussion of this section was to present an informal introduction to the problem of evaluating the "goodness" of decision criteria. In Sections 10.2 and 10.3 we shall treat this problem in a more general and in a more rigorous fashion.

EXERCISES

1. Suppose that, in the example discussed in the text, the criterion is changed so that the old appeal is retained if the number of responses produced by the 1,000 mailings is below 125 and the new appeal is adopted if the number of responses is 125 or more.
    (a) Find the probability that this modified criterion will lead to the adoption of the new appeal even though it is only as good as the

old one. (*Hint:* Use the normal curve approximation to find the probability of getting 125 or more responses when p equals 0.10.)

(b) Find the probability that this modified criterion will lead to the retention of the old appeal even though the new appeal increases the average percentage of responses to 14 per cent. (*Hint:* Use the normal curve approximation to find the probability of getting fewer than 125 responses when p equals 0.14.)

2. Suppose that in the example in the text the criterion is modified so that the old appeal is retained if the number of responses produced by the 1,000 mailings is below 115 and the new appeal is adopted if the number of responses is 115 or more.

(a) Find the probability that this modified criterion will lead to the adoption of the new appeal even though it is only as good as the old one.

(b) Find the probability that this modified criterion will lead to the retention of the old appeal even though the new appeal increases the average percentage of responses to 14 per cent.

3. Suppose that in the example in the text the sample size is increased to 4,000, the old appeal is retained if the number of responses is below 440, and the new appeal is adopted if the number of responses is 440 or more.

(a) Find the probability that this new criterion will lead to the adoption of the new appeal even though it is only as good as the old one.

(b) Find the probability that this new criterion will lead to the retention of the old appeal even though the new appeal increases the average percentage of responses to 12 per cent.

4. On page 213 we remarked that if the new appeal is *worse* than the old one, the chances are that the old appeal will be retained. Calculate the probability that the old appeal will be retained with the criterion given on page 213, if the new appeal *decreases* the average percentage of responses to $9\frac{1}{2}$ per cent. (*Hint:* Use the normal curve approximation to find the probability of getting fewer than 120 responses when p equals 0.095.)

10.2 Type I and Type II Errors

To study the problem which we introduced in Section 10.1 in a more rigorous fashion, let us now refer to the hypothesis "*the new appeal is exactly as effective as the old one*" as "*Hypothesis H.*" Since effectiveness was measured in terms of percentage response and the old appeal yielded an average response of 10 per cent, we can also write

$$\textit{Hypothesis H: } p = 0.10,$$

where p is the probability that any person chosen from the company's mailing list will respond to the new appeal. To state it differently,

p is the proportion of people who in the long run will respond to the new appeal.

Now, clearly, hypothesis H is either true or false and it will either be accepted or it will be rejected. If H is true and we accept it *or* if H is false and we reject it, our decision will be the right one. If H is true and we reject it, we will be making an error which, in statistics, is referred to as a *Type I error. Generally speaking, a Type I error is committed when we reject a hypothesis which should be accepted.* In our example, the error of switching to the new appeal when actually the new appeal is exactly as effective as the old one, is a Type I error.

If H is false and we accept it, we will be making an error which is called a Type II error. *In general, a Type II error is committed when we accept a hypothesis which should be rejected.* In our example, the error of retaining the old appeal if actually the new one is better is a Type II error.

We thus find ourselves in the situation which is described by the following table:

	Accept H	*Reject H*
H is true	correct decision	Type I error
H is false	Type II error	correct decision

This analysis of the Type I and Type II errors is by no means limited to the special hypothesis which we formulated for the mail appeal illustration. H could be the hypothesis that a shipment of steel will arrive on time, it could be the hypothesis that two companies will merge, it could be the hypothesis that the price of coffee will remain the same over a given period of time, it could be the hypothesis that a new business venture will succeed—to name but a few among countless possibilities.

Suppose that H is the hypothesis that *Mr. Jones, who is being considered for a responsible position in a bank, is an honest man.* If, for some reason, we decide that he cannot be trusted though, actually, he is a man of unquestionable integrity, we will be making a Type I error. If we decide that he is honest and give him the job, we will realize that we have committed a Type II error if he absconds with the company's funds.

It should be clear from our illustrations that whether an error is a Type I error or a Type II error depends on how we formulate Hypothesis H. If in the last example we had formulated the hypothesis that Mr. Jones is *dishonest,* our Type I error would have been a Type II error and vice versa.

In Section 10.1 we evaluated the "goodness" of our decision criterion, the one given on page 213, by calculating the probabilities of committing Type I and Type II errors.   We saw that the probability of committing the Type I error of *switching to the new appeal although, actually, the new appeal is no better than the old one,* is 0.02.   (This is a Type I error because we are rejecting a hypothesis which should be accepted, namely, the hypothesis that $p = 0.10$.   As before, p is the probability that any one person will respond to the new appeal.)†

We also saw that the probability of committing the Type II error of *retaining the old appeal when p is greater than 0.10* depends on the actual value of p.   We showed that when $p = 0.14$, the probability

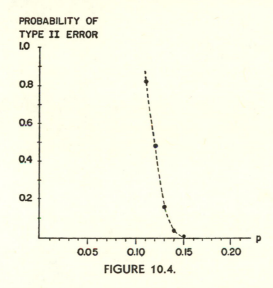

FIGURE 10.4.

of committing a Type II error is 0.03 and that when $p = 0.12$, the probability of committing a Type II error is 0.48.   Using identical methods we could also show that when $p = 0.11$, 0.13, and 0.15, the probabilities of committing Type II errors are, roughly, 0.83, 0.16, and 0.004.

If we now plot these probabilities as shown in Figure 10.4, we obtain a diagram which might reasonably be called a "performance chart" of the decision criterion.   It shows the probabilities of committing Type II errors for various alternative values of p.

Figure 10.4 actually does not show *all* the risks to which we are exposed by the decision criterion, it does not indicate the probability

---

† It is customary, in statistical literature, to use the Greek letters α and β (*alpha* and *beta*) to denote the probabilities of committing Type I and Type II errors.

of committing a Type I error.   This may be taken care of as follows: Since a Type II error is committed when we accept a hypothesis H which should be rejected, let us label the vertical scale of Figure 10.4 "probability of accepting H" instead of "probability of committing Type II error."   This does not change anything so far as the points of Figure 10.4 are concerned, but we can now add a point for $p = 0.10$. When $p = 0.10$ (in our example), the probability of accepting H is the probability of making the *right* decision or

1 — Probability of committing Type I error.

Since the probability of committing a Type I error was shown to be 0.02, the probability of *accepting H when H is true* (when $p = 0.10$) is

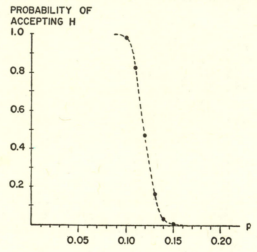

FIGURE 10.5.   Operating Characteristic Curve.

$1 - 0.02 = 0.98$.   Plotting this value for $p = 0.10$, we finally obtain the curve shown in Figure 10.5.   It is called the *Operating Characteristic Curve* of the criterion or, simply, its *OC-curve*.   If we plotted the probabilities of *rejecting H* instead of the probabilities of *accepting H*, we would obtain a curve which is called the *power function* of the decision criterion.

An operating characteristic curve gives the probabilities of committing Type II errors for various values of a parameter (in our example, the parameter p) with one exception.   For the value assumed under the hypothesis, it gives the probability of making the right decision, namely, 1 *minus* the probability of committing a Type I error. An OC-curve provides us, thus, with a picture of the various risks to which we are exposed by a decision criterion.

A thorough study of operating characteristic curves would go far beyond the scope of this text. The main purpose of our example was to provide an illustration of how statistical methods can be used to analyze and describe various risks connected with a decision criterion.

EXERCISES

1. Suppose that a marketing expert is asked to test the hypothesis H that *at least one million units of a new toy will be sold.* Explain under what conditions this marketing expert would be committing a Type I error and under what conditions he would be committing a Type II error.
2. Suppose that on the basis of a sample we want to test the hypothesis that *a shipment of nails meets the specifications as to length.* Explain under what conditions we would be committing a Type I error and under what conditions we would be committing a Type II error.
3. Explain with reference to Exercise 2 why the probability of committing a Type I error may be referred to here as the *producer's risk* and the probability of committing a Type II error as the *consumer's risk.*
4. Calculate the probability that the criterion of Exercise 3 on page 217 will lead to a Type II error, if (a) $p = 0.11$ and (b) $p = 0.13$.
5. Use the results of Exercise 3 on page 217 and Exercise 4 above to plot the operating characteristic curve of the decision criterion.

10.3 Null Hypotheses and Significance Tests

In the example treated in the preceding section we had fewer difficulties in the analysis of Type I errors than in the analysis of Type II errors. This was due to the simple fact that we formulated our hypothesis in such a way that the probability of committing a Type I error could immediately be calculated. Had we formulated the hypothesis that *the new appeal is better than the old one,* that p is greater than 0.10, we would not have been able to calculate the probability of committing a Type I error. At least, we could not have calculated it without specifying "how much better," that is, without specifying a particular value of p.

In choosing the hypothesis which we used in the mail appeal illustration, we followed the widely accepted rule of *always formulating hypotheses in such a way that we know what to expect if they are true.* By formulating the hypothesis $p = 0.10$, we *know* that on the average we can expect to get 100 responses to 1,000 mailings, we *know* that the sampling distribution of the number of responses to 1,000 mailings has the standard deviation $\sqrt{np(1 - p)} = \sqrt{1000(0.10)(0.90)} = 9.5$, and we *know* that this sampling distribution can be approximated closely with a normal curve. *Hence, we have sufficient knowledge to calculate the probability of committing a Type I error.* If we formulated the hypothesis $p > 0.10$, we would *not* know exactly what to expect and we would *not* be able to calculate the probability of committing a Type I error. In other words, the hypothesis must be *specific*, which $p > 0.10$ is not.

To follow the rule given in the last paragraph, we often have to assume the exact opposite of what we are trying to prove. If we wanted to show that Fertilizer *A* will produce a higher (or lower) yield of corn than Fertilizer *B*, we would formulate the hypothesis that there is *no difference* in the effectiveness of the two fertilizers. Similarly, if we wanted to show that a new drug reduces the mortality rate of a given disease we would formulate the hypothesis that it does *not;* and if we wanted to show that there is more absenteeism in one industry than there is another, we would formulate the hypothesis that their rates of absenteeism are the *same*. The fact that we assumed that there is *no difference* in the effectiveness of the two fertilizers, that there is *no difference* regardless of whether or not we use the drug, and that there is *no difference* between the rates of absenteeism of the two plants, explains why hypotheses like these are referred to as *null hypotheses*.

Although we may be avoiding one kind of difficulty by always formulating hypotheses so that we can calculate the probability of committing a Type I error, this will not help so far as problems connected with Type II errors are concerned. The only time that there will be no problem in calculating the probability of committing a Type II error is when we test a specific hypothesis against a specific alternative. For instance, if in the mail appeal example we tested the hypothesis $p = 0.10$ against the alternative hypothesis $p = 0.14$, it would be as easy to calculate the probability of committing a Type II error as it would be to calculate the probability of committing a Type I error.

One possible escape from problems connected with the evaluation of Type II errors is to avoid making Type II errors altogether. To illustrate how this can be done, let us suppose that we want to test the hypothesis that *a given coin is balanced* (that the probability for heads is 0.50), and that we want to base our decision on the number of heads observed in 100 tosses of the coin. If we use the criterion

reject the hypothesis if the number of heads obtained in the 100 tosses of the coin exceeds 60 or is less than 40; reserve judgment if there are anywhere from 40 to 60 heads,

it can easily be shown that the probability of committing a Type I error is 0.0358. There is no need to calculate the probability of committing a Type II error; we never really accept the hypothesis and we, therefore, cannot possibly make the mistake of accepting a false hypothesis.

If we obtained, say, 64 heads, we could *reject* the hypothesis that the coin is balanced.   If we obtained 53, we would say that *the difference between what we expected and what we got, namely, the difference between 50 and 53, is not significant and may reasonably be attributed to chance.*

The procedure we have outlined in this illustration is referred to as a *test of significance.*   If the difference between what we expect and what we get is so *large* that it cannot reasonably be attributed to chance, we reject the hypothesis on which our expectation is based. If the difference between what we expect and what we get is so *small* that it may reasonably be attributed to chance, we say that our results are *not (statistically) significant.*   We can then accept the hypothesis or reserve judgment, depending on whether a definite decision (a definite action) is required.

To give an example in which "reserving judgment" would be appropriate, let us suppose that we want to test the effectiveness of a new drug in the treatment of a disease for which the mortality rate is 0.20.   Formulating the (null) hypothesis that the new drug is *not* effective (that even with the new drug the mortality rate is still 0.20), let us consider the following criterion:

> *the new drug is to be given to 100 patients having the given disease, and the null hypothesis is to be rejected only if fewer than 12 of these patients die.*

Suppose now that 15 of the patients die.   On the basis of the given criterion we *cannot* reject the hypothesis that the drug is *not* effective, *we cannot decide that it is effective.*   On the other hand, there were fewer deaths than expected and it would be quite dangerous to rule out the possibility that the drug might nevertheless be effective.   It would be appropriate to report that, *statistically speaking, the effect of the drug was not significant, but the results of the experiment indicate that further investigations should be conducted.*   In other words, we should be reluctant to decide on the basis of this experiment that the drug is *not effective* (that the original hypothesis should be *accepted*).†

If we refer again to the mail-appeal illustration, we can convert the criterion given on page 213 into that of a significance test by writing

---

† Perhaps a more reasonable criterion would be to *reject* the hypothesis if fewer than 12 of the patients die, to *accept* it if 20 or more die, and to *reserve judgment* if 12 to 19 die.

reject the hypothesis (that $p = 0.10$) if there are 120 or more responses to the 1,000 mailings; reserve judgment if the number of responses is less than 120.

Comparing this criterion with the one given on page 113, we find that the rule for rejecting the hypothesis is still the same and that, therefore, the probability of committing a Type I error is still 0.02. However, so far as the acceptance of the hypothesis is concerned, we are now playing it safe by reserving judgment and not making a decision.

This raises the very important question of whether we can afford the luxury of "reserving judgment" in this example. Since the purpose of sending out the 1,000 mailings was to decide whether to switch to the new appeal or to retain the old one, we really have no choice but to go back to the original criterion. Since we must reach a decision one way or the other, we cannot "reserve judgment" and we cannot escape the possibility of committing a Type II error. We will have to calculate the probabilities of committing Type II errors for various alternative values of p and, if necessary, investigate the entire OC-curve.

Since the general problem of testing hypotheses and constructing decision criteria is quite complicated, we shall simplify our work in the remainder of this book by using the following procedure:

(i) *We formulate a (null) hypothesis in such a way that the probability of committing a Type I error can be calculated. We also formulate an alternative hypothesis so that the rejection of the null hypothesis is equivalent to the acceptance of the alternative hypothesis.*

In the mail appeal illustration the null hypothesis was $p = 0.10$, where p is the probability of getting a response to the new appeal, and the alternative hypothesis was $p > 0.10$. In the coin tossing example on page 222, we assumed under the null hypothesis that $p = 0.50$ and under the alternative hypothesis that $p \neq 0.50$. In the drug testing example on page 223, we assumed under the null hypothesis that $p = 0.20$ and under the alternative hypothesis that $p < 0.20$.

(ii) *We specify the probability of committing a Type I error.*

This probability is also called the *level of significance* at which the test is being conducted. Although the choice of a level of significance is essentially arbitrary, it is customary to use 0.05 or 0.01.

(iii) *We use statistical theory to construct a criterion for testing the hypothesis formulated in (i) at the level of significance specified in (ii).*

The construction of such a criterion will depend on the particular *statistic* on which we will base our decision and on its sampling distribution. A considerable part of the remainder of this book will be devoted to the construction of such criteria.

(iv) *We specify whether the alternative to rejecting the null hypothesis is (a) to accept it or (b) to reserve judgment.*

As we saw in our example, this will have to depend on the nature of the decisions that have to be reached and on the consequences and risks that are involved. *We shall often choose (a) and accept the null hypothesis with the tacit hope that we are not exposing ourselves to too high a risk of committing a serious Type II error.* Of course, if necessary we can always calculate the probabilities of committing Type II errors for various alternate values of the parameter in question and study the resulting OC-curve.

Before going into the actual problem of constructing various kinds of decision criteria, let us point out that the discussion of this and the preceding sections is *not* limited to tests or decisions concerning proportions. The various concepts we have introduced apply equally well to hypotheses about means or standard deviations, about the randomness of samples, about trends in time series, about the relationships of several variables, and so forth.

10.4 Tests Concerning Proportions

To illustrate the four steps outlined in the preceding section, let us suppose that a manufacturer of a suntan lotion claims that his product is 80 per cent effective in preventing sunburn; in other words, he claims that 80 per cent of the people who use his product while spending a good deal of time out in the sun will not get sunburnt. To investigate this claim, a consumer testing service proposes to try the lotion on a random sample of 100 persons and to reach its decision at a level of significance of 0.05.

Letting p stand for the probability that a person using this lotion will receive adequate protection, the hypothesis to be tested is

$$Hypothesis: p = 0.80$$

and the alternative hypothesis is

$$Alternative: p < 0.80.$$

(The alternative hypothesis is $p < 0.80$ since the testing service is interested primarily in testing the manufacturer's claim against the alternative that the product is *not as good as claimed.*)

It stands to reason that the acceptance or rejection of the hypothesis which we have formulated should be based on the number of "successes" or, in other words, on the number of persons in our sample who receive adequate protection. The only question that remains is "Where do we draw the line?" To answer this question, let us consider Figure 10.6, which represents the sampling distribution of x, the number of persons in our sample who receive adequate protection. Since the level of significance (the probability of committing a Type I error) was given as 0.05, the dividing line of the criterion must be such that the shaded area of Figure 10.6 is equal to 0.05. According to Table I, 5 per cent of the area under a normal curve lies to the right of

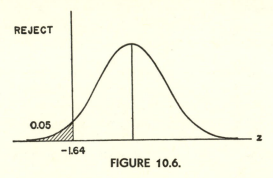

FIGURE 10.6.

$z = 1.64$, or to the left of $z = -1.64$, and we can, therefore, formulate our criterion as follows:†

reject the hypothesis (and accept the alternative) if $z < -1.64$; accept the hypothesis if $z \geqslant -1.64$, where z is to be calculated by means of the formula

$$z = \frac{x - np}{\sqrt{np(1 - p)}}$$ (10.4.1)★

with $n = 100$ and $p = 0.80$.

(We are again using the normal curve approximation to the binomial distribution. The formula for z is obtained by subtracting from x, the number of "successes," the mean of its sampling distribution and then dividing by its standard deviation.)

If the experiment is actually performed and 72 persons get adequate protection while 28 get sunburnt, we obtain

$$z = \frac{72 - 100(0.80)}{\sqrt{100(0.80)(0.20)}} = -2.00$$

---

† We are assuming that the consumer testing service wants to reach a decision one way or the other. Otherwise, we could have written the criterion as

reject the hypothesis if $z < -1.64$ and reserve judgment if $z \geqslant -1.64$.

Since this is less than -1.64, we can *reject* the hypothesis—we can conclude that *the manufacturer's claims are not valid.*

The criterion which we used in this example is referred to as a *one-tail test.* The hypothesis is rejected only if the observed value of x falls into the left-hand tail of the sampling distribution (see Figure 10.6). We used this criterion because the alternative to the hypothesis was the *one-sided alternative $p < 0.80$.* Clearly, there would have been no sense in criticizing the manufacturer for playing down his own product, and we considered his advertising to be misleading only for p less than 0.80. Had we wanted to test the hypothesis $p = 0.80$ against the *two-sided alternative $p \neq 0.80$* (p is less than 0.80 *or* greater than 0.80) we would have had to reject the hypothesis for very small

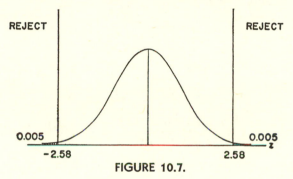

FIGURE 10.7.

as well as very large values of x. In other words, we would have had to use a *two-tail test.*

To give an example in which we would use a *two-tail test,* let us suppose that an advertising executive claims to a sponsor that 40 per cent of the people who saw a certain big "spectacular" television program will remember the name of the product advertised 24 hours after they have seen the show. If, in a sample survey conducted 24 hours after this show was on the air, 152 of 400 viewers remembered the name of the product advertised, test the above claim at a level of significance of 0.01.

Having no idea whether 40 per cent might be correct, too high, or too low, we shall test the hypothesis that the percentage is 40 per cent against the alternative that it is not. Formally,

$$Hypothesis:\ p = 0.40$$

$$Alternative:\ p \neq 0.40$$

where p is the probability that any person who saw the show will remember the name of the product advertised 24 hours later.

Since we shall want to base our decision on that fact that 152 of 400 viewers remembered the name of the product, let us consider the normal curve shown in Figure 10.7 which approximates the sampling distribution of the number of "successes" in 400 trials with $p = 0.40$. The probability of committing a Type I error (the level of significance) is given by the shaded area of Figure 10.7 and, as was specified above, it equals 0.01. Putting half of this area into each tail of the sampling distribution, the z values of the dividing lines of the criterion must be such that 99 per cent of the area lies between $-z$ and z. As we saw on page 191, 99 per cent of the area under a normal curve lies between $z = -2.58$ and $z = 2.58$ and we, therefore have the following criterion:

reject the hypothesis (and accept the alternative) if $z < -2.58$ or $z > 2.58$; accept the hypothesis if $-2.58 \leqslant z \leqslant 2.58$, where z is to be calculated by means of the formula

$$z = \frac{x - np}{\sqrt{np(1 - p)}}$$

with $n = 400$ and $p = 0.40$.

Substituting $x = 152$, $n = 400$, and $p = 0.40$ into this formula for z, we get

$$z = \frac{152 - 400(0.40)}{\sqrt{400(0.40)(0.60)}} = -0.82$$

Since this value lies between -2.58 and 2.58, we shall *accept* the hypothesis that $p = 0.40$, and *we shall accept the advertising executive's claim.* (We might add, that we shall also keep our fingers crossed that the above criterion is not exposing us to too high a probability of committing a serious Type II error.)

Although the choice of the level of significance is really arbitrary, it has become common practice to use either 0.05 or 0.01. If a level of significance of 0.01 is used in a one-tail test (like the one discussed earlier in this section), the dividing line of the criterion is $z = -2.33$ or $z = 2.33$, depending on whether the hypothesis is rejected for values falling into the left-hand tail or the right-hand tail of the distribution. If a level of significance of 0.05 is used in a two-tail test like the one described above, the dividing lines of the criterion have the z values -1.96 and 1.96. *Let us repeat that a one-tail test is used when the alternative hypothesis is one-sided and a two-tail test is used when the alternative hypothesis is two-sided.*

10.5 Differences Between Proportions

A question which must frequently be decided is whether an observed difference between two sample proportions is significant or whether it may reasonably be attributed to chance. For instance, if a sample check reveals that one clerk has misfiled 7 of 200 reports while another clerk has misfiled 10 of 400, we may want to decide whether the difference between $\frac{7}{200} = 0.035$ and $\frac{10}{400} = 0.025$ may reasonably be attributed to chance or whether it implies that on the average the two clerks do *not* make the same number (proportion) of mistakes. Similarly, if one manufacturing process produces 16 defective pieces in a sample of size 400 while another process produces 24 defective pieces in a sample of size 300, it may be of interest to know whether the difference between $\frac{16}{400} = 0.04$ and $\frac{24}{300} = 0.08$ may reasonably be attributed to chance.

Problems of this kind are usually treated with the use of the following theory: *if x_1 and x_2 are the number of "successes" observed in large random samples of size n_1 and n_2 taken from populations having the proportions p_1 and p_2, the sampling distribution of the statistic*

$$\frac{x_1}{n_1} - \frac{x_2}{n_2}$$

can be approximated closely with a normal curve whose mean and standard deviation are

$$\mu = p_1 - p_2 \tag{10.5.1}$$

and
$$\sigma_{\frac{x_1}{n_1} - \frac{x_2}{n_2}} = \sqrt{\frac{p_1(1 - p_1)}{n_1} + \frac{p_2(1 - p_2)}{n_2}} \tag{10.5.2}$$

(For a derivation of formulas (10.5.1) and (10.5.2), see Bibliography on page 244.)

To illustrate what is meant by the sampling distribution of the difference of two proportions, let us suppose that p_1 is the actual proportion of male voters in Detroit who favor Candidate A and that p_2 is the corresponding proportion of female voters. If we sent out a large number of field workers, telling each one to interview random samples of n_1 male voters and n_2 female voters in Detroit, we should expect that the figures they get for x_1 and x_2, the number of male and female votes cast for Candidate A, will not all be the same, and neither will be the values they get for $(x_1/n_1) - (x_2/n_2)$. In fact, if we group the various values which the field workers will get for the difference between x_1/n_1 and x_2/n_2 we will get an *experimental sampling distribution* of the statistic $(x_1/n_1) - (x_2/n_2)$. The sampling distribution referred to above is the corresponding *theoretical sampling distribution*.†

---

† This example also serves to illustrate the fact that the two samples must be *independent*, they must be selected *separately* and at random. It stands to reason that we might get very misleading results if we interviewed, for instance, married couples, whose political views are apt to be the same.

Using the terminology introduced in Chapter 8, we shall refer to the standard deviation given in (10.5.2) as the *standard error of the difference between two proportions*. It is the standard deviation of the sampling distribution of such differences.

Since we are usually interested in testing whether $p_1 = p_2$, we shall formulate the *null hypothesis* $p_1 = p_2$ $(= p)$ and the *two-sided alternative* $p_1 \neq p_2$. Substituting p for p_1 and p_2 in (10.5.1) and (10.5.2), we find that under this null hypothesis the formulas for the mean and

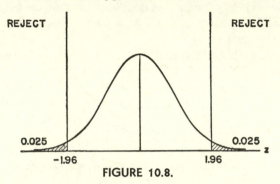

FIGURE 10.8.

standard deviation of the sampling distribution of $(x_1/n_1) - (x_2/n_2)$ become

$$\mu = 0 \tag{10.5.3}\star$$

and

$$\sigma_{\frac{x_1}{n_1} - \frac{x_2}{n_2}} = \sqrt{p(1 - p)\left(\frac{1}{n_1} + \frac{1}{n_2}\right)} \tag{10.5.4}\star$$

Since the last formula cannot be calculated without knowledge of p and p is usually unknown, we shall approximate it by substituting for p the proportion of "successes" in the two samples combined, namely,

$$\frac{x_1 + x_2}{n_1 + n_2} \tag{10.5.5}\star$$

In addition to the fact that we are using a normal curve approximation, this is another reason why the above theory should be used only for *large samples*.

To demonstrate how the theory we have discussed may be used to test hypotheses about *two* proportions, let us refer to the first example mentioned on page 229, namely, the one dealing with the two clerks who misfiled reports. Letting p_1 and p_2 stand for the probabilities that the two given clerks will misfile a report, our hypotheses are

Null hypothesis: $p_1 = p_2$

Alternative: $p_1 \neq p_2$

Let us also specify that the level of significance is to be 0.05.

If we now consider the normal curve of Figure 10.8 which represents the sampling distribution of $(x_1/n_1) - (x_2/n_2)$, we can formulate the following criterion:

reject the null hypothesis if $z < -1.96$ or $z > 1.96$; accept the null hypothesis if $-1.96 \leqslant z \leqslant 1.96$, where z is to be calculated by means of the formula

$$z = \frac{\dfrac{x_1}{n_1} - \dfrac{x_2}{n_2}}{\sqrt{p(1-p)\left(\dfrac{1}{n_1} + \dfrac{1}{n_2}\right)}} \qquad (10.5.6)\star$$

with $(x_1 + x_2)/(n_1 + n_2)$ substituted for p.

It should be noted that this formula for z was obtained by subtracting from the observed difference between the two sample proportions the mean of its sampling distribution as given by (10.5.3) and then dividing by the standard deviation of its sampling distribution as given by (10.5.4).

Substituting the numerical values given on page 229, we find that for $x_1 = 7$, $n_1 = 200$, $x_2 = 10$, and $n_2 = 400$, formulas (10.5.5) and (10.5.6) yield

$$p = \frac{7 + 10}{200 + 400} = 0.028$$

and

$$z = \frac{\frac{7}{200} - \frac{10}{400}}{\sqrt{(0.028)(0.972)(\frac{1}{200} + \frac{1}{400})}} = 0.71$$

Since this value falls between -1.96 and 1.96, we shall *accept* the null hypothesis that $p_1 = p_2$, namely, the hypothesis that the proportions of misfilings made on the average by the two clerks are the same. If we did not want to expose ourselves to the risk of committing a Type II error, we could merely state that *the difference between the two sample proportions is not statistically significant.* This is meant to say that the difference between 0.035 and 0.025 is not big enough in this case to conclude that one clerk is actually better than the other. It does not say that the two clerks are necessarily equally good.

The above example is typical of tests concerning two proportions or, as they are usually called, tests of the significance of the difference between two proportions. A test which serves to compare more than two proportions will be discussed later in Section 11.1.

EXERCISES

1. It has been claimed that 25 per cent of all students entering college drop out during their first year. Test this claim against the alterna-

tive that the percentage is less than 25 per cent, if a random sample of 500 students, who entered college in 1968 contains 149 who dropped out during or after the first year.   Use a level of significance of 0.05.

2. An airline claims that 60 per cent of all single women hired as airline stewardesses get married and leave their jobs within two years after they are hired.   Test this hypothesis against the alternative that $p \neq 0.60$, if among 300 such stewardesses 168 got married within two years after they were hired and left their jobs.   Use a level of significance of 0.05.

3. An opinion research organization is asked to test the hypothesis that Candidate X will receive 55 per cent of the vote against the alternative that he will receive less than 55 per cent of the vote.   What decision should this research organization reach at a level of significance of 0.01 if a random sample of 1,000 voters contained 506 who favored Candidate X and 494 who favored his opponent?

4. It has been claimed that 50 per cent of all families move at least once every 3 years.   Test this claim against the alternative that $p > 0.50$, if a sample survey showed that among 400 families interviewed 218 had moved at least once during the past 3 years.   Use a level of significance of 0.05.

5. A car manufacturer claims that 35 per cent of all cars built by his firm in 1958 were still in running condition in 1968. Show that this claim cannot be rejected at a level of significance of 0.01 (the alternative hypothesis is $p \neq 0.35$) if in a random sample of 800 such cars 257 were still in running condition in 1968.   How would you phrase your result so that you would not be exposed to the risk of committing a Type II error?

6. A large hardware manufacturer believes from experience that at least 70 per cent of all shoppers prefer to purchase small parts which are packaged in clear plastic containers.   In connection with studies of a different type of package the company conducts a market survey in which a sample of 300 shoppers are interviewed.   Of them 203 are found to prefer the clear plastic container.   Test the hypothesis $p = 0.70$ against the alternative hypothesis $p < 0.70$ at a level of significance of 0.05.

7. On page 229 we mentioned two manufacturing processes producing 16 and 24 defective pieces in samples of size 400 and 300, respectively. Use a level of significance of 0.05 to test whether the difference between the two sample proportions of defectives is significant.

8. A marketing study conducted in Philadelphia showed that in a random sample of 300 housewives 213 preferred breakfast cereal A to breakfast cereal B.   In a similar study made in St. Louis, a random sample of 100 housewives showed that 68 preferred cereal A to cereal B.   Use a level of significance of 0.01 to test whether the difference between the two proportions of preferences for cereal A is significant.

9. The following are data obtained by the management of a department store in a study of delinquent time-payment accounts: In a sample of 600 time-payment accounts opened by individuals who had resided

in the community for more than 5 years, 58 had become delinquent at one time or another. In a sample of 400 time-payment accounts opened by individuals who had resided in the community for less than 5 years, 26 had become delinquent. Use a level of significance of 0.05 to test whether the difference between the two proportions of delinquent accounts is significant.

10. A sample of 600 buy orders for stock on one regional exchange showed that 360 were executed by women, while a sample of 520 such orders on another regional exchange showed that 300 were executed by women. Is this evidence that the proportion of women customers is the same in the two exchanges for the time period involved? Use a level of significance of 0.05.

10.6 Tests Concerning Means

Once the fundamental ideas underlying tests of hypotheses are understood, the various tests which we shall study in this and in later sections should not present any further difficulties. In principle they are all the same. We assume a (null) hypothesis for which we can calculate the probability of committing a Type I error, an alternative hypothesis, and then we construct a decision criterion on the basis of the sampling distribution of an appropriate statistic.

The test which we shall discuss in this section applies to problems in which we want to decide on the basis of a sample whether the mean of a population equals some assumed value μ. We may be interested, for example, in testing whether the average distance required to stop a car going 20 miles per hour is 25 feet, whether the average price of two-bedroom houses in a certain area is $15,000, whether the average age of bank presidents is 57, or whether the average hourly wage paid to workers in a certain industry is $2.89.

To illustrate the approach used in problems of this type, let us test the hypothesis that *the average length of local telephone calls made in a certain community is 4 minutes*, if a random sample of 100 such calls had a mean of $\bar{x} = 3.4$ minutes and a standard deviation of $s = 2.8$ minutes.

Letting μ stand for the true average length of local calls made in the given community, the hypothesis to be tested and the alternative hypothesis are

$$Hypothesis:\ \mu = 4$$

$$Alternative:\ \mu \neq 4$$

Let us specify, furthermore, that the level of significance is to be 0.05.

In the customary language of tests of significance, we shall want to decide whether the difference between the observed value of 3.4

and the assumed value of 4 is *significant*, namely, large enough to reject the hypothesis that $\mu = 4$. If the difference is *not significant*, we have the choice between reserving judgment or accepting the original hypothesis. As we said before, this choice will have to depend on the nature of the problem and the risks (consequences) that are involved.

Having already studied the sampling distribution of $\bar{x}$ in Chapters 8 and 9, we know that for large n it can be approximated closely with a normal curve having the mean μ and the standard deviation $\sigma_{\bar{x}} = \sigma/\sqrt{n}$, where μ and σ are the mean and standard deviation of the population from which we are sampling and n is the size of the sample. Considering the normal curve of Figure 10.9 which represents such a

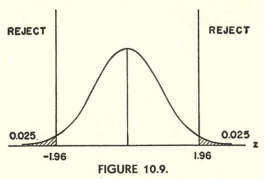

FIGURE 10.9.

sampling distribution of $\bar{x}$, we can now formulate the following criterion:

reject the hypothesis if $z < -1.96$ or $z > 1.96$; accept the hypothesis (or reserve judgment) if $-1.96 \leqslant z \leqslant 1.96$, where

$$z = \frac{\bar{x} - \mu}{s/\sqrt{n}} \qquad (10.6.1)\star$$

and $\mu = 4$.

We obtained this formula for z by subtracting from $\bar{x}$ the *mean* of its sampling distribution and then dividing by its *standard deviation*. In the latter we substituted s for σ, and this is one reason why the above criterion should be used only for *large samples*, namely, when $n \geqslant 30$. Another reason is that we are using the normal curve approximation to the sampling distribution of $\bar{x}$ and this is generally not permissible when n is *very small*.†

---

† If σ is known (see, for instance, Exercise 3 on page 242) it is unnecessary to substitute s for σ and we can use the above criterion with (10.6.1) written as

$$z = \frac{\bar{x} - \mu}{\sigma/\sqrt{n}}$$

In that case the criterion applies approximately, even when n is small.

If we now substitute the numerical values given in our example, we get

$$z = \frac{3.4 - 4}{2.8/\sqrt{100}} = \frac{-0.6}{0.28} = -2.14$$

and since this value is less than -1.96, we can *reject* the hypothesis that $\mu = 4$. Hence, we conclude that the average length of local calls in the given community is *not* 4 minutes.

In the above example we used a two-tail test because we originally formulated the two-sided alternative $\mu \neq 4$. To give an example in which we would use a one-sided alternative and, correspondingly, a one-tail test, let us suppose that some educator claims that *the average I.Q. of American college students is at most 110.* In a study made to test this claim, 150 American college students, selected at random, had an average I.Q. of 111.2 with a standard deviation of 7.2.

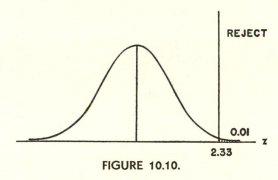

FIGURE 10.10.

As we have formulated this problem, we should really test the hypothesis $\mu \leqslant 110$ against the alternative hypothesis $\mu > 110$. However, since this would not give us a unique value for the probability of committing a Type I error, we shall write instead

Hypothesis: $\mu = 110$

Alternative: $\mu > 110$

If we now consider Figure 10.10, it can easily be seen that the one-tail criterion for testing this hypothesis at a level of significance of 0.01 is:

reject the hypothesis if $z > 2.33$; accept the hypothesis (or reserve judgment) if $z \leqslant 2.33$, where

$$z = \frac{\bar{x} - 110}{s/\sqrt{n}}$$

Substituting $n = 150$, $\bar{x} = 111.2$, $s = 7.2$, we get

$$z = \frac{111.2 - 110}{7.2/\sqrt{150}} = 2.03$$

and since this value does not exceed 2.33, *we cannot reject the educator's claim.* Depending on whether or not a definite decision is required, we shall have to accept the educator's claim or reserve judgment.

It is interesting to note that if we had used a level of significance of 0.05 in this example, our result would have been significant and we would have been able to reject the educator's claim. *This illustrates the very important point that the level of significance should always be specified before any statistical tests are made.* This will save us from the temptation of later on choosing a level of significance which happens to suit our objectives.

It is important to remember that the method which we have discussed should be used only for large samples. If n is *small*, less than 30, we shall proceed as in Section 9.3 and use the statistic

$$t = \frac{\bar{x} - \mu}{s} \sqrt{n} \tag{9.3.1}$$

whose sampling distribution is the Student-t distribution with $n - 1$ degrees of freedom (provided that the population from which we are sampling can be approximated very closely with a normal curve). Since the above formula for t is identical with the formula for z given in (10.6.1), the only difference between the small-sample criterion and the large-sample criterion is that we shall use $t_{.05}$, $t_{.025}$, $t_{.01}$, and $t_{.005}$ as given in Table II instead of the z values 1.64, 1.96, 2.33, and 2.58.

To illustrate the small-sample technique, let us suppose that an oil company claims that (under standard conditions) its gasoline yields on the average 20 miles per gallon. If 10 cars, driven under specified road conditions with 1 gallon of this gas, averaged 19.4 miles with a standard deviation of 0.9 miles, let us test the *hypothesis* $\mu = 20$ against the *alternative hypothesis* $\mu < 20$ at a level of significance of 0.05.

As can easily be seen from the graph of Figure 10.11, which represents the Student-t distribution with $n - 1 = 9$ degrees of freedom, the criterion for testing the given hypothesis may be written as

reject the hypothesis if $t < -t_{.05}$; accept the hypothesis (or reserve judgment) if $t \geqslant -t_{.05}$, where

$$t = \frac{\bar{x} - \mu}{s} \sqrt{n} \tag{10.6.2}\star$$

with $\mu = 20$ and the number of degrees of freedom equal to $n - 1$.

If we now substitute $\bar{x} = 19.4$, $s = 0.9$, and $n = 10$, we get

$$t = \frac{19.4 - 20}{0.9} \sqrt{10} = -2.11$$

and since this value is less than -1.833, *we can reject the hypothesis that $\mu = 20$, i.e., we can reject the oil company's claim.* (The reader may wish to check for himself that 1.833 is the value given in Table II for $t_{.05}$ with 9 degrees of freedom.)

Although we have shown that the difference between the observed average of 19.4 miles per gallon and the assumed value of 20 miles per gallon is *significant*, this does not necessarily mean that the difference is "*commercially*" *significant; in other words, the difference may be statistically significant but it may also be too small to worry about from a purely practical point of view.*

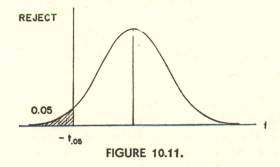

FIGURE 10.11.

Had we wanted to test the oil company's claim against the two-sided alternative $\mu \neq 20$, our criterion would have read

reject the hypothesis if $t < -t_{.025}$ or $t > t_{.025}$; accept the hypothesis (or reserve judgment) if $-t_{.025} \leqslant t \leqslant t_{.025}$.

As can easily be verified, this modified criterion would not have enabled us to reject the oil company's claim.

10.7 Differences Between Means (Large Samples)

Another important test of significance concerns the question whether an observed difference between two sample means may be attributed to chance or whether the difference is large enough to enable us to conclude that the two samples came from populations with unequal means. We may want to decide, for example, whether there actually is a difference in the strength of two kinds of steel if a sample of one kind showed an average strength of 57,000 psi while a sample of another kind showed an average strength of 55,600 psi. Similarly, we may want to decide whether there actually is a difference in the speed with which men and women can perform a certain task, if a sample of 100 men took on the average 6 minutes 35 seconds while a sample of 80 women took on the average 6 minutes 42 seconds.

In both of these problems we are interested in deciding whether an observed difference $\bar{x}_1 - \bar{x}_2$ is significant or whether it may reasonably be attributed to chance. The technique which is generally used in problems of this type is based on the following theory: *if $\bar{x}_1$ and $\bar{x}_2$ are the means of two large (independent)† random samples of size n_1 and n_2, the theoretical sampling distribution of the statistic $\bar{x}_1 - \bar{x}_2$ can be approximated closely with a normal curve whose mean and standard deviation are*

$$\mu = \mu_1 - \mu_2 \tag{10.7.1}$$

and
$$\sigma_{\bar{x}_1 - \bar{x}_2} = \sqrt{\frac{\sigma_1^2}{n_1} + \frac{\sigma_2^2}{n_2}} \tag{10.7.2}$$

where μ_1 and μ_2 are the means of the populations from which the two samples were obtained and σ_1 and σ_2 are their standard deviations. (For a derivation of formulas (10.7.1) and (10.7.2), see Bibliography on page 244.)

The sampling distribution referred to above is the distribution which we could expect if we grouped values of $\bar{x}_1 - \bar{x}_2$ obtained from a large number of repeated samples from two populations. Using the terminology introduced in Chapter 8, we shall refer to $\sigma_{\bar{x}_1 - \bar{x}_2}$, the standard deviation of this sampling distribution, as the *standard error of the difference of two means.*

If the population standard deviations are unknown, which is usually the case, we shall approximate (10.7.2) by substituting the sample standard deviations s_1 and s_2 for σ_1 and σ_2. We shall thus write

$$\sqrt{\frac{s_1^2}{n_1} + \frac{s_2^2}{n_2}} \tag{10.7.3}\star$$

instead of (10.7.2), *with the provision that this formula is to be used only for large samples.*

To illustrate how the above theory is applied, let us suppose that we want to compare two kinds of electric light bulbs on the basis of an experiment in which 100 light bulbs made by Company A had an average lifetime of 952 hours with a standard deviation of 85 hours, while 50 light bulbs made by Company B had an average lifetime of

---

† By "independent" we mean that the selection of one sample is in no way affected by the selection of the other. The test described in this section does *not* apply, for example, to "before and after" kinds of comparisons. If $\bar{x}_1$ is the average weight of a number of people before starting a certain diet and $\bar{x}_2$ is their average weight after the diet, the two samples are clearly *not independent.* A special method for handling problems of this kind is referred to on page 244.

987 hours with a standard deviation of 92 hours. Symbolically, we can write the information supplied by this experiment as follows:

$$n_1 = 100, \ \bar{x}_1 = 952, \ s_1 = 85$$
$$n_2 = 50, \ \bar{x}_2 = 987, \ s_2 = 92$$

Letting μ_1 and μ_2 stand for the true average lifetimes of the two kinds of light bulbs, the hypothesis to be tested and the alternative hypothesis are

Null Hypothesis: $\mu_1 = \mu_2$

Alternative: $\mu_1 \neq \mu_2$

Let us specify, furthermore, that the level of significance is to be 0.05.

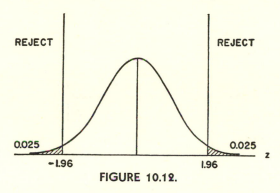

FIGURE 10.12.

Basing our argument on the sampling distribution of $\bar{x}_1 - \bar{x}_2$ as approximated by means of the normal curve of Figure 10.12, we can now formulate the following criterion:

reject the null hypothesis if $z < -1.96$ or $z > 1.96$; accept the null hypothesis (or reserve judgment) if $-1.96 \leqslant z \leqslant 1.96$, where

$$z = \frac{\bar{x}_1 - \bar{x}_2}{\sqrt{\dfrac{s_1^2}{n_1} + \dfrac{s_2^2}{n_2}}} \qquad (10.7.4)\star$$

It should be noted that under the null hypothesis the mean of the sampling distribution of $\bar{x}_1 - \bar{x}_2$, as given by (10.7.1), is equal to 0. The formula for z was obtained by subtracting from $\bar{x}_1 - \bar{x}_2$ the mean of its sampling distribution, namely, 0 and then dividing by (10.7.3).

Substituting the numerical values given for the experiment involving the two kinds of light bulbs, we get

$$z = \frac{952 - 987}{\sqrt{\dfrac{85^2}{100} + \dfrac{92^2}{50}}} = \frac{-35}{15.5} = -2.26$$

Since this value is less than -1.96, we conclude that the observed difference between 952 and 987 is *significant* and that the null hypothesis is to be *rejected*. (Whether we would actually buy the second kind of light bulb in preference to the first is another matter; it would have to depend on their price and, perhaps, other factors. All we have shown in our test is that on the average the two do not last equally long.)

10.8 Differences Between Means (Small Samples)

If our samples come from populations that can be approximated closely with normal curves and if, furthermore, $\sigma_1 = \sigma_2 \ (=\sigma)$, a *small*

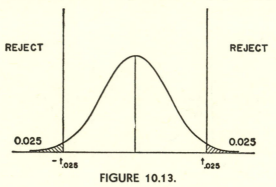

REJECT   REJECT

0.025   0.025

$-t_{.025}$ $t_{.025}$

FIGURE 10.13.

sample test of the significance of the difference between two sample means may be based on the Student-t distribution. Making use of the fact that the sampling distribution of the statistic†

$$t = \frac{\bar{x}_1 - \bar{x}_2}{\sqrt{\dfrac{(n_1 - 1)s_1^2 + (n_2 - 1)s_2^2}{n_1 + n_2 - 2}} \cdot \sqrt{\dfrac{1}{n_1} + \dfrac{1}{n_2}}} \qquad (10.8.1)\star$$

---

† Letting $\sigma_1 = \sigma_2 = \sigma$, formula (10.7.2) becomes

$$\sigma_{\bar{x}_1 - \bar{x}_2} = \sigma \sqrt{\frac{1}{n_1} + \frac{1}{n_2}} \qquad (10.8.2)$$

If we then *estimate* the σ appearing in this formula with the statistic

$$\sqrt{\frac{(n_1 - 1)s_1^2 + (n_2 - 1)s_2^2}{n_1 + n_2 - 2}} \qquad (10.8.3)$$

we obtain the expression given in the denominator of (10.8.1). It should be noted that the expression inside the radical of (10.8.3) is a *weighted average* of s_1^2 and s_2^2. Since $n_1 - 1$ of the deviations from the mean are *independent* in s_1^2 (see page 98) and $n_2 - 1$ of the deviations from the mean are *independent* in s_2^2, we have altogether $(n_1 - 1) + (n_2 - 1) = n_1 + n_2 - 2$ *independent* deviations from the mean on which to base our estimate of σ and, hence, that many *degrees of freedom*.

is the *Student-t distribution* with $n_1 + n_2 - 2$ degrees of freedom, we can test the hypothesis

$$Null\ hypothesis:\ \mu_1 = \mu_2$$

against the alternative hypothesis

$$Alternative:\ \mu_1 \neq \mu_2$$

with the criterion shown in Figure 10.13.   Using a level of significance of 0.05, our criterion will read

> *reject the null hypothesis if $t < -t_{.025}$ or $t > t_{.025}$; accept the null hypothesis (or reserve judgment) if $-t_{.025} \leqslant t \leqslant t_{.025}$, where t is to be calculated according to (10.8.1) and the number of degrees of freedom equals $n_1 + n_2 - 2$.*

To illustrate this procedure, let us suppose that a study has been made to compare the nicotine content of two brands of cigarettes and that 10 cigarettes of Brand A had an average nicotine content of 23.1 milligrams with a standard deviation of 1.5 milligrams, while 8 cigarettes of Brand B had an average nicotine content of 22.7 milligrams with a standard deviation of 1.7 milligrams.   (We shall assume that these two samples are random samples from normal populations and that, furthermore, since the sample standard deviations are fairly close, the two samples come from populations with equal standard deviations.)   The level of significance to be used is 0.05.

Substituting $n_1 = 10$, $\bar{x}_1 = 23.1$, $s_1 = 1.5$, $n_2 = 8$, $\bar{x}_2 = 22.7$, and $s_2 = 1.7$ into (10.8.1), we obtain

$$t = \frac{23.1 - 22.7}{\sqrt{\dfrac{9(1.5)^2 + 7(1.7)^2}{16}} \cdot \sqrt{\dfrac{1}{10} + \dfrac{1}{8}}} = 0.53$$

Since $t_{.025} = 2.120$ for $n_1 + n_2 - 2 = 10 + 8 - 2 = 16$ degrees of freedom (see Table II), we find that *the observed difference between the average nicotine contents of the two brands of cigarettes is not significant.*

A test which will enable us to decide whether observed differences between *several* sample means are significant will be presented in Section 11.4.   Applications of the various tests discussed in this Chapter to problems of *quality control* are treated briefly in Appendix II.

EXERCISES

1. A random sample of 30 cans of tomato soup, each labeled "10.75 ounces net weight," has an average net weight of 10.6 ounces with a standard deviation of 0.3 ounces.   Does this information enable us to reject the hypothesis $\mu = 10.75$ ounces, if we test it against the alternative hypothesis $\mu < 10.75$ ounces at a level of significance of 0.01?

2. Test the hypothesis that private passenger cars, which are not used for business purposes, are driven on the average 12,000 miles a year, if a random sample of 100 such cars averaged 12,750 miles with a standard deviation of 2,500 miles. Use a two-sided alternative and a level of significance of 0.05.

3. A company has manufactured millions of hooks which are designed to have a breaking strength of 100 pounds. The variability in the process as measured by the standard deviation is stable and, on the basis of thousands of tests, well established at 20 pounds. The company ships the hooks in very large lots and the decision as to whether the average breaking strength is 100 pounds is made by subjecting 400 randomly selected hooks to appropriate breaking tests, and testing the hypothesis $\mu = 100$ against the alternative hypothesis $\mu \neq 100$ at a level of significance of 0.05. What will be the company's decision about a lot whose sample showed an average strength of 96.9 pounds? (*Hint:* See footnote on page 234.)

4. A paint manufacturer claims that on the average a gallon of his paint will cover 500 square feet. Test this claim against the alternative $\mu < 500$, if a random sample of 12 gallon cans of this paint covered on the average 486 square feet with a standard deviation of 32 square feet. Use a level of significance of 0.05.

5. A random sample of 8 steel beams has an average compressive strength of 56,598 psi with a standard deviation of 564 psi. Test the hypothesis that the true average strength of the steel beams from which this sample was taken is 56,000 psi. Use a two-sided alternative and a level of significance of 0.05. Would your conclusion be the same if you used a level of significance of 0.01?

6. A bank president made a speech in which he asserted that, after taking account of recent trends in prices and building practices, the average price of all homes sold in a certain city during 1968 was $22,500. To test this claim, an investigator took a random sample of 20 sales contracts and found an average price of $17,750 with a standard deviation of $4,250. Test, at a level of significance of 0.05, whether it seems reasonable to claim that the average price of all homes sold was $22,500.

7. An identical mechanical aptitude test was given to random samples of size 50 at two different schools. Given that the results were

$$n_1 = 50 \qquad \bar{x}_1 = 89 \qquad s_1 = 4$$
$$n_2 = 50 \qquad \bar{x}_2 = 92 \qquad s_2 = 3$$

test at a level of significance of 0.05 whether the difference between the two sample means is significant.

8. A random sample of 100 strips of type A carpeting and a random sample of 80 strips of type B carpeting were tested for wear in the

corridors of a university. Type A had an average lifetime of 72 months with a standard deviation of 7.0 months. Type B had an average lifetime of 75 months with a standard deviation of 6.8 months. Use a level of significance of 0.01 to test whether the observed difference between the two average lifetimes is significant.

9. In a survey of buying habits, 400 women shoppers are chosen at random in Supermarket A located in a certain section of the city. Their average weekly food expenditure is $50 with a standard deviation of $15. For 400 women shoppers chosen at random in Supermarket B in another section of the city, the average weekly food expenditure is $38 with a standard deviation of $18. Test, at a level of significance of 0.05, whether the average weekly food expenditures of the two populations of shoppers from which the samples were obtained are equal.

10. An agricultural experiment showed that 6 test plots planted with one variety of corn yielded on the average 85.3 bushels per acre with a standard deviation of 5.8 bushels per acre while 6 test plots planted with another variety of corn yielded on the average 92.7 bushels per acre with a standard deviation of 6.1 bushels per acre. Show that at a level of significance of 0.05 the difference between the average yields is not significant.

11. A random sample of 8 discount stores in New York City showed markdown percentages of 8.6, 9.4, 10.7, 7.8, 10.3, 7.9, 8.3, and 11.4, while a random sample of 10 Los Angeles stores showed markdown percentages of 10.2, 7.8, 11.0, 11.5, 9.5, 12.2, 10.3, 8.8, 8.9, and 9.8. Is this evidence (at a level of significance of 0.05) that the average markdowns in discount stores in the two cities differ significantly?

BIBLIOGRAPHY

Informal discussions of the two types of errors, tests of significance, null hypotheses, operating characteristic curves, and the various tests discussed in this chapter, may also be found in

Croxton, F. E., Cowden, D. J., and Klein, S., *Applied General Statistics*, 3rd ed., Englewood Cliffs, N.J.: Prentice-Hall, Inc., 1967, Chaps. 24 and 25.

Moroney, M. J., *Facts from Figures*. London: Penguin Books, 1956, Chap. 13.

Neter, J., and Wasserman, W., *Fundamental Statistics for Business and Economics*, 3rd ed. Boston: Allyn and Bacon, Inc., 1966, Chap. 10.

Paden, D. W., and Lindquist, E. F., *Statistics for Economics and Business*, 2nd ed. New York: McGraw-Hill, 1956, Chap. 11.

Wallis, W. A., and Roberts, H. V., *Statistics: A New Approach*. Glencoe, Ill.: Free Press, 1956, Chaps. 12 and 13.

and in many other elementary texts.

Derivations of the formulas for the standard error of the difference between two proportions and the standard error of the difference between two means are given in

Hoel, P. G., *Introduction to Mathematical Statistics*, 3rd ed. New York: John Wiley, 1962, Chap. 6.

A test of the significance between two sample means based on *dependent* samples is given in the book (page 654) by Croxton, Cowden, and Klein mentioned above, and among others, in

Lewis, E. E., *Methods of Statistical Analysis in* Economics and Business, 2nd ed. Boston: Houghton Mifflin Co., 1965, Chap. 6.

A discussion of the *small sample t-test* for the comparison of two means may be found in the book by Hoel mentioned above (page 227) and in most textbooks of mathematical statistics. See also

Freund, J. E., *Mathematical Statistics*. Englewood Cliffs, N.J.: Prentice-Hall, Inc., 1962.

The above-mentioned book contains a derivation of the formula for the standard error of the difference between two means.

CHAPTER 11

Further Tests of Hypotheses*

11.1 Tests Concerning *k* Proportions

To generalize the work of Section 10.5, let us now consider problems in which we are interested in deciding whether observed differences between *more than two* sample proportions may reasonably be attributed to chance. For example, we may want to test whether a production process is "in control," whether the true proportion of defectives remains constant, if on 3 consecutive days there were 12 defectives in a sample of size 100, 15 defectives in a sample of size 120, and 6 defectives in a sample of size 80. Similarly, we may want to know whether there really is a difference in opinion concerning a certain piece of legislation, if random samples of size 420, 500, 300, and 380 of the members of four different unions contained 312, 348, 243, and 297 unfavorable votes.

In the first of these two examples we are interested in testing whether the difference between $\frac{12}{100} = 0.12$, $\frac{15}{120} = 0.125$, and $\frac{6}{80} = 0.075$ is *significant* and in the second example we are interested in deciding the same question with regard to the differences between $\frac{312}{420} = 0.74$, $\frac{348}{500} = 0.70$, $\frac{243}{300} = 0.81$, and $\frac{297}{380} = 0.78$.

To illustrate the procedure used in problems of this kind, let us refer to the first of these two examples and let us rewrite the given information in the following table:

---

* This chapter may be omitted without loss of continuity. Also, the material covered in Sections 11.1, 11.2, and 11.3 is not a prerequisite for that covered in Sections 11.4 and 11.5.

	1st day	2nd day	3rd day	*Totals*
Number of defectives	12	15	6	33
Number of non-defectives	88	105	74	267
Totals	100	120	80	300

Since this table contains 2 rows and 3 columns, it is referred to as a *2 by 3 table*.

Denoting the true proportions of defectives produced by the process on the given three days as p_1, p_2, and p_3, we shall want to test the hypothesis

$$Null\ hypothesis:\ p_1 = p_2 = p_3\ (=p)$$

against the alternative hypothesis that the three p's are *not all the same*.

Assuming that p is unknown, we shall proceed as on page 230 and estimate it as the proportion of defectives observed in the three samples combined, namely, as

$$\frac{12 + 15 + 6}{100 + 120 + 80} = \frac{33}{300} = 0.11$$

We can now ask for the number of defectives that we could have expected in each of the 3 samples if the null hypothesis were true and p equalled 0.11. Clearly, in a sample of size 100 we could have expected $100(0.11)$ = 11.0 defectives, in a sample of size 120 we could have expected $120(0.11)$ = 13.2, and in a sample of size 80 we could have expected $80(0.11)$ = 8.8. Writing the *expected frequencies*, the expected numbers of defectives and non-defectives, below the corresponding entries of the 2 by 3 table, we get

	1st day	2nd day	3rd day
Number of defectives	12 (11.0)	15 (13.2)	6 (8.8)
Number of non-defectives	88 (89.0)	105 (106.8)	74 (71.2)

where the expected numbers of non-defectives were obtained by subtracting the expected numbers of defectives from the respective totals of the 3 samples.

In order to test the hypothesis formulated above, we can now compare the expected frequencies shown in this table with the frequencies which were actually observed. *It stands to reason that the null hypothesis should be accepted if these two sets of frequencies are very much alike. After all, we would then have obtained almost exactly what we could have expected if the null hypothesis were true. If the discrepancies between the two sets of frequencies are large, the observed frequencies do not agree with what we could have expected and we shall conclude that our expectations and, hence, the null hypothesis must have been false.*

Before we introduce the criterion used for tests of this kind, let us first formulate what we have done in a more general fashion. Let us suppose that we have k random sample from as many populations and that the observed numbers of "successes" and "failures" are as shown in the following *2 by k table:*

	1st sample	2nd sample	. . . .	kth sample	*Totals*
Number of successes	n_{11}	n_{12}	. . . .	n_{1k}	$n_{1.}$
Number of failures	n_{21}	n_{22}	. . . .	n_{2k}	$n_{2.}$
Totals	$n_{.1}$	$n_{.2}$	. . . .	$n_{.k}$	n

In the notation used here, n_{ij} stands for the entry given in the cell which belongs to the ith row and jth column. The first subscript is 1 or 2, depending on whether we are referring to "successes" or "failures"; the second subscript represents the number of the sample. Furthermore, the total numbers of successes and failures for the k samples combined, the *row totals*, are written as $n_{1.}$ and $n_{2.}$, while the total number of trials in the respective samples, *the column totals,* are written as $n_{.1}, n_{.2}, \ldots$, and $n_{.k}$. The total number of trials for all k samples combined is written as n.†

If we now let $p_1, p_2, \ldots$, and p_k stand for the true proportions of successes in the populations from which the k samples were obtained, we shall want to test the hypothesis

Null hypothesis: $p_1 = p_2 = \cdots = p_k \ (=p)$

against the alternative hypothesis that these p's are not all the same.

Assuming that p, the actual proportion of successes (which under the null hypothesis is supposed to be the same for all k populations), is *unknown*, we shall proceed as before and estimate it as the proportion of successes in the k samples combined, namely, as

$$\frac{n_{11} + n_{12} + \cdots + n_{1k}}{n} = \frac{n_{1.}}{n} \qquad (11.1.1)\star$$

As in our numerical example, *we can now ask for the numbers of successes and failures that could have been expected in each of the k samples if the null hypothesis were true and p equalled $n_{1.}/n$.*

---

† This symbolism may seem unnecessarily complicated for problems of this type, but it has the advantage that it can be used also for more general kinds of problems, such as those discussed in Section 11.2.

Multiplying the estimate of p given by (11.1.1) by the number of trials in each of the k samples, we find that in the first sample the *expected* number of successes is $n_{.1}(n_{1.}/n)$, in the second sample the *expected number* of successes is $n_{.2}(n_{1.}/n)$, . . . , and in the kth sample the *expected* number of successes is $n_{.k}(n_{1.}/n)$. The *expected* numbers of "failures" may again be obtained by subtracting the expected numbers of successes from the totals of the respective samples. To simplify our notation, we shall write the expected frequency corresponding to n_{ij} as e_{ij}. (In other words, e_{ij} is the expected frequency belonging to the ith row and jth column of the 2 by k table.)

We have now reached the point at which we interrupted our numerical example; *we have shown how to calculate the expected frequencies e_{ij} under the assumption that the null hypothesis is true and that p equals the value given by (11.1.1)*. Our next step will be to test whether the discrepancies between the observed frequencies n_{ij} and the expected frequencies e_{ij} are significant or whether they may reasonably be attributed to chance. The criterion that is generally used for this purpose is based on the statistic

$$\chi^2 = \sum \frac{(n_{ij} - e_{ij})^2}{e_{ij}} \qquad (11.1.2)\star$$

which is called "chi-square" and whose symbol is the Greek letter *chi* with the exponent 2. It owes its name to the fact that (if the null hypothesis is true) its sampling distribution can be approximated very closely with a theoretical distribution called the "*chi-square distribution.*" The summation in (11.1.2) extends over all cells of the 2 by k table.† In other words, we must calculate

$$\frac{(n_{ij} - e_{ij})^2}{e_{ij}}$$

separately for each cell of the 2 by k table and then *add* the values obtained.

If there is a close agreement between the observed and expected frequencies, the differences $n_{ij} - e_{ij}$ and χ^2 will be *small;* if the agreement is poor, some of the differences $n_{ij} - e_{ij}$ and χ^2 will be *large.*

Since χ^2 is calculated on the basis of samples, we must allow for the fact that, like any other statistic, it is subject to chance variations

---

† To be more explicit, we could use a *double summation* and write the formula for χ^2, as it will be used in this section, as

$$\chi^2 = \sum_{i=1}^{2} \sum_{j=1}^{k} \frac{(n_{ij} - e_{ij})^2}{e_{ij}}$$

as characterized by a certain (theoretical) sampling distribution.   This is the distribution which we could *expect* to obtain for values of χ^2 that are calculated according to (11.1.2) for a large number of repeated samples from the k populations.

As we mentioned above, the theoretical sampling distribution of χ^2 can be approximated very closely with a theoretical distribution which is called the *chi-square distribution* (see Figure 11.1).   Areas under this curve are given in Table III on page 503, where they are tabulated in the same way in which we tabulated areas under the Student-t distribution.   Analogous to $t_{.05}$, $t_{.025}$, $t_{.01}$, and $t_{.005}$, we shall now write $\chi^2_{.05}$, $\chi^2_{.025}$, $\chi^2_{.01}$, and $\chi^2_{.005}$ for values which are such that 5,

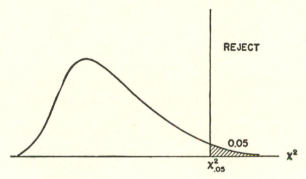

FIGURE 11.1.   A Chi-Square Distribution.

$2\frac{1}{2}$, 1, and $\frac{1}{2}$ per cent of the area under the chi-square distribution lies to their right.

The chi-square distribution, like the Student-t distribution, depends on a quantity referred to as *the number of degrees of freedom*.   When χ^2 is calculated for a 2 by k table, the number of degrees of freedom equals $k - 1$ and this may be explained as follows: *Once the expected numbers of successes have been calculated with the use of (11.1.1) for any $k - 1$ of the samples, all of the other e_{ij} are automatically fixed, that is, they may be obtained by subtraction from appropriate row and column totals.*   It will be left as an exercise (Exercise 6 on page 253) to show that if the e_{ij} are calculated with the use of (11.1.1), the sum of the *expected* numbers of successes (or failures) must equal the sum of the *observed* numbers of successes (or failures).

Using χ^2, we can now test the null hypothesis formulated on page 247 with the following criterion (see Figure 11.1):

reject the null hypothesis if $\chi^2 > \chi^2_{.05}$; accept the null hypothesis (or reserve judgment) if $\chi^2 \leqslant \chi^2_{.05}$, where χ^2 is to be calculated by means of formula (11.1.2) and the number of degrees of freedom equals $k - 1$.

If we wanted to use a level of significance of 0.01, we would only have to substitute $\chi^2_{.01}$ for $\chi^2_{.05}$.

To illustrate the use of this criterion, let us now continue the numerical example which we studied earlier in this section and let us calculate χ^2 for the observed and expected frequencies shown in the 2 by 3 table on page 246. Substituting the given frequencies into (11.1.2), we obtain

$$\chi^2 = \frac{(12 - 11.0)^2}{11.0} + \frac{(15 - 13.2)^2}{13.2} + \frac{(6 - 8.8)^2}{8.8} + \frac{(88 - 89.0)^2}{89.0}$$
$$+ \frac{(105 - 106.8)^2}{106.8} + \frac{(74 - 71.2)^2}{71.2}$$
$$= 1.379$$

Since this is less than 5.991, the value given in Table III for $\chi^2_{.05}$ with $3 - 1 = 2$ degrees of freedom, *the null hypothesis cannot be rejected.* The discrepancies between the proportions of defectives may be attributed to chance and we shall conclude that the production process was under control, in other words, that the actual proportion of defectives remained constant.

To give another illustration of the technique presented in this section, let us consider the second example mentioned on page 245 and let us arrange the given information in the following *2 by 4 table:*

	Union A	Union B	Union C	Union D	Totals
Unfavorable votes	312	348	243	297	1200
Favorable votes	108	152	57	83	400
Totals	420	500	300	380	1600

Letting p_1, p_2, p_3, and p_4 stand for the actual proportions of members of Unions A, B, C, and D who are against the proposal, we shall want to test the hypothesis

Null hypothesis: $p_1 = p_2 = p_3 = p_4 \ (= p)$

against the alternative hypothesis that the four p's are not all the same.

Using (11.1.1) to obtain an estimate of p, the true proportion of unfavorable votes which under the null hypothesis is supposed to be the same for all 4 unions, we get

$$\frac{312 + 348 + 243 + 297}{1600} = \frac{1200}{1600} = 0.75$$

Multiplying this value by the sizes of the 4 samples, we find that the *expected numbers of unfavorable votes* are $420(0.75) = 315$, $500(0.75) = 375$, $300(0.75) = 225$, and $380(0.75) = 285$, respectively. (The fourth of these expected frequencies could also have been obtained by subtracting $315 + 375 + 225$ from 1200.) Subtracting the expected numbers of unfavorable votes from the totals of the respective samples, we find that the *expected numbers of favorable votes* are $420 - 315 = 105, 500 - 375 = 125, 300 - 225 = 75$, and $380 - 285 = 95$. Writing the e_{ij} which we have just calculated in parentheses below the corresponding n_{ij}, we get

	Union A	Union B	Union C	Union D
Unfavorable votes	312 (315)	348 (375)	243 (225)	297 (285)
Favorable votes	108 (105)	152 (125)	57 (75)	83 (95)

If we now substitute these observed and expected frequencies into formula (11.1.2), we get

$$\chi^2 = \frac{(312 - 315)^2}{315} + \frac{(348 - 375)^2}{375} + \frac{(243 - 225)^2}{225} + \frac{(297 - 285)^2}{285}$$
$$+ \frac{(108 - 105)^2}{105} + \frac{(152 - 125)^2}{125} + \frac{(57 - 75)^2}{75} + \frac{(83 - 95)^2}{95}$$

$$= 15.572$$

Assuming that the level of significance is to be 0.01, we find that $\chi^2_{.01}$ for $4 - 1 = 3$ degrees of freedom equals 11.345 (see Table III on page 503. Since the value which we obtained for χ_2 exceeds 11.345, the null hypothesis must be rejected and we shall conclude that the actual proportions of unfavorable votes for the given piece of legislation are *not* equal. In other words, we have shown that there is a *significant difference* between the four sample proportions, which were 0.74, 0.70, 0.81, and 0.78.

It may have occurred to the reader that for $k = 2$, the method discussed in this section provides an alternate criterion for testing the significance of the difference between *two* sample proportions. Actually, for $k = 2$ the method discussed here and the method discussed in Section 10.5 are *equivalent*. It can be shown that when $k = 2$ the value obtained for χ^2 with formula (11.1.2) equals the *square* of the z value obtained with formula (10.5.6). (See Exercises 7 and 8 on page 253.)

When calculating the expected frequencies of a 2 by k table, it is customary to round off to the nearest integer or to one decimal. The entries in Table III are given to three decimals, but there is seldom any need to carry more than two decimals when calculating χ^2, itself.

Since the sampling distribution of χ^2, as defined by means of formula (11.1.2), is only approximately the theoretical distribution on which Table III is based, the criterion given on page 249 should not be used when the expected frequencies are very small. *A relatively safe rule followed by many statisticians is to use the chi-square criterion only when none of the expected frequencies are less than 5.* If the chi-square criterion is to be used when one (or more) of the expected frequencies is less than 5, we can obey this rule by combining cells, i.e., by adding their respective expected and observed frequencies before calculating χ^2 (see page 258). If 2 cells are thus treated as one, it will also be necessary to subtract 1 from the number of degrees of freedom.

EXERCISES

1. Smith, Brown, and Jones perform the same task on a production line, soldering electrical connections. A sample of 400 soldered connections made by Smith yielded 38 defects, a sample of 400 soldered connections made by Brown yielded 56 defects, and a sample of 400 soldered connections made by Jones yielded 50 defects. Test the null hypothesis that there is no difference in the effectiveness of the three workers. Use a level of significance of 0.05.

2. To test whether the proportions of sons taking up the occupations of their fathers are equal for a selected group of occupations, a research organization took a random sample and obtained the following results:

	Accountants	Dentists	Professors	Musicians	Fireman
Same occupation	47	31	28	32	26
Different occupation	153	139	72	118	154

Find the expected cell frequencies under the null hypothesis that the true proportions of sons taking up the occupations of their fathers are the same for the given occupations, calculate χ^2, and test the null hypothesis at a level of significance of 0.01.

3. A research organization, interested in testing whether a new TV program has equal appeal for different age groups, took a random telephone sample of TV viewers after the first showing of the program and obtained the following results:

	Age Groups			
	Under 20	20–39	40–59	60 and over
Liked the TV show	153	72	46	36
Did not like the TV show	47	28	44	24

Find the expected frequencies under the null hypothesis that the true proportions of favorable opinions are the same, calculate χ^2, and test the null hypothesis at a level of significance of 0.01.

4. Repeat Exercise 7 on page 232, basing the test of significance on the χ^2 criterion. Compare the value of χ^2 obtained by means of formula (11.1.2) with the *square* of the z value obtained previously by means of formula (10.5.6).

5. Repeat Exercise 8 on page 232, basing the test of significance on the χ^2 criterion. Compare χ^2 with the *square* of the z value obtained previously with formula (10.5.6).

6. (*Theoretical Exercise*) Show that if the expected number of "successes" for the jth sample is calculated by means of the formula $n_{.j}(n_{1.}/n)$, the total of the *expected* numbers of successes will have to equal the total of the *observed* numbers of successes.

7. (*Theoretical Exercise*) If random samples of size n_1 and n_2 yield x_1 and x_2 successes, show that χ^2, calculated according to the method studied in this section, may be written in the form of

$$\chi^2 = \frac{(n_1 + n_2)(n_2 x_1 - n_1 x_2)^2}{n_1 n_2 (x_1 + x_2)[(n_1 + n_2) - (x_1 + x_2)]}$$

8. (*Theoretical Exercise*) Show that the *square* of the expression given for z in (10.5.6) equals the expression given for χ^2 in Exercise 7.

11.2 The Analysis of an r by k Table

The binomial distribution applies to situations in which each trial must result in one of two possible outcomes which, traditionally, have been labeled "success" and "failure." The methods we shall study in this section pertain to situations in which each trial must result in one of *more than two* possible outcomes. For example, a person who is interviewed by an opinion poll may be given the choice of being (1) for Candidate A, (2) for Candidate B, or (3) undecided; a piece of equipment may be classified as being (1) in perfect condition, (2) slightly defective but usable, or (3) defective and not usable; a consumer testing service may classify a product as (1) superior, (2) good, (3) average, or (4) poor.

Without going into a detailed study of the theory underlying situations such as those described above, we shall investigate only the special kind of problem illustrated by the following example: Let us suppose that the management of a firm wants to know how their employees feel about working conditions, particularly whether there are differences in sentiment between various departments, and that a study based on random samples of the employees of 4 departments yielded the results shown in the following *3 by 4 table:*

	Dept. A	Dept. B	Dept. C	Dept. D	Totals
Working conditions are very good	65	112	85	80	342
Working conditions are average	27	67	60	44	198
Working conditions are poor	8	21	15	16	60
Totals	100	200	160	140	600

The null hypothesis which we shall want to test is that *the distribution of the proportions of employees who think that working conditions are very good, average, and poor is the same for all four departments.*† The alternative hypothesis is that *the respective proportions are not all the same.*

To test this hypothesis we shall proceed as in Section 11.1. First, we shall find the *expected* frequencies under the assumption that the null hypothesis is true and then we shall calculate χ^2 and compare it with a suitable value of $\chi^2_{.05}$ obtained from Table III.

Duplicating the steps used on page 246, let us combine the four samples and estimate the actual proportions of employees who feel that working conditions are very good, average, and poor as

$$\frac{65 + 112 + 85 + 80}{600} = \frac{342}{600} = 0.57$$

$$\frac{27 + 67 + 60 + 44}{600} = \frac{198}{600} = 0.33$$

and $$\frac{8 + 21 + 15 + 16}{600} = \frac{60}{600} = 0.10$$

Multiplying these proportions by the totals of the four columns (the sizes of the four samples), we find that the *expected frequencies for the first row of our 3 by 4 table* are $100(0.57) = 57.0$, $200(0.57) = 114.0$, $160(0.57) = 91.2$, and $140(0.57) = 79.8$; that the *expected frequencies for the second row* are $100(0.33) = 33.0$, $200(0.33) = 66.0$, $160(0.33) = 52.8$, and $140(0.33) = 46.2$; and that the *expected frequencies for*

---

† To formulate this hypothesis more precisely, let us write p_{ij} for the *actual* proportion of the employees of Department j who choose alternative i. Here j equals 1, 2, 3, or 4, depending on whether we are referring to Department A, B, C, or D; and i equals 1, 2, or 3, depending on whether we are referring to the proportion of employees who think that working conditions are very good, average, or poor. With this notation we can write

Null hypothesis: $p_{11} = p_{12} = p_{13} = p_{14}$

$$p_{21} = p_{22} = p_{23} = p_{24}$$

(Since the 3 proportions for each department must add up to 1, it follows automatically that under the null hypothesis also $p_{31} = p_{32} = p_{33} = p_{34}$.)

the third row are $100(0.10) = 10.0$, $200(0.10) = 20.0$, $160(0.10) = 16.0$, and $140(0.10) = 14.0$.   Writing these expected frequencies below the corresponding observed frequencies of the 3 by 4 table, we now have

	Dept. A	Dept. B	Dept. C	Dept. D
Working conditions are very good	65 (57.0)	112 (114.0)	85 (91.2)	80 (79.8)
Working conditions are average	27 (33.0)	67 (66.0)	60 (52.8)	44 (46.2)
Working conditions are poor	8 (10.0)	21 (20.0)	15 (16.0)	16 (14.0)

The method which we used to determine the expected frequencies of the 3 by 4 table is identical with that used in Section 11.1.   In fact, if we had k samples and each "trial" permitted r possible outcomes (alternatives), the expected frequencies of the resulting r by k table could be calculated in the same way.   To find e_{ij}, the expected frequency for alternative i in sample j (for row i and column j), we would *estimate* the true proportion of cases falling into alternative i, *which under the null hypothesis is supposed to be the same for all k populations*, as $n_{i.}/n$.   Multiplying this proportion by $n_{.j}$, the number of trials in the jth sample, we find that

$$e_{ij} = n_{.j}\left(\frac{n_{i.}}{n}\right) = \frac{(n_{.j})(n_{i.})}{n} \tag{11.2.1}\star$$

It should be noted that any one e_{ij} may thus be found by multiplying the total of Row i by the total of Column j and then dividing by the grand total.   For example, in our numerical example we could have written

$$e_{23} = \frac{160 \cdot 198}{600} = 52.8, \qquad e_{14} = \frac{140 \cdot 342}{600} = 79.8, \ldots$$

To test the null hypothesis that *the distribution of the proportions associated with the r alternatives is the same for all k populations*, we shall again calculate χ^2 and use the following criterion:

reject the null hypothesis if $\chi^2 > \chi^2_{.05}$; accept the null hypothesis (or reserve judgment) if $\chi^2 \leqslant \chi^2_{.05}$, where χ^2 is to be calculated by means of formula (11.1.2) with the summation extending over all cells of the r by k table.   The number of degrees of freedom equals $(r - 1)(k - 1)$.†

---

† To explain this formula for the number of degrees of freedom, let us point out first that if the e_{ij} are calculated with (11.2.1), *the sum of the expected frequencies of any row or column must equal the sum of the corresponding observed frequencies*. It can then be shown that if $(r - 1)(k - 1)$ of the expected frequencies have been calculated with formula (11.2.1), all of the remaining e_{ij} may be obtained by subtraction from the totals of appropriate rows or columns.

If we wanted to change the level of significance from 0.05 to 0.01, we would only have to substitute $\chi^2_{.01}$ for $\chi^2_{.05}$.

Let us now apply this criterion to the numerical example in which we were interested in determining whether there is a difference in sentiment about working conditions between four departments of a given firm. Substituting the observed and expected frequencies of the 3 by 4 table shown on page 255 into formula (11.1.2), we get

$$\chi^2 = \frac{(65 - 57.0)^2}{57.0} + \frac{(112 - 114.0)^2}{114.0} + \frac{(85 - 91.2)^2}{91.2} + \frac{(80 - 79.8)^2}{79.8}$$
$$+ \frac{(27 - 33.0)^2}{33.0} + \frac{(67 - 66.0)^2}{66.0} + \frac{(60 - 52.8)^2}{52.8} + \frac{(44 - 46.2)^2}{46.2}$$
$$+ \frac{(8 - 10.0)^2}{10.0} + \frac{(21 - 20.0)^2}{20.0} + \frac{(15 - 16.0)^2}{16.0} + \frac{(16 - 14.0)^2}{14.0}$$
$$= 4.570$$

Using a level of significance of 0.05, we find that $\chi^2_{.05}$ for $(r - 1)(k - 1) = (3 - 1)(4 - 1) = 6$ degrees of freedom equals 12.592 (see Table III). Since the value which we obtained in our experiment was *less* than 12.592, we cannot reject the null hypothesis and we shall conclude that *there is no difference in sentiment about working conditions between the four departments*.

Since the analysis of an r by k table is practically identical with that of a 2 by k table, as discussed in Section 11.1, we shall not consider additional examples at this time. Further applications of the method discussed in this section will be treated later in Chapter 15.

EXERCISES

1. An attorney was asked to collect overdue accounts randomly selected from the files of a department store, a credit jeweler, and a furniture store. His collection efforts produced the following results:

	Department Store	Jeweler	Furniture Store
Collected completely	25	50	35
Received part payment	25	35	40
Could not collect	50	15	25

   Test the null hypothesis that the distribution of the true proportion of the items falling into the three categories is the same for the department store, jeweler, and furniture store. Use a level of significance of 0.05.

2. Random samples of eligible voters in 4 cities produced the following results:

	St. Louis	*Detroit*	*Baltimore*	*Hartford*
For Candidate A	80	75	55	40
For Candidate B	40	50	30	15
Undecided	30	40	25	20

Test the null hypothesis that the distribution of the true proportion of votes falling into the three categories is the same for all four cities. Use a level of significance of 0.05.

3. In late 1969 a bureau of business research polled a sample of 400 men engaged in different fields of business in a certain state, to determine whether there were any differences in attitude towards the prospects for over-all business activity in the coming year. The results of this poll were as follows:

	Retailing	*Services*	*Utilities*	*Manufacturing*
Increased activity	40	41	24	16
Decreased activity	32	80	61	52
No appreciable change	8	19	15	12

Test the null hypothesis that the distribution of the true proportions corresponding to the 3 categories is the same for all 4 kinds of businessmen. Use a level of significance of 0.05.

11.3 Tests of "Goodness of Fit"

The chi-square criterion, which we studied in the two previous sections, served as a measure of the *compatibility* of two sets of frequencies. We used it to test whether a set of observed frequencies differed sufficiently from a set of expected frequencies to reject the hypothesis under which the expected frequencies were obtained. Although we made this comparison only in the analysis of an r by k table, the chi-square criterion applies to many other situations in which we want to investigate whether differences between observed frequencies and expected frequencies may be attributed to chance.

For example, the chi-square criterion may be used to test whether it is justifiable to approximate an observed distribution by means of a normal curve. The reader may recall that in Section 7.8 we considered a distribution of the weights of 300 army recruits and calculated a set of frequencies which we referred to as "expected normal curve frequencies." These were the frequencies we could have expected to find in the various classes if we actually had a normal distribution with the same mean and standard deviation as our observed distribution. The values we obtained were

Weights (in pounds)	Observed Frequencies	Expected Normal Curve Frequencies
150–158	9	9.0
159–167	24	25.4
168–176	51	51.5
177–185	66	71.2
186–194	72	67.8
195–203	48	44.6
204–212	21	20.2
213–221	6 $\Big\}$ 9	6.3 $\Big\}$ 7.7
222–230	3	1.4

On page 164 we compared these two sets of frequencies by looking at the respective histograms; now we shall compare them on the basis of χ^2 as calculated by means of the formula

$$\chi^2 = \sum \frac{(n_i - e_i)^2}{e_i} \qquad (11.3.1)\star$$

This formula differs slightly from (11.1.2) inasmuch as we are now comparing two sets of frequencies arranged in *single columns* instead of the rectangular arrays of r by k tables. We still square the difference between each pair of frequencies, divide by the respective expected frequencies, and then determine the sum of the quotients $(n_i - e_i)^2/e_i$.

Substituting the observed frequencies and the expected normal curve frequencies of the weight distribution into (11.3.1), we get

$$\chi^2 = \frac{(9 - 9.0)^2}{9.0} + \frac{(24 - 25.4)^2}{25.4} + \frac{(51 - 51.5)^2}{51.5} + \frac{(66 - 71.2)^2}{71.2}$$
$$+ \frac{(72 - 67.8)^2}{67.8} + \frac{(48 - 44.6)^2}{44.6} + \frac{(21 - 20.2)^2}{20.2} + \frac{(9 - 7.7)^2}{7.7}$$
$$= 1.232$$

It should be noted that we combined the last two classes of the frequency table in accordance with the rule that the chi-square criterion should not be used unless each of the expected frequencies is at least equal to 5.

Using the value which we obtained for χ^2, we can now decide whether the normal curve provides a reasonably good fit to the original distribution. *If χ^2 is very small the fit is good, if it is large the fit is bad.* Actually, we shall test the hypothesis that *the observed distribution constitutes a sample from a population having a normal distribution* with the following criterion:

reject the hypothesis (or state that the fit is poor) if $\chi^2 > \chi^2_{.05}$; accept the hypothesis (or state that the fit is good) if $\chi^2 \leqslant \chi^2_{.05}$, where χ^2 is *to be calculated by means of formula (11.3.1) and the number of degrees of freedom equals k − 3. Here k is the number of terms in (11.3.1).*

In our numerical example $k - 3 = 8 - 3 = 5$ and, hence, $\chi^2_{.05} = 11.070$ (see Table III on page 503). Since 1.232 is less than this tabular value of $\chi^2_{.05}$, we conclude that the difference between the observed frequencies and the expected frequencies is not significant. *In other words, the normal curve provides a very good fit.*

To explain the formula for the number of degrees of freedom, let us point out that the expected frequencies we calculated in Section 7.8 had to satisfy 3 conditions. The sum of the observed frequencies had to equal the sum of the expected frequencies and the mean and standard deviation of the normal curve had to equal the mean and standard deviation of the observed distribution. *Generally speaking, in a chi-square test of goodness of fit the number of degrees of freedom equals the number of pairs of frequencies for which we compute $(n_i - e_i)^2/e_i$ minus the number of quantities, determined from the observed data, which are used in the calculation of the expected frequencies.*

To give another example in which chi-square is used to test goodness of fit, let us suppose that in Exercise 1 on page 130 we obtained the following distribution for 160 tosses of 4 coins:

Number of Heads	Observed Frequencies
0	17
1	52
2	54
3	31
4	6

Using the formula for the binomial distribution with $n = 4$ and $p = 0.50$ (see page 134), we find that under the assumption that the coins are balanced, the probabilities of getting 0, 1, 2, 3, or 4 heads are $\frac{1}{16}$, $\frac{4}{16}$, $\frac{6}{16}$, $\frac{4}{16}$, and $\frac{1}{16}$. If we now multiply each of these probabilities by the total frequency, 160, we obtain the *expected* frequencies shown in the following table:

Number of Heads	Observed Frequencies	Expected Frequencies
0	17	10
1	52	40
2	54	60
3	31	40
4	6	10

Substituting these frequencies into (11.3.1), we get

$$\chi^2 = \frac{(17 - 10)^2}{10} + \frac{(52 - 40)^2}{40} + \frac{(54 - 60)^2}{60} + \frac{(31 - 40)^2}{40} + \frac{(6 - 10)^2}{10}$$
$$= 12.725$$

and this measures how well the binomial distribution fits to our observed data. Actually, we shall test the hypothesis that *the tosses of the coins are random and that the probability of getting heads is 0.50 for each coin.* The alternative hypothesis is that *either the coins were not properly tossed or that at least one of the coins is not balanced.* The criterion which we shall use reads as follows:

reject the hypothesis (or state that the fit is poor) if $\chi^2 > \chi^2_{.05}$; accept the hypothesis (or state that the fit is good) if $\chi^2 \leqslant \chi^2_{.05}$, where χ^2 is to be calculated by means of formula (11.3.1) and the number of degrees of freedom equals $k - 1$. Again, k is the number of terms in (11.3.1).

It should be noted that this formula for the number of degrees of freedom agrees with the rule given on page 259. The total frequency was the only quantity, obtained from the observed distribution, which we used in the calculation of the expected frequencies.

Since the value which we obtained for χ^2 in our numerical example *exceeds* $\chi^2_{.05} = 9.488$, the value given in Table III for $\chi^2_{.05}$ with $k - 1 = 5 - 1 = 4$ degrees of freedom, we conclude that *the binomial distribution does not provide a good fit.* We shall accept the alternative hypothesis that either one or more of the coins are not balanced or that the coins were not properly flipped. It will be left to the reader to show that if we had used a level of significance of 0.01, we would not have been able to reject the hypothesis. We would then have stated that the difference between the two sets of frequencies is not significant or that *the binomial distribution provides a (fairly) good fit.*

EXERCISES

1. Test the goodness of the fit of the normal curve which was fitted to the mail-order distribution in Exercise 4 on page 164. Use a level of significance of 0.05.
2. Test the goodness of the fit of the normal curve which was fitted in Exercise 5 on page 165. Use a level of significance of 0.05.
3. Test the goodness of the fit of the normal curve fitted to the distribution of the number of physicians per 100,000 population. In Exercise 6 on page 165. Use a level of significance of 0.05.

4. Assuming that the expected normal curve frequencies given below were calculated according to the method of Section 7.8, test for goodness of fit at a level of significance of 0.05:

Observed Frequencies	Expected Normal Curve Frequencies
29	25
160	156
314	312
202	215
42	40
3	2

5. Use the χ^2 criterion to compare the observed and expected frequencies of Exercise 1 on page 130.   Test for goodness of fit at a level of significance of 0.05.

6. Use the χ^2 criterion to compare the observed and expected frequencies of Exercise 3 on page 130.   Test for goodness of fit at a level of significance of 0.01.

7. The following table contains a distribution obtained for 320 tosses of 5 coins.   The corresponding expected frequencies were calculated with the formula for the binomial distribution with $n = 5$ and $p = \frac{1}{2}$. Test for goodness of fit at a level of significance of 0.05.   State the hypothesis which is being tested and the conclusions reached.

Number of Heads	Observed Frequencies	Expected Frequencies
0	13	10
1	49	50
2	87	100
3	109	100
4	56	50
5	6	10

11.4 Tests Concerning k Means

To generalize the work of Section 10.7, let us now consider problems in which we must decide whether observed differences between *more than two* sample means may reasonably be attributed to chance. For example, we may wish to decide whether there really is a difference between 3 kinds of tires, if 5 tires made by Company A lasted on the average 22,150 miles, 5 tires made by Company B lasted on the average 21,400 miles, and 5 tires made by Company C lasted on the average 20,745 miles.   Similarly, we may wish to test whether there really is a difference in the yield of 4 varieties of corn if 6 test plots planted with

Variety A yielded on the average 81.7 bushels per acre, 6 test plots planted with Variety B yielded on the average 72.5 bushels per acre, 6 test plots planted with Variety C yielded on the average 78.7 bushels per acre, and 6 test plots planted with Variety D yielded on the average 83.2 bushels per acre. In both of these examples we are interested in deciding whether differences between several $\bar{x}$'s are significant or whether they may reasonably be attributed to chance.

To illustrate the procedure used in problems of this kind, let us suppose that we want to compare 3 brands of tires and that we are interested in particular in their average braking distances at 30 miles per hour, that is, the average distances required to stop cars going at that speed. Let us suppose, furthermore, that actual tests produced the following results (data given in feet):

	Brand A	Brand B	Brand C
	22	22	25
	21	25	29
	26	24	28
	23	25	30
Means	23	24	28

In this table we have 3 random samples of size 4 with *means* of 23, 24, and 28 feet. *What we would like to know is whether the difference between these means is significant.*

Letting μ_1, μ_2, and μ_3 stand for the *true* average braking distances (at 30 miles per hour) for the 3 kinds of tires, we shall want to test the hypothesis

Null hypothesis: $\mu_1 = \mu_2 = \mu_3$

against the alternative hypothesis that the three μ's are *not* all the same.

In order to obtain a measure of the size of the discrepancies between the three $\bar{x}$'s, let us calculate their *variance*, the square of their standard deviation. Using formula (4.4.3) given on page 85, we find that $s_{\bar{x}}^2$, the variance of the 3 means, equals

$$s_{\bar{x}}^2 = \frac{(23 - 25)^2 + (24 - 25)^2 + (28 - 25)^2}{3 - 1} = 7$$

This quantity will play an important role in our decision to accept or reject the null hypothesis formulated above. It stands to reason that the null hypothesis should be *rejected* if the difference between the $\bar{x}$'s and, hence, $s_{\bar{x}}^2$ is *very large*. On the other hand, the hypothesis should *not* be rejected if the difference between the $\bar{x}$'s is *very small*. The question that remains is "Where do we draw the line?"

Since the technique which we shall discuss would otherwise not apply, let us first make the assumption that our 3 samples came from populations having normal distributions with the same standard deviation σ. This means that we shall assume that if we tested many tires of each of the 3 brands, the distributions we would get for the braking distances would be close to normal curves having the same standard deviation.

Combining the assumptions made in the last paragraph with the assumption that the null hypothesis is true, we can look upon our 3 samples as samples from one and the same population. Consequently, $s_{\bar{x}}^2$ may be looked upon as an *estimate* of $\sigma_{\bar{x}}^2 = \sigma^2/n$ (see Chapter 8) and $n \cdot s_{\bar{x}}^2$ may be looked upon as an estimate of σ^2.

If σ^2 were known, we could compare $n \cdot s_{\bar{x}}^2$ with σ^2 and reject the null hypothesis if $n \cdot s_{\bar{x}}^2$ were much larger than σ^2. However, in our example (as in most practical problems) σ^2 is unknown and we have no choice but to estimate it on the basis of the given data. The estimate which we shall use for this purpose is simply *the average of the squares of the standard deviations of the individual samples.* Using formula (4.4.3) on page 85 for each of the 3 samples and averaging the results, we get

$$\frac{s_1^2 + s_2^2 + s_3^2}{3}$$

$$= \frac{1}{3}\left[\frac{(22-23)^2 + (21-23)^2 + (26-23)^2 + (23-23)^2}{3}\right.$$

$$+ \frac{(22-24)^2 + (25-24)^2 + (24-24)^2 + (25-24)^2}{3}$$

$$\left.+ \frac{(25-28)^2 + (29-28)^2 + (28-28)^2 + (30-28)^2}{3}\right]$$

$$= 3.78$$

where s_1, s_2, and s_3 are the standard deviations of the 3 samples.

We now have *two* estimates of σ^2. The first is

$$n \cdot s_{\bar{x}}^2 = 4 \cdot 7 = 28$$

and the second is

$$\frac{s_1^2 + s_2^2 + s_3^2}{3} = 3.78$$

Whereas the first estimate is based on the variation *between* the sample means, the second is based on the variation *within* the 3 samples and it may be looked upon as a measure of *chance variation. Since the first estimate we obtained is much larger than the second, it would seem reasonable to say that the variation between the means is too large to be accounted for by chance.*

Before we introduce the criterion on which the test of the original null hypothesis will be based, let us formulate what we have done in a more general fashion. Let us suppose that we have k random samples of size n from as many populations and that our observations, written as x_{ij}, are as shown in the following table:

1st Sample	2nd Sample	$\cdots$	kth Sample
x_{11}	x_{12}	$\cdots$	x_{1k}
x_{21}	x_{22}	$\cdots$	x_{2k}
x_{31}	x_{32}	$\cdots$	x_{3k}
$\cdots$	$\cdots$	$\cdots$	$\cdots$
x_{n1}	x_{n2}	$\cdots$	x_{nk}
Means　$\bar{x}_1$	$\bar{x}_2$	$\cdots$	$\bar{x}_k$

In this notation x_{ij} stands for the ith observation of the jth sample, that is, the value given in the ith row and the jth column. The first subscript is 1, 2, 3, . . . , or n, depending on whether we are referring to the first, second, third, or nth observation of the respective sample; the second subscript simply specifies the sample. Furthermore, the means of the k samples are written as $\bar{x}_1$, $\bar{x}_2$, $\bar{x}_3$, . . . , $\bar{x}_k$ and the *over-all mean*, the mean of all of the observations, is written as $\bar{x}$.

If we now write the means of the k *populations* as μ_1, μ_2, . . . , and μ_k, the hypothesis which we shall want to test is

Null hypothesis: $\mu_1 = \mu_2 = \cdots = \mu_k$

and the alternative hypothesis is that *these μ's are not all the same.* As in the numerical example, we shall assume that our k random samples came from populations having normal distributions with the same standard deviation σ.

Measuring, as before, the variation between the means in terms of their *variance* (the square of their standard deviation), we get

$$s_{\bar{x}}^2 = \frac{\sum\limits_{j=1}^{k} (\bar{x}_j - \bar{x})^2}{k - 1} \qquad (11.4.1)$$

according to (4.4.3). Since we are assuming that our samples came from populations having equal means and standard deviations—and, for that matter, equal normal distributions—we can look upon $s_{\bar{x}}^2$ as an estimate of $\sigma_{\bar{x}}^2 = \sigma^2/n$ or upon $n \cdot s_{\bar{x}}^2$ as an estimate of σ^2. If we now multiply $s_{\bar{x}}^2$ as given in (11.4.1) by n, we get

$$\frac{n \cdot \sum\limits_{j=1}^{k} (\bar{x}_j - \bar{x})^2}{k - 1} \qquad (11.4.2)\star$$

and we shall use this as our *first* estimate of σ^2. As was explained on page 98, the quantity $k - 1$ by which we divide in this formula is referred to as the *number of degrees of freedom.*

In our numerical example we based the second estimate of σ^2 on the variation *within* the samples, calculating it simply as the average of the squares of the standard deviations of the individual samples. Repeating this here, we shall write our second estimate of σ^2 as

$$\frac{s_1^2 + s_2^2 + \cdots + s_k^2}{k} \qquad (11.4.3)$$

where $s_1, s_2, \ldots$, and s_k are the standard deviations of the k samples. Substituting for each of these standard deviations an appropriate expression according to (4.4.3), it is easy to show that (11.4.3) may be written as

$$\frac{\sum\limits_{i=1}^{n} \sum\limits_{j=1}^{k} (x_{ij} - \bar{x}_j)^2}{k(n-1)} \qquad (11.4.4)\star$$

It should be noted that in this formula, which represents our *second* estimate of σ^2, we subtract $\bar{x}_j$, *the mean of the jth sample*, from x_{ij}, *the ith observation of the jth sample*, and then sum the squares of these deviations from the mean over all values of i and j. Since $n - 1$ of the deviations from the mean are *independent* in each of the k samples (see page 98), we have altogether k *times* $n - 1$ independent deviations, or $k(n - 1)$ *degrees of freedom*.

If the null hypothesis is true, the variation between the k means must be due to chance and (11.4.2) should be about as large as (11.4.4). *If (11.4.2) is much larger than (11.4.4), it would seem reasonable to conclude that the variation between the k means cannot be attributed to chance and that the null hypothesis must be rejected.* The reader will recall that (11.4.2) and (11.4.4) are both estimates of σ^2, but that the first describes the variation *between* the k means while the second describes the variation *within* the samples, that is, chance variation.

To put the comparison of (11.4.2) and (11.4.4) on a precise basis, let us introduce the statistic F, which is defined as the *ratio* of our two estimates of σ^2, namely, as

$$F = \frac{\text{Value obtained with (11.4.2)}}{\text{Value obtained with (11.4.4)}} \qquad (11.4.5)\star$$

The theoretical sampling distribution of this statistic is called the F-distribution. In case the reader has some difficulty in understanding the double subscripts and double summation in (11.4.4), he may put (11.4.3), *the mean of the squared sample standard deviations*, into the denominator of F.

Since the null hypothesis formulated on page 264 is to be rejected when (11.4.2) is significantly greater than (11.4.4), *when the variation between the means is significantly greater than it should be if it were due*

only to chance, we shall reject the null hypothesis when F is *large*.
Depending on the level of significance, we shall thus be interested in
$F_{.05}$ or $F_{.01}$, namely, in values which are such that 5 per cent or 1 per
cent of the area under the F-distribution lies to their right.   These
values, which will provide the dividing lines of our criterion, are given
in Tables IVa and IVb on pages 504 and 505.

Tables IVa and IVb differ from the corresponding tables for χ^2
and t inasmuch as their entries depend on *two quantities*, the number
of degrees of freedom for the numerator of F and the number of degrees
of freedom for its denominator.   In view of what we said above, the
number of degrees of freedom for (11.4.2) appearing in the numerator

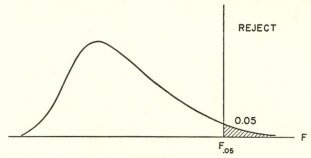

FIGURE 11.2.   An F-Distribution.

of F is $k - 1$ and the number of degrees of freedom for (11.4.4) appearing in the denominator of F is $k(n - 1)$.

Using the theory discussed above, we shall test the null hypothesis
$\mu_1 = \mu_2 = \cdots = \mu_k$ with the following criterion (see also Figure
11.2):

> *reject the null hypothesis if $F > F_{.05}$; accept the null hypothesis (or
> reserve judgment) if $F \leqslant F_{.05}$, where F is to be calculated by means
> of formula (11.4.5) and the number of degrees of freedom for the
> numerator and denominator of F equal $k - 1$ and $k(n - 1)$,
> respectively.*

If we wanted to change the level of significance from 0.05 to 0.01, we
would only have to substitute $F_{.01}$ for $F_{.05}$.

Returning now to our numerical example, the one dealing with the
braking distances of the 3 kinds of tires, we find that (11.4.2) and
(11.4.4) equalled 28 and 3.78 and that, therefore,

$$F = \frac{28}{3.78} = 7.41$$

Since this value *exceeds* 4.26, the value given in Table IVa for $F_{.05}$ with
$k - 1 = 3 - 1 = 2$ and $k(n - 1) = 3(4 - 1) = 9$ degrees of free-

dom, we find that the null hypothesis must be rejected. *This means that there is a significant difference between the 3 sample means and that we shall conclude that the true average braking distances for the 3 kinds of tires are not the same.*

In practice, the calculation of (11.4.2) and (11.4.4) may be simplified by using short-cut formulas analogous to those which we used for standard deviations. References to such formulas for (11.4.2) and (11.4.4) are given in the Bibliography on page 269. *As a word of caution, let us add that the technique discussed in this section should be used only if it is reasonable to assume that our samples come from populations which have equal standard deviations and which can be approximated closely with normal curves.*

EXERCISES

1. The following are 3 different weeks' earnings (in dollars) of 5 piece-rate production workers employed by a certain company:

Jones	Smith	Brown	Miller	Black
126	148	204	119	156
152	171	182	125	184
139	176	172	155	161

   Calculate F and, assuming that the necessary assumptions can be met, test at a level of significance of 0.05 whether the differences between the average earnings of these workers over the given period of time may be attributed to chance.

2. In order to compare the mileage yields of 4 different kinds of gasoline, several tests were run using the gasoline in a motor scooter. The following results were obtained: (each figure represents the number of miles obtained with a gallon of the respective gasoline.)

Gasoline A	Gasoline B	Gasoline C	Gasoline D
62	38	62	58
42	44	57	70
48	58	68	62
56	58	60	60
62	52	58	65

   Calculate F and, assuming that the necessary assumptions can be met, test at a level of significance of 0.05 whether the observed differences between the means obtained for the 4 kinds of gasoline may be attributed to chance.

3. Random samples of ten students taken from 4 very large sections of a course in marketing contain the following grades:

Section 1	Section 2	Section 3	Section 4
55	86	72	58
74	99	85	76
92	73	67	85
58	84	53	49
72	76	74	66
67	83	70	71
51	94	81	64
89	95	74	88
72	61	63	92
80	79	51	41

Calculate F and, assuming that the necessary assumptions can be met, test at a level of significance of 0.01 whether the differences between the means of the 4 samples may be attributed to chance.

4. (*Theoretical Exercise*) Show, symbolically, that when $k = 2$, the value of F obtained according to the method of this section equals the *square* of the value given for t by formula (10.7.5) with $n_1 = n_2 = n$. This shows that when $k = 2$ and the two samples are of the same size, the small sample t-test discussed in Section 10.7 and the F-test of Section 11.4 are *equivalent*.

11.5 Analysis of Variance

The method which we studied in the preceding section belongs to a very important part of statistics called the *analysis of variance*. To explain the meaning of this term, let us look upon the observations x_{ij} as *one* sample of size $k \cdot n$ and let us write its *variance* (the square of its standard deviation) as

$$s^2 = \frac{\sum\limits_{i=1}^{n} \sum\limits_{j=1}^{k} (x_{ij} - \bar{x})^2}{kn - 1} \tag{11.5.1}$$

Rearranging the numerator of (11.5.1), it can be shown that

$$\sum_{i=1}^{n} \sum_{j=1}^{k} (x_{ij} - \bar{x})^2 = \sum_{i=1}^{n} \sum_{j=1}^{k} (x_{ij} - \bar{x}_j)^2 + n \cdot \sum_{j=1}^{k} (\bar{x}_j - \bar{x})^2 \tag{11.5.2}$$

and we can, thus, express the numerator of (11.5.1) as the *sum* of the numerators of (11.4.2) and (11.4.4). In other words, in (11.5.2) we analyze the numerator of (11.5.1), measuring the *total variation* of our data, into *two* quantities, the first being a measure of *chance variation* (see page 265) and the second being a measure of the *variation between the means*.

The analysis of variance provides extremely powerful tools for the analysis of experimental data. It enables us to analyze the total variation of our data into components which may be attributed to various "sources" or "causes" of variation. For instance, if we had planned the experiment involving the braking distances in a more elaborate fashion, we might have been able to analyze the total variation of our data into different components caused by, say, differences between the 3 brands of tires, differences between cars used in the test, differences in road condition, and perhaps differences in the reflexes of the drivers. References to fairly elementary introductions to the subject of analysis of variance are given below.

BIBLIOGRAPHY

An informal treatment of the chi-square criterion and its various applications may be found in

Moroney, M. J., *Facts from Figures*. London: Penguin Books, 1956, Chap. 15.

A more formal treatment of the methods of Sections 11.1 and 11.2 may be found in

Siegel, S., *Nonparametric Statistics for the Behavioral Sciences*. New York: McGraw-Hill, 1956, Chaps. 6 and 8.

Elementary introductions to the analysis of variance and related problems of experimental design are given in the book (chapter 19) by Moroney mentioned above and, among others, in

Dixon, W. J., and Massey, F. J., *Introduction to Statistical Analysis*, 2nd ed. New York: McGraw-Hill, 1957, Chap. 10.
Ferber, R., *Statistical Techniques in Market Research*. New York: Mc Graw-Hill, 1949, Chap. 10.

Short-cut formulas for the computation of the sums of squares in (11.4.2) and (11.4.4) may be found in each of the books listed above.

For a more advanced treatment of the theory underlying the chi-square and F-distribution see

Freund, John E., *Mathematical Statistics*. Englewood Cliffs, N.J.: Prentice-Hall, Inc., 1962.
Hoel, P. G., *Introduction to Mathematical Statistics*, 3rd ed. New York: John Wiley, 1962.

CHAPTER 12

Problems of Sampling

12.1 Introduction

All of the methods which we studied in the last few chapters were based on the assumption of *randomness*. In everyday language randomness means a certain haphazardness or lack of bias which is supposed to assure that no single item or groups of items belonging to a population are preferred, avoided, or distorted during the process of sampling. *In this sense randomness really refers to methods which we use in obtaining a sample, particularly, to the things which we must avoid so that we can rightfully say that we have a random sample.*

In Chapter 8 we defined randomness, at least for samples from *finite* populations, by saying that each item contained in the population must have an equal chance of being included in the sample. To this end we suggested the use of gambling devices or, preferably, random numbers. A sample which is selected in such a way that each item in the population has a known (not necessarily equal) chance of being chosen is appropriately called a *probability sample*.

In contrast to probability samples, we shall refer to a sample as a *judgment sample*, if in addition to (or instead of) chance, *personal judgment* plays a role in the selection of the sample. For instance, if a market research organization feels that a certain town in New Jersey is typical of communities of its size in the United States, a survey based solely on this town would constitute a judgment sample *the moment generalizations were made beyond this particular town.*

Although there may be situations where, for practical reasons, an element of judgment must replace probability in the selection of a sample, such samples have the undesirable feature that we cannot apply any of the methods which we have studied to evaluate the accuracy of estimates or to calculate the probabilities of making various kinds of errors. The evaluation of the "goodness" of estimates or decisions based on judgment samples is again a matter of personal judgment.

The question of randomness is one factor which must always be considered in the selection of a sample. *Another factor is whether we are sampling from the correct population.* To illustrate what is meant here by the word "correct," let us consider the classical example of the poll made by the now defunct *Literary Digest* in the 1936 presidential election. The *Literary Digest* predicted a Republican victory on the basis of a very large sample taken from names listed in telephone directories and from lists of automobile registrations. Although the results obtained by this poll may have been quite indicative of voter preferences in the population from which the sample was drawn, it was definitely not random with respect to the population of all eligible voters. It so happened that President Roosevelt's support came largely from the lower income groups which were not adequately represented among the owners of telephones and automobiles. Consequently, the *Literary Digest* poll was biased, it was not a random sample from the correct population, namely, the population of all eligible voters, and the effect of this error of sampling was disastrous to the magazine.

The same problem presents itself in a different way in the analysis of mail questionnaires. Let us suppose, for instance, that a manufacturer of electric ranges encloses with the guarantee of each range a self-addressed card and requests that the new owner check on this card whether the new range replaces another electric range or one using natural gas, bottled gas, wood, or coal. Let us suppose, furthermore, that of the 5,420 cards distributed 1,562 were returned and, of these, 683, or 44 per cent, showed that the new range replaced an old electric range. This raises the question whether a percentage based on returned questionnaires can be used as an estimate of the corresponding percentage for the entire population, in this case the population of all the ranges that were replaced with an electric range made by the given firm. A question like this is difficult to answer, particularly if a mail questionnaire involves issues that might be affected by snobbery, prejudice, embarassment, or by issues that are apt to arouse emotional reactions. Persons who are vitally interested in an issue touched upon in a mail questionnaire are more likely to return the questionnaire and this can easily introduce a bias. In such a situation we have the choice of either conducting a follow-up study by writing again or making personal visits to some of the individuals who did not return the questionnaires or relying on personal judgment.

Let us always remember that *predictions, estimations, and decisions that are made on the basis of samples apply only to the particular popula-*

tions from which the samples were obtained.   Any generalizations to
other populations are either pure guesswork or they must be preceded
by an extremely careful analysis.   In either case, theory based on the
assumptions of probability sampling would no longer apply.

12.2 A Test of Randomness

So far we have discussed the problem of randomness only in connection with the selection of samples drawn from *finite populations.*
The definition given on page 170 does not apply to samples from
infinite populations; it does not apply to a sample of, say, 100 flips
from the infinite population consisting of all possible flips of a given
coin.

The question of randomness also raises difficulties in problems in
which we have little or no control over the selection of our data.   If
we wanted to predict a store's volume of Christmas sales, we would
have no choice but to use data based on previous years and, perhaps,
collateral information about general economic conditions.   None of
this information constitutes a random sample in the sense that it was
obtained with the use of random numbers or other probability schemes.
Similarly, we have no choice but to rely on available records if we want
to estimate the mortality rate of a disease, if we want to make long-
range predictions of the weather, or if we want to evaluate the merits
of a stock or bond.   In situations like these it is often very difficult
to decide whether available information may be treated as if it constitutes a random sample.

In the remainder of this section we shall limit our discussion to
specific kinds of *bias or nonrandomness* that can be detected by studying
the *internal structure* of a sample.   To illustrate what is meant here by
"internal structure," let us suppose that a market research organization sends out 3 investigators, instructing them to interview samples
of 30 housewives and to ask, among other things, whether they prefer
to buy a certain product in round or in square bottles.   If each investigator merely records the number of preferences he obtains for each
type of bottle, there is no way in which we can check whether there is
any reason to suspect that the samples were not random.   However,
if each investigator lists the *order* in which his preferences for round
and square bottles were obtained, there is the possibility that we might
detect certain patterns or regularities which would make it unreasonable to refer to their samples as random.   To study the internal structure of a sample, it will thus be necessary to record the individual items
(measurements or observations) that constitute a sample in the precise
order in which they were obtained.

Letting r denote a preference for round bottles and s a preference for square ones, let us suppose that the worksheets of the 3 investigators showed the following raw data:

Investigator 1

r r r r r r r r r r r r r r r s s s s s s s s s s s s s s s

Investigator 2

r s r s r s r s r s r s r s r s r s r s r s r s r s r s r s

Investigator 3

r r s r s s s r r s r r r s s r s s s s r r s r s s r s r r

We can now ask whether any features of these r's and s's might be interpreted as evidence that any or all of these samples are not random.

The sample of the first investigator has the very unusual feature that the first 15 housewives he interviewed preferred round bottles while the last 15 preferred square ones. Although this could be due to chance, we might suspect that the investigator began his sample in a neighborhood where women shop mostly in a store which sells the product in one kind of bottle and that he then moved to a neighborhood where women shop in a store which sells the product in the other.

The data obtained by the second investigator are even more difficult to understand or explain. In any case, it would seem quite unreasonable to accept this regular alternation of r's and s's as a random sample or a random arrangement.

Finally, the data obtained by the third investigator appear to agree with the intuitive notion of randomness. At least, there is no obvious reason why we should suspect that this arrangement of r's and s's is not random.

Several techniques have been developed in recent years to test hypotheses about the randomness of samples on the basis of the order in which the individual items constituting the samples were obtained. The method which we shall treat in this section is based on the so-called *theory of runs.* Alternate methods are referred to in the Bibliography on page 284.

A run is defined as a succession of identical letters which is followed and preceded by a different letter or by no letter at all. Taking the results of Investigator 3, for example, we find that he first had a run of *two* r's, then a run of *one s*, then a run of *one r*, then a run of *three s*'s, then a run of *two r*'s, then a run of *one s*, and so on. Using braces to combine the letters which constitute the various runs, it can easily be seen that the sample of Investigator 3 has the following 17 runs:

rr	s	r	sss	rr	s	rrr	ss	r	ssss	rr	s	r	ss	r	s	rr
1	2	3	4	5	6	7	8	9	10	11	12	13	14	15	16	17

The total number of runs appearing in a sample of this kind is often a good indication of a possible lack of randomness. If there are *too few runs*, we might suspect a significant grouping or clustering of like elements or, perhaps, a trend, as will be explained later in Section 12.3. If there are *too many runs*, we might suspect that there is a cyclic pattern (see page 278).

To construct a criterion on the basis of which we will be able to decide whether u, the observed number of runs, is too large or too small, let us suppose that a sample contains n_1 letters of one kind and n_2 of another. (Which we refer to as n_1 and which as n_2 is immaterial.) We can then study the *sampling distribution of the statistic u*, namely, the distribution which we could expect to obtain for the total number of runs if we repeatedly used random methods to order n_1 letters of one kind and n_2 of another. To construct an experimental sampling distribution of u for given values of n_1 and n_2, we could label n_1 slips of paper with the letter a, n_2 slips with the letter b, repeatedly draw them one at a time out of a box, goldfish bowl, or urn, and in each case count the total number of runs.

Although we shall not attempt to prove it, let us use the fact that the *mean* of the theoretical sampling distribution of the total number of runs in a random arrangement of n_1 letters of one kind and n_2 of another is

$$\mu = \frac{2n_1 n_2}{n_1 + n_2} + 1 \qquad (12.2.1)\star$$

and that its *standard deviation* is

$$\sigma_u = \sqrt{\frac{2n_1 n_2 (2n_1 n_2 - n_1 - n_2)}{(n_1 + n_2)^2 (n_1 + n_2 - 1)}} \qquad (12.2.2)\star$$

It has also been shown that unless n_1 or n_2 is very small the sampling distribution of u can be approximated closely with a normal curve and we can, therefore, test the null hypothesis that a sample is random against the alternative hypothesis that it is not random with the following criterion:

reject the null hypothesis if z > 1.96 or z < −1.96; accept the null hypothesis (or reserve judgment) if −1.96 ⩽ z ⩽ 1.96, where

$$z = \frac{u - \mu}{\sigma_u} \qquad (12.2.3)\star$$

and μ and σ_u are to be calculated by means of formulas (12.2.1) and (12.2.2).

This criterion provides a *two-tail test* at a level of significance of 0.05. To change the level of significance to, say, 0.01, we would only have to replace 1.96 with 2.58.

It is difficult to make an exact statement as to how large n_1 and n_2 must be before the normal curve approximation and the above criterion may be used. Tables which make it possible to conduct an *exact* test when n_1 and n_2 are both less than or equal to 20 are referred to in the Bibliography on page 285. We shall use the normal curve approximation so long as neither n_1 nor n_2 is less than 10.

Returning now to the sample of Investigator 3 (see page 273), we find that $n_1 = 15$, $n_2 = 15$, $u = 17$ and that according to (12.2.1) and (12.2.2)

$$\mu = \frac{2(15)(15)}{15 + 15} + 1 = 16$$

and

$$\sigma_u = \sqrt{\frac{2(15)(15)[2(15)(15) - 15 - 15]}{(15 + 15)^2(15 + 15 - 1)}} = 2.69$$

Substituting these values into (12.2.3), we get

$$z = \frac{17 - 16}{2.69} = 0.37$$

and since this value falls between -1.96 and 1.96, *there is no reason to suspect that the sample of Investigator 3 is not random.*

If we applied the same test to the samples of Investigators 1 and 2, we would get $n_1 = 15$, $n_2 = 15$, $u = 2$ for the first; $n_1 = 15$, $n_2 = 15$, $u = 30$ for the second; and we would find that both samples are *significantly nonrandom.*

To give another illustration, let us consider the following arrangement of successive orders for buying (B) and selling (S) received at a certain stock exchange:

S S B S S S S S B B S S B S S S S
(cont.) S S B B S S S B S S S S B S S S B
(cont.) B S S S S S S B S S S B B S S S S
(cont.) B S S S S B S S S B B B S S S S S
(cont.) B B B S S B B B B S B B B S S S B
(cont.) B B B B B B S B B S B S S B B

Since $n_1 = 60$ and $n_2 = 40$, formulas (12.2.1) and (12.2.2) yield

$$\mu = \frac{2(60)(40)}{60 + 40} + 1 = 49$$

and

$$\sigma_u = \sqrt{\frac{2(60)(40)[2(60)(40) - 60 - 40]}{(60 + 40)^2(60 + 40 - 1)}} = 4.77$$

Substituting these values together with $u = 38$ into (12.2.3), we get

$$z = \frac{38 - 49}{4.77} = -2.31$$

and we can *reject* the null hypothesis of randomness at a level of significance of 0.05. A careful study of the data shows that the lack of randomness may very well be due to the fact that there was a trend from selling to buying. At first most orders were S's and later on most were B's.

EXERCISES

1. At a meeting of the yacht club, the owners of Sailboats and owners of Motorboats seated themselves in the following arrangement:

```
        S  S  S    M  M  M  M  M  M  M  S  S  S  S  S  M  S
(cont.) S  S  S    M  M  M  M  M  M  S   S  S  S  M  M  S  S
(cont.) S  S  M    M  M  M  M  S  M  M   M  M  S  S  S  S
```

Use a level of significance of 0.05 to test whether this arrangement is random.

2. The following is the arrangement of a number of men and women lined up at the box office of the college athletic stadium:

```
        w  w  m  m  m  m  w  w  m  w  m  w  m  w  m  w
(cont.) m  m  w  m  m  m  w  m  w  w  m  m  m  w  m  w
(cont.) m  w  m  w  m  m  m  w  m  w  m  m  m  w  m  m
(cont.) w  m
```

Test at a level of significance of 0.05 whether this arrangement is random.

3. Reading successive rows from left to right, the following is the order in which defective (D) and non-defective (N) pieces were produced by a machine during a certain period of time:

```
N N N N N N N N N D D D N N N N N N N N
N N N D N N N N N N N N N D D N N D D N N
N N N N N N N N N N N N N D D D N N D N
D N N D N N N N N N N N D D N N N D N N
N N N N N N N N D D D N N N N N N N N N
```

Test at a level of significance of 0.05 whether this arrangement is random.

4. Flip a coin 100 times and record an H for each head and a T for each tail. Then test at a level of significance of 0.01 whether this arrangement of H's and T's is random.

5. The following is the order in which codfish and mackerel were caught from a fishing boat:

```
        c  c  c  m  c  c  c   c  c  m  m  m  c  c   c  c  c
(cont.) c  c  c  c  c  c  m  m  m  c   c  c  c  m  m  m  m
```

Test for randomness at a level of significance of 0.05.

6. The theory which we have discussed in this section may also be used to test whether two samples come from populations with equal means, that is, *it may be used as a substitute for testing the significance of the difference between two sample means.* Given samples from two populations, we simply combine the two samples and rank (arrange) *all* of the observations in an increasing or decreasing order. Writing an *A* below observations belonging to the first sample and a *B* below those belonging to the second, we can then test whether this arrangement of *A*'s and *B*'s is random. If there are *too few runs*, we shall use this as an indication that the samples come from populations with unequal means. In other words, we shall reject the null hypothesis $\mu_1 = \mu_2$ at a level of significance of 0.05 if z, as calculated by means of (12.2.3), is less than -1.64. If 12 test runs with a gallon of one kind of gasoline yielded

> 18.2, 19.8, 17.4, 17.7, 20.3, 19.9,
> 18.4, 20.6, 19.6, 17.5, 20.2, and 20.5 miles,

while 12 test runs with a gallon of a second kind of gasoline yielded

> 19.3, 20.0, 21.7, 23.5, 20.9, 22.0,
> 23.1, 19.4, 21.0, 22.3, 19.0, and 21.3 miles,

test at a level of significance of 0.05 whether the true average mileage yields of the two kinds of gasoline are the same. (For a more detailed treatment of this application of the theory of runs see Bibliography on page 285.)

12.3 Runs Above and Below the Median

The usefulness of the method discussed in the last section is not limited to testing the randomness of series of attributes such as the *r*'s and *s*'s, *B*'s and *S*'s of our two illustrations. Any sample that consists of numerical measurements or observations can be treated similarly by denoting each number falling *below the median of the sample* with the letter *b* and each number falling *above the median* with the letter *a*. Numbers that are equal to the median are omitted and the resulting series of *a*'s and *b*'s can then be tested for randomness with the method of Section 12.2.

To illustrate this technique, let us consider the following data constituting the number of defective pieces in samples of size 400 taken from large lots of machine parts produced on 40 consecutive days: 6, 9, 12, 11, 5, 9, 8, 10, 4, 1, 7, 10, 6, 6, 14, 7, 8, 9, 10, 2, 3, 5, 9, 12, 11, 11, 4, 10, 13, 9, 7, 8, 7, 9, 4, 5, 2, 11, 3, and 6. The median of these numbers is 8 and we can, therefore, write

> b a a a b a a b b b a b b a b a a b b
> (cont.) b a a a a b a a a b b a b b b a b b

Since $n_1 = 19$, $n_2 = 18$, and $u = 19$, substitution into (12.2.1) and (12.2.2) yields

$$\mu = \frac{2(19)(18)}{19 + 18} + 1 = 19.5$$

and
$$\sigma_u = \sqrt{\frac{2(19)(18)[2(19)(18) - 19 - 18]}{(19 + 18)^2 (19 + 18 - 1)}} = 3.00$$

and (12.2.3) becomes

$$z = \frac{19 - 19.5}{3.00} = -0.17$$

Since this value lies between -1.96 and 1.96, the critical values for a level of significance of 0.05, we cannot reject the hypothesis that the given arrangement of a's and b's is random. Hence, we shall conclude that the numbers of defectives produced on the 40 days may be looked upon as a random sample.

The method of runs above and below the median is particularly useful in the detection of *significant trends or cyclic patterns*. If there is a trend, there will first be mostly a's and later mostly b's (or vice versa) and if there is a cyclic pattern there will generally be a definite alternation of a's and b's. In the first case the number of runs will be too small and in the second case it will be too large. We thus have a useful technique for testing whether an apparent trend or cyclic pattern may be attributed to chance. (If we wanted to test the null hypothesis of randomness against the specific alternative that *there is a trend*, we would use a *one-tail test* and reject the null hypothesis at a level of significance of 0.05 if u is too small, namely, if the corresponding z value is less than -1.64. If the alternative hypothesis is that *there is a cyclic pattern*, we would reject the null hypothesis if z exceeds 1.64.)

EXERCISES

1. A fishing vessel recorded the following catches, in pounds, for 30 consecutive fishing days:

    739, 760, 752, 735, 717, 692, 745, 751, 743, 741,
    (cont.) 718, 720, 764, 725, 745, 754, 761, 682, 714, 729,
    (cont.) 705, 731, 726, 719, 703, 694, 738, 734, 674, 726

    Test at a level of significance of 0.05 whether these numbers may be looked upon as a random sample.

2. Use the method of runs above and below the median to test whether it is reasonable to treat the "downtimes" of Exercise 8 on page 29 as a random sample. Read successive rows across and use a level of significance of 0.05. (This median was found in Exercise 5 on page 65.)

3. Use the method of runs above and below the median to test whether it is reasonable to treat the number of physicians per 100,000 population in Exercise 12 on page 31 as a random sample. Read successive rows and use a level of significance of 0.05.

4. The following are the yearly averages of the *Wholesale Price Index* (1957–59 = 100) (for all commodities) as reported by the Bureau of Labor Statistics for the years 1937–1966: 47.2, 43.0, 42.2, 43.0, 47.8, 54.0, 56.5, 56.9, 57.9, 66.1, 81.2, 87.9, 83.5, 86.8, 96.7, 94.0, 92.7, 92.9, 93.2, 96.2, 99.0, 100.4, 100.6, 100.7, 100.5, 100.6, 100.3, 100.5, 102.5, 105.9. Test at a level of significance of 0.05 whether this arrangement is random or whether there is a significant trend.

12.4 Sample Designs

The methods which we have discussed in Chapters 8 to 11 were always based on the assumption that we are dealing with random samples. Although, for problems involving finite populations, we have considered only *simple random* sampling, that is, sampling in which each item of the population has an equal chance of being included, there are many instances, particularly in surveys, where such sampling is very difficult and highly impractical. Logically speaking, we could sample with random numbers after assigning a number to each farm in the United States, to each eligible voter in California, to each housewife in Chicago, to each tree in the Rockies, . . . , but it is hardly necessary to point out that such a procedure would be extremely costly and, perhaps, even physically impossible. In situations like these it is desirable and usually necessary to replace simple random sampling with alternate sample designs. *A sample design is a definite plan, completely determined before any data are collected, for obtaining a sample from a given population.*

Since the subject of sample designs covers a considerable amount of material, we shall merely attempt to present a brief introduction to such special sampling schemes as *stratified sampling, quota sampling, cluster sampling,* and *systematic sampling.* More detailed treatments of this very important part of statistics are mentioned in the Bibliography on page 285.

In *stratified sampling* we divide the population into sub-populations, *strata,* to which we allocate specific portions of the total sample. Such sampling can actually improve the accuracy of estimates and predictions since we are making sure that the various strata that make up the total population are all represented in our sample. If the items selected from each stratum constitute a simple random sample, this entire procedure, first stratification and then simple random selection, is called *stratified random sampling.* In many instances, of course, it is not feasible to make the ultimate choice of items by simple random selection, in which case alternate methods, to be discussed below, will have to be used.

An important question that arises in stratified sampling is how to allocate the sample to the various strata. Let us suppose, for instance, that income level is a factor which has an important bearing on public opinion concerning a given election and that it is known that 40 per cent of the voters have low incomes, 50 per cent have medium incomes, while 10 per cent have high incomes. If this is the case, we can improve the accuracy of our prediction of the outcome of the election by allocating 40 per cent of the total sample to members of the low-income group, 50 per cent to the medium-income group, and the remaining 10 per cent to the high-income group. Such sampling is referred to as *proportional stratified sampling:* we use subsamples which are proportional in size to the size of the strata. If, in the above example, we wanted to take a total sample of 400 individuals, we would select $400(0.40) = 160$ from the low-income group, $400(0.50) = 200$ from the medium-income group, and $400(0.10) = 40$ from the high-income group.

Let us now suppose that we want to take a poll to predict a national election and that we have decided to stratify by states. Although we could allocate our sample proportionally, that is, proportional to the number of voters in each state, it would seem only reasonable to allocate a larger portion to a state in which the election is apt to be close than to a state in which the outcome is a foregone conclusion. In a situation like this it may be desirable to use what is called *optimum allocation* of a stratified sample. If we wanted to estimate the percentage of the total vote that Candidate X will get in a national election, we would be using optimum allocation if we determine the sample size for each state, or stratum, so that the standard error of the statistic which we use as our final estimate will be minimized.

A further refinement in stratified sampling consists of stratifying the population with respect to *several* relevant characteristics. Had we known, for instance, that nationality background and geographic location are additional factors having an important bearing on public opinion concerning the above-mentioned election, we could have allocated a portion of our sample to persons with low incomes who belong to nationality group A and live in the South, another portion to persons with high incomes who belong to nationality group B and live in the West, and so forth. Such a process is referred to as *cross-stratification.*

It is important to remember that when we use stratified sampling, regardless of the method of allocation, all of the standard error formulas which we studied in previous chapters must be modified. Suitable references to the necessary modifications are given in the Bibliography on page 285.

Another form of sampling used sometimes in public opinion polls and market surveys goes by the name of *quota sampling*. In a quota sample, each interviewer is asked to question specified numbers of individuals having given characteristics. For instance, a person working for a public opinion poll may be asked to interview 5 retail merchants of German origin who own homes, 10 wage earners of Anglo-Saxon origin who live in rented apartments, 2 retired persons who live in trailers, and the like. A quota sample differs from a stratified sample inasmuch as the "filling of the quota" is usually left entirely at the interviewer's discretion. Since interviewers are apt to select individuals who are easy to reach, perhaps, individuals who work in the same building, shop in the same store, or reside in the same general area, *quota samples are essentially judgement samples and, therefore, do not lend themselves to any sort of statistical evaluation.*

To illustrate another kind of sampling, let us suppose that we want to study family expenditures in the Los Angeles area and that we have decided to interview a total of 4,000 families. In a problem like this, the use of simple random sampling poses difficulties, for suitable lists are usually hard to obtain and the cost of visiting families scattered over a wide area can be very high. A more appropriate method of sampling would be to divide the total area in which we are interested into smaller areas, say, city blocks, and then to interview all (or samples of) families in a number of randomly selected blocks. Such sampling is referred to as *cluster sampling;* we divide the total population into relatively small subdivisions, or clusters, and then use random numbers to choose the particular clusters on which the sample is to be based. If the clusters are geographic subdivisions, as they were in our example, cluster sampling is also referred to as *area sampling.*

To give further illustrations of cluster sampling, let us suppose that the management of a large chain-store wants to interview a sample of its employees to determine their opinion about a pension plan. If they used random methods to select, say, 3 or 5 of their stores in which to interview employees, the resulting sample would be a cluster sample. Also, if the dean of a university wanted to know how fraternity men at his institution feel about a certain regulation, he would obtain a cluster sample if he interviewed the members of a number of randomly selected fraternities.

Although estimates based on cluster samples are generally not as reliable as estimates based on simple random samples of the same size, they are usually more reliable *per unit cost.* Referring again to the survey of family expenditures in the Los Angeles area, it is easy to see that it may well be possible to obtain a cluster sample several times the size of a simple random sample at the same expense. It is much cheaper to visit and interview families living close together in clusters than families selected at random over a wide area.

The question of subdividing a population into strata or clusters is one important aspect of sample design. Another important aspect is how to select a sample from each stratum, from each chosen cluster, or, for that matter, from the entire population. Although we could always use simple random sampling, there are many instances in which the most practical method of selecting a sample is to take every 5th, 10th, 20th, . . . , item of a numbered list or other arrangement that represents the population (or subpopulation). Such sampling is referred to as *systematic sampling*. Although a systematic sample is not a random sample as defined in Section 8.1, it is usually quite reasonable to use statistical theory based on random sampling if a sample includes, say, every 10th voucher in a file, every 20th name in a telephone directory, every 50th piece coming off an assembly line, and so forth.

One random element in systematic sampling is the choice of the number with which to start. For instance, if we wanted to obtain a 5 per cent sample of the names on a certain mailing list, we would first use random numbers or some other gambling device to pick (with equal probabilities) a number from 1 to 20. If, by chance, we selected 7, we would then include in our systematic sample the 7th, 27th, 47th, . . . , item on the list.

Whether or not a systematic sample may be treated as if it were a simple random sample depends entirely on the structure (order) of the population. In many instances systematic sampling actually provides an improvement over simple random sampling inasmuch as the sample is spread "more evenly" over the entire population. The main danger in systematic sampling lies in *hidden periodicities*. Our results would be quite biased, for instance, if we inspected every 10th piece coming off an assembly line and it happened that owing to a defect in the machine every 10th piece has imperfections. Similarly, a systematic sample might yield misleading results if we interviewed the residents of every 10th house along a certain route and it happened that each 10th house is in a choice location, say, on a corner lot.

In practice, it is often desirable to use in the same survey several of the methods which we have discussed. For instance, if government statisticians wanted to study living conditions on American farms, they might first stratify by states or some other geographic subdivision. To obtain a sample for each stratum, they might then use cluster sampling, subdividing each stratum into a number of small geographic units; and, finally, to obtain samples from the chosen clusters, they might use 100 per cent (exhaustive) sampling, simple random sampling, or systematic sampling.

12.5 Double, Multiple, and Sequential Sampling

The term *single sampling* is used when we make an estimate or test a hypothesis on the basis of one sample whose size is determined before any data are actually collected. The disadvantage of single sampling lies in the fact that the predetermined sample size is often unnecessarily large. Let us suppose, for instance, that a public opinion poll wants to predict a gubernatorial election and that it is decided to use a sample of size 1,000. *If the contest turned out to be very one-sided, it could well be that a sample of size 200 or 400 would have sufficed.*

The sampling plans we shall discuss in this section have the advantage that, *since the data are collected in several stages, we can often reach decisions on the basis of smaller samples and, hence, at a lower cost.* First, let us consider *double sampling.* To illustrate the meaning of this term, let us suppose that we want to inspect lots of manufactured products and that we decide to use the following scheme: we first take a random sample of size 50, accepting the lot if the number of defectives is 2 or less, rejecting the lot if the number of defectives is 7 or more, and taking an additional sample of size 100 if the number of defectives is 3, 4, 5, or 6. If the second sample is needed, we combine the two samples, accepting the lot if the total number of defectives is 6 or less, rejecting it if the total number of defectives is 7 or more.

In double sampling we thus begin with a relatively small sample and if the results are not decisive we supplement it with another. The advantage of this procedure is that if the quality of a lot is *very high or very low,* this will immediately become apparent on the basis of a relatively small first sample and the second sample will not be needed.

In a handbook of industrial statistics, the double-sampling criterion which we described might be presented, formally, in the following fashion:

Double Sampling Plan

Sample	Sample size	Combined Samples		
		Size	Acceptance number	Rejection number
First	50	50	2	7
Second	100	150	6	7

A further refinement which can produce even greater savings consists of beginning with a small sample and adding further samples until a decision can be reached. Such a procedure is referred to as *multiple sampling.* In the industrial example used above, we might use the following multiple sampling scheme:

Multiple Sampling Plan

Sample	Sample size	Combined Samples		
		Size	Acceptance number	Rejection number
First	20	20		3
Second	20	40	1	4
Third	20	60	2	5
Fourth	20	80	3	6
Fifth	20	100	5	7
Sixth	20	120	6	8
Seventh	20	140	7	8

We begin with a sample of size 20, rejecting the lot if the number of defectives is 3 or more and continuing with a second sample if the number of defectives is less than 3. If the second sample is needed, we combine the two samples, accepting the lot if the total number of defectives is 1 or less, rejecting it if the total number of defectives is 4 or more, and continuing with a third sample if the total number of defectives is 2 or 3. We thus proceed until the lot is either accepted or rejected.

Carrying the ideas presented in this section one step further, we could take observations *one at a time,* deciding after each observation whether to accept the hypothesis which we are testing, whether to reject it, or whether to continue sampling. Such a procedure is referred to as *sequential sampling.* An appropriate sequential sampling plan for our industrial example would be similar in form to the double and multiple sampling plans given above, although we would have to give acceptance and rejection numbers corresponding to each additional observation. More detailed treatments of the sampling plans discussed in this section are referred to below. Any of the plans discussed in this section may be set up in such a way that a hypothesis is tested against a given alternative with specified probabilities of committing Type I and Type II errors.

BIBLIOGRAPHY

Tests of randomness which may be used instead of the one based on the total number of runs are given in

Hoel, P. G., *Introduction to Mathematical Statistics*, 3rd ed.   New York: John Wiley, 1962.

Sprowls, R. C., *Elementary Statistics for Students of Social Science and Business*.   New York: McGraw-Hill, 1955, Chap. 11.

Wallis, W. A., and Roberts, H. V., *Statistics: A New Approach*.   Glencoe, Ill.: Free Press, 1956, Chap. 18.

Tables which make it possible to conduct an exact test for randomness based on the total number of runs are given in

Freund, J. E., and Williams, F. J., *Elementary Business Statistics.   The Modern Approach*.   Englewood Cliffs, N.J.: Prentice-Hall, Inc., p. 445.

for n_1 and n_2 both less than or equal to 20.

The application of theory of runs to testing the hypothesis that two samples come from populations with equal means is treated in more detail in

Siegel, S., *Nonparametric Statistics for the Behavioral Sciences*.   New York: McGraw-Hill, 1956, Chap. 6.

Informal introductions to the subject of sample design may be found in

Neter, J., and Wasserman, W., *Fundamental Statistics for Business and Economics*, 3rd ed. Boston: Allyn and Bacon, Inc. 1966, Chap. 8.

Spurr, W. A., and Bonini, C. P., *Statistical Analysis for Business Decisions*. Homewood, Ill.: Richard Irwin, 1967, Chap. 11.

in the above-mentioned book by Sprowls (Chap. 6), and in many other introductory texts.   More advanced treatments are given in

Cochran, W. G., *Sampling Techniques*, 2nd ed.   New York: John Wiley, 1963.

Deming, W. E., *Some Theory of Sampling*.   New York: Dover Publications, 1966.

Hansen, M. H., Hurwitz, W. N., and Madow, W. G., *Sample Survey Methods and Theory* (2 volumes).   New York: John Wiley, 1953.

These books contain detailed discussions of the reliability of estimates and decisions based on various kinds of sample designs.

For further information about double, multiple, and sequential sampling see, for instance,

Bowker, A. H., and Lieberman, G. J., *Handbook of Industrial Statistics*. Englewood Cliffs, N.J.: Prentice-Hall, Inc., 1955.

Dodge, H. F., and Romig, H. G., *Sampling Inspection Tables*, 2nd ed. New York: John Wiley, 1959.

Statistical Research Group, Columbia University, *Sampling Inspection*. New York: McGraw-Hill, 1948.

CHAPTER 13

Linear Regression

13.1 Introduction

The foremost objective of many statistical investigations in business and economics is to predict—that is, to forecast such things as the potential market for a new product, the future value of a piece of property, the growth of an industry, or over-all economic conditions. Although it would be desirable to make *infallible* predictions, forecasts based on statistical data will seldom fit this description. Basing their predictions on statistical information, businessmen and economists can at best formulate their forecasts in terms of probabilities, being satisfied if they are right a high percentage of the time or if their predictions are *on the average* reasonably close.

Whenever possible, scientists strive to express (or approximate) relationships between variables, namely, relationships between quantities that are known and quantities that are to be predicted, in terms of mathematical equations. This approach has been very successful in the physical sciences, where it is known, for instance, that (at constant temperature) the relationship between the volume (y) of a gas and its pressure (x) is given by the formula

$$y = \frac{k}{x}$$

where k is a numerical constant. Similarly, in biological science it has been discovered that the relationship between the size of a culture of bacteria and the time that it has been exposed to certain favorable conditions may be written as

$$y = ab^x$$

where y stands for the size of the culture, x for time, while a and b are numerical constants.

Although we introduced the two equations of the previous paragraph with reference to problems in physics and biology, they apply equally well to describe relationships in other fields. Businessmen and economists have borrowed and continue to borrow liberally from the tools of the natural sciences. Among others, the two aforementioned equations are often used to describe relationships between such things as the total consumption and the price of a commodity, a company's sales and the number of years it has been in operation, the size of an order and unit cost, and so on.

One of the simplest and most widely used equations for expressing relationships in various fields in the linear equation† which is of the form

$$y = a + bx \qquad (13.1.1)$$

Here a and b are numerical constants and once they are known we can calculate a predicted value of y for any given value of x by direct substitution. *Linear equations are useful and important not only because there exist many relationships that are actually of this form, but also because they often provide close approximations to complicated relationships which would otherwise be difficult to describe.*

In this chapter we shall limit our discussion to relationships which, mathematically speaking, can be represented by means of an equation of the form $y = a + bx$. More complicated kinds of relationships and their equations, among others the demand and growth curves mentioned above, will be treated briefly in Sections 18.5, 18.6, and 18.7.

The term "linear equation" owes its name to the fact that, when plotted on ordinary graph paper, all pairs of values of x and y which satisfy an equation of the form $y = a + bx$ will fall on a straight line. To illustrate, let us consider the equation

$$y = 21.9 + 3.3x$$

whose graph is shown in Figure 13.1. To give it some meaning, let us suppose that x stands for the July rainfall in inches (in a certain location) and that y stands for the corresponding yield of corn in bushels per acre. If in a given July there are 5 inches of rain, we find by substitution that the (predicted) yield of corn for that year is $y = 21.9 + 3.3(5) = 38.4$ bushels per acre. Similarly, if the July rainfall is 3 inches we get by substitution a predicted yield of corn of

---

† More correctly, $y = a + bx$ should be referred to as a *linear equation in two unknowns*. Linear equations in more than two unknowns will be mentioned briefly in Section 13.5.

$y = 21.9 + 3.3(3) = 31.8$ bushels per acre, and if the July rainfall is
1 inch we get a predicted yield of $y = 21.9 + 3.3(1) = 25.2$ bushels
per acre.   Taking many such values of x, calculating the correspond-

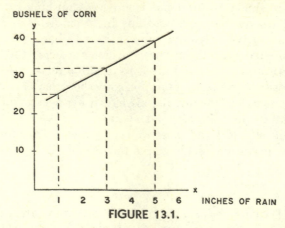

BUSHELS OF CORN

FIGURE 13.1.

ing values of y, and plotting the corresponding points, the reader can
easily check for himself that all these points will fall on the line of
Figure 13.1.

13.2 The Method of Least Squares

To illustrate the general problem of *linear curve fitting*—in other
words, the problem of fitting a straight line to a set of data consisting
of paired observations of two variables x and y—let us consider the
following concrete example: let us suppose that a firm wants to predict
new salesmen's total first year sales on the basis of a special aptitude
test prepared by its personnel department.   Let us suppose, furthermore, that the following data constitute a random sample from the
company's files:

	Test Score	First Year Sales (thousands of dollars)
	x	y
Mr. Brown	48	312
Mr. Smith	32	164
Mr. Jones	40	280
Mr. Taylor	34	196
Mr. Black	30	200
Mr. Miller	50	288
Mr. Green	26	146
Mr. Roberts	50	361
Mr. White	22	149
Mr. Baker	43	252

Plotting the points corresponding to these 10 salesmen as we have done in Figure 13.2, it is apparent that although the points do not actually fall on a straight line, they are reasonably close to the dotted line shown in that diagram.   Deciding that a straight line will give a fairly good description of the relationship between a salesman's test score and his first year sales, *we now face the problem of finding the equation of the line which in some sense provides the best possible fit to our data and, perhaps, later on will yield the best possible predictions.*

Anyone who has had the experience of analyzing data which were plotted as points on a piece of graph paper has probably felt the urge

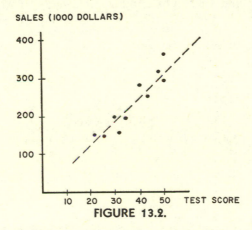

SALES (1000 DOLLARS)

FIGURE 13.2.

to take a ruler, juggle it around, and find in this fashion a straight line which presents a fairly good fit.   Of course, there is no law to prevent our doing this, but it is highly questionable whether we should trust our eyesight and aesthetic judgment in choosing a line which is supposed to yield the best possible predictions.   Another argument against freehand curve fitting is that it is largely subjective and, hence, that there is no way in which we will be able to evaluate the merits of subsequent predictions.

Since there is, logically speaking, no limit to the number of lines which one can draw on a piece of paper, *we will need some criterion on the basis of which we will be able to single out one line as the one that provides the best fit to our data.*   The choice of such a criterion is self-evident only in the very special case where all points actually fall on a straight line; but we can hardly expect this to happen very often when dealing with statistical data.   In general we will have to be satisfied with a straight line which, although it does not pass through all the points, will have some less perfect yet still desirable properties.

The method which nowadays is used almost universally for fitting straight lines (and other curves) to numerical data was proposed early in the nineteenth century by the French mathematician Adrien Legendre and it is known as the *method of least squares*. The criterion of least squares, as it is to be used in this chapter, demands that the line which we fit to our data be such that *the sum of the squares of the vertical deviations (distances) from the points to the line be a minimum*. With reference to our numerical example, the method of least squares requires that for the line which we fit to our data the sum of the squares of the solid line segments of Figure 13.3 be a minimum. The logic behind this least squares approach may be explained as follows: Mr. Baker, one of the 10 salesmen mentioned above, had a test score of 43 and total first year sales of $252 thousand. If we used the line of Figure 13.3 to "predict" his first year sales, we could do so by

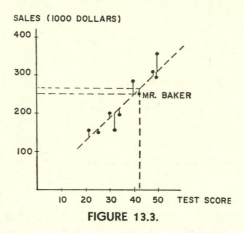

FIGURE 13.3.

either substituting $x = 43$ into the equation of the line or by reading his predicted sales directly off the diagram. As can be seen from Figure 13.3, *the vertical distance from the point representing Mr. Baker to the dotted line is the difference between his actual first year sales and his predicted first year sales and it, thus, measures the error of this prediction.*

The reason we minimize the sum of the *squares* of the vertical deviations (or errors) and not their sum is that some of the differences between the observed values of y and the corresponding values on the line are positive while others are negative. If we considered the sum of the deviations we would run into the difficulties which we encountered with the standard deviation, namely, that this sum can be zero even though the deviations are large.

Since it makes no difference whether we call one variable x and the other y or vice versa, *we shall agree to reserve y for the variable which is to be predicted in terms of the other*. (If we wanted to predict x in terms of y, we would have to apply the method of least squares differently and minimize the sum of squares of the *horizontal deviations* from the line, namely, the sum of squares between the observed values of x and the corresponding values on the line.   (See Exercises 7 and 8 on page 296.)

To show how a least squares line may be fitted to a set of data, let us consider n pairs of numbers (x_1, y_1), (x_2, y_2), . . . , and (x_n, y_n), which might, for example, represent the heights and weights of n persons, stockmarket averages and department stores sales at n different times, temperature and attendance at n picnics, or automobile registration and total gasoline sales in n different towns.   Let us suppose, furthermore, that the line which we fit to these data has the equation

$$y' = a + bx$$

using the symbol y' to differentiate between *observed* values of y and the corresponding values calculated by means of the equation of the

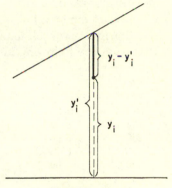

FIGURE 13.4.

line.   For each given value x_i ($i = 1, 2, . . . ,$ or n) we thus have a *given* value y_i and a *calculated* value y'_i, obtained by substituting x_i into $y'_i = a + bx_i$.

If the line given by this equation were found by the method of least squares, the numerical values of a and b would have to be such that after substituting the x_i and calculating the y'_i

$$\sum_{i=1}^{n} (y_i - y'_i)^2 \qquad\qquad (13.2.1)$$

is as small as possible (see Figure 13.4).   Since the derivation of the formulas which the method of least squares yields for a and b is somewhat involved, requiring either calculus or a great deal of algebra, let us merely state the result as

$$b = \frac{n \left(\sum\limits_{i=1}^{n} x_i y_i \right) - \left(\sum\limits_{i=1}^{n} x_i \right) \left(\sum\limits_{i=1}^{n} y_i \right)}{n \left(\sum\limits_{i=1}^{n} x_i^2 \right) - \left(\sum\limits_{i=1}^{n} x_i \right)^2} \qquad (13.2.2)\star$$

$$a = \frac{\sum\limits_{i=1}^{n} y_i - b \cdot \sum\limits_{i=1}^{n} x_i}{n} \qquad (13.2.3)\star$$

Here, n is the number of pairs of observations, Σx_i and Σy_i are the sums of the given x's and y's, Σx_i^2 is the sum of the squares of the x's, and $\Sigma x_i y_i$ is the sum of products of the corresponding x's and y's. It should be noted that by using these formulas we first calculate b with (13.2.2) and then use this result in (13.2.3) to find a. (Derivations of the two formulas are referred to in the Bibliography on page 305.)

Returning to our numerical example, let us now use formulas (13.2.2) and (13.2.3) to fit a least squares line to the test scores and first year sales. Copying the first two columns from page 288, we get the needed summations from the following kind of table:

x	y	x^2	xy
48	312	2,304	14,976
32	164	1,024	5,248
40	280	1,600	11,200
34	196	1,156	6,664
30	200	900	6,000
50	288	2,500	14,400
26	146	676	3,796
50	361	2,500	18,050
22	149	484	3,278
43	252	1,849	10,836
375	2,348	14,993	94,448

This gives us $n = 10$, $\Sigma x_i = 375$, $\Sigma y_i = 2,348$, $\Sigma x_i^2 = 14,993$, $\Sigma x_i y_i = 94,448$, and upon substituting these values into (13.2.2) we get

$$b = \frac{10(94,448) - (375)(2,348)}{10(14,993) - (375)^2} = \frac{63,980}{9,305} = 6.88$$

Substituting this value (which was rounded to two decimals), together with $n = 10$, $\Sigma x_i = 375$, and $\Sigma y_i = 2,348$, into (13.2.3), we get

$$a = \frac{2,348 - 6.88(375)}{10} = -23.20$$

and we can finally write the equation of the line as

$$y' = -23.20 + 6.88x$$

By using this formula we can now forecast a new salesman's first year sales on the basis of his score on the aptitude test. For instance, if a new applicant receives a score of 45, we shall predict that his first year sales will be

$$y' = -23.20 + 6.88(45)$$

or \$286.4 thousand. How such a prediction is to be interpreted and how its "goodness" may be evaluated will be discussed in the next two sections.

Instead of calculating a and b by means of (13.2.2) and (13.2.3), these constants may also be obtained by solving the following two equations, called the *normal equations:*

$$\sum_{i=1}^{n} y_i = na + b \sum_{i=1}^{n} x_i \qquad (13.2.4)^\star$$

$$\sum_{i=1}^{n} x_i y_i = a \sum_{i=1}^{n} x_i + b \sum_{i=1}^{n} x_i^2 \qquad (13.2.5)^\star$$

Had we used these equations in our numerical example, we would have had to solve the following two simultaneous linear equations

$$2{,}348 = 10a + 375b$$

$$94{,}448 = 375a + 14{,}993b$$

Using the method of elimination or any one of the other methods discussed in algebra texts, we would again get $a = -23.20$ and $b = 6.88$. If the reader is not familiar with solving simultaneous linear equations, he should find it easier to use (13.2.2) and (13.2.3). Actually, the two sets of formulas will yield identical results.

The two normal equations are fairly easy to remember, easier than (13.2.2) and (13.2.3). *We have only to remember that (13.2.4) may be obtained by putting a summation sign before each term of $y = a + bx$, which gives us $\Sigma y = \Sigma a + \Sigma bx$, and, according to the rules of Chapter 1, this may be rewritten as $\Sigma y = na + b \cdot \Sigma x$. Similarly, (13.2.5) may be obtained by first multiplying both sides of $y = a + bx$ by x and then putting a summation sign in front of each term.*

The calculation of a and b can sometimes be simplified by changing the scale of x so that the sum of the new x's is equal to zero. How this is done in the special problem of fitting least squares trend lines to time series will be discussed in Section 18.3.

EXERCISES

1. The following are data (in millions of short tons) on *shipments of steel for converting and processing* and *shipments of bolts, nuts, rivets and screws.* (Source: *Statistical Abstract of the United States 1967.*)

Shipments of steel	x	1.6	2.7	3.9	3.8	2.9
Shipments of bolts, etc.	y	0.7	1.1	1.4	1.5	1.0

   (a) Use formulas (13.2.2) and (13.2.3) to find a and b for the equation of the line which will enable the prediction (estimate) of the amount of *shipments of bolts,* etc., in terms of *shipments of steel.*

   (b) Construct suitable scales for x and y and plot the original data as well as the line obtained in (a) in one diagram.

   (c) Estimate the *shipments of bolts,* etc., for a year in which the *shipments of steel* are 3.0 (*i*) by substituting $x = 3.0$ into the equation of the line and (*ii*) by reading the estimate off the diagram obtained in (b).

2. Repeat (a) of Exercise 1, using the two normal equations to find a and b.

3. An educational association wished to predict a community's percentage of professional men and women on the basis of the community's median income. For this purpose, the association conducted a random sample of 8 communities in a 15-state area. The results of this sample were as follows:

Per Cent of Professional Men and Women	Median Community Income
6.3	$3,816
17.0	7,020
6.0	4,438
10.2	5,387
6.7	4,915
7.2	4,244
12.5	6,215
23.9	7,566

   (a) Use formulas (13.2.2) and (13.2.3) to find the equation of a line which will enable us to predict a community's median income on the basis of the percentage of its population that are professional men and women.

   (b) Use the equation obtained in (a) to predict a community's median income if 12 per cent of its population are professional men and women.

4. In a shelter study made by a state welfare bureau for the years 1967-68, it is desired to find the relationship between the county assessed

valuation of dwellings (x) and the property upkeep per month (y). A random sample of 10 homes is taken, yielding the following data:

Assessed Valuation	Property Upkeep per Month
$ 3,000	$23.00
11,000	52.00
7,000	31.00
9,000	41.00
10,000	54.00
17,000	76.00
5,000	24.00
12,000	55.00
10,000	45.00
6,000	23.00

(a) Use formulas (13.2.2) and (13.2.3) to compute a and b from these data and write the equation of the least squares line.

(b) Plot the original data and the line on one diagram.

(c) Use the least squares equation to estimate the monthly property upkeep for a house assessed at $6,000.

5. The following are the ages and second-hand prices charged for a certain make two-door sedan:

Age (years)	Price (dollars)
1	2,125
6	1,495
2	1,795
4	1,485
8	1,345
5	1,295
10	795
1	2,295

Use formulas (13.2.2) and (13.2.3) or the two normal equations to find the equation of the least squares line which will enable us to predict the second-hand price of a two-door sedan of the given make on the basis of its age. What value would we obtain with this equation if we wanted to estimate the second-hand price of a 3 year old model of this kind of car? Explain why it would be unreasonable to use this equation for $x = 20$, that is, for a twenty year old car?

6. A manufacturer of farm tools and supplies wishes to study the relationship between his sales and the income level of farmers in a certain area. Taking a sample of 11 counties, recording for each the median income of farm families (x) in 1968 and his corresponding sales (y), both in thousands of dollars, he gets:

Median Income	Sales
13	28
9	19
14	32
10	24
8	17
9	20
13	30
12	26
8	15
14	33
7	16

(a) Use either formulas (13.2.2) and (13.2.3) or the normal equations to find the equation of the straight line which best fits these data.

(b) Plot the original data as well as the line on one diagram.

(c) Estimate this manufacturer's sales of farm tools and supplies for a county in which the median farm family income is $11,000.

7. (a) Use formulas (13.2.2) and (13.2.3) to fit a least squares line of the form $y = a + bx$ to the following data:

x	1	2	3	4	5
y	18	9	13	18	27

(b) Suppose that on the basis of the data given in (a) we wanted to predict x in terms of y by means of an equation of the form $x = c + dy$. Applying the method of least squares *so that the sum of the squares of the horizontal deviations from the line is minimized,* find c and d by using formulas (13.2.2) and (13.2.3), respectively, with x replaced by y and y replaced by x wherever these symbols occur in the two formulas.

(c) Show that if the equation obtained in (a) is solved for x it will not be the same as the equation obtained in (b). Plot both lines as well as the original data on one diagram.

8. Suppose that in Exercise 3 we wanted to predict median community income in terms of per cent of professional men and women by means of an equation of the form $x = c + dy$. Find c and d with formulas (13.2.2) and (13.2.3), respectively, substituting x for y and y for x wherever these symbols occur in the two formulas. Measuring x as usual along the horizontal scale and y along the vertical scale, plot the original data as well as the line obtained in this exercise, and indicate the deviations, the sum of whose squares we minimized by applying the method of least squares.

13.3 Linear Regression

Let us suppose that the personnel department of the firm referred to in our example uses the equation $y' = -23.20 + 6.88x$

and predicts that an applicant who scores 35 in the aptitude test will sell $y' = -23.20 + 6.88(35) = 217.6$ or \$217,600 worth of merchandise during the first year and that someone who scores 50 will sell $y' = -23.20 + 6.88(50) = 320.8$ or \$320,800.

As we pointed out in the beginning of this chapter, predictions based on statistical information are seldom infallible and it would certainly be unreasonable to expect that each applicant who scores 35 will have total first year sales of \$217,600 and that each applicant

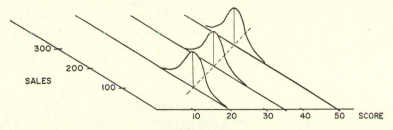

FIGURE 13.5. Conditional Sales Distributions for Fixed Test Scores.

who scores 50 will have total first year sales of \$320,800. As a matter of fact, we find on page 288 that Mr. Miller, who scored 50 on the test, had total first year sales of \$288,000 and that Mr. Roberts, who also scored 50, had total first year sales of \$361,000.

In view of the fact that we cannot expect predictions based on equations like the one used in our example to be perfect, it would seem desirable, at least, to be able to say that applicants who score 35 will *on the average* have first year sales of \$217,600 and that applicants who score 50 will *on the average* have first year sales of \$320,800. We could then assert our predictions by saying that we *expect* a man who scores 35 in the test to have total first year sales of \$217,600, interpreting the word "expect" as a mathematical expectation, namely, as an average.

To illustrate this situation graphically, let us consider Figure 13.5 in which we have shown (in perspective) several distribution curves. One represents the distribution of the total first year sales of *all* applicants who score 35 on the test, a second represents the distribution of the total first year sales of *all* applicants who score 50 on the test, a third represents the distribution of the total first year sales of *all* applicants who score 20 on the test, and the reader should be able to visualize similar distributions for other values of x. These distributions are usually referred to as the *conditional distributions* of y (total first year sales) for fixed values of x (test scores). (By "all applicants" we mean the population of all past, future, or potential applicants who obtain the respective scores in the company's aptitude test.)

The curve that is formed by the *means* of the distributions illustrated in Figure 13.5 might, appropriately, be called a *curve of means* (see Figure 13.5), but it is customarily referred to as a *regression curve*.† In particular, if it is a straight line as in Figure 13.5, it is called a *regression line*. To indicate that it contains the means of y for fixed values of x, it is referred to more specifically as a *regression line of y on x.*

In this chapter we shall consider only problems in which the *true* regression curve (the curve containing the true means of y for fixed values of x) is a straight line, and we shall write its equation as

$$y = \alpha + \beta x$$

In the same sense in which we previously used μ and σ for the true means and standard deviations of populations and $\bar{x}$ and s for the means and standard deviations of samples, we shall now use $y = \alpha + \beta x$ for the equation of the line containing the true means of y and $y' = a + bx$ for the *estimated regression line* obtained with the method of least squares. Thus the value which formula (13.2.2) yields for a is only an estimate of α (alpha) and the value which formula (13.2.3) yields for b is only an estimate of β (beta). This symbolism is conventional, but it is unfortunate in the sense that beginners have been known to confuse these symbols with the ones used for the probabilities of committing Type I and Type II errors (see page 219).

It stands to reason that if we duplicated the example which we used in this chapter with different data, the aptitude scores and sales records of different salesmen, we could hardly expect to end up with the same line, that is, the same values for a and b. These values are only estimates based on samples and therefore, subject to chance variation like all the other statistics discussed in previous chapters.

The constants a and b which we obtain with the method of least squares are usually referred to as the *estimated regression coefficients;* the quantities which they are supposed to estimate, namely, α and β, are called the (true) *regression coefficients.*

Returning now to our numerical example, we will have to take into account that $y' = -23.20 + 6.88x$, the line which we obtained with the method of least squares, is only an *estimate* of the line containing the true average first year sales of individuals obtaining given scores

---

† The origin of the term "regression" as it is used here is explained in the book by Sprowl referred to on page 306.

on the test.   Before using it for purposes of prediction, it might be well to investigate any one or all of the following questions:

(1) Since a and b are only estimates of α and β, what can we say about the "goodness" or "precision" of these estimates?†
(2) If we used the equation to estimate (predict) *average first year sales* of individuals scoring, say, 35 on the test, what could we say about the "goodness" of this estimate?
(3) How can we obtain *limits* (two numbers) between which we can expect a man's first year sales to fall, say, with a probability of 0.95, if he scores 35 on the test?

In reply to the first two questions, we could construct *confidence intervals for α and β and a confidence interval for the true average first year sales of individuals scoring 35 on the test.*   In principle, this would not differ from our work in Chapter 9, but since the underlying theory is fairly advanced and the resulting formulas are quite complicated we shall not take it up at this time.   Suitable references to the construction of such confidence intervals are given on page 306.   To answer the third question we could construct so-called *limits of prediction* and since this introduces a new concept we shall go into it briefly in Section 13.4.   *It is worth noting that none of the above questions could be answered if we fitted a freehand line.   They can be answered (with some qualification) if we use the method of least squares.*

13.4 Limits of Prediction‡

In Chapters 9 and 10 we learned that to evaluate the accuracy of an estimate or the "goodness" of a decision we must know something about the *variability* of the population or populations from which our samples are obtained.   (All of our methods involved in some way an estimate of a population standard deviation.)   To answer any one of the questions posed in the previous section, we will similarly have to obtain some information about the variability of the population or populations from which our sample is obtained.   *This variability is usually measured by the standard deviations of the distributions of Figure 13.5, namely, the standard deviations of the conditional distributions of y for fixed values of x.*

---

† In some studies the estimation of β is of utmost importance as it is a measure of the *change in y* associated with a *unit change in x.*

‡ The technique discussed in this section is of a somewhat more advanced nature and it may be omitted without loss of continuity.

Since the data that are used in problems of this kind usually consist of *very few* values of y for any given value of x, we would have difficulties getting anywhere in estimating these standard deviations unless we assumed that they are all the same. *We shall, therefore, assume that the distributions illustrated in Figure 13.5, the conditional distributions of y for fixed values of x, all have the same standard deviation.* Whether or not this assumption is defensible in any given problem is often difficult to decide; since we would have to go considerably beyond the scope of this book, we shall not consider this question at this time.

Having made the assumption that the true standard deviation of the y's is the same for each value of x, we can look upon the vertical deviations from the line (shown again in Figure 13.6) as *deviations*

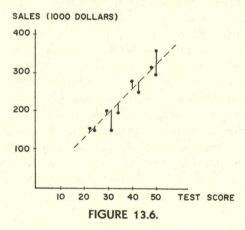

FIGURE 13.6.

from the mean or, better, as *deviations from the respective means.* Using a formula similar to that of a sample standard deviation, let us estimate σ, the standard deviation of the distributions illustrated in Figure 13.5, as

$$s_{y \cdot x} = \sqrt{\frac{\sum_{i=1}^{n} (y_i - y_i')^2}{n - 2}} \qquad (13.4.1)\star$$

Here the subscript "$y \cdot x$" is used to indicate that we estimate the true standard deviation of the y's for fixed values of x. It is customary to refer to $s_{y \cdot x}$ as the *standard error of estimate*, although it should be remembered that it is based on a sample and, therefore, itself only an estimate.

In (13.4.1) the sum of the squares is divided by $n - 2$ instead of the $n - 1$ which we might have expected in line with the definition of a sample standard deviation. The usual explanation given for this division by $n - 2$ is that the two constants a and b were calculated on the basis of the original data and we, thus, lose *two degrees of freedom*. To justify this theoretically is beyond the scope of this text.

To illustrate the calculation of $s_{y \cdot x}$, let us again refer to the aptitude scores and first year sales and let us arrange the necessary work in the following kind of table:

x	y	y^2	y'	$y - y'$	$(y - y')^2$
48	312	97,344	307.04	4.96	24.6016
32	164	26,896	196.96	−32.96	1,086.3616
40	280	78,400	252.00	28.00	784.0000
34	196	38,416	210.72	−14.72	216.6784
30	200	40,000	183.20	16.80	282.2400
50	288	82,944	320.80	−32.80	1,075.8400
26	146	21,316	155.68	− 9.68	93.7024
50	361	130,321	320.80	40.20	1,616.0400
22	149	22,201	128.16	20.84	434.3056
43	252	63,504	272.64	−20.64	426.0096
		601,342			6,039.7792

The first two columns were copied from page 288, the entries in the fourth column were obtained by substituting the given x's into $y' = -23.20 + 6.88x$, the fifth column contains the differences between y and y', and the last column contains the squares of the differences between y and y'. Substituting the rounded-off total of the last column into (13.4.1), we get

$$s_{y \cdot x} = \sqrt{\frac{6,040}{10 - 2}} = \sqrt{755}$$

or approximately $s_{y \cdot x} = 27.5$ (thousand dollars). This figure tells us how much variation there is in the first year sales of individuals obtaining a given score in the company's aptitude test. It is an estimate of the standard deviation of the conditional distributions of first year sales illustrated in Figure 13.5.

The standard error of estimate can also be calculated directly by means of the formula

$$s_{y \cdot x} = \sqrt{\frac{\Sigma y^2 - a \cdot \Sigma y - b \cdot \Sigma xy}{n - 2}} \qquad (13.4.2)^\star$$

where a and b are the values obtained from (13.2.2) and (13.2.3). The advantage of this formula is that it gives $s_{y \cdot x}$ without actually having to find the calculated values of y, the differences $y_i - y'_i$, and their squares. It will be left as an exercise for the reader to recalculate $s_{y \cdot x}$ for the test scores and first year sales data with the use of (13.4.2). The quantity Σy^2 is given in the table above.

To illustrate how a calculated value of $s_{y \cdot x}$ may be used in formulating predictions, let us suppose that a new applicant for a sales position in the given company scores 35 on the test and that we wish to predict his total first year sales. Substituting $x = 35$ into $y' = -23.20 + 6.88x$, we get $y' = 217.6$ and, as we pointed out before, this means that we shall estimate *average* first year sales of applicants who score 35 as being \$217,600. Although it may be of interest to know such an estimated average, *let us suppose that we would prefer to have two numbers so that we can predict with a probability of, say, 0.95 that the new applicant's first year sales will fall between the two.*

Assuming that the distributions of y for fixed values of x (the conditional distributions illustrated in Figure 13.5) are normal curves, two such numbers, called *limits of prediction*, may be obtained with the following technique:

for a given value of x we can predict with a probability of 0.95 that the corresponding y will lie between $y' - A$ and $y' + A$, where

$$A = t_{.025} \cdot s_{y \cdot x} \sqrt{\frac{n+1}{n} + \frac{(x - \bar{x})^2}{\Sigma x^2 - n \cdot \bar{x}^2}} \qquad (13.4.3)\star$$

and $t_{.025}$ is to be obtained from Table II with $n - 2$ degrees of freedom.

Here x is the value for which we want to predict the corresponding y, y' is the value obtained by substituting this x into the equation of the least squares line, while n, $\bar{x}$, and Σx^2 pertain to the data for which the line was originally obtained. If we wanted to change the probability from 0.95 to 0.98 or 0.99, we would only have to replace $t_{.025}$ by $t_{.01}$ or $t_{.005}$ in (13.4.3).

Returning to our numerical example, we already have found that $s_{y \cdot x} = 27.5$ and that $y' = 217.6$ for $x = 35$. From page 292 we know that $n = 10$, $\bar{x} = \frac{375}{10} = 37.5$, $\Sigma x^2 = 14{,}993$, and from Table II on page 502 we find that for $10 - 2 = 8$ degrees of freedom $t_{.025}$ equals 2.306. Substituting these values into (13.4.3) we get

$$A = (2.306)(27.5) \sqrt{\frac{10 + 1}{10} + \frac{(35 - 37.5)^2}{14{,}993 - 10(37.5)^2}}$$

$$= (2.306)(27.5) \sqrt{1.107}$$

$$= 66.8 \text{ (approx.)}$$

and our limits of prediction become $y' - A = 217.6 - 66.8 = 150.8$ and $y' + A = 217.6 + 66.8 = 284.4$. *We can thus predict with a probability of 0.95 that the first year sales of the new applicant who scored 35 will fall between \$150,800 and \$284,400.*

There are two reasons this interval is so wide that it may be relatively useless for purposes of prediction: *first*, we obtained our results on the basis of as few as ten points, the sales records and aptitude scores of ten salesmen; and *second*, $s_{y \cdot x}$ was \$27,500, which indicated that there is considerable variation in total first year sales even among salesmen having identical aptitude scores. The first of these two causes may be remedied by taking a larger sample, but the second cannot be controlled. If there is a large variation in the total first year sales of individuals having identical aptitude scores, we may have to consider other relevant factors such as past employment record, age, intelligence, and the like, in order to obtain a narrower interval of prediction.

The technique which we have introduced in this section must be used with considerable caution. To begin with, we assumed that the true regression curve is a straight line, that the conditional distributions of y for fixed values of x are normal curves, and that they all have equal standard deviations. In addition, it is important to note that *the probability of 0.95 applies to the entire process of obtaining the data to which the method of least squares is applied, calculating the limits, and drawing the additional observation whose y value we are trying to predict.* Another factor, which is often overlooked, is that the probability of 0.95 applies only for *one* prediction based on a given set of data. If a regression line is to serve for *several* predictions, which is often the case, the limits of prediction described in this section cannot be applied. This is a serious limitation and, as we said before, it is often overlooked.

Finally, it must be understood that the methods of prediction discussed in this chapter apply only if all relevant conditions remain unchanged. Clearly, it would be nonsensical to use the equation which we obtained in our example if the company had changed its prices since the original data were obtained or if there had been a change in the demand for the company's product.

EXERCISES

1. Recalculate $s_{y \cdot x}$ for the aptitude scores and first year sales data given on page 288 with the use of formula (13.4.2). Use the values of a and b obtained in the text.

2. Use the results of Exercise 1 on page 294 to construct 0.95 limits of prediction for the shipments of bolts when shipments of steel are 3.0.

3. Use the results of Exercise 3 on page 294 to construct 0.95 limits of prediction for median community income if 12 per cent of its population are professional men and women.

4. Use the results of Exercise 4 on page 294 to construct 0.99 limits of prediction for the monthly property upkeep of a house assessed at \$9,000.

5. Use the results of Exercise 5 on page 295 to construct 0.95 limits of prediction for the second-hand price of a 3 year old model of the given kind of car.

6. Use the results of Exercise 6 on page 295 to construct 0.95 prediction limits for the manufacturer's sale of farm tools and supplies in a county with median farm income of $12,000.

7. Explain on the basis of (13.4.3) why the limits of prediction will become *wider* if the x for which we want to make the prediction is further away from $\bar{x}$.

13.5 Multiple Linear Regression

Although there are many situations in which fairly accurate predictions can be made of one variable in terms of another, it stands to reason that these predictions should be improved if they took into account additional relevant information. For instance, we should be able to make better predictions of the demand for coffee if we considered not only its price per pound but also the price of tea. Similarly, we should be able to make better predictions of a cow's yield of milk if we considered the amount of grain fed to it each day, in addition to its age, and we should be able to make better predictions of a community's demand for new dwelling units if we considered the size of its population, the number of available units, and, perhaps, average monthly rent.

Among the many equations that can be used to express relationships between more than two variables, the most widely used in statistical work are *linear equations* of the form

$$y = a + bx_1 + cx_2 + dx_3 + \cdots \qquad (13.5.1)$$

where y is the variable which is to be predicted while $x_1, x_2, x_3, \ldots,$ are the known variables on which the prediction is to be based. Of the latter there may be two, three, four, or more, depending on the nature of each individual problem.

To give an example of a *multiple linear regression equation*, as these equations are called, let us consider the following, which arose in a study of the demand for beef and veal:[†]

$$y = 3.4892 - 0.0899x_1 + 0.0637x_2 + 0.0187x_3$$

Here y stands for the total consumption of federally inspected beef and veal in millions of pounds; x_1 stands for a composite retail price of beef in cents per pound; x_2 represents a composite retail price of

---

† H. Schultz, *The Theory and Measurement of Demand* (Chicago: University of Chicago Press, 1938), pp. 582 ff.

pork in cents per pound; and x_3 stands for income as measured by a certain payroll index. With this equation it is possible to forecast the total consumption of federally inspected beef and veal on the basis of known or estimated values of x_1, x_2, and x_3. This equation also shows how much the total consumption of beef and veal is decreased by an *increase* in the price of beef or a *decrease* in the price of pork.

The major problem in fitting a linear equation in more than two unknowns to a given set of data is that of finding numerical values for a, b, c, d, . . . , so that the resulting equation will yield the best possible predictions. As in the two-variable case treated in Section 13.2, this problem is generally solved by using the method of least squares, which provides the values of a, b, c, d, . . . , that make $\Sigma(y_i - y_i')^2$ a minimum. In this notation, y_i stands for an observed value of y while y_i' stands for the corresponding value calculated with the linear equation (13.5.1).

In principle, the problem of finding predicting equations of the form $y = a + bx_1 + cx_2 + dx_3 + \cdots$, is no different from that of fitting lines of the form $y = a + bx$. In practice, it is much more tedious because the method of least squares yields a set of *normal equations* which consists of three or more linear equations in as many unknowns. Suitable references to these normal equations and illustrations of their application are given below.

BIBLIOGRAPHY

Other informal introductions to the problem of least squares curve fitting and regression may be found in

Moroney, M. J., *Facts From Figures*. London: Penguin Books, 1956, Chap. 16.

Neter, J., and Wasserman, W., *Fundamental Statistics for Business and Economics*, 3rd ed. Boston: Allyn and Bacon, Inc., 1966.

Wallis, W. A., and Roberts, H. V., *Statistics: A New Approach*. Glencoe, Ill.: Free Press, 1956, Chap. 17.

Derivations of equations (13.2.2) and (13.2.3) by means of calculus may be found in most textbooks of mathematical statistics; for example, in

Hoel, P. G., *Introduction to Mathematical Statistics*, 3rd ed. New York: John Wiley, 1962, Chap. 7.

The origin of the term "regression" as it is used in statistics is explained briefly in

Sprowls, R. C., *Elementary Statistics for Students of Social Science and Business*. New York: McGraw-Hill, 1955, Chap. 10.

Confidence intervals for regression coefficients are treated in the book by Hoel mentioned above and confidence intervals for the true average values of y for fixed values of x are given in, among others, the books by Neter and Wasserman *and* Wallis and Roberts. The theory underlying the formula we gave for limits of prediction is treated in

Mood, A. M., and Graybill, F. A., *Introduction to the Theory of Statistics*, 2nd ed. New York: McGraw-Hill, 1963, Chap. 13.

For a brief discussion of multiple linear regression and the normal equations needed to calculate the coefficients, see

Croxton, F. E., and Cowden, D. J., *Practical Business Statistics*. Englewood Cliffs, N.J.: Prentice-Hall, Inc., 1960, Chap. 20.

Ezekiel, M., and Fox, K. A., *Methods of Correlation and Regression Analysis*, 2nd ed. New York: John Wiley, 1959.

Ferber R., *Statistical Techniques in Market Research*. New York: McGraw-Hill, 1949, Chap. 12.

Lewis, E. E., *Methods of Statistical Analysis in Economics and Business*. Boston: Houghton Mifflin, 1953, Chap. 14.

CHAPTER 14

Correlation

14.1 The Coefficient of Correlation

Having seen that there are formulas to describe almost any feature of a set of numerical data, it would be surprising if statisticians had not developed some way of measuring the goodness of the fit of a regression line. Since the method of least squares defined "goodness of fit" in terms of the sum of the squares of the vertical deviations from the line (see Figure 13.3), it would only seem reasonable to measure this goodness of fit in terms of the quantity

$$\sum_{i=1}^{n} (y_i - y_i')^2 \qquad (14.1.1)$$

If the differences between the observed and calculated values of y are small, (14.1.1) will be small, and if the differences are large, it will be large.

Although this would seem to be the natural way of measuring "goodness of fit" when using the method of least squares, it has the unfortunate feature that the magnitude of (14.1.1) depends on the scale of measurement—on the units of y. In our example of the aptitude scores and first year sales we found that this quantity equals approximately 6,040, the y's being given in thousands of dollars. Had the y's been given in dollars, the sum of the squares of the differences between the observed and calculated values of y would have equalled 6,040,000,000 and had they been given in cents, it would have equalled 60,400,000,000,000.

To eliminate this difficulty we shall measure the goodness of the fit of a regression line not merely in terms of $\Sigma(y_i - y_i')^2$, but by comparing this quantity with the sum of the squares of the deviations of the y's from their *mean*, namely, with $\Sigma(y_i - \bar{y})^2$. To illustrate, let us first consider Figures 14.1 and 14.2, whose points represent the number of customers who visited a certain restaurant on 12 consecutive days and the number of steak dinners served on the same days. The diagram on the left shows the vertical deviations of the y's from

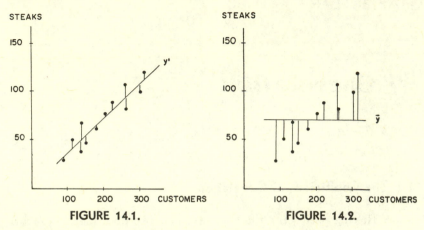

FIGURE 14.1.   FIGURE 14.2.

the least squares line; the one on the right shows the deviations of the y's from their mean. *It is apparent that in these two diagrams* $\Sigma(y_i - y_i')^2$ *is much smaller than* $\Sigma(y_i - \bar{y})^2$. To consider another example, let us take Figures 14.3 and 14.4; whose points represent annual savings increases of households in Mutual Savings Banks for

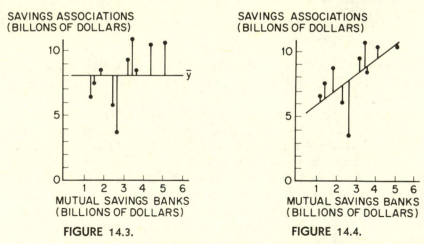

FIGURE 14.3.   FIGURE 14.4.

the ten year period 1958 through 1967, and annual savings increases of households in Savings Associations during the same years.

Again the diagram on the left shows the vertical deviations of the y's from the least squares line and the one on the right shows the deviations of the y's from their mean, *but this time* $\Sigma(y_i - y_i')^2$ *is just about as large as* $\Sigma(y_i - \bar{y})^2$. We can thus see that in the first example, where the fit of the least squares line is good, $\Sigma(y_i - y_i')^2$ is much smaller than $\Sigma(y_i - \bar{y})^2$, while in the second example, where the points are widely scattered and the fit is poor, the two sums of squares are almost the same size.

To put this comparison on a precise basis, it is customary to use the statistic

$$r = \pm \sqrt{1 - \frac{\sum_{i=1}^{n} (y_i - y_i')^2}{\sum_{i=1}^{n} (y_i - \bar{y})^2}} \qquad (14.1.2)$$

which is called the *coefficient of correlation*. If the fit is *poor*, the ratio of the two sums of squares will be close to 1 and r, *the coefficient of correlation, will be close to 0.*† On the other hand, if the fit is *good*, the ratio of the two sums of squares will be close to zero and *the coefficient of correlation will be close to $+1$ or -1.* (The significance of the sign of r will be discussed below.)

The statistic we have just defined by means of (14.1.2) is the most widely used measure of the *strength of linear relationships between two variables*. It measures the goodness of the fit of the least squares line, and this, in turn, tells us whether or not it is reasonable to say that there exists a linear relationship (correlation) between x and y. If r is close to 0, the fit is poor and we shall say that *the relationship is weak or nonexistent;* if r is close to $+1$ or -1, we shall say that *there is a strong correlation*, with the tacit understanding that we are referring to a *linear* relationship and nothing else. The statistic r is sometimes referred to somewhat more elaborately as the *Pearson product-moment coefficient of correlation*.

To illustrate the calculation of r with (14.1.2), let us again use the aptitude score and first year sales data of Chapter 13. Having already found on page 301 that $\sum_{i=1}^{n} (y_i - y_i')^2$ is approximately 6,040, we will only have to calculate the sum of the squares of the deviations of the y's from their mean and substitute into (14.1.2). Calculating this sum of squares for the first year sales data given on page 288, we get

$$\sum_{i=1}^{n} (y_i - \bar{y})^2 = \sum_{i=1}^{n} y_i^2 - n \cdot \bar{y}^2$$

$$= 601,342 - 10(234.8)^2$$

$$= 50,032$$

---

† Actually $\Sigma(y_i - \bar{y})^2$ *cannot be less than* $\Sigma(y_i - y_i')^2$. Since the line was fitted with the method of least squares, the sum of the squares of the vertical deviations from any other line, even the horizontal line through $\bar{y}$, cannot be less than $\Sigma(y_i - y_i')^2$.

and substitution into (14.1.2) yields

$$r = \sqrt{1 - \frac{6{,}040}{50{,}032}} = 0.94.$$

We shall explain later why we have taken the positive rather than the negative square root. Incidentally, if we computed the coefficient of correlation for the restaurant data shown in Figures 14.1 and 14.2 we would get $r = 0.91$ and if we calculated it for the savings and work stoppages data of Figures 14.3 and 14.4 we would get 0.62.

Although formula (14.1.2) serves to *define* the coefficient of correlation, it is seldom, if ever, used in practice. Its use is rather involved, requiring that we first find the equation of the least squares line, substitute the given values of x to obtain the calculated values of y, determine the two sums of squares, and finally substitute into (14.1.2). A great deal of this work may be avoided if we use an alternate formula which, mathematically speaking, is equivalent to (14.1.2). Without actually giving its derivation (see Bibliography on page 336), let us merely state this formula as

$$r = \frac{n \cdot \sum_{i=1}^{n} x_i y_i - \left(\sum_{i=1}^{n} x_i \right) \left(\sum_{i=1}^{n} y_i \right)}{\sqrt{n \cdot \sum_{i=1}^{n} x_i^2 - \left(\sum_{i=1}^{n} x_i \right)^2} \sqrt{n \cdot \sum_{i=1}^{n} y_i^2 - \left(\sum_{i=1}^{n} y_i \right)^2}} \qquad (14.1.3)\star$$

Although this expression may look quite formidable, the reader will find after working a few examples that it is actually easy to use. Since (14.1.3) requires less work than (14.1.2) and will produce the identical results, it is recommended that it always be used for the calculation of r from ungrouped data.

To find a coefficient of correlation we have only to calculate the five sums Σx_i, Σy_i, Σx_i^2, Σy_i^2, $\Sigma x_i y_i$ and substitute them together with n, the number of pairs of observations, into (14.1.3). To illustrate, let us recalculate r for the aptitude scores and first year sales data. Since we already know from page 292 that $n = 10$, $\Sigma x_i = 375$, $\Sigma y_i = 2{,}348$, $\Sigma x_i^2 = 14{,}993$, and $\Sigma x_i y_i = 94{,}448$, we will only have to determine Σy_i^2. Squaring the values given in the second column on page 292, we get $\Sigma y_i^2 = 601{,}342$ and substituting the various sums into (14.1.3), we obtain

$$r = \frac{10(94{,}448) - (375)(2{,}348)}{\sqrt{10(14{,}993) - (375)^2} \sqrt{10(601{,}342) - (2{,}348)^2}}$$

$$= 0.94$$

This is identical with the result which we got earlier with (14.1.2).

To give another example, let us calculate r, the degree of relationship between x the annual per capita consumption of chewing tobacco (by males 18 years and over) in the United States from 1959 through 1966, and y, production of coal in the United States during the same years.   Writing the original data in the first two columns, the necessary calculations may be arranged as in the following table:†

	x (Pounds)	y (Hundred Millions Tons)	x^2	y^2	xy
1959	1.20	4.12	1.4400	16.9744	4.9440
1960	1.13	4.16	1.2769	17.3056	4.7008
1961	1.13	4.03	1.2769	16.2409	4.5539
1962	1.10	4.22	1.2100	17.8084	4.6420
1963	1.11	4.59	1.2321	21.0681	5.0949
1964	1.11	4.87	1.2321	23.7169	5.4057
1965	1.07	5.12	1.1449	26.2144	5.4784
1966	1.05	5.34	1.1025	28.5156	5.6070
	8.90	36.45	9.9154	167.8443	40.4267

Substituting these values together with $n = 8$ into (14.1.3), we get

$$r = \frac{8(40.4267) - (8.90)(36.45)}{\sqrt{8(9.9154) - (8.90)^2} \; \sqrt{8(167.8443) - (36.45)^2}}$$

$$= -.78$$

Having obtained a *negative r*, let us now see what significance there is attached to the sign of the coefficient of correlation.   Plotting the chewing tobacco and coal production as we have done in Figure 14.5, it may be seen that *large values of x go with small values of y while small values of x go with large values of y*.   This, indeed, characterizes what is called a *negative correlation*.   If small values of x go with small values of y and large values of x go with large values of y, r will be positive and there is said to be *positive correlation*.   Geometrically speaking, a correlation is positive when, going from left to right, the regression line slopes *upward* and it is negative when, going from left to right, the regression line slopes *downward* (see Figure 14.6).   As can be seen from our two examples, *the sign of r is automatically taken care of in formula (14.1.3)*.

Since we obtained a fairly strong negative correlation in our last example, we might be led to believe that there exists a "real" relationship between the consumption of chewing tobacco and the production of coal.   Offhand, this does not make sense and we shall have more to

---

† If these calculations are performed with a machine, the totals of the various columns may be accumulated directly without having to write down the individual squares and products.

say about it in Section 14.2.   Until then let us be careful to interpret r only as a quantity which measures how closely a regression line fits a given set of data.

There exist a number of tricks which will simplify the calculation of correlation coefficients.   Since r does not depend on the scale of

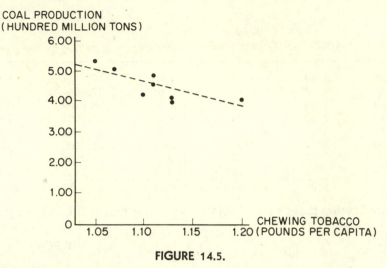

FIGURE 14.5.

x or y, it is based only on the *ratio* of $\Sigma(y_i - y_i')^2$ and $\Sigma(y_i - \bar{y})^2$, we can often simplify the numbers with which we have to work by adding an arbitrary positive or negative number to each x, each y, or both; or by multiplying each x, each y, or both by arbitrary constants.   For

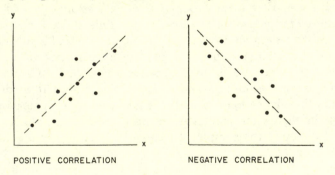

FIGURE 14.6.

instance, in the last example we could have multiplied each x by 100 and then subtracted 111, substituting 9, 2, 2, -1, 0, 0, -4, and -6 into the first column.   The reader may wish to verify for himself that the resulting value of r would again have been -0.78.   To simplify things further we could also have multiplied each y value by 100 and then subtracted 450 (or some other number) from each y.   The numbers which we may want to add or subtract or by which we may want to multiply or divide will of course, have to depend on the numerical values of our data. The important thing to keep in mind is that the purpose of these manipulations is to simplify our work as much as possible.

EXERCISES

1. Calculate r for the shipments of steel and shipments of bolts data of Exercise 1 on page 294.
2. Calculate r for the per cent of professional men and women, and median community income data on page 294.
3. Calculate r for the valuations and cost of upkeep of Exercise 4 on page 294.
4. Calculate r for the median income and sales data of Exercise 6 on page 296.
5. If you were asked to calculate the coefficient of correlation for *each* of the following two sets of data

x	y	x	y
141	7	17	89
113	5	54	32

would you be surprised if your answers were $r = 1$ and $r = -1$, respectively? Explain.
6. Recalculate r for the chewing tobacco and coal production data on page 311 after multiplying each x by 100 and subtracting 111, and multiplying each y by 100 and subtracting 450.
7. Using a calculating machine and formula (14.1.3), find r for the 48 valuations and costs of upkeep on page 320.
8. Calculate r for the following data, which represents the 1966 *Consumer Price Index* for food and apparel in 15 cities: (1957–59 = 100)

	Food	Apparel
Atlanta	112.9	111.0
Baltimore	115.9	111.4
Chicago	114.6	105.8
Cincinnati	111.8	110.7
Detroit	112.2	110.5
Houston	115.4	108.6
Kansas City	117.2	112.9
Los Angeles	113.3	109.4
New York	115.1	112.7
Philadelphia	113.1	115.0
Pittsburgh	111.8	112.1
St. Louis	117.8	110.9
San Francisco	114.2	113.3
Seattle	114.1	111.8
Washington, D.C.	114.0	112.2

9. State in each case whether you would expect to find a positive correlation, a negative correlation, or no correlation:
   (a) The density of the population and the crime rate.
   (b) Spraying of insecticide and number of mosquitos.
   (c) Head size and musical talent.
   (d) Economic growth and unemployment.
   (e) Application of fertilizer and size of crop.

14.2 The Interpretation of r

There is no difficulty in explaining the meaning of a correlation coefficient which equals 0, +1, or −1. If it is 0, the fit of the regression line is so poor that our knowledge of x will not aid in the prediction (forecasting) of y. If it is +1 or −1, all the points lie on a straight line and it stands to reason that we should be able to make very close predictions with the equation of the line. Values of r falling between 0 and 1 or between 0 and −1 are more difficult to explain: someone who has no knowledge of statistics might easily be led to the erroneous idea that a correlation of 0.40 is "half as good" as a correlation of 0.80 or that a correlation of 0.90 is "three times as good" as a correlation of 0.30.

To clarify the meaning of such intermediate values of r, let us again look at the aptitude test and first year sales data which we introduced first on page 288. Noting that there are considerable differences between the first year sales (the smallest value is $146,000 and the largest is $361,000) let us see whether we cannot find out why there is such a large variation. One factor which must surely affect a salesman's first year sales is his aptitude for the job. Others might include his appearance, the territory that he is assigned, and probably quite a bit of luck. To illustrate our point, let us group these factors determining first year sales into (1) aptitude as measured by the company's test and (2) all others, including luck. *This raises the question as to how much of the total variation of the first year sales may be attributed to differences in aptitude and how much may be attributed to all other factors.*

Measuring the total variation of the first year sales by means of the quantity $\Sigma(y_i - \bar{y})^2$, *the sum of the squares of their deviations from the mean*, rather complicated algebraic manipulations will show that this sum of squares can be written as†

$$\Sigma(y_i - \bar{y})^2 = \Sigma(y_i - y_i')^2 + \Sigma(y_i' - \bar{y})^2 \qquad (14.2.1)$$

† What we are doing here actually goes under the heading of "analysis of variance." We are decomposing the total variation of the y's as measured by $\Sigma(y_i - \bar{y})^2$ into two parts which we attribute to different causes or, better, to different *sources* of variation. See also Bibliography on page 336.

Considering the first term on the right-hand side of (14.2.1), we find that it consists of the sum of the squares of the vertical deviations from the regression line and, as we pointed out on page 301, it measures the variation between the first year sales of salesmen having *identical* aptitude scores. *Hence, this term gives us the portion of the total variation of the first year sales that is caused by factors other than aptitude for the job.*

The second term on the right-hand side of (14.2.1) is illustrated in Figure 14.7, and it may be seen from this diagram that $\Sigma(y_i' - \bar{y})^2$ *measures the amount of variation there would have been in the first year*

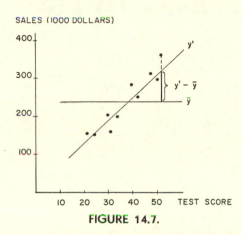

FIGURE 14.7.

sales if differences in aptitude had been the only contributing factor or, in other words, if the points had all fallen on the line.

Expressing this last term as a percentage of the *total* variation of the first year sales, we get

$$\frac{\Sigma(y_i' - \bar{y})^2}{\Sigma(y_i - \bar{y})^2} \cdot 100 \qquad (14.2.2)$$

and with the use of (14.2.1) and (14.1.2) it can easily be shown that this quantity equals $100 \cdot r^2$. (See Exercise 6 on page 320.) In our numerical example we can thus say that $100(0.94)^2 = 88$ per cent of the variation in first year sales is due to differences in aptitude.

More generally, *if the coefficient of correlation for observations on two variables x and y is equal to r, then $100 \cdot r^2$ per cent of the variation of the y's may be accounted for by the relationship with the variable x.* If in a given experiment r equals 0.70, then 49 per cent of the variation of the y's is accounted for (perhaps caused) by differences in the variable x; if r is equal to 0.30, then only 9 per cent of the variation of the y's is due to the relationship with x. In the sense of "percentage of variation accounted for," we can thus say that a correlation of 0.80

is *four times* as strong as a correlation of 0.40 or that a correlation of 0.90 is *nine times* as strong as a correlation of 0.30.

In the preceding discussion we did not refer to the sign of r, because this had no bearing on the strength of the relationship. The distinction between values such as $r = 0.80$ and $r = -0.80$ lies in the fact that one regression line slopes upward while the other slopes downward, and this has nothing to do with the goodness of the fit or the strength of the relationship between x and y.

The coefficient of correlation is one of the most widely used and also one of the most widely *abused* of statistical measures. It is abused in the sense that one sometimes overlooks the fact that r measures

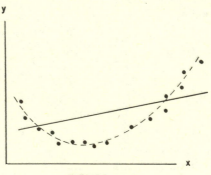

FIGURE 14.8.

nothing but the strength of *linear* relationships and that it does *not necessarily* imply a cause-effect relationship.

If r is calculated indiscriminately, for example, for the data of Figure 14.8, a small value of r does not mean that the two variables are not related. As a matter of fact, the dotted curve provides an excellent fit (even though the straight line does not), and we can say that there is a strong *curvilinear* correlation. Let us remember, therefore, that r measures only the strength of linear relationships.

The fallacy of interpreting a high correlation as necessarily implying a cause-effect relationship is best explained with a few examples. A classical example is the high positive correlation obtained for data pertaining to teachers' salaries and figures on the consumption of liquor. Clearly, this is not a cause and effect relationship; the correlation results from the fact that both variables are effects of a common cause, namely, the over-all standard of living. Another classical example is the strong positive correlation obtained for the number of storks seen nesting in the chimneys of various English villages and towns and the number of child births recorded in the same communities. We shall leave it to the reader's ingenuity to figure out why there is such a high correlation in this example without there being a direct cause-effect relationship between storks and babies. A third example is the strong negative correlation which we obtained in Section 14.1 for the production of coal and the per capita consumption of chewing tobacco. In this case the high correlation is due to the dependence of both variables on economic and social patterns changing with time. These examples demonstrate why it is safer to interpret correlation coefficients as measures of *association* rather than *causation*. For further information about such *spurious* correlations see Bibliography on page 336.

14.3 A Significance Test for r

Let us suppose that we take a pair of dice, one red and one green, and that rolling them together five times we get the following results.

Red Die *x*	Green Die *y*
1	2
4	6
5	4
6	4
2	2

Calculating *r* for these numbers, we get the surprisingly high value of 0.67, and this raises the question whether there is anything wrong with the natural assumption that there should be no relationship between the results produced by the two dice. After all, one die does not know what the other one is doing.

To answer this question we shall have to see whether we could reasonably have obtained this high value of *r by chance* even though there is really no relationship between the performances of the two dice.

Whenever *r* is calculated on the basis of a sample, the value which we obtain for *r* is only an estimate of ρ (rho), the *true* correlation coefficient which we would obtain if we calculated it for the entire population. (In our example, the population, hypothetical as it is, consists of all possible rolls of the given pair of dice.) To test the hypothesis that there is no correlation, that $\rho = 0$, we shall as always have to consider the sampling distribution of our statistic, in this case the sampling distribution of *r*. Such a sampling distribution could be obtained experimentally for the above example by repeatedly rolling the pair of dice 5 times, calculating *r* in each case, and grouping the results into an appropriate distribution.

Since the theory underlying sampling distributions of *r* is considerably beyond the scope of this text, let us merely state that *under the null hypothesis of no correlation the sampling distribution of r can be approximated with a normal curve having the mean 0 and the standard deviation* $1/\sqrt{n-1}$, *provided that n is large and that the observations x and y may be looked upon as samples from normal populations.* Using the terminology of Section 9.5, we shall refer to the standard deviation of this sampling distribution as the *standard error of r* and write it as

$$\sigma_r = \frac{1}{\sqrt{n-1}} \qquad (14.3.1)^\star$$

It should be noted that this formula applies only when the null hypothesis of no correlation is true.

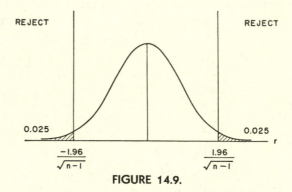

FIGURE 14.9.

Referring to Figure 14.9, whose normal curve represents the sampling distribution of r under the null hypothesis of *no correlation*, we can test this hypothesis (at a level of significance of 0.05) with the following criterion:

reject the null hypothesis of no correlation if $r < \dfrac{-1.96}{\sqrt{n-1}}$ *or* $r >$ $\dfrac{1.96}{\sqrt{n-1}}$; *reserve judgment (or accept the hypothesis) if* $\dfrac{-1.96}{\sqrt{n-1}}$ $\leqslant r \leqslant \dfrac{1.96}{\sqrt{n-1}}.$

If we wanted to use a level of significance of 0.02 or 0.01, we would only have to substitute 2.33 or 2.58 for 1.96.

To illustrate the use of this test, let us suppose that a sample of 50 paired observations produced a correlation of $r = 0.40$. Since $1.96/\sqrt{n-1} = 1.96/\sqrt{50-1} = 0.28$, we can *reject* the hypothesis of no correlation or, as is customary, we can say that there is a *signifi-*

cant correlation. Had the observed value of r been -0.15, which lies between -0.28 and 0.28, we would have said that the correlation is *not significant* or, in other words, that this value may reasonably be attributed to chance.

Although the test we have presented here is based on a normal curve approximation and the assumption that n is large, it is *safe* to use it even when n is small. If we used the exact sampling distribution of r for samples from normal populations to evaluate the probability of getting an r that is less than $-1.96/\sqrt{n-1}$ or greater than $1.96/\sqrt{n-1}$, we would get a value slightly less than 0.05. Hence, if our criterion allows us to say that a correlation coefficient is *significant* at a level of significance of 0.05, an exact test based on the actual sampling distribution of r would do the same. (An exact small sample test based on the Student-t distribution is referred to in the Bibliography on page 336.)

It is also worth noting that if a correlation coefficient is significant by the above criterion, this does not necessarily imply that it is also significant in the sense of being *meaningful*. The correlation which we obtained for the chewing tobacco and coal production data on page 311 was -0.78, and this is significant for $n = 8$ at a level of significance of 0.05 (provided that the assumptions underlying the test can be met). Nevertheless, the correlation is *spurious;* it is not indicative of a direct cause-effect relationship.

EXERCISES

1. Assuming that the conditions underlying the test on page 318 can be met, test at a level of significance of 0.05 whether the following correlation coefficients are significant:

   (a) $n = 50$ and $r = -.32$
   (b) $n = 82$ and $r = .40$
   (c) $n = 101$ and $r = -.17$
   (d) $n = 122$ and $r = .20$
   (e) $n = 145$ and $r = .75$
   (f) $n = 626$ and $r = -.02$

2. Repeat Exercise 1, using a level of significance of 0.01.
3. Test, at a level of significance of 0.05, whether the correlation coefficient calculated in Exercise 3 on page 313 is significant.
4. Test, at a level of significance of 0.05, whether the correlation coefficient calculated in Exercise 4 on page 313 is significant.

5. (Group Exercise) Using a pair of dice, one red and one green (or otherwise distinguishable), take a large number of samples, each consisting of 5 rolls of the dice. Calculate r separately for each sample, group these data into a frequency distribution, and compare the standard deviation of this experimental sampling distribution with the value which we should have expected according to (14.3.1).

6. (*Theoretical Exercise*) Solve (14.2.1) for $\Sigma(y_i' - \bar{y})^2$ and, substituting your result into (14.2.2), show that this quantity equals $100 \cdot r^2$.

14.4 The Calculation of r from Grouped Data

In Chapters 3 and 4 we saw that the calculation of means and standard deviations of large sets of data can be simplified considerably by first grouping the data into a frequency table. Since the calculation of r can be quite tedious, particularly when n is large and the work has to be done without a machine, let us now demonstrate how r can be calculated from grouped data.

To illustrate the steps needed to group paired observations and to calculate r on the basis of the resulting frequency table, let us refer to a study designed to investigate the relationship between county assessed valuations of dwellings in a certain area and the cost of their upkeep. The following data, in which x stands for county assessed valuation and y for cost of upkeep per month (both in dollars), constitute a random sample of 48 homes in the area under consideration:

x	y	x	y	x	y	x	y
3,000	23.00	11,000	54.00	9,000	31.00	6,000	31.00
11,000	36.00	2,000	22.00	5,000	24.00	7,000	41.00
9,000	51.00	6,000	42.00	9,000	43.00	3,000	33.00
7,000	36.00	6,000	35.00	13,000	48.00	12,000	60.00
7,000	50.00	12,000	46.00	10,000	50.00	11,000	45.00
9,000	42.00	15,000	71.00	15,000	68.00	11,000	47.00
11,000	61.00	11,000	49.00	3,000	31.00	15,000	66.00
16,000	56.00	17,000	71.00	15,000	56.00	12,000	51.00
6,000	23.00	10,000	33.00	14,000	53.00	3,000	40.00
7,000	37.00	7,000	31.00	7,000	41.00	11,000	56.00
14,000	66.00	10,000	55.00	19,000	72.00	11,000	58.00
11,000	45.00	13,000	61.00	5,000	32.00	17,000	68.00

The problem of grouping paired observations is almost the same as that of constructing an ordinary frequency distribution. *We have to decide how many classes to use for each of the two variables and from where to where each class is to go.* Choosing the five classes $0–$3,999

$4,000–$7,999, $8,000–$11,999, $12,000–$15,999, and $16,000–$19,999, for the valuations and the six classes $20.00–$29.99, $30.00–$39.99, $40.00–$49.99, $50.00–$59.99, $60.00–$69.99, and $70.00–$79.99 for the monthly upkeeps, we obtain the following *two-way table:*

	Valuations (x)				
Upkeep (y)	0–3,999	4,000–7,999	8,000–11,999	12,000–15,999	16,000–19,999
20.00–29.99					
30.00–39.99					
40.00–49.99					
50.00–59.99					
60.00–69.99					
70.00–79.99					

Having constructed this table, the next step is to tally the data by putting a check into the appropriate cell for each pair of observations. For instance, $x = 3000$ and $y = 23.00$ goes into the *first row* and *first column*, $x = 11,000$ and $y = 36.00$ goes into the *second row* and *third column*, and $x = 9000$, and $y = 51.00$ goes into the *fourth row* and third column. After completing the tally for our data and counting the number of checks in each cell, we get the following *two-way frequency distribution:*

	Valuations (x)				
Upkeep (y)	0–3,999	4,000–7,999	8,000–11,999	12,000–15,999	16,000–19,999
20.00–29.99	2	2			
30.00–39.99	2	6	3		
40.00–49.99	1	3	6	2	
50.00–59.99		1	6	3	1
60.00–69.99			1	5	1
70.00–79.99				1	2

In the construction of this table we followed the customary procedure of writing the x-classification on top of the table, increasing from left to right, and the y-classification on the left-hand side, increasing downward. If it is desired to present a two-way distribution graphically, this may be done by drawing a three-dimensional histogram like the one shown in Figure 14.10. Here the *heights* of the blocks represent the frequencies of the cells on which they stand just as the heights of the rectangles in an ordinary histogram represent the frequencies of the various classes.

Having taken care of the first part of our problem, namely, that of

grouping paired data, it still remains to be seen how r is to be calculated on the basis of a two-way frequency table.   To be able to do
so we shall assume, as we did for the mean and standard deviation,
that all measurements contained in a class are located at its class mark.
We shall, thus, say that the two items falling into the first row and
first column of our table have $x = \$1,999.50$ and $y = \$24.995$, that
the six items falling into the second row and second column have
$x = \$5,999.50$ and $y = \$34.995$ and so forth.   Having made the same

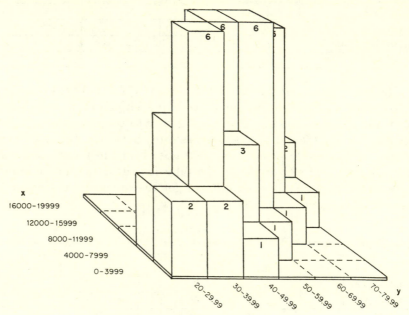

FIGURE 14.10.   A Three-Dimensional Histogram.

assumption as in the case of the mean and standard deviation, let us
also use the same simplification of changing the scale.   Since we now
have two variables x and y, we shall replace the x-scale by a u-scale
and the y-scale by a v-scale, doing this as before by numbering the
successive class marks $-3, -2, -1, 0, 1, 2, 3, \ldots$ In our example
we will, thus, get

	-2	-1	0	1	2	u-scale
-2	2	2				
-1	2	6	3			
0	1	3	6	2		
1		1	6	3	1	
2			1	5	1	
3				1	2	

v-scale

As before, the choice of the zero of each new scale is arbitrary and it should be chosen so as to simplify subsequent calculations as much as possible.

On page 312 we remarked that the coefficient of correlation will not change if we add fixed numbers to our observations (not necessarily the same for x and y) or if we multiply or divide the observations by any constants. *This implies that if we calculate r in terms of the u's and v's, we will get precisely the same result as if we had used the class marks in the original scales.* Consequently, substituting u's and v's for the x's and y's of (14.1.3) and allowing for the fact that we are now dealing with grouped data, the *formula for calculating r from grouped data becomes*

$$r = \frac{n(\Sigma uvf) - (\Sigma uf_u)(\Sigma vf_v)}{\sqrt{n(\Sigma u^2 f_u) - (\Sigma uf_u)^2}\ \sqrt{n(\Sigma v^2 f_v) - (\Sigma vf_v)^2}} \qquad (14.4.1)\star$$

It should be noted that this short-cut formula for r for grouped data can be used only if the original classifications of x and y, respectively, have class intervals of equal width.

The various sums that are needed for substitution into (14.4.1) are most conveniently obtained by arranging the calculations in the following kind of table:

					(1) v	(2) f_v	(3) vf_v	(4) v^2f_v	(5) uvf
2	2				-2	4	-8	16	12
2	6	3			-1	11	-11	11	10
1	3	6	2		0	12	0	0	0
	1	6	3	1	1	11	11	11	4
		1	5	1	2	7	14	28	14
			1	2	3	3	9	27	15
						48	15	93	55
(6) u	-2	-1	0	1	2	48			
(7) f_u	5	12	16	11	4	48			
(8) uf_u	-10	-12	0	11	8	-3			
(9) u^2f_u	20	12	0	11	16	59			
(10) uvf	12	9	0	16	18	55			

To simplify the over-all picture, we put the u and v scales at the bottom and right-hand side of this *correlation table*, labeling them row (6) and column (1), respectively. Column (2) contains the frequencies f_v, the frequencies corresponding to the different values of v, and they are obtained by adding the frequencies in the respective rows. Column (3) contains the products vf_v of the corresponding entries of columns (1) and (2), and column (4) contains the products v^2f_v, which may be obtained by either squaring each entry of column (1) and multiplying by the corresponding entry of column (2) or by multiplying the corresponding entries of columns (1) and (3). The figures shown in rows (7), (8), and (9) are obtained by performing the same operations on the u's and f_u's.

The totals of columns (2), (3), and (4) provide n, Σvf_v, and $\Sigma v^2 f_v$, while the totals of rows (7), (8), and (9) provide n, Σuf_u, and $\Sigma u^2 f_u$. To calculate r with (14.4.1) we still lack Σuvf, which stands for the sum of the terms obtained by individually multiplying each cell frequency by the corresponding u and v. To simplify the calculation of Σuvf, we shall work on each row separately, first multiplying each cell frequency by the corresponding u and then multiplying the sum of these products by the corresponding v. For example, for the second row we get $2(-2)$, $6(-1)$, $3(0)$, whose sum is -10, and multiplying this sum by $v = -1$ we get $(-10)(-1) = 10$. This is the entry shown in column (5). Similarly, for the fifth row we get $1(0)$, $5(1)$, $1(2)$, and multiplying the sum of these products by $v = 2$ we get 14.

Interchanging u and v, we can also calculate Σuvf by working separately on each column, multiplying each cell frequency by the corresponding v, and then multiplying the sum of these products by the corresponding u. For the first column we will thus get $2(-2)$, $2(-1)$, $1(0)$, and multiplying the sum of these products by $u = -2$ we get 12, namely, the first entry of row (10). The sum of the products uvf is thus given by the total of column (5) as well as that of row (10), and although it is not necessary to calculate them both, it is advisable to do so simply as a check.

Substituting the totals of the appropriate rows and columns of our correlation table into (14.4.1), we finally get

$$r = \frac{48(55) - (-3)(15)}{\sqrt{48(59) - (-3)^2} \; \sqrt{48(93) - (15)^2}}$$
$$= 0.80$$

and this measures the strength of the relationship between the assessed valuations and the monthly upkeeps of the given dwellings.

Having seen only this one illustration, the reader may feel that the calculation of r from grouped data is quite involved. However, after having worked a few problems it will become apparent that the use of (14.4.1) can save a considerable amount of work, at least, so long as the calculations are done without a machine. Incidentally, had we calculated r in this example without first grouping our data (see Exercise 7 on page 313), we would have obtained 0.87. This goes to show that, as we have pointed out before, by working with grouped data we lose some information.

EXERCISES

1. The following are the populations of 38 selected Metropolitan Areas (1960 census) having populations under 250,000 and the number of new private housing units authorized in each, in 1966: (Source: *Statistical Abstract of the United States 1967*)

	Population (*in thousands*)	New Private Housing Units Authorized by Building Permits
Ann Arbor, Mich.	172	3936
Atlantic City, N.J.	161	1386
Augusta, Ga.	217	1550
Austin, Tex.	212	3396
Bay City, Mich.	107	417
Billings, Mont.	79	267
Cedar Rapids, Iowa	137	1051
Colorado Springs, Colo.	144	1864
Decatur, Ill.	118	481
Eugene, Oreg.	163	2222
Gadsden, Ala.	97	189
Galveston, Tex.	140	382
Hamilton-Middletown, Ohio	199	1225
Huntsville, Ala.	154	1354
Jackson, Mich.	132	568
Jackson, Miss.	221	1004
Kalamazoo, Mich.	170	2202
Lafayette, La.	85	355
Lake Charles, La.	145	320
Laredo, Tex.	65	167
Macon, Ga.	180	1154
Mansfield, Ohio	118	382
Meriden, Conn.	52	306
Midland, Tex.	68	413
Muncie, Ind.	111	64
New Britain, Conn.	129	1504
Norwalk, Conn.	97	665
Ogden, Utah	111	386
Pensacola, Fla.	203	1486
Pine Bluff, Ark.	81	318
Pueblo, Colo.	119	521
Raleigh, N.C.	169	2030
Sioux Falls, S.Dak.	87	321
Springfield, Mo.	126	491
Waterloo, Iowa	122	763
Wheeling, W.Va.	180	162
Wichita Falls, Tex.	130	314
Wilmington, N.C.	92	501

   Group these data into a suitable two-way table and use formula (14.4.1) to calculate r.

2. Calculate r from the following two-way table, with x representing the average weekly wages of families selected at random in a given area

and y representing their average weekly expenditure for entertainment (movies, magazines, etc.):

Average Weekly Expenditure for Entertainment (in dollars)	Average Weekly Wages (in dollars)						
	60–79	80–99	100–119	120–139	140–159	160–179	180–199
0.00– 1.99	1						
2.00– 3.99	2	3	1				
4.00– 5.99	1	2	10	2			
6.00– 7.99		5	6	5	1	1	1
8.00– 9.99			2	4	3	2	1
10.00–11.99				1	10	6	2
12.00–13.99				2	5	2	2
14.00–15.99						1	1

3. Group the following data into a frequency table and calculate r with formula (14.4.1). This historical data constitutes a sample of 50 counties in the southern part of the United States taken in a certain year to determine the relationship between the income of farm families and the level of urban-industrial development. The variable y stand for a county's percentage of non-farm population and x stands for the median income of its farm families and unrelated individuals.

x	y	x	y	x	y	x	y	x	y
$ 878	49	$ 923	71	$1403	66	$ 819	50	$1073	61
1690	94	850	42	1605	88	1219	49	1858	86
1318	56	738	60	842	58	890	48	1612	65
818	30	1087	55	768	49	699	65	726	47
867	61	758	31	2048	87	808	51	1509	85
978	64	799	35	1388	82	746	52	680	47
1045	39	753	17	672	39	753	48	1088	60
763	67	1263	60	915	51	904	42	1106	45
1194	70	1295	57	440	26	1219	72	1454	96
874	38	957	37	1958	83	424	39	815	57

4. Calculate r from the following correlation table with x representing grades in an economics test and y representing grades in a finance examination:

	1–20	21–40	41–60	61–80	81–100	x-scale
1–20	2	4	2			
21–40		8	6	4		
41–60		2	10	14	4	
61–80			4	6	6	
81–100			2	4	8	

y-scale

14.5 Rank Correlation

At times it is convenient to measure the strength of a relationship between paired observations on the basis of their *ranks* within the respective samples instead of their actual numerical values. The advantage of this is that the so-called *rank correlation coefficient* is generally easier to find.

To illustrate, let us consider a study whose purpose is to determine whether there is a relationship between the number of motor vehicles registered in different states and the mileage of their rural and municipal roads. The first two columns of the following table contain the 1966 figures on motor vehicle registration and total road mileage as reported in the *Statistical Abstract of the United States 1967* for the 11 Mountain and Pacific States:

	Number of Motor Vehicles (in thousands) x	Road Mileage (in 1000 miles) y	Rank of x	Rank of y
Arizona	863	39.1	5	10
California	10,347	164.2	1	1
Colorado	1,201	80.1	3	3
Idaho	446	53.3	8	8
Montana	439	72.7	9	5
Nevada	279	46.7	10	9
New Mexico	549	66.4	6	7
Oregon	1,167	83.6	4	2
Utah	544	37.7	7	11
Washington	1,756	71.7	2	6
Wyoming	224	76.3	11	4

In the third column we ranked the x's giving Rank 1 to the highest value which is 10,347, Rank 2 to 1756, Rank 3 to 1201, and so on. In the fourth column we similarly ranked the y's giving Rank 1 to 164.2, Rank 2 to 83.6, Rank 3 to 80.1 . . . , and Rank 11 to 37.7, which is smallest.

Proceeding from here we could calculate r with formula (14.1.3) for the ranks instead of the actual numbers, but we would get the *same* result faster and more easily if we used the formula

$$r' = 1 - \frac{6(\Sigma d_i^2)}{n(n^2 - 1)} \qquad (14.5.1)^\star$$

which defines the *coefficient of rank correlation*.† In this formula the

---

† We shall use the symbol r', but there is little consistency in this respect. The reader will find that in other books the coefficient of rank correlation is referred to by the symbol r' as well as r_S r_r, S, and ρ (rho).

d_i stands for the differences between the ranks of the corresponding x s and y's and n, as always, for the size of the sample, that is, for the number of pairs of values given for x and y. In using this formula it does not matter whether we give Rank 1 to the *highest* value and rank the others in decreasing order of magnitude 2, 3, 4, . . . , or give Rank 1 to the *lowest* and rank the other values in increasing order of magnitude 2, 3, 4, ... We will have to be consistent, however, using the same type of ranking for both x and y.

The sum of the squares of the d's, which is needed in (14.5.1), can easily be found as follows:

Rank of x	Rank of y	d	d^2
5	10	−5	25
1	1	0	0
3	3	0	0
8	8	0	0
9	5	+4	16
10	9	+1	1
6	7	−1	1
4	2	+2	4
7	11	−4	16
2	6	−4	16
11	4	+7	49
			128

Substituting $n = 11$ and $\Sigma d_i^2 = 128$ into the formula for r', we get

$$r' = 1 - \frac{6(128)}{11(11^2 - 1)} = 0.58$$

as a measure of the strength of the relationship between vehicle registration and road mileage in different states.

If we tried to calculate r' for the aptitude test and first year sales data used as an illustration in this and the preceding chapter, we would run into the problem of ties in rank. Considering the data on page 288, we would find that so far as test scores are concerned, Mr. Miller and Mr. Roberts are tied for the lowest (or highest) rank. *It is customary, in a situation like this, to give each item, each value, the average of the ranks which they jointly occupy.* Hence, since Mr. Miller's and Mr. Robert's aptitude scores occupy Ranks 1 and 2, we assign each the rank $(1 + 2)/2 = 1.5$. Similarly, if three values were tied for ranks 5, 6, and 7, we would assign each the rank $(5 + 6 + 7)/3 = 6$; and if two values were tied for ranks 8 and 9, we would assign each the rank $(8 + 9)/2 = 8.5$. If there are ties in rank, formula (14.5.1) will generally not equal r, the ordinary correlation coefficient calculated on the basis of the ranks, but if the number of ties is small, the difference between r and r' will generally not be very large.

In addition to the fact that the calculation of r' is much simpler than that of r, the coefficient of rank correlation has the added advantage that it can be used in problems where items can be ranked even though they cannot be measured on a numerical scale. We could use r', for example, to measure the correlation between the rankings given by two judges to various exhibits at an industrial fair or we could use it to measure the relationship between two persons' preferences (rankings) for various kinds of foods. In either case, formula (14.5.1) provides an easy to use measure of correlation although, strictly speaking, we are not dealing with numerical data.

The coefficient of rank correlation may be used to test the null hypothesis of no correlation with the same criterion given for r in Section 14.3. The standard error of r' is also $1/\sqrt{n-1}$, and when n is large we can say that there is a significant correlation at a level of significance of 0.05 if r' is less than $-1.96/\sqrt{n-1}$ or if it exceeds $1.96/\sqrt{n-1}$. One advantage of using r' instead of r in such a test of significance is that we no longer have to assume that our samples come from normal populations.

EXERCISES

1. Compute r' for the aptitude scores and first year sales data given on page 288.
2. Calculate r' for the per cent of professional men and women, and median community income of Exercise 3 on page 294.
3. Calculate r' for the assessed valuation and cost of upkeep data of Exercise 4 on page 294.
4. Calculate r' for the farm income and sales data of Exercise 6 on page 295.
5. Calculate r' for the median incomes and percentages of Exercise 3 on page 326.
6. Two housewives, asked to express their preference for different kinds of detergents, gave the following replies:

	Mrs. Brown	Mrs. Green
Detergent A	4	5
Detergent B	3	4
Detergent C	1	1
Detergent D	5	2
Detergent E	2	6
Detergent F	6	11
Detergent G	12	10
Detergent H	11	9
Detergent I	10	7
Detergent J	8	8
Detergent K	9	12
Detergent L	7	3

Calculate r' as an indication of the *consistency* of these two sets of ratings.

14.6 Multiple and Partial Correlation

The study of multiple and partial correlation is usually omitted in introductory courses in statistics because there is already enough material to be covered without going into these more advanced topics. Nevertheless, the underlying concepts of multiple and partial correlation are quite easy to grasp and we shall discuss them briefly so that the reader will at least know what these terms mean.

In Section 14.1 we defined the coefficient of correlation as a measure of the goodness of the fit of the least squares line $y' = a + bx$. Without going into any detail, we shall now extend this measure of goodness of fit to apply to problems in which y is predicted by means of a *multiple regression equation* of the form

$$y' = a + bx_1 + cx_2 + dx_3 + \cdots \qquad (14.6.1)$$

On page 304 we illustrated such an equation with reference to the problem of predicting the consumption of beef and veal on the basis of the price of beef, the price of pork, and personal income.

To measure how closely a set of observed values of y agrees with the corresponding calculated values $y' = a + bx_1 + cx_2 + dx_3 + \cdots$, we shall use the same formula as that which on page 309 served to define r, namely,

$$\sqrt{1 - \frac{\Sigma(y_i - y_i')^2}{\Sigma(y_i - \bar{y})^2}} \qquad (14.6.2)$$

Now this formula defines the *multiple correlation coefficient*, provided, of course, that y' is calculated with a *multiple regression equation* instead of a simple regression equation of the form $y' = a + bx$. [Although (14.6.2) serves to *define* what is meant by a multiple correlation coefficient, it is rarely used in practice, because there exist alternate formulas which are much easier to apply.]

In contrast to r, we cannot say that a multiple correlation coefficient measures the goodness of the fit of a regression *line*. Conceptually speaking, things are more complicated now because $y' = a + bx_1 + cx_2$ is the equation of a *plane*, whereas $y' = a + bx_1 + cx_2 + dx_3$, $y' = a + bx_1 + cx_2 + dx_3 + ex_4$, etc., are the equations of *multidimensional hyperplanes*, that is, geometrical configurations that cannot even be visualized.

To illustrate the simplest case, let us suppose that we are given data on the consumption of coffee (y), the price of coffee (x_1), and the price of tea (x_2), and that we want to predict the consumption of coffee by means of an equation of the form $y' = a + bx_1 + cx_2$. Geometrically, this equation is represented by the plane of Figure 14.11 and the goodness of the fit is measured, essentially, by the sum of the squares of the vertical deviations shown in this diagram. *The multiple correlation coefficient thus measures how closely the plane fits to the points representing the given set of data or, in other words, how closely the consumption of coffee is related to both the price of coffee and the price of tea.*

Now let us say a few words about *partial correlation*. When we mentioned the problem of "correlation and causation," we pointed out that a high correlation between two variables can sometimes be

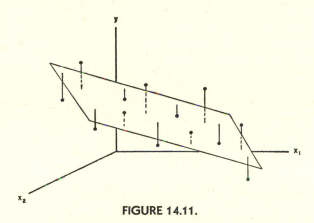

FIGURE 14.11.

ascribed to their separate relationships to a third variable, which is, so to speak, their common cause. On page 316 we used the example of birth registrations and the prevalance of storks, and now, to give another example, let us suppose that we are interested in determining the relationship between x_1, the weekly amount of hot chocolate sold by a refreshment stand in a summer resort, and x_2, the weekly number of tourists visiting this resort. Furthermore, let us suppose that on the basis of a relatively large set of weekly data we found that for the two given variables the ordinary correlation coefficient equals -0.30. This result is startling, to say the least, for we would expect higher sales of hot chocolate when there are more tourists and, hence, a *positive* correlation.

Going into this more thoroughly, we surmise that the negative correlation could be accounted for by the fact that the sales of hot chocolate as well as the number of tourists visiting the resort might be related to a third variable x_3, the average weekly temperature at the resort. *If the temperature is high there will be many visitors, but they will prefer cold drinks to hot chocolate; if the temperature is low there will be fewer visitors, and they will prefer hot chocolate to cold drinks.* Getting data on the resort's average temperature for the same weeks for which we previously calculated the correlation between x_1 and x_2, let us suppose that the correlation coefficient for x_1 and x_3 is -0.70 while that for x_2 and x_3 is 0.80. These values seem reasonable since high temperature should go with low sales of hot chocolate, low temperatures with high sales, while high temperatures should go with many tourists, low temperatures with fewer.

In order to study the actual effect of the number of tourists on the sale of hot chocolate, we will have to investigate the relationship between x_1 and x_2 with all other factors, primarily temperature, held fixed. Since it is seldom possible to make a separate study of two variables when all other relevant factors are fixed (we seldom have sufficient data for that type of analysis) it has been found that a statistic called the *partial correlation coefficient* does a fair job of eliminating the interference of other variables. Writing the ordinary correlation coefficient for x_1 and x_2 as r_{12}, that for x_1 and x_3 as r_{13}, and that for x_2 and x_3 as r_{23}, *the partial correlation coefficient measuring the strength of the correlation between x_1 and x_2 while x_3 is held fixed is defined as*

$$r_{12.3} = \frac{r_{12} - (r_{13})(r_{23})}{\sqrt{(1 - r_{13}^2)(1 - r_{23}^2)}}. \qquad (14.6.3)\star$$

Substituting the values given above, we get

$$r_{12.3} = \frac{(-0.30) - (-0.70)(0.80)}{\sqrt{[1 - (-0.70)^2][1 - (0.80)^2]}}$$
$$= 0.62$$

and this shows that, as we should have expected, there is a *positive* correlation between the number of tourists visiting the resort and the sales of hot chocolate when the temperature is the same.

The above example has been given mainly to illustrate what is meant by partial correlation. At the same time it has served to emphasize again that an ordinary correlation coefficient can lead to very misleading conclusions unless it is interpreted with great care.

Formula (14.6.3) provides a measure of the strength of the correlation between two variables x_1 and x_2 when a third variable x_3 is held fixed. It can easily be generalized to apply to situations in which we want to measure the strength of the correlation between two variables when *several* other variables are held fixed. More detailed treatments of multiple and partial correlation, the necessary formulas and computational techniques, are referred to in the Bibliography on page 336.

14.7 The Correlation of Qualitative Data†

Although we did not mention it explicitly, it must be apparent from the formula for r that the coefficient of correlation applies only to *quantitative* data. It cannot be used if one or both of the variables are *qualitative*. For example, it cannot be used to measure the strength of whatever relationship there might exist between consumer preferences and geographic location, between temperament and the speed of professional advancement, between the occupations of fathers and the occupations of sons, political affiliation and education, and the like. Since problems like these are of considerable interest to persons engaged in market research, personnel work, and management in general, let us demonstrate briefly how the technique which we discussed in Section 11.2 may be applied to situations of this kind.

Let us suppose that we want to determine whether there exists a relationship between the income of one-car families and the kind of car that they drive, classifying income into low, medium, and high and cars into sedans, hardtops, and convertibles. (Although income can be expressed numerically, the cars cannot.) Let us suppose, furthermore, that a random sample of 400 one-car families residing in a certain area yielded the information presented in the following table:

		Sedans	Hardtops	Convertibles
	Low	77	13	8
Income	Medium	145	58	27
	High	21	32	19

If we formulate the hypothesis that there is *no relationship* between a family's income and the style of its car, we mean that the proportions of families driving sedans, hardtops, and convertibles *is the same* for each income group. Checking on page 254, the reader will find that this is precisely the kind of hypothesis which we tested with reference to an r by k table. Hence, we can test the null hypothesis of no correlation by first finding the *expected* cell frequencies with (11.2.1), calculating χ^2 with formula (11.1.2), and then using the criterion given on page 255.

Calculating the expected cell frequencies with (11.2.1), namely, by multiplying each cell's row total by its column total and then divid-

---

† The technique treated in this section is based on Section 11.2, and it should be omitted unless the reader has studied the χ^2 test as it applies to the analysis of an r by k table.

ing by the grand total, we get the values shown in parentheses in the following table:

		Sedans	Hardtops	Convertibles
	Low	77 (59.5)	13 (25.2)	8 (13.3)
Income	Medium	145 (139.7)	58 (59.2)	27 (31.1)
	High	21 (43.8)	32 (18.6)	19 (9.6)

Substituting these observed and expected frequencies into (11.1.2), we get

$$\chi^2 = \frac{(77 - 59.5)^2}{59.5} + \frac{(13 - 25.2)^2}{25.2} + \frac{(8 - 13.3)^2}{13.3} + \frac{(145 - 139.7)^2}{139.7}$$

$$+ \frac{(58 - 59.2)^2}{59.2} + \frac{(27 - 31.1)^2}{31.1} + \frac{(21 - 43.8)^2}{43.8} + \frac{(32 - 18.6)^2}{18.6}$$

$$+ \frac{(19 - 9.6)^2}{9.6} = 44.65$$

and to test the null hypothesis of no correlation at a level of significance of 0.05 we have only to compare this value with $\chi^2_{.05}$ as given in Table III on page 503. Since the number of degrees of freedom is $(3 - 1)(3 - 1) = 4$ (see page 255), we get $\chi^2_{.05} = 9.488$, and since this value is exceeded by the χ^2 which we calculated for our data, the null hypothesis of no correlation can be rejected. *If the value of χ^2 which we obtain for our data is large, so large that the null hypothesis can be rejected, we shall say, as before, that there is a significant correlation. If χ^2 is small, smaller than the value of $\chi^2_{.05}$ looked up in Table III, we shall say that the relationship is not significant.*

Using the method of Section 11.2 we can, thus, test whether there exists a significant relationship between *qualitative* variables. The only distinction between the over-all methods described here and in Section 11.2 lies in the selection of the data. In Section 11.2 we assumed that we had k separate samples from as many populations; now we look upon the entire table as *one* sample and the totals of the columns, instead of being fixed, depend on chance. The 3 by 3 table used in our example and, more generally, r by k tables used in problems of this type are also referred to as *contingency tables.*

Once we have shown by means of the χ^2 criterion that a correlation between two qualitative variables is significant, it is often desirable to obtain some measure of the strength of the relationship. Let us.

therefore, use the following formula to define a new measure of correlation, the so-called *contingency coefficient:*

$$C = \sqrt{\frac{\chi^2}{\chi^2 + n}} \qquad (14.7.1)\star$$

Here n is the grand total of the frequencies of the contingency table while χ^2 is the value obtained with formula (11.1.2).

Contingency coefficients are in many respects similar to ordinary correlation coefficients, being close to 0 when there is no correlation and close to 1 when the relationship is strong. To interpret C correctly, it is worth noting that for contingency tables having relatively few rows and columns the maximum value of C is actually less than 1. For a 2 by 2 table a "perfect" correlation would yield $C = 0.707$ and for a 3 by 3 table $C = 0.816$.

Had we computed the contingency coefficient to measure the strength of the relationship between a family's income and the style of its car, we would have obtained

$$C = \sqrt{\frac{44.65}{44.65 + 400}} = 0.32$$

This indicates a significant though not particularly strong correlation between the two variables. The fact that it is significant was shown previously by means of the χ^2 criterion.

EXERCISES

1. Use the following table to test at a level of significance of 0.05 whether there exists a significant relationship between aggressiveness and the speed of professional advancement. If there is a significant relationship, calculate C.

| | Speed of Advancement | | | |
	Slow	Average	Fast	Very fast
Very aggressive	48	51	43	35
Average	22	42	59	57
Very passive	15	22	33	13

2. A study to determine whether there exists a relationship between intelligence (as measured by a certain objective test) and salesmanship (as measured by volume of sales) produced the results shown in the following table:

		Intelligence	
	Below average	Average	Above average

		Below average	Average	Above average
Volume of sales	Low	18	28	14
	Average	37	63	30
	High	15	29	16

Test at a level of significance of 0.05 whether there exists a significant relationship and, if so, calculate C.

3. A study made by a bookstore in a particular community to determine whether there exists a relationship between education and the frequency of purchase of books yielded the results shown in the following table:

	Frequency of Book Purchases		
	Almost never	Occasionally	Frequently
Did not finish high school	82	65	12
Finished high school	59	112	24
At least one year of college	37	94	42

Test at a level of significance of 0.01 whether there exists a significant correlation and, if so, calculate C.

BIBLIOGRAPHY

Similar treatments of the problem of correlation and further illustrations and examples may be found in most elementary statistics texts.

See, for example,

Bryant, E. C., *Statistical Analysis*, rev. ed. New York: McGraw-Hill, 1966, Chaps. 7 and 10.

Croxton, F. E., and Cowden, D. J., *Practical Business Statistics*, 3rd ed. Englewood Cliffs, N.J.: Prentice-Hall, 1960, Chaps. 24 to 27.

A derivation of (14.1.3) may be found in

Richardson, C. H., *An Introduction to Statistical Analysis*. New York: Harcourt, Brace, 1944, Chap. 8.

Various types of spurious correlations are discussed in

Johnson, P. O., and Jackson, R. W., *Introduction to Statistical Methods*. Englewood Cliffs, N.J.: Prentice-Hall, Inc., 1953, Chap. 11.

and a theoretical treatment of this problem may be found in

Simon, H. A., "Spurious Correlation: A Causal Interpretation," *Journal of the American Statistical Association*, Vol. 49, 1954, No. 267.

An exact small-sample test of the significance of r based on the t-distribution is given in

Neiswanger, W. A., *Elementary Statistical Methods as Applied to Business and Economic Data*, rev. ed.   New York: Macmillan, 1956, Chap. 19.

and an alternate test based on the so-called z-transformation is mentioned in

Mills, F. C., *Introduction to Statistics*.   New York: Holt, Rinehart & Winston, 1956.

A detailed study of multiple and partial correlation is given in

Ezekiel, M., and Fox, K. A., *Methods of Correlation and Regression Analysis*, 3rd ed.   New York: John Wiley, 1959.

CHAPTER 15

Index Numbers:
Basic Concepts

15.1 Introduction

High on the list of statistical measures that are the most useful to the typical businessman are those telling him how much quantities have changed or how one quantity compares with another.   He may want to know, for example, that if the production of corn for grain in the United States has increased from 2,764 million bushels in 1952 to 4,103 million bushels in 1966, this marks an increase of about 48 per cent, or, better, that the 1966 production of corn for grain is about 148 per cent of what it was in 1950.   Similarly, if a housewife paid 50 cents for a pound of butter in 1944 and 82 cents in 1968, she may justifiably complain that the 1968 price is 82/50 or 164 per cent of what it was in 1944.   And if the population of the United States has increased from 132,164,569 in 1940 to 179,323,175 in 1960, we can express this by saying that the size of the population in 1960 is 135.7 per cent of what it was in 1940.

In each of these illustrations we compared two things: first we compared two production totals, then two prices, and finally two population sizes.   Such *binary* comparisons, that is, comparisons of *two* things, in which a relative change is expressed by giving one number as a percentage of another are the most primitive of a large set of statistical measures called *index numbers*.   Although index numbers are, by definition, statistical measures that express *binary* comparisons, they are often used for what is called a *comparison in series*.   Instead of taking a single index number and considering it as expressing a comparison of two things, one may take a whole set, or series, of index numbers and compare them with one another.   For example, we might construct individual index numbers by expressing each year's production of corn as a percentage of that of 1950 and we could then study changes in the production of corn by comparing these index numbers instead of the actual totals.   For that matter, we could also compare them with similar series of index numbers for *other* commodities; this would be more meaningful than a direct comparison of the original data because each series of index numbers represents *percentage changes.*

The usefulness of index numbers is by no means limited to measuring changes in the prices or production of single commodities.   They are widely used to express changes in such complex economic phenomena as the cost of living, total industrial production, and business cycles.   This, of course, involves a process of combining many prices or quantities in such a way that a single number can be used to indicate over-all changes.

Although index numbers are commonly associated with the evaluation of business conditions, economic conditions, and the like, they are also used in other fields.   Sociologists may speak of population indexes, making comparisons like the one given in the first paragraph of this section; psychologists measure intelligence quotients, which are essentially index numbers comparing a person's intelligence score with that of an average for his or her age; health authorities prepare indexes to display changes in the adequacy of hospital facilities; and educational research organizations have devised formulas to measure changes in the effectiveness of school systems.   In view of our primary interests in this book, we shall limit our discussion to business data and we shall, furthermore, consider only index numbers measuring changes relative to *time*.   In contrast, there also are index numbers which compare economic conditions of different locations, different industries, different cities, or different countries, but since the basic problems are essentially the same and since most of the important index numbers published by the government and private research organizations do refer to data collected at different *times*, it will be convenient to make this restriction.

During the past few years the use of index numbers has extended to many fields; they no longer are tools used only by economists or highly trained business analysts.   They are now of vital interest to millions of workers whose wages automatically go up or down depending on the *Consumer Price Index* prepared by the Bureau of Labor Statistics; they are of great concern to all farmers whose subsidies depend on the *Parity Index* of the federal government; and they are no less important to business firms and individuals for whom they provide actual insurance against changing prices.   Although index numbers have been used for quite some time by government agencies, private firms, and individuals as *one* of many factors considered in the formulation of over-all policies, it is only very recently that they have been written into laws and contracts as "trigger fingers" that automatically determine changes in wages, subsidies, prices, and credit. The widely used and much publicized *Consumer Price Index* which, prior to 1945 was known as the *Cost of Living Index*, has for years aided businessmen all over the world in determining prices and analyzing markets, but in addition it now determines whether or not raises are given to millions of wage earners whose services are covered by wage contracts including so-called "escalator clauses." For example, the United Automobile, Aerospace, and Agricultural Implement Workers International Union (U.A.W.) negotiates local agreements with employers which result in an automatic 1¢ increase in wages for every .4 per cent increase in The Consumer Price Index. This same index is even found in alimony agreements and trust fund payments, which can thus be made to vary with the value of the dollar. Other index numbers serve much the same purpose with respect to the rental or lease of both real estate and machinery.

The increasingly significant role that index numbers have been assuming in business planning and in the formulation of executive decisions not only puts a tremendous burden on the statisticians who are responsible for their construction, but it also presents the businessman who uses them with the responsibility of using them intelligently in full awareness of what, through their strength, they show and, through their inherent weaknesses and limitations, they fail to show.

Among the problems that arise in the construction and application of index numbers, the following are the most basic and they will be discussed below in some detail: *1. Purpose of the Index, 2. Availability and Comparability of Data, 3. Selection of Items to Be Included, 4. Choice of the Base Period, 5. Choice of the Weights, and 6. Methods of Construction.*

15.2 Purpose of the Index

The purpose for which an index is to be used is generally specified before any attempt is made to construct it, although we might add that indexes often find important uses for which they were not originally intended. A clear and precise statement of the purpose of an index will often settle automatically some of the other problems listed above. Suppose, for instance, that we are asked to construct an index whose purpose is to measure the annual changes in sales and shipments of the U.S. Steel Corporation for the years 1960 through 1966 *relative to* 1959.   As the problem is stated, there is only one item to be included in the index and it is specifically mentioned.   The data are available and, upon examination, are found to be comparable (in the same unit, recorded by the same organization, and so forth).   The *base year*, that is, the year which constitutes the basis of the comparison, is specified as 1959 and for the *weights*, about which we shall have a great deal to say later on, this problem need not be considered inasmuch as we are including only one item.

An index of the sort suggested here will lead to a comparison in *series*.   To see how this comparison might be made, let us take the production data from the (1959–1966) Annual Reports of the United States Steel Corporation and reproduce them in the following table:

	Sales and Shipments in Millions of Tons	Index
1959	18.1	100
1960	18.7	103
1961	16.8	93
1962	17.8	98
1963	18.9	104
1964	21.2	117
1965	22.5	124
1966	21.6	119

The values given in the third column make up the index of Sales and Shipments of steel for these data relative to 1959.   They were obtained by dividing each year's sales and shipment by 18.1, the 1959 sales and shipments total, and then multiplying by 100.   In this index number series each figure, which is in fact a *quantity index* of the simple form called a *quantity relative*, relates the sales and shipments of that year to the 1959 total.

Using this series, we can now make various kinds of comparisons. For instance, we can say that the 1966 figure was 119 per cent of, or 19 per cent higher than, the 1959 shipments and sales.   We can also say, for instance, that the 1961 index is 7 "points" lower than the 1959 index, that the 1965 index is 31 "points" higher than the 1961 index, or that the 1965 index is 7 "points" higher than the 1964 index.

In contrast to the above example in which the statement of the purpose of the index solved most of our problems, there are many situations in which the statement of the problem raises complex questions of the sort enumerated above and to be discussed below.

15.3 Availability and Comparability of Data

It is needless to say that it is impossible to make appropriate comparisons unless the necessary statistical data can be obtained.   Many index number workers have been frustrated by the fact that essential information was tabulated by counties, whereas actually they needed it by townships, they have run into difficulties because sales data were available only by type of merchandise and not by brand, and they have been moved to despair upon discovering that insurance losses were given per risk while no information was available about the sizes of individual claims. Unfortunately, there often is a considerable lack of uniformity in the methods used by various agencies in the reporting of statistical data, and it is, of course, only seldom that an index number worker can go out and collect his own data. Since large-scale surveys are very expensive and since the construction of index number series demands the repeated collection of data at frequent intervals of time, most of the important indexes are constructed by the federal government. Rather the exception than the rule are some widely used and respected indexes reported by non-government organizations of one kind or another.

Regardless of who attempts to construct an index number series, the problem of the availability of data can create a serious dilemma years after a series has been started. Consider, for example, the case of a statistician who wants to use a certain formula to construct an index that is supposed to measure changes in the prices of commodities playing a significant role in the average family's budget. Suppose, furthermore, that he decides to compare in this index each year's prices with average prices from 1910 to 1914. Looking around him, he now finds the widespread use of frozen and packaged foods, millions of television sets, and numerous small and large appliances of various sorts. Since these items represent a considerable portion of many budgets, he feels that they must be considered in the construction of his index. This brings him face to face with the unpleasant fact that many of these things did not sell commercially from 1910 to 1914 and, to name but one, television had not even been invented. He is thus left with two alternatives, neither of which is very attractive: he can change the entire method of construction on which his index is based, or he can invent fictitious prices representing what these things might have cost if they had been on the market in the period from 1910 to 1914. It is easy to see that problems of this sort may constantly arise in view of the ever changing buying habits of the general public and the availability of different and new commercial products.

The problem of the *comparability* of data used in an index can also be quite troublesome. In congressional hearings the representatives of some labor organizations complained that the government price indexes of certain commodities included in the *Consumer Price Index* are too low—that they reflect deterioration of quality instead of an actual change in prices. It does not matter at this point whether this criticism is valid, but arguments of this type are frequently raised against index numbers, because it is an exceedingly difficult problem to make sure that prices are actually comparable, that is, that they really refer to goods and services that are identical in quality. The common error of mistaking a change in quality for a change in price can be illustrated with the following example: a man who has just bought a new pair of shoes complains that he had to pay $22.95 instead of the $16.95 which he paid for his last pair three years ago. If the two pairs of shoes were made by the same manufacturer and were identical in every respect, his criticism would be valid. If, on the other hand, they were made by different manufacturers, it could well be that a large portion of the difference in price might be attributed to a difference in quality.

The comparability of statistical data may also be questioned if parts of the data were collected by different agencies. It is very confusing, for example, to note that the *Bureau of Census* reports that the total value of the 1939 manganese ore production in the United States was $916,000 although the corresponding figure published by the *Bureau of Mines* reads only $794,000. Such differences can arise due to variations in the methods that are used in the collection and compilation of the data. In our example they arise from the fact that the *Bureau of Census* figures are those submitted by producers while the *Bureau of Mines* figures are those given by purchasers and transportation companies.

If someone wanted to construct an index comparing the total June, 1968, employment in the United States with that of June, 1967, he would be wrong to divide the 1968 figure of 77.3 millions as reported by the *Bureau of Census* by the 1967 figure of 66.5 millions as reported by the *Bureau of Labor Statistics*. Even though both these figures are assertedly national employment totals, they are not comparable because, among other things, the *Bureau of Census* includes persons who are self-employed while the *Bureau of Labor Statistics* does not.

Mistakes in the selection of data that are really not comparable can also be made at times due to the carelessness of the person constructing an index. The two terms "life insurance in force in the U.S." and "life insurance in force in U.S. companies" sound very much alike, but it may be seen from the 1967 *Life Insurance Fact Book* that, for instance, in 1966 there is a difference of $67 billion. The two figures refer to entirely different things since residents of the United States can buy insurance in foreign companies while U.S. companies do, of course, issue policies to individuals living abroad.

To summarize, it is important to keep in mind that, insofar as is possible, data which are used in the construction of an index number must be comparable in the sense that if one wants to compare prices one is not really comparing quality. Furthermore, the goods or services to which the prices or quantities refer must adhere to uniform definitions, that is, rigorous specifications. How to achieve these goals in practice is a problem that has never been solved entirely to everyone's satisfaction.

15.4 Selection of the Items

One of the most difficult problems in index number construction is the choice of the items that are to be included in the comparison. If an index is designed for the special purpose of comparing the prices of a commodity at two different times or the quantities of a commodity produced at two different seasons, then there is no question as to what figures are to be included. The problem is clear cut, for instance, in the construction of an index designed to measure the relative change in the total production of electric energy in the United States from 1955 to 1965. According to reports of the *Federal Power Commission* these figures are approximately 629 billion Kw. hrs. in 1955, 1248 Kw. hrs. in 1965, and the desired index is 1248/629 = 1.98 or 198 per cent. If the purpose of the index had been to measure the change in the amount of electric power produced by *publicly owned* utilities, these figures could not have served, but there is little doubt as to which ones would have been required.

The situation is entirely different in the construction of so-called *general purpose indexes* such as those designed to measure general changes in wholesale or consumer prices. It must be clear that it is physically impossible, or at least highly impractical, to include in such an index all commodities from aspirin to zithers, shoelaces to pianos, and to include, furthermore, *all* prices at which these items are traded in every single transaction throughout the entire country. The only feasible alternative is to take *samples* (of items as well as transactions) in such a way that it may reasonably be presumed that the items and transactions which are included adequately reflect, or indicate, the over-all picture.

For example, an index of *Wholesale Food Prices*, published by *Dun & Bradstreet* since 1916, includes an assumedly typical "market basket" of 31 food commodities, among them milk, butter, and ham. It is obvious that such an index cannot apply to persons whose diet consists mainly of champagne and caviar but, on the other hand, it is a fairly good indicator for those persons whose tastes are more ordinary. Also, a general index of *Wholesale Prices* has been published by the Bureau of Labor Statistics since 1890, and it is currently based on the prices of a sample of about 2,300 commodities chosen from what must literally be millions of commodities. This index is published weekly and monthly, and it has been supplemented since 1939 by a daily index of *Spot Market Prices* based, since 1951, on 22 commodities.

The important *Consumer Price Index* is also a "market basket" index comparing the prices of about 400 items, goods, and services that play a significant role in the average budget of persons belonging to a certain population group. The prices considered in this index are samples with respect to the goods and services that are included and also with respect to the stores and cities that are canvassed in the necessary surveys.

The fact that most general purpose indexes are based on samples of specific items must not be overlooked in the application of such an index to the solution of practical business problems. Another factor to be considered is that, for example, the *Consumer Price Index* really measures only changes in the retail prices of certain specific goods and services that appear in the budgets of wage-earners and clerical-worker families residing in cities. Any generalizations to other population groups are questionable at best and should, whenever possible, be avoided.

Let us also point out that the sampling methods used in selecting the commodities to be included in an index frequently come under the heading of what in Chapter 12 was referred to as "judgment sampling." This means that statisticians who construct these indexes usually do *not* write the names of all possible commodities on slips of paper and draw some of them at random (or select them by means of random numbers) but that they select items which in their professional judgment are considered to be the most representative of the general situation the index is supposed to describe. They thus use judgment to decide whether a commodity is really basic, whether its quality is kept at a constant level, and so on.

15.5 Choice of the Base Period

If an index number is designed for the special purpose of comparing 1968 figures with those of some other year, say, 1959, it is customary to refer to 1968 as the *given year*, to 1959 as the *base year*, and the latter is usually expressed by writing 1959 = 100. Although this is nonsensical as it stands, it is simply meant to imply that the 1959 figures represent 100 per cent. In general, the year or period which one wants to compare is called the *given year* or *given period*, while the year or period relative to which the comparison is made is called the *base year* or the *base period*.

The choice of the base year or base period does not offer any particular difficulties if a special purpose index is to be constructed for a single binary comparison.   Evidently, if one wants to compare coffee prices in 1968 with those in 1958 the formulation of the problem automatically specifies the base year as being 1958.   Similarly, if one wants to compare coffee prices in 1967 with those during World War II, the basis of the comparison would, by definition, have to be some average of the five-year period 1941–1945.

If an index number is constructed for comparisons in series, the base period is often not specified in the formulation of the problem and, fundamentally speaking, it is essentially arbitrary.   Nevertheless, it should not be difficult to see that it is generally desirable to base comparisons on a period of relative *economic stability* as well as a period that is *not too distant in the past*.   The first stipulation is important because during times of abnormal economic conditions there may be no free trading of certain commodities, for example, no new passenger cars were available during World War II, prices may be controlled by law, black markets may arise, and the buying habits of the public may be irregular due to the lack or shortage of products that would ordinarily figure in the average budget.

It is preferable to select a base period that is not too distant in the past because index numbers which are published periodically, that is, in series, often serve for the investigation of month to month or year to year changes.   A base period that is far in the past raises difficulties not unlike those facing an art critic who, instead of judging two paintings by holding them next to one another, is forced to compare them separately with the *Mona Lisa* and then compare the two individual comparisons.   Clearly, direct comparisons are likely to be more accurate and more meaningful, *and* if direct comparisons are desirable, as they are in the case of index number series, the base period must not be too remote.   Another reason is that rapid changes in the availability of commercial products make it difficult, sometimes impossible, to find suitable prices for periods dating back a decade or more.   We have already pointed out the trouble connected with the problem of incorporating the prices of television sets into an index whose basis is the period from 1910 to 1914.

Currently   most   government   indexes   have   1957–1959 = 100, although this period can hardly be described as one of economic stability in the classical sense.   However, this period has the advantage that it is relatively recent and it might be argued that it is at least as stable as any period within the last decade.   Some index numbers still have as their basis the pre-World War I period from 1910 to 1914 and this is due primarily to the fact that they are tied to this base period by law.

15.6 Choice of the Weights

If the factors which we have so far discussed were the only ones that have to be considered in the construction of an index, our task would be relatively easy and suitable index numbers could readily be defined by means of any one of the formulas which will be treated in Section 16.2. Unfortunately, we will not get very far in our work before we will realize that in many instances the various prices or quantities that are to be considered in an index cannot simply be added. Let us suppose, for example, that we want to construct an index comparing the 1958 and 1968 *wholesale prices* of certain wood products used in the home and that we have arbitrarily decided to include the two items *toothpicks* and *furniture*. Let us suppose, furthermore, that the index (the ratio of the 1968 to the 1958 prices multiplied by 100) for *furniture alone* is 150 per cent, while that for *toothpicks alone* is 120. The question now arises as to how these two figures can be combined. We could go ahead and compute the (arithmetic) mean of the two indexes, getting $(150 + 120)/2 = 135$, and on the basis of this result it would seem that the prices of these wood products have increased on the average by 35 per cent. Mathematically speaking, this answer is correct, but it should not take much to convince anyone that it is also quite useless. *The difficulty which we have to face is that the numbers we are trying to average are not of equal importance.* To get a meaningful result we would have to weight them in some way to account for their significance in the over-all phenomenon that the index is supposed to describe. This is the identical problem we first met in our discussion of the *weighted mean* in Chapter 3, where we pointed out that the weights which we assign to the various items must be measures of their relative importance and should be chosen with great care to avoid biased and misleading results.

The problem of choosing suitable weights is usually not an easy one. It depends on whether we want to average prices, quantities, or indexes pertaining to individual commodities, and we shall defer further discussion of this matter until we come to specific index number formulas in Chapter 16.

15.7 Methods of Construction

Just as we were able to describe the "average" of a set of data by the use of the mean, median, mode, and still other measures of central location, we shall see that we can similarly describe relative changes

by employing any one of a number of formulas, all of which, by definition, provide us with index numbers. In the following chapter we
shall discuss the nature of many of these formulas and some of the
factors that must be taken into account in choosing an appropriate
one for a given problem. After having read Chapter 16, it should be
clear to the reader that any such choice will ultimately have to be
based not only on the mathematical niceties of the formulas but also
on practical considerations.

The symbolism used in the literature of index numbers can hardly
be described as uniform. Most people refer to every index by the
letter I; some distinguish between price, quantity, and value indexes
by representing them with the letters P, Q, and V; and some indicate
the base year with the letter o, the given year with the letter n, and
write the corresponding index as $I_{o,n}$.

To avoid the arbitrary introduction of new symbols, we shall refer
to most index numbers as I, except for a few special cases in which
different symbols are necessary or more appropriate. Unless otherwise stated, all formulas in the next chapter will refer to *price indexes*,
although, as we shall see, they can readily be changed to *quantity
indexes* by substituting p's (prices) for q's (quantities) and q's for p's
wherever these symbols occur.

In order to distinguish between the prices of the base year (or
base period) and those of the given year, we shall write the first as
p_o and the latter as p_n. Similarly, the quantities corresponding to the
base year and the given year will be written as q_o and q_n, respectively.
If one wants to distinguish between the base year prices of k different
commodities, one can write them as p_o', p_o'', p_o''' . . . , $p_o^{(k)}$; and the
same symbolism can also be used for given-year prices, base-year
quantities, and given-year quantities. In Chapter 16 we shall omit
the primes; instead, we shall write, for instance, Σp_o for the sum of the
base-year prices, Σq_n, for the sum of the given-year quantities, and
$\Sigma p_o q_n$ for the sum of the products obtained by multiplying base-year
prices by the corresponding given-year quantities.

BIBLIOGRAPHY

Further details on the fundamental problems of index number construction
may be found in the following books and pamphlets:

Doody, F. S., *Introduction to the Use of Economic Indicators*. New York:
    Random House, 1965.
Fisher, I., *The Making of Index Numbers*. Boston: Houghton Mifflin, 1923.
King, W. I., *Index Numbers Elucidated*. New York: Longmans Green,
    1930.

Mitchell, W. C., *The Making and Using of Index Numbers*. Bulletin 656, Bureau of Labor Statistics, Washington, D.C., 1938.

Morrell, A. J. H., *Introduction to Index Numbers*. London: H.F.L. Publishers, Ltd., 1948.

Mudgett, B. D., *Index Numbers*. New York: John Wiley, 1951.

Persons, W. M., *The Construction of Index Numbers*. Boston: Houghton Mifflin, 1928.

Books and articles dealing with the construction and use of specific index numbers are referred to in the Bibliography on page 380.

CHAPTER 16

Index Numbers: Theory and Application

16.1 Introduction

In the preceding chapter we introduced the idea of index numbers, directing attention to their usefulness and wide applicability, and we also discussed the nature of some of the basic problems that arise in their construction. It should not be surprising that many methods have been proposed for handling these problems; the history of many other fields shows that when someone feels that a "standard" method of expression does not fit his particular needs, he sets out to find a new, different, and sometimes better method. In this chapter we shall discuss some of the most useful ways of combining data into index numbers which represent changes in things or groups of things that happen to be of interest in particular problems.

16.2 Unweighted Index Numbers

Let us first take up those index numbers which are generally referred to as *unweighted*. In order to introduce such indexes, let us suppose that we are faced with the problem of measuring, for a particular locality, the change in the retail prices of certain foods from 1956 to 1968 and that, on the basis of a sample of stores and transactions, we have the following information (all prices being given in cents per pound):

350

	1956	1968
Pork chops	79.4	109.0
Sliced bacon	67.2	98.0
Roasting chicken	48.5	79.0
Butter	81.9	80.0
Wheat flour	10.4	15.0
Potatoes	4.8	14.0
Sugar	10.1	12.0

If we add the 1968 prices and divide their *sum* by that of the corresponding 1956 prices, we obtain

$$I = \frac{109.0 + 98.0 + 79.0 + 80.0 + 15.0 + 14.0 + 12.0}{79.4 + 67.2 + 48.5 + 81.9 + 10.4 + 4.8 + 10.1}$$

$$= 1.3463$$

and this tells us that the combined 1968 prices are 135 per cent of what they were in 1956, in other words, the index stands at 135 per cent. The method which we employed in this comparison of the two sets of prices is called the *simple aggregative method* and the resulting index number is accordingly called a *simple aggregative index*. In general, a *simple aggregative index* is computed by means of the formula

$$I = \frac{\Sigma p_n}{\Sigma p_0} \cdot 100 \qquad\qquad (16.2.1)^\star$$

in which the quotient of the sum of the given-year prices over the sum of the base-year prices is multiplied by 100 to express the index as a percentage.

The greatest inherent weakness of a simple aggregative index is that it can produce vastly divergent results if the various items and their prices are quoted in different units. A simple aggregative index fails what is often referred to as the "units test." To illustrate, let us consider the problem of calculating an index comparing the 1968 cost of construction with that of 1956 and, for the sake of argument, let us include only two items, the cost of labor and the price of cement. Using the following figures

	1956	1968
Average *hourly* wage paid to construction workers	$2.74	$3.85
One barrel of cement	$5.20	$6.60

and substituting them into (16.2.1), we get

$$I = \frac{3.85 + 6.60}{2.74 + 5.20} \cdot 100 = 132 \text{ per cent}$$

As the problem has been stated, there is no reason this index should be based on hourly rather than on weekly wages, and assuming a 35 hour week, we could rewrite our figures as

	1956	*1968*
Average *weekly* wage paid to construction workers	$95.90	$134.75
One barrel of cement	$ 5.20	$ 6.60

Substituting these new figures into (16.2.1), we get

$$I = \frac{134.75 + 6.60}{95.90 + 5.20} \cdot 100 = 140 \text{ per cent}$$

The reader can easily check for himself that if we had used hourly wages together with the prices of 1000 barrels of cement, the resulting index would have been $I = 127$ per cent.

Results of this sort seem to support the old aphorism that "you can prove anything with statistics." Using the identical data we showed that from 1956 to 1968 construction costs increased by 32, 40, or 27 per cent. It is largely for this reason that the simple aggregative index has never gained any great degree of acceptance. Any index whose value can be manipulated more or less to suit oneself by quoting prices per pound rather than per ton, per gallon rather than per pint, per package rather than per carton, and so forth, can hardly be used as an *objective* measure.

Among the few published indexes that are still of the simple aggregative type, perhaps the most widely known is the *Dun & Bradstreet Wholesale Food Price Index* which has appeared monthly in *Dun's Statistical Review* since 1916. This index is based on the prices of one pound each of 31 basic commodities and it is released as the sum Σp_n, which is not really a ratio of two aggregates but which can easily be changed into one if we divide by some base figure Σp_o. For instance, for the week ending July 24, 1968, Dun & Bradstreet reported that the index stood at $6.68 as compared to $6.73 for the preceding week and $6.59 for the corresponding week in 1967. Using the 1967 week as a base, the index for the week ending July 24, 1968, could be written as $6.68/6.59 \cdot 100 = 101$ per cent. The principal reason for the continued use of this technique in the construction of the index is to preserve the continuity of the series.

An alternate way of comparing the two sets of prices which we gave on page 351, would be to calculate first a separate index for each of the seven foods and then to average the ratios p_n/p_o, also called the individual *price relatives*, using any one of the measures of central location which we discussed in Chapter 3. Writing the price relatives of the seven foods as percentages, we obtain

Price Relatives

Pork chops $\quad \dfrac{109.0}{79.4} \cdot 100 = 137$ per cent

Sliced bacon $\quad \dfrac{98.0}{67.2} \cdot 100 = 146$ per cent

Roasting chicken $\quad \dfrac{79.0}{48.5} \cdot 100 = 163$ per cent

Butter $\quad \dfrac{80.0}{81.9} \cdot 100 = 98$ per cent

Wheat flour $\quad \dfrac{15.0}{10.4} \cdot 100 = 144$ per cent

Potatoes $\quad \dfrac{14.0}{4.8} \cdot 100 = 292$ per cent

Sugar $\quad \dfrac{12.0}{10.1} \cdot 100 = 119$ per cent

where each 1968 price is divided by the corresponding 1956 price and their ratio is multiplied by 100. To construct an over-all index combining these seven commodities, we can now take the mean, median, mode, geometric mean, or some other "average" of the price relatives.

If we choose to average the price relatives by calculating their arithmetic mean, we get

$$I = \frac{137 + 146 + 163 + 98 + 144 + 292 + 119}{7} = \frac{1099}{7} = 157$$

and this index is appropriately called an *arithmetic mean of price relatives*. Symbolically, the formula for this index is

$$I = \frac{\Sigma \dfrac{p_n}{p_0} \cdot 100}{k} \qquad (16.2.2)\star$$

where k is the number of items (commodities) whose price relatives are thus averaged. Inasmuch as we have written each price relative as a percentage, a value of I computed with (16.2.2) is already a percentage and the result must *not* be multiplied by another factor 100.

Among the other averages which we might have used, the *median* would have led to $I = 144$, the *mid-range* (the mean of the smallest and largest) would have led to $I = (98 + 292)/2 = 195$, and the *geometric mean* would have led to $I = 149$. Although any measure of central location could serve to define an over-all index, price relatives are usually averaged with either the arithmetic or geometric mean.

Using the more convenient logarithmic form, see (3.7.2) on page 67, the formula for the index consisting of the *geometric mean of the price* relatives can be written as

$$\log I = \frac{\Sigma \log \dfrac{p_n}{p_0} \cdot 100}{k} \qquad (16.2.3)\star$$

Substituting into this formula the logarithms of the seven price relatives (which may be obtained with the use of Table VIII), we have

$$\log I =$$
$$\frac{2.1367 + 2.1644 + 2.2122 + 1.9912 + 2.1584 + 2.4654 + 2.0755}{7}$$

$$= 2.1720$$

Since the number whose logarithm is closest to 2.1720 is 149 (see Table VIII), we have thus found that the geometric mean of the price relatives is 149 per cent.

We now have four values, 157, 144, 195, and 149, corresponding to the mean, median, mid-range, and geometric mean of the seven price relatives and it is only natural to ask which one gives the best description of the over-all change in the prices of these foods. To answer this question we will have to investigate, as in Chapter 3, the various desirable and undesirable properties of the respective measures of central location.

Some economists, notably F. Y. Edgeworth, have preferred to use the median which, as we pointed out earlier, is not affected by a single extreme value. Since this argument is important only when an index is based on a very small number of commodities, it generally does not carry too much weight and the median is seldom used in actual practice. The mid-range is generally considered to be a rather poor average, for it is based only on the two extreme values and does not account in any way for the remainder of the data. Although the arithmetic and geometric means have both been used, the arithmetic mean is often preferred because it is easier to compute and much better known.

It is a matter of historical interest to note that the earliest index number on record is an arithmetic mean of price relatives. In the middle of the eighteenth century G. R. Carli, an Italian, calculated the effect of the import of silver from America on the value of money, using a formula analogous to (16.2.2) to compare the 1750 prices of oil, grain, and wine, with those of the year 1500. Among the first to use the geometric mean in the construction of an index number was W. S. Jevons, an Englishman, who in 1863 constructed an index designed to measure the effect of the discovery of gold in California on the value of gold. This index included 39 commodities and compared the 1862 prices with average prices for 1845–1850.

Today very few index numbers are actually calculated with formulas as simple as (16.2.1), (16.2.2), or (16.2.3), because the need for the employment of weights has been almost universally accepted. Among the important government indexes only the daily *Index of*

Spot Market Prices is still calculated as a simple geometric mean of price relatives. Prior to 1914 the *Wholesale Price Index* of the Bureau of Labor Statistics was an arithmetic mean of the price relatives of about 250 commodities. It was changed to a weighted index after the well-known study by W. C. Mitchell (see Bibliography on page 348), which since its publication in 1915 has had a pronounced effect on index number construction.

Formulas (16.2.1), (16.2.2), as well as (16.2.3), define price indexes, but they are readily adaptable to comparisons of other types of data. Any one of the three formulas could be transformed into a *quantity index* by changing the p's to q's without altering the subscripts. If it were desired to compare prices in two different geographic locations instead of at two different times, the p_n's could be taken as prices quoted, say, in Chicago and the p_o's as prices quoted in New York. It should also be pointed out that in contrast to (16.2.1), the index numbers defined by formulas (16.2.2) and (16.2.3) are *not* affected by changes in the units.

EXERCISES

1. The following are the average prices paid by farmers in the United States for selected foods in cents

	Sept. 15 1966	Dec. 15 1966	Mar. 15 1967	June 15 1967	Sept. 15 1967
Round steak	99.3	97.9	98.4	98.3	101.0
Bacon, sliced	91.6	76.5	73.5	82.5	78.4
Chicken, frying, per pound	40.7	37.8	38.6	37.7	37.9
Milk, fluid, per quart	27.0	27.0	26.8	27.1	27.5

   (a) Find for Dec. 15, 1966; Mar. 15, 1967; June 15, 1967; and Sept. 15, 1967, simple aggregative indexes of the prices of the given foods, using Sept. 15, 1966 = 100.

   (b) Find the simple aggregative index of the Sept. 15, 1967, prices of the given foods relative to those of Dec. 15, 1966.

   (c) Find for Dec. 15, 1966 and Sept. 15, 1967, the arithmetic mean of the price relatives of the given foods, using Sept. 15, 1966 = 100.

   (d) Find the arithmetic mean of the relatives comparing the Sept. 15, 1967, prices with those of Mar. 15, 1967.

2. Using the data of Exercise 1 above, calculate the geometric mean of the relatives comparing the Sept. 15, 1967 prices with those of March 15, 1967.

3. The total July output of selected durable goods (in thousands of units) in 1965 and 1966 was:

	July 1965	July 1966
Passenger cars	754.0	488.4
Radio sets	1,757.0	1,233.7
Television sets	596.3	585.9
Ranges	148.5	157.0
Vacuum cleaners	329.2	414.6

(a) Calculate a simple aggregative index comparing the July, 1966, production of these goods with that of July, 1965.

(b) Find the arithmetic mean of the relatives comparing the July, 1966, production of these goods with that of July, 1965.

4. The following are the average prices (in dollars) paid by farmers for certain supplies:

	Average 1957–59	Sept. 1965	Sept. 1966
Axes, with handle, each	6.03	6.16	6.26
New baskets, round stave, bu., with cover, doz.	4.67	5.37	5.56
Milk pails, 12 quarts, each	2.21	2.55	2.56
Hoes, 7 inch, blade, each	2.38	2.71	2.80

(a) Calculate for September, 1965, and September, 1966, a simple aggregative index of the prices of these supplies, using 1957–59 = 100.

(b) Calculate for September, 1965, and September, 1966, the arithmetic mean of the price relatives, using 1957–59 = 100.

(c) Calculate the geometric mean of the relatives comparing the September, 1966, prices of these supplies with those of September, 1965.

16.3 Weighted Index Numbers

Although we referred to the index numbers of the previous section as *unweighted*, they are in reality weighted, the weights being implicit rather than explicit. The example on page 351 illustrated how we can increase the relative importance of the cost of labor in the construction of a certain index by substituting weekly wages for hourly wages and how we can similarly increase the relative importance of cement by quoting its price per 1,000 barrels rather than per barrel. *We can thus change the importance of different items even in an unweighted index by quoting prices relative to different units.* It should be clear that such an implicit weighting is far from realistic in all but a few cases and that the construction of useful index numbers requires a conscious effort to assign to each commodity a weight in accordance with its importance in the total phenomenon that the index is supposed to describe.

We have often stressed the basic arbitrariness in the selection of statistical formulas, and in the assigning of weights we are again faced by a more or less arbitrary decision. Of course, this choice is arbitrary only within certain limits because, as we stated above, the weights must fulfill a well-defined function.

To construct an index that is similar in form to the simple aggregative index but which involves some sort of weighting of the prices before they are summed, we could write in general

$$I = \frac{\Sigma p_n \cdot w}{\Sigma p_0 \cdot w} \cdot 100 \qquad (16.3.1)$$

where the w's are the weights which we assign to the various prices. Inasmuch as the importance of the price of a commodity in the over-all picture described by an index is generally determined by its *quantity* produced, consumed, bought, or sold, the w's of formula (16.3.1) should preferably be *quantity weights* and we are left with the choice of taking the quantities of the given year, those of the base year, or some other set of quantities whose use can reasonably be justified.

Let us consider first two price indexes using as weights the base-year quantities and given-year quantities, respectively. Using the base-year quantities q_0, we obtain a *weighted aggregative index* which, since it was first suggested by Laspeyres, will be represented by the letter L. Symbolically, we get

$$L = \frac{\Sigma p_n q_0}{\Sigma p_0 q_0} \cdot 100 \qquad (16.3.2)\star$$

If, instead of the base-year quantities, we use the given-year quantities q_n, a second weighted aggregative index is given by the formula

$$P = \frac{\Sigma p_n q_n}{\Sigma p_0 q_n} \cdot 100 \qquad (16.3.3)\star$$

where P stands for Paasche, the statistician who first suggested its use.[†]

To illustrate the use of these two new formulas, let us consider the problem of constructing an index number comparing the prices paid to farmers for five of their major crops in 1966 with those received for the same commodities in 1960. In the following table the prices given

---

[†] It should be noted that it is *not* permissible to use given-year quantities in the numerator and base-year quantities in the denominator of (16.3.1). The resulting index would be a *value index* instead of a price index, because $\Sigma p_n q_n$ is the total value of the commodities in the given year and $\Sigma p_0 q_0$ is their total value in the base year.

are averages for the marketing season in cents per bushel and the quantities are in millions of bushels.

	Prices		Quantities	
	1960	1966	1960	1966
Wheat	174.0	163.0	1,355	1,311
Oats	60.0	67.0	1,153	798
Rye	88.0	107.0	33	28
Barley	84.0	106.0	429	390
Corn for Grain	100.0	129.0	3,907	4,103

Weighting the 1960 and 1966 prices with the 1960 (or base-year) quantities, substitution into (16.3.2) yields

$$L = \frac{163.0(1355) + 67.0(1153) + 107.0(33) + 106.0(429) + 129.0(3907)}{174.0(1355) + 60.0(1153) + 88.0(33) + 84.0(429) + 100.0(3907)} \cdot 100$$

$$= 115.9 \text{ per cent}$$

Weighting the same prices with the 1966 (or given-year) quantities gives

$$P = \frac{163.0(1311) + 67.0(798) + 107.0(28) + 106.0(390) + 129.0(4103)}{174.0(1311) + 60.0(798) + 88.0(28) + 84.0(390) + 100.0(4103)} \cdot 100$$

$$= 116.5 \text{ per cent}$$

In the first of these indexes a constant aggregate of goods, that of the 1960 base period, was priced at 1960 prices, again at 1966 prices, and the index reflects only changes in price since the quantities of goods priced were the same. For the same reason, the second index also measures only changes in price although we priced a different aggregate of goods, that of the year 1966.

The difference between the results which we obtained for L and P is evidently slight in this particular example; but it can be large, and this raises the question as to which of the two formulas should be preferred. Evidently such a question can only be answered by investigating some of the basic properties of these index numbers. To begin with, neither of the two indexes is subject to the criticism raised earlier against the simple (unweighted) aggregative index, *they are not affected by changes in the units to which the prices refer*. The reason for this is that whenever the units are changed, the quantity weights assigned to the prices also change and the two changes compensate for one another. In other words, the products p times q remain the same. A man selling 20 dozen pencils at 60 cents a dozen could also be reported as selling 240 pencils at 5 cents each. The products of price and quantity are evidently the same—60 times 20 in the first case, and 5 times 240 in the second.

From a practical point of view, Laspeyres' index is often preferred to Paasche's because the weights q_o are base-year quantities and do not change from one year to the next. On the other hand, the use of Paasche's index requires the continuous use of *new* quantity weights q_n for each period considered, and in most cases these weights are difficult and expensive to obtain. The problem of fixed vs. flexible (changing) weights is of considerable importance and it will be touched upon again later on in this chapter.

An interesting property of the index numbers given by (16.3.2) and (16.3.3) is that Laspeyres' index can generally be expected to *overestimate* or to have an *upward bias*, while Paasche's index will generally do the exact opposite. When prices increase there is usually a reduction in the consumption of those items for which the increase has been the most pronounced and, hence, by using base-year quantities we will be giving too much weight to the prices that have increased the most and the numerator of (16.3.2) will be too large. Similarly, when prices go down, consumers will shift their preference to those items which have declined the most and, hence, by using base-period weights in the numerator of L we will not be giving sufficient weight to the prices that have gone down the most and the numerator of (16.3.2) will again be too large. We shall leave it to the reader to find similar reasons for Paasche's index's *underestimating*, that is, for its having a *downward bias*. Incidentally, the above argument does not imply that L must necessarily always be larger than P.

Unless drastic changes have taken place between the base year and the given year, the difference between L and P will generally be small and either could serve as a satisfactory measure. If the difference is large it may be that both will have to be considered as wholly unsatisfactory. If it is impossible to choose between the two on theoretical grounds in a situation in which the two indexes agree quite well, a practical compromise may be reached by either averaging the two indexes or by averaging the weights.

Considering first the alternative of averaging L and P, let us define the following two index numbers by forming their arithmetic and geometric means:

$$I = \frac{L + P}{2} = \frac{\dfrac{\Sigma p_n q_o}{\Sigma p_o q_o} + \dfrac{\Sigma p_n q_n}{\Sigma p_o q_n}}{2} \cdot 100 \qquad (16.3.4)\star$$

and

$$I = \sqrt{L \cdot P} = \sqrt{\frac{\Sigma p_n q_o}{\Sigma p_o q_o} \cdot \frac{\Sigma p_n q_n}{\Sigma p_o q_n}} \cdot 100 \qquad (16.3.5)\star$$

The first of these formulas is named after Drobisch, who suggested its use as early as 1871, and the second is known as the *Ideal Index*, a name given to it by Irving Fisher in 1920. Although there is seldom a big difference between the indexes given by (16.3.4) and (16.3.5), Fisher's Ideal Index is generally preferred because it satisfies mathematical criteria which will be discussed in Section 16.8.

Had we computed formulas (16.3.4) and (16.3.5) for the data given on page 358, our results would have been $I = \dfrac{(115.9 + 116.5)}{2} = 116.2$ and $I = \sqrt{115.9 \cdot 116.5} = 116.2$.

Considering the second alternative of averaging the weights, we could substitute for the w's in (16.3.1) the arithmetic or geometric means of the base-year and given-year quantities, namely, $(q_o + q_n)/2$ or $\sqrt{q_o q_n}$, and we would get

$$I = \frac{\Sigma p_n (q_o + q_n)}{\Sigma p_o (q_o + q_n)} \cdot 100 \tag{16.3.6)$\star$}$$

and

$$I = \frac{\Sigma p_n \sqrt{q_o q_n}}{\Sigma p_o \sqrt{q_o q_n}} \cdot 100 \tag{16.3.7)$\star$}$$

The factor 2 has been cancelled in the numerator and denominator of (16.3.6).

To calculate the first of these new index numbers, the so-called Edgeworth formula, for the data given on page 358, we would first have to find the sums of the base-year and given-year quantities *separately* for each commodity and then multiply the respective prices by these weights. It will be left to the reader to show that this index will also equal 116.2.

A method which is currently in great favor in the construction of index number series is the *fixed-weight aggregative* index that is defined by the formula

$$I = \frac{\Sigma p_n q_a}{\Sigma p_o q_a} \cdot 100 \tag{16.3.8)$\star$}$$

Here the weights are quantities referring to some period other than the base year o or the given year n. An important advantage of this formula is that like Laspeyres' index it does not demand yearly changes in the weights and, in addition, the base period can be changed without necessitating corresponding changes in the weights. This is very important because the construction of appropriate quantity weights for a general purpose index usually requires a considerable amount of work. Weights can, thus, be kept constant until new census (or other survey) data become available to revise the index.

One of the most important fixed weight aggregative indexes is the *Wholesale Price Index* published by the Bureau of Labor Statistics. This index is currently based on about 2,300 commodities divided into 15 major groups (farm products, processed foods, textiles, metals and metal products, for example) which in turn are divided into 98 subgroups, for all of which separate indexes are periodically calculated.† The base used in this index is the three-year period 1957–1959. Effective January 1967, the weights q_a are the values of net shipments of commodities in 1963. The choice of the year a, namely 1963, reflects the continuing updating of the weight period which has been revised four times since the year 1955.

We shall not illustrate the use of formula (16.3.8) with a numerical example, because the calculations are identical with those of formulas (16.3.2) and (16.3.3), the only difference being that the quantity weights refer to some year other than the base year or the given year.

Let us also point out that if we substituted p's for q's and q's for p's in any one of the formulas given in this section, the resulting formula would represent the corresponding *quantity index* in which the quantities of the different commodities are weighted by their prices. There would be no particular advantage in using (16.3.8) in that case since there is generally no problem in obtaining prices. The much more difficult question is that of obtaining the quantities that are to be compared in a quantity index.

EXERCISES

1. The following table contains stumpage prices and production totals of three selected softwoods, prices being given in dollars per thousand board feet, and production figures in billions of board feet:

	Prices				Quantities			
	1963	*1964*	*1965*	*1966*	*1963*	*1964*	*1965*	*1966*
Douglas Fir	27.9	38.1	42.6	51.6	8.6	9.0	9.2	8.4
Southern Pine	25.1	27.8	31.7	38.5	6.1	6.4	6.6	6.7
Western Pine	15.8	19.0	19.8	19.6	9.3	10.4	10.3	10.4

   (a) Using 1963 quantities as weights and 1963 = 100, find weighted aggregative indexes for the 1964, 1965, and 1966 prices of these woods.
   (b) Calculate an aggregative index number comparing the 1966 prices of the three woods with those of 1963, using as weights the 1966 quantities.
   (c) Calculate an aggregative index number comparing the 1966 prices of the three woods with those of 1963 using as weights the averages of the 1963 and 1966 quantities.

---

† The Bureau of Labor Statistics also constructs wholesale price indexes by stage of processing for which commodities are separated into three categories: (1) raw or crude materials for further processing, (2) intermediate materials, supplies, and components, and (3) finished goods.

(d) Use the results of (a) and (b) to calculate an Ideal Index comparing the 1966 prices of the three woods with those of 1963.
2. Apply the Edgeworth formula, formula (16.3.6), to the data given on page 358 and verify that the index number thus obtained is 116.2.
3. The following are the 1940, 1950 and 1966 wholesale prices (in cents per gallon) and production (in hundreds of millions of barrels) of three fuels:

| | Prices | | | Quantities | | |
	1940	1950	1966	1940	1950	1966
Gas	4.6	10.0	11.4	6.7	10.2	17.9
Kerosene	5.4	9.4	10.4	.7	1.2	1.0
Distillate fuel oil	4.7	8.3	9.4	1.8	4.0	7.9

(a) Using the 1940 quantities as weights, construct aggregative indexes comparing the 1950 and 1966 prices, respectively, with those of 1940.
(b) Using the 1950 quantities as weights, construct aggregative indexes comparing the 1950 and 1966 prices, respectively, with those of 1940.
(c) Using 1966 quantities as weights, construct an aggregative index comparing the 1966 prices with those of 1940.
(d) Basing your calculations on the results of (a) and (c), calculate the Ideal Index for the 1966 prices, with 1940 = 100.
(e) Using the means of the 1940 and 1966 quantities as weights, construct an aggregative index comparing the 1966 prices with those of 1940.
4. Using the data of Exercise 3 above, the 1950 quantities as weights, and 1950 = 100, construct aggregative indexes for the prices of the given commodities in 1940 and 1966.

16.4 Weighted Averages of Price Relatives

In Section 16.2 we showed how index numbers can be constructed by first calculating separate indexes, or relatives, for each individual commodity and then combining the relatives into some over-all index. In this section we shall do precisely the same, but instead of computing arithmetic and geometric means with formulas such as (16.2.2) and (16.2.3), we shall use similar formulas involving weights.

Taking the formula for the weighted mean as given by (3.8.1), we can write the formula for a *weighted arithmetic mean of price relatives* as

$$I = \frac{\Sigma(p_n/p_o) \cdot w}{\Sigma w} \cdot 100 \qquad (16.4.1)$$

where the w's are the weights assigned to the individual price relatives. In this formula we left the relatives p_n/p_o as *proportions*, multiplying their weighted average by 100 so that the index number itself is expressed as a percentage. Instead we could also multiply each relative by 100, as we did in (16.2.2), and we would then have omitted the 100 shown in (16.4.1).

Since the importance of a relative change in the price of a commodity is most adequately measured by the *total amount of money* that is spent on it, it is customary to use *value weights* for the w's of (16.4.1). This raises the question whether one should use the values (prices times quantities) of the base year, those of the given year, or perhaps some other fixed-value weights. If we substituted base year values $v_o = p_o q_o$ for the w's of (16.4.1), we would get

$$I = \frac{\Sigma(p_n/p_o)(p_o q_o)}{\Sigma p_o q_o} \cdot 100 \qquad (16.4.2)$$

and it can easily be seen that we would *not* have arrived at a new index. The p_o's in the numerator cancel, leaving $\Sigma p_n q_o$, and the formula reduces to (16.3.2), namely, that of the Laspeyres index. If the necessary prices and quantities are available, it is generally easier to use (16.3.2) than (16.4.2). However, if the price data are already given in the form of relatives or if the raw data are given as values rather than quantities, it may be easier to use (16.4.2).

Substituting *given year* values $v_n = p_n q_n$ for the w's of (16.4.1), we get

$$I = \frac{\Sigma(p_n/p_o)(p_n q_n)}{\Sigma p_n q_n} \cdot 100 \qquad (16.4.3)\star$$

as an alternative formula for a *weighted arithmetic mean of price relatives*. This index has been criticized by economists on the grounds that while the denominator of the index constitutes a *real* value, the numerator has no real meaning in a physical sense. The index is, furthermore, not very practical because it requires given-year values or quantities.

To illustrate the calculation of a weighted arithmetic mean of price relatives in which the weights are given-year values, let us refer again to the five farm commodities considered on page 358. Specifically, we shall weight the relatives $p_n/p_o = p_{66}/p_{60}$ by the values $p_{66}q_{66}$. (In this notation, the subscripts on the p's and q's refer to the years 1960 and 1966.) The table given below contains the 1960 and 1966 prices of the five commodities in cents per bushel and the corresponding quantities in millions of bushels. The fifth column contains the relatives p_{66}/p_{60}; and the sixth, the value weights $p_{66}q_{66}$.

| | Prices | | Quantities | | Relatives | Values |
	1960	1966	1960	1966	p_{66}/p_{60}	$p_{66}q_{66}$
Wheat	174.0	163.0	1,355	1,311	0.94	213,693
Oats	60.0	67.0	1,153	798	1.12	53,466
Rye	88.0	107.0	33	28	1.22	2,996
Barley	84.0	106.0	429	390	1.26	41,340
Corn	100.0	129.0	3,907	4,103	1.29	529,287

Substituting the figures given in the last two columns into (16.4.3), we obtain

$$I = \frac{0.94(213{,}693) + 1.12(53{,}466) + 1.22(2{,}996) + 1.26(41{,}340) + 1.29(529{,}287)}{213{,}693 + 53{,}466 + 2{,}996 + 41{,}340 + 529{,}287} \cdot 100$$

$$= 119 \text{ per cent}$$

It is important to note that if all the weights in formulas (16.4.2) and (16.4.3) are multiplied or divided by an arbitrary fixed number, *the value of the index will not change.* Hence, it is quite common to multiply or divide value weights by a suitable number to simplify calculations. For example, if three value weights are $1,000, $2,000, and $5,000, we could just as well use the weights 1, 2, and 5. Another modification that is frequently used is to give each weight as a *percentage* of the total of the weights assigned to all of the commodities. For the above figures the sum of the weights is $1,000 + 2,000 + 5,000 = 8,000$, of which the first weight is $1,000/8,000 = 12.5$ per cent, the second is $2,000/8,000 = 25$ per cent, and the third is $5,000/8,000 = 62.5$ per cent. Instead of the original weights we could, thus, use the percentage weights 12.5, 25, and 62.5.

Value weights that belong neither to the base year nor to the given year are used, for instance, in the *Index of Industrial Production* prepared by the Federal Reserve Board. This index is a quantity index rather than a price index, and its formula is obtained by substituting q's for p's in (16.4.1). In this quantity index, published monthly in the *Federal Reserve Bulletin*, the average daily production figures for a given month are divided by the corresponding average monthly production figures for the base 1957–59 and these *quantity relatives* are then weighted by means of fixed value weights that are essentially the value added by manufacture in the year 1957. The main reason for the use of this particular method in the calculation of the *Index of Industrial Production* is that quantity relatives as well as price relatives do not depend on the units to which quantities and prices refer. Since production figures are sometimes given in man-hours of work and sometimes in production volumes, it would be difficult to justify a direct addition of the various quantities.

For the sake of completeness let us also mention that we could compute a *weighted geometric mean of price relatives* using, as before, value weights. However, such indexes are rarely used; whatever theoretical advantages they might have are outweighed by far by the troublesome calculations required in their construction.

EXERCISES

1. Using the data of Exercise 1 on page 361 and formula (16.4.3), construct an index comparing the 1966 prices of the three woods with those of 1963.
2. Using the data of Exercise 1 on page 361 and formula (16.4.1) with 1964 value weights, calculate an index comparing the 1965 prices of the three woods with those of 1963.
3. Using the data of Exercise 3 on page 362 and formula (16.4.3), construct an index comparing the 1966 prices of the three fuels with those of 1940.
4. Using the data of Exercise 3 on page 362 and formula (16.4.1) with 1950 value weights, calculate an index comparing the 1966 prices of the three fuels with those of 1940.
5. Using the data of Exercise 1 on page 361 and formula (16.4.3) with p's replaced by q's and vice versa, construct an index comparing the 1966 production of the three woods with that of 1964.

16.5 Chain Index Numbers

Most published index numbers serve the dual purpose of providing individual long-range comparisons as well as comparisons in series. For instance, a contractor may be interested in knowing that the July 1966 wholesale price index for lumber and wood products was 110.5 (1957–1959 = 100), but it may be even more significant to him to know how much these prices have changed since the previous month.

If an index number is used for the comparison of successive days, weeks, months, or years, the reference to a fixed base period is really unnecessary. It could be avoided altogether if, instead of comparing each set of annual, monthly, weekly, or daily figures with those of a fixed base, we were to compare them with those of the preceding year, month, week, or day and construct in this manner a so-called *chain index*. Since there is no difference in the formulas and methods used for day to day, week to week, month to month, and year to year comparisons, we shall limit our discussion chiefly to comparisons made on an annual basis.

There are various reasons why chain indexes are sometimes preferable to the indexes which we have already discussed. Chain index methods permit the introduction of new commodities, or reporting units, and the deletion of old ones without necessitating either the recalculation of the entire series or other drastic changes. This flexibility is very important in the construction of general purpose indexes. The main feature of a chain index, however, is that it makes it possible to adjust the weights as frequently as necessary.

The symbolism that is commonly used for a chain index comparing the prices (quantities or values) of some year i with those of the preceding year $i - 1$ is $I_{i-1,i}$. Dropping as before the first two digits, we shall write a chain index providing a comparison between 1967 and 1968 as $I_{67,68}$.

The calculation of a chain index can be based on any one of the methods which we have already discussed and, of course, on others. To illustrate the general procedure we shall restrict ourselves to the fixed-weight aggregative type. Omitting the factor 100, (16.3.8) yields the following formula for the individual comparisons of a chain index, that is, for the comparison of any year i with the preceding year $i - 1$:

$$I_{i-1,i} = \frac{\Sigma p_i q_a}{\Sigma p_{i-1} q_a} \qquad (16.5.1)\star$$

Having two successive comparisons of a chain index, say $I_{66,67}$ and $I_{67,68}$, one might ask whether it is possible to combine the two in some way so as to obtain a comparison between 1966 and 1968. Indeed, it would seem logical that $I_{66,68}$, the index comparing 1966 and 1968 should be the *product* of $I_{66,67}$ and $I_{67,68}$. After all, if the 1967 price of a commodity is twice what it was in 1966 while its 1968 price is three times what it was in 1967, then its 1968 price is $2 \cdot 3 = 6$ times what it was in 1966. Writing

$$I_{66,68} = I_{66,67} \cdot I_{67,68} = \frac{\Sigma p_{67} q_a}{\Sigma p_{66} q_a} \cdot \frac{\Sigma p_{68} q_a}{\Sigma p_{67} q_a} = \frac{\Sigma p_{68} q_a}{\Sigma p_{66} q_a}$$

we find that the product of $I_{66,67}$ and $I_{67,68}$ indeed gives the appropriate fixed-weight aggregative index comparing 1966 and 1968.

It would seem natural to extend this method to a comparison of any two years, say 1962 and 1967. If we try it and write

$$I_{62,67} = I_{62,63} \cdot I_{63,64} \cdot I_{64,65} \cdot I_{65,66} \cdot I_{66,67}$$

we will find that after making the necessary substitutions and cancellations we get

$$I_{62,67} = \frac{\Sigma p_{67} q_a}{\Sigma p_{62} q_a}$$

and this is again the appropriate fixed-weight aggregative index comparing 1962 and 1967. It should be noted that the above argument assumes that each individual comparison of the chain index is computed with (16.5.1) *and* that the identical commodities are included each year. If the same commodities were not included each year, the various sums would not cancel and the product of the individual comparisons would not reduce to a fixed-weight aggregative index.

Incidentally, what we have done here would lead to confusion if we had written the individual comparisons as percentages rather than proportions; this is why the factor 100 is omitted in (16.5.1). The reader may also wish to check that if the individual comparisons were calculated as Laspeyres or Paasche indexes (with the 100 omitted), the long-range comparison obtained by multiplying successive chain indexes would *not* be of the same type. The sums will not cancel and the expression will not simplify.

If chain index numbers are multiplied by one another for long-range comparisons, it can happen that minor errors accumulate into serious proportions. For instance, if an index comparing price changes from January to February were incorrectly given as 120 instead of 110 and if the index giving the price change from February to March were 150, the index comparing January and March would be given as 180 instead of 165. Whereas the error was only 10 "points" in the January-February comparison, it is 15 "points" in the January-March comparison. The use of chain indexes has sometimes been attacked on these grounds but by and large this criticism does not seem too serious.

In computing the *Consumer Price Index*, its various group and subindexes, and other major indexes, the Bureau of Labor Statistics uses formula (16.5.1) as well as the equivalent formula

$$I_{i-1,i} = \frac{\Sigma(p_i/p_{i-1})(p_{i-1}q_a)}{\Sigma p_{i-1}q_a} \qquad (16.5.2)\star$$

Since the p_{i-1} appearing in the numerator of (16.5.2) can be cancelled, the two indexes are clearly the same. Both formulas represent fixed-weight aggregative indexes and they differ only in the mechanics of calculation.

The *Consumer Price Index* is computed as a month to month chain index and it is currently based on about 400 items whose prices are gathered periodically in 56 cities, and metropolitan areas, varying in size from very small towns to the largest city in the country. The CPI may be defined as a statistical measure of changes in prices of goods and services bought by urban wage earners and clerical workers, including families and single persons. The prices entering the index are drawn from six major groups, namely, (1) food, (2) housing, (3) apparel and upkeep, (4) transportation, (5) health and recreation, (6) and miscellaneous. Group indexes are calculated for 23 of the largest cities and metropolitan areas and these indexes are combined into over-all city indexes, group indexes for all cities combined, and one index covering all groups and all cities combined. When the groups are combined into a city index, each group is assigned a weight, differing from one city to another, intended to represent the "relative importance" of the group in the average budget, or family expenditure, of those people to whom the index refers. The national index, called the *Consumer Price Index*, is constructed from the data of the various cities, giving the results for each city a weight proportionate to the wage-earner and clerical-worker population it represents.

The value weights used in (16.5.2) are the products of the prices p_{i-1}, and the fixed quantities, q_a, of the items comprising the so-called "index market basket." This "market basket" is made up of some 400 items—meats, dairy products, residential rents, appliances, men's suits, slacks, and shirts, women's coats and dresses, gasoline and oil, prescriptions, drugs, haircuts, razor blades, TV sets, newspapers, cigarettes, beer, and the like—which are of major importance in family purchases and, in fact, represent the greater part of family spending. In the value weights, $p_{i-1}q_a$, the prices, p_{i-1}, vary from one time period to another, but the fixed quantities, q_a, do not. The original value weights, $v_a = p_a q_a$, were established from a survey conducted in 1950 of the consumer expenditure pattern of 7000 wage-earner and clerical-worker families in 91 cities, with the data of that year being adjusted to 1952 prices and consumption levels. At the present time, the *Consumer Price Index* is reported with 1957–59 as the base period while the period a used for the quantity weights is 1960–1961.

To illustrate the use of (16.5.2) as well as the calculation of the value weights $p_{i-1}q_a$, let us suppose that a survey conducted in 1959 showed that the average person living in a given community spent annually \$24.32 for eggs, \$8.83 for butter, and \$41.83 for milk. We are asked to calculate an index of the *Consumer Price Index* type of the three commodities for 1968 (1960 = 100) on the basis of the above values and the following data:

| | Prices in Cents | | | |
	1959	*1960*	*1967*	*1968*
Eggs, per dozen	69.6	72.3	60.4	73.7
Butter, per pound	80.5	86.7	72.9	81.9
Milk, per quart	19.6	21.8	20.6	23.1

If all we wanted was $I_{60,68}$, we could obtain this index directly by substituting into

$$I_{60,68} = \frac{\Sigma p_{68} q_{59}}{\Sigma p_{60} q_{59}} \cdot 100$$

after having found the 1959 quantities by dividing the 1959 values by the 1959 prices. To illustrate the chain index technique, let us instead calculate $I_{67,68}$ and then multiply by $I_{60,67}$ to get $I_{60,68}$. (In the calculations which follow we shall assume that $I_{60,67}$ is known to equal 89.9 per cent. See also Exercise 3 on page 371.)

To find $I_{67,68}$ with formula (16.5.2), we will first have to calculate the value weights $p_{67}q_{59}$. Since the 1959 quantities were not given explicitly we shall use the identity

$$p_{i-1}q_a = \frac{p_{i-1}}{p_a}(p_a q_a) = \frac{p_{i-1}}{p_a}(v_a) \qquad (16.5.3)\star$$

whose validity can easily be verified. Substituting $p_{i-1} = p_{67} = 60.4$, $p_a = p_{59} = 69.6$, and $v_a = v_{59} = \$24.32$, we find that the 1967 value weight for eggs is

$$p_{67}q_{59} = \frac{60.4}{69.6} \cdot 24.32 = \$21.10$$

Similarly, the 1967 value weights for butter and milk are $(72.9/80.5) \cdot 8.83 = \$8.00$ and $(20.6/19.6) \cdot 41.83 = \$43.96$, respectively. Substituting these value weights together with the 1967 and 1968 prices into (16.5.2), we get

$$
I_{67,68} = \frac{\dfrac{73.7}{60.4} \cdot 21.10 + \dfrac{81.9}{72.9} \cdot 8.00 + \dfrac{23.1}{20.6} \cdot 43.96}{21.10 + 8.00 + 43.96}
$$

$$
= \frac{\$25.75 + \$8.99 + \$49.29}{\$21.10 + \$8.00 + \$43.96} = 115.0 \text{ per cent}
$$

The last expression shows very clearly that this chain index may also be looked upon as a fixed-weight aggregative index. Whereas \$24.32 was the amount spent for a certain number of eggs in 1959, \$21.10 and \$25.75 are the amounts that would have to be spent in 1967 and 1968 for the *same* number of eggs. The quantities q_a are held constant throughout the calculations of the index, and it is for this reason that the index is said to refer to a fixed market basket of goods. Finally, to get $I_{60,68}$, namely, the index which relates the 1968 prices to those of 1960, we have only to write

$$I_{60,68} = I_{60,67} \cdot I_{67,68} = (.899)(1.15) = 103.4 \text{ per cent}$$

This illustrates the general procedure used in the construction of the *Consumer Price Index*.

It is important to realize and to see how the "relative importance" weights change from year to year, or month to month. Returning to our last illustration, we stated on page 368 that in 1959 the average person in the community under investigation spent \$24.32 on eggs, \$8.83 on butter, and \$41.83 on milk. Adding these values we find that the average person in this community spent in 1959 a total of $74.98 on the three food products. Dividing each of the three values by 74.98 and multipling by 100, we find that 32.4 per cent of the $74.98 was spent on eggs, 11.8 per cent was spent on butter, while 55.8 per cent was spent on milk. Applying the same calculations to the 1967 and 1968 values $p_{67}q_{59}$ and $p_{68}q_{59}$, which appear in the denominator and numerator of the expression which we obtained above for $I_{67,68}$, we find that the respective "relative importance" percentages are

	1967	1968
Eggs	28.9	30.6
Butter	10.9	10.7
Milk	60.2	58.7

Incidentally, we could have used the 1967 "relative importance" percentage weights in our calculation of $I_{67,68}$ and, as we pointed out earlier, the result would have been the same.

In the *Consumer Price Index* the "relative importance" weights also change with time. For example, for December 1965 and December 1966 they were:

	Relative Importance Percentages	
	December, 1965	December, 1966
Food	22.83	22.94
Housing	32.98	32.89
Apparel and upkeep	10.50	10.54
Transportation	13.80	13.70
Medical care	5.80	5.97
Personal care	2.70	2.69
Reading and recreation	5.87	5.80
Other goods and services	5.14	5.09
Miscellaneous	.38	.38
All items	100.00	100.00

These changes in relative importance indicate how families would spend their money if they continued buying the same quantities and qualities of items as during the most recent survey of consumer expenditure patterns. Thus the index measures the effect of changes in the prices of a fixed quantity of goods and services and it does not reflect changes in the level at which families live.

The fact that the importance played by various items in an individual's spending pattern varies as prices change, while the quantities that constitute the market basket are kept the same, is often used as an argument against the validity of fixed quantity indexes. Actually, when prices change and when new and different items appear on the market, people do shift their purchases in certain ways, for instance, to those commodities which have increased less than others, to those which have declined more than others, or to lower grades of the same commodities. Moreover, individuals sometimes gradually or abruptly change the level at which they live as the result of hard work, fortuitous circumstance, or both. Naturally, none of these changes are reflected in any weighting system based on the presumption that families will continue to buy the same quantities of goods and services they bought in some earlier period.

In normal times the relative importance of family expenditures for various categories of goods and services probably changes fairly slowly so that the weights need only infrequent revision. However, the relative importance of items in the *Consumer Price Index* coincides with the actual distribution of family expenditures only at the particular points where revised weights are introduced. It may be assumed that the December 1966 relative importance weights approximated the distribution of family expenditures at that time, but the growth of television, frozen foods, and migration to the suburbs has certainly brought about changes in family spending patterns that must be recognized. As additional expenditure and value data become available from new surveys, the index weights will, if necessary, be changed.

EXERCISES

1. Suppose that in the example used in the text (see page 368) we also knew that the 1966 price of eggs was 58.5 cents per dozen, that of butter was 72.4 cents per pound, and that of milk was 23.0 cents per quart. Duplicating the method used in the text, find $I_{66,67}$. (This index is again to be a fixed-weight aggregative index with 1959 quantity weights.)

2. Use formula (16.5.1) to verify directly that for the example used in the text $I_{60,68}$ equals 103.4 per cent.

3. Use formula (16.5.1) to verify the value of 89.9 per cent given for $I_{60,67}$ on page 369.

4. Suppose that in the example used in the text (see page 368) we also knew that the 1969 price of eggs was 78.4 cents per dozen, that of butter was 72.0 cents per pound, and that of milk was 26.0 cents per quart. Duplicating the method used in the text, find $I_{68,69}$. (This index is again to be a fixed-weight aggregative index with 1959 quantity weights.)

5. Using the value of $I_{60,68}$ obtained in the text and the result of the previous exercise, calculate $I_{60,69}$.

6. Verify the 1967 and 1968 relative importance percentages of eggs, butter, and milk, which we gave on page 370.

16.6 Shifting the Base

Under certain circumstances it may be advisable or necessary to change the point of reference, that is, the base, of an index from one period to another; such a change is usually referred to as *shifting the base*. To illustrate how this is done, let us suppose that a research foundation studies the business and economic conditions of a certain community and that, among other indexes, it constructs one purporting to indicate changes in the cost of living of typical wage-earner families residing in this community. This index is published with 1960 = 100 and, for the sake of argument, let us suppose that an investigator wants to shift the base to 1964 = 100 so that he can compare cost of living changes in the community with those of another region for which the corresponding index is given with the base year 1964.

In the table given below, the first column contains the values of the index as it was originally reported with the base 1960 and the second column contains the corresponding values of the index after the base has been shifted to 1964:

	I $1960 = 100$	I $1964 = 100$
1959	98.0	90.7
1960	100.0	92.5
1961	102.2	94.5
1962	101.8	94.2
1963	104.5	96.7
1964	108.1	100.0
1965	108.9	100.1
1966	110.0	101.8
1967	112.3	103.9
1968	113.1	104.6

The entries of the second column were obtained very simply by dividing each entry of the first column by 108.1, namely, by the original value of the index for 1964, and multiplying by 100. This method is based on the argument (discussed previously with reference to chain indexes) that, for example, $I_{60,67} = I_{60,64} \cdot I_{64,67}$ or

$$I_{64,67} = \frac{I_{60,67}}{I_{60,64}}$$

More generally, for any year i

$$I_{64,i} = \frac{I_{60,i}}{I_{60,64}} \qquad\qquad (16.6.1)$$

provided that these indexes are given as proportions. It is important to note that (16.6.1) is *not* valid for all index numbers. It
does not apply, for example, to a Laspeyres index or to one calculated
as an arithmetic mean of price relatives. Hence, the above method of
shifting the base will not necessarily coincide with the method in
which we start anew with the original data and recompute the whole
series with the new base. It all depends on how the index is constructed and what weights are being used. Nevertheless, since it is
sometimes impossible to do otherwise in practice, the simple method
illustrated above is often employed regardless of whether a complete
recomputation of the index would produce the identical results.

16.7 The Use of Index Numbers in Deflating

We often hear statements such as "compared to 1960 the *construction dollar* is worth only 83 cents," "compared to 1950 the *food dollar*
is worth only 43 cents," or "compared to 1955 the *rent dollar* is worth
only 66 cents." This is meant to say that what used to buy a certain
kind of house now pays for only 83 per cent of it, the amount of money
which used to feed a family for 100 days now feeds them for only 43,
and what used to pay a year's rent now pays for roughly 8 months.
·In each of these statements we reflected a change in prices in terms of a
so-called change in the value of the dollar.

As long as we refer to a single commodity, it is easy to illustrate
how such a "value" of the dollar is obtained. If a certain restaurant
charged $1.75 for a steak dinner in 1961 and $3.50 in 1967, the value
of the 1967 "steak dinner dollar" is only 50 cents compared to 1961.
This simply means that the price has doubled and that the value or
the *purchasing power* of the "steak dinner dollar" has been cut in half.
It is easy to see that the value (purchasing power) of a dollar is simply
the reciprocal of an appropriate price index written as a proportion.
If prices increase by 50 per cent, the price index is 1.50 (150 per cent),
and what a dollar will buy is only $1/1.50 = \frac{2}{3}$ of what it used to buy.
In other words, the purchasing power is $\frac{2}{3}$ of what it was or approximately 67 cents. Similarly, if prices increase by 25 per cent, the price
index is 1.25 (125 per cent), and the purchasing power of the dollar is
$1/1.25 = 0.80$ or 80 cents.

The problem becomes more complicated when we speak of a construction dollar, a food dollar, and the like, which evidently do not
refer to single commodities. To illustrate the general procedure used,
let us refer to the example of the previous section and let us suppose
that the research foundation also compiles annual figures on average
wages paid in the community in various fields of activity. The

trouble with wages reported in this way is that gains in money wages often turn out to be more apparent than real. *What the wage earner really wants to know is not how much money he earns but how much his wages will buy.* Clearly, if wages and prices both double, the wage earner can buy just as much as before even though he does get a bigger pay check.

To calculate so-called *real* wages, we shall proceed as above, multiplying cash wages by a quantity measuring the purchasing power of the dollar or, better, *dividing by an appropriate price index.* This process is referred to as *deflating.*† In principle it is simple enough, but in practice it is often difficult to find an appropriate index to deflate a given set of values.

In our illustration we will use the *Consumer Price Index*, assuming that it is appropriately computed for the same individuals on which the wage averages are based. In the table shown below, the first column contains average weekly wages in dollars for all manufacturing industries as they are reported by the United States Department of Labor, the second column contains the *Consumer Price Index*, while the third column contains the corresponding *real wages* obtained by dividing each entry of the first column by the corresponding entry of the second converted into a proportion.

	Average Weekly Wages (All Manufacturing Industries)	Consumer Price Index 1957–59 = 100	Real Wages
1957	$ 81.59	98.0	$83.25
1958	82.71	100.7	82.14
1959	88.26	101.5	86.96
1960	89.72	103.1	87.02
1961	92.34	104.2	88.62
1962	96.56	105.4	91.61
1963	99.63	106.7	93.37
1964	102.97	108.1	95.25
1965	107.53	109.9	97.84
1966	111.92	113.1	98.96

Several interesting things can be read from this table or the graphs of cash wages and real wages shown in Figure 16.1. Using the years 1958

---

† When we *deflate* the 1967 value of a single commodity to, say, 1960 prices, we divide its value, v_{67}, by the index p_{67}/p_{60}, getting

$$\frac{v_{67}}{p_{67}/p_{50}} = \frac{p_{67} \cdot q_{67}}{p_{67}/p_{60}} = p_{60}q_{67}$$

and we are thus calculating the value of q_{67} at 1960 prices. Although this argument does not apply strictly when we deflate an aggregate of the values of several commodities, we are in a sense estimating the total value of the *same* goods at base year prices.

and 1966 as an example, *money wages* increased by (111.92 − 82.71)/
82.71 = 0.35 or 35 per cent, which certainly looks nice in the pay
envelope, but changes in prices have reduced the quantities which
can be purchased with these higher wages and *real wages* are up only
(98.96 − 82.14)/82.14 = 0.20 or 20 per cent. Expressed in another
way, what could have been bought for $98.96 during the base period
1957–1959, cost 111.92 in 1966.

The method which we have illustrated here is frequently used to
deflate individual values, value series (as in our example), or value
indexes. It is applied in problems dealing with such diversified things

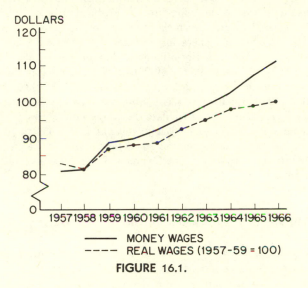

MONEY WAGES
REAL WAGES (1957-59 = 100)

FIGURE 16.1.

as dollar sales, dollar inventories of manufacturers, wholesalers, and
retailers, total values of construction contracts, incomes, wages, and
so forth. The main difficulty always consists of finding appropriate
indexes, or as they are here also called, appropriate *deflators*.

EXERCISES

1. Shift the base of the cost of living index given on page 372 to the year
   1962.
2. The following are the 1957 to 1965 values of a cost of living index
   published in Sweden, with 1958 = 100: 95, 100, 101, 105, 107, 112, 115,
   119, and 125. Shift the base of this index to 1964.
3. The following are the 1958 to 1967 values of the *Manufacturing
   Production-worker Employment Index* of the U.S. Department of
   Labor, with 1957–59 = 100: 95.2, 100.1, 99.9, 95.9, 99.1, 99.7, 101.5,
   106.7, 113.3, 112.9. Shift the base of this index to the year 1965.

4. The average weekly wages (in dollars) for all manufacturing industries in the U.S. for the 14 months March, 1966, through April, 1967, were 110.95, 111.24, 112.05, 112.74, 111.11, 111.78, 113.71, 113.85, 113.99, 114.40, 113.42, 111.48, 112.44, and 112.56.   For the same months the *Consumer Price Index* equalled 112.0, 112.5, 112.6, 112.9, 113.3, 113.8, 114.1, 114.5, 114.6, 114.7, 114.7, 114.8, 115.0, and 115.3.   Use these values of the *Consumer Price Index* to deflate the above wages.   Also calculate the actual percentage increase in these wages from March, 1966, to April, 1967, and compare it with the corresponding percentage change in the "real" wages.

5. Using the data of the previous exercise, calculate an index of the purchasing power of the dollar for the given months on the basis of the *Consumer Price Index*, which, incidentally, has 1957–59 = 100.

6. The 1958 to 1966 operating revenues received by Class I line-haul railroads in the U.S. (in millions of dollars) were 9,565; 9,825; 9,517; 9,187; 9,440; 9,560; 9,857; 10,208; and 10,655.   For the same years the *Consumer Price Index* equalled 100.7, 101.5, 103.1, 104.7, 105.4, 106.7, 108.1, 109.9, and 113.1.   Shift the base of this index to 1964 and use the resulting index to deflate the above value series.   Also calculate the actual percentage decrease in revenue from 1958 to 1960 and compare it with the corresponding percentage change in the "real" revenues. Discuss the adequacy of the *Consumer Price Index* as a deflator in this exercise.

16.8 Mathematical Properties of Index Numbers

We already pointed out a number of reasons some index number formulas are preferred to others.   Several indexes are more or less neglected because of their computational complexity while others are disliked because they require difficult to obtain quantity or value weights.   We also mentioned another criterion for choosing between index numbers, the so-called *units test*, which requires that an index be independent of the units in which, or for which, prices and quantities are quoted.   Of all the index numbers that we have met only the simple (unweighted) aggregative index failed to meet this condition.

Statisticians and economists have developed a number of other theoretical criteria, the most interesting of which are the *time reversal* and the *factor reversal* tests.   The *time reversal test* expresses the intuitive idea that if an index number for 1968 with 1950 = 100 is 200 per cent, the same index for 1950 with 1968 = 100 should be 50 per cent. In other words, if one object is twice as big as another, the second must be half as big as the first.   In spite of the evident desirability of this property, it will probably come as a surprise that the time reversal test is *not* satisfied by many of the major index numbers discussed in this chapter.

Symbolically, the time reversal test demands that $I_{o,n}$ be the reciprocal of $I_{n,o}$ or, which is the same, that

$$I_{o,n} \cdot I_{n,o} = 1 \qquad\qquad (16.8.1)\star$$

where both index numbers must be given as *proportions*. If we want to determine whether an index satisfies the time reversal test, we have only to interchange the subscripts o and n wherever they appear in the formula of the index and substitute into (16.8.1). For example, formula (16.2.1) meets the test because

$$\frac{\Sigma p_n}{\Sigma p_o} \cdot \frac{\Sigma p_o}{\Sigma p_n} = 1$$

while formula (16.3.2), the Laspeyres index, does *not* because

$$\frac{\Sigma p_n q_o}{\Sigma p_o q_o} \cdot \frac{\Sigma p_o q_n}{\Sigma p_n q_n}$$

generally is not equal to 1. It will be left as an exercise for the reader to check which of the index numbers that we have defined do or do not satisfy the time reversal test.

We pointed out earlier that the various formulas defined in this chapter can be converted into *quantity indexes* by the simple process of replacing p's with q's and q's with p's. Using this relationship between the formula of a price index and that of the corresponding quantity index, the *factor reversal test* demands that the product of the two equal the *value index*

$$V = \frac{\Sigma p_n q_n}{\Sigma p_o q_o} \qquad\qquad (16.8.2)$$

This criterion is, of course, satisfied if we compare the prices, quantities, and values of a *single* commodity taken at times o and n. The price index (or relative) is p_n/p_o, the corresponding quantity index (or relative) is q_n/q_o, and the product of the two is $p_n q_n/p_o q_o$. After all, if the price of a commodity has *doubled* and we sell *three times* as much, the total amount of money which we will receive must be $2 \cdot 3 = 6$ times as much.

The factor reversal criterion, formulated originally by Irving Fisher, is satisfied by the Ideal Index and some others, but it is not satisfied by any of the other formulas which we have given in this chapter. Writing (16.3.5), the Ideal Index, as

$$I_p = \sqrt{\frac{\Sigma p_n q_o}{\Sigma p_o q_o} \cdot \frac{\Sigma p_n q_n}{\Sigma p_o q_n}}$$

the corresponding quantity index obtained by interchanging p's and q's is

$$I_q = \sqrt{\frac{\Sigma q_n p_o}{\Sigma q_o p_o} \cdot \frac{\Sigma q_n p_n}{\Sigma q_o p_n}}$$

and it can easily be seen that

$$I_p \cdot I_q = \sqrt{\frac{\Sigma p_n q_o}{\Sigma p_o q_o} \cdot \frac{\Sigma p_n q_n}{\Sigma p_o q_n} \cdot \frac{\Sigma q_n p_o}{\Sigma q_o p_o} \cdot \frac{\Sigma q_n p_n}{\Sigma q_o p_n}} = \frac{\Sigma p_n q_n}{\Sigma p_o q_o} = V$$

It would be erroneous to assume that the criteria mentioned in this section provide an absolute yardstick by which one can measure the relative merits of index numbers. Some of them are undoubtedly of more academic than practical interest. Logically, all such tests should be considered only as supplementary to practical considerations that arise in the construction of an index. When practical advantages clash with theoretical considerations, practical needs are usually given the most attention. Although the Ideal Index, for example, is an excellent index so far as the theoretical tests are concerned, the fact that it requires up to date quantity weights q_n makes it difficult to use for a general purpose index that is published every day, week, or month.

EXERCISES

1. Check whether indexes (16.2.3), (16.3.3), (16.3.4), (16.3.5), (16.3.6), and (16.3.7) satisfy the time reversal test.
2. Show that the factor reversal test is *not* satisfied by indexes (16.2.1) and (16.2.2).

16.9 Current Problems

Statisticians and economists directly connected with the construction of general purpose index numbers can be divided into two schools of thought: there are those who continuously clamor for bigger, better, and more general indexes, and those who feel that the value of an index number decreases if it is intended to describe too general a situation, in other words, a phenomenon that is too wide in scope. Some persons have criticized the *Consumer Price Index* because it covers only urban wage-earner and clerical-worker families and have asked for a general price index covering *all* goods and services. Such an over-all index would be quite useful, perhaps, in the development of general economic theory, but, as is always true with descriptions that are too general, it would be essentially useless in many practical situations. In the earlier parts of this book we stressed the idea that whenever we take a large set of data and represent it by a single statistic, a single description, we lose a tremendous amount of information. Accordingly, there are statisticians who feel that the phenomena that are described in general purpose indexes, for example, wholesale prices, retail prices, production, and the like, are much too wide to be described by means of a single number.

In view of the fact that most useful index numbers actually refer to relatively limited phenomena it is very important that great care be exercised in their use and interpretation. Erroneously, the *Consumer Price Index* is often considered to represent, or indicate, the *cost of living* of all individuals in the United States. This it does not do even for the families to which it is meant to apply, for it does not consider, among other things, taxes paid to federal and state governments. The Bureau of Labor Statistics takes great pains to refer to the index now as the *Consumer Price Index for Urban Wage Earners and Clerical Workers.*

Problems related to the choice of base periods, the selection of weights, the sampling of items, as well as many others, are constantly on the minds of statisticians engaged in the field of index number construction. Changes of base period and weights have produced increasingly difficult problems in recent years because the method of construction of some index numbers is specified by law and written into contracts or labor-management agreements. Inasmuch as a change in the base year or weights can produce noticeable differences in an index, one of the contracting parties is likely to be aggrieved and may object violently to the revision of the index. If wage agreements are tied to "point" changes of an index, this can lead to difficult problems; a change in the base may mean that a change of one "point" in the old index will equal a change of only half a point or less in the new one.

Other problems that usually arise in the construction of a general purpose index are those concerning the sampling of the commodities or items to be included, the maintenance of uniform quality and standards, and the recognition of bargain prices. In view of the fact that this discussion of index numbers is only part of a general introductory course in statistics, we have tried to present to the reader some of the basic methods and problems and to show that the construction of index numbers is by no means a cut and dried proposition.

The views expressed in this chapter may be summarized quite well by quoting the following "Letter to the Editor" by I. H. Siegel,† formerly on the staff of the President's Council of Economic Advisers:

---

† *The American Statistician*, February, 1952.

It ought to be conceded that index numbers are essentially arbitrary. Being at best rearrangements of data wrenched out of original market and technological contexts, they strictly have no economic meaning. Changes in tastes, technology, population composition, etc., over time increase their arbitrariness. But, of course, there is no bar to the use of indexes "as if" they did have some unequivocal meaning provided that users remember that they themselves made up the game and do not threaten to "kill the umpire" when the figures contradict expectations.

BIBLIOGRAPHY

Further information about index numbers, their theory, construction, and application, may be found in the books listed in the Bibliography on page 348. A detailed treatment of the major index numbers published by the federal government as well as other agencies and organizations is given in

Hauser, P. M., and Leonard, W. R., *Government Statistics for Business Use.* New York: John Wiley, 1956.

Snyder, R. M., *Measuring Business Changes.* New York: John Wiley, 1955.

Many articles on current problems in index number theory and practice appear regularly in statistical and business journals—among others, in the publications of the *American Statistical Association*, in the *Monthly Labor Review*, and in other government publications.

Some recent articles that will be of interest to those seriously interested in the subject are

U.S. Dept. of Labor, Bureau of Labor Statistics, *The Consumer Price Index, A Short Description.* Washington, D.C., 1967.

U.S. Department of Labor, Bureau of Labor Statistics, *The Relative Importance of Items in the Consumer Price Index.* Washington, D.C., December, 1966.

U.S. Department of Labor, Bureau of Labor Statistics, *Wholesale Prices and Prices Indexes.* Washington, D.C., September, 1967.

U.S. Department of Labor, Bureau of Labor Statistics, *Prices and Price Indexes Group I—Farm Products.* Washington, D.C., 1963 (A Reprint from BLS Bulletin No. 1513).

U.S. Department of Labor, Bureau of Labor Statistics, *Prices and Price Indexes Group II—Processed Foods.* Washington, D.C., 1963 (A Reprint from BLS Bulletin No. 1513).

U.S. Department of Labor, *Major BLS Programs—A Summary of Their Characteristics.* Washington, D.C., 1966.

Federal Reserve Board, *Industrial Production Measurement in the United States: Concepts, Uses and Compilation Practices.* Washington, D.C.: Board of Governors of the Federal Reserve System, 1964.

U.S. Bureau of The Budget, *Statistical Services of the United States Government Part II.*, rev. ed. Washington, D.C., 1963.

CHAPTER 17

Time Series Analysis:
Basic Concepts

17.1 Introduction

One of the principal tasks facing the manager of a business or any other organization is that of planning for the future. This discipline requires looking, to the best of one's ability, into an area where, regrettably, no one sees quite as clearly as he would like. However, if business and economic planning is to proceed, someone must take the responsibility for looking ahead and making the *forecasts* of future levels of activity upon which all intelligent planning necessarily depends. Since there can be no doubt that planning *must* proceed, *a large part of the effort devoted to running an organization is spent in planning for the future.*

Whatever misgivings they might have, businessmen and government officials at all levels are forced to make plans for both the near and distant future which often commit them to spend vast sums of money. Hardly a day goes by on which the financial news fails to carry some tangible evidence of this kind of forward thinking. For example, in one week in July, 1968, The International Bank for Reconstruction and Development announced plans to sell $900,000,000 of securities, with the proceeds to be used to expand investments in a large number of countries throughout the world. The State of California sold $100,000,000 of bonds for the purpose of developing water resources. In the same week, the Standard Oil Co. of New Jersey sold $250,000,000 of securities, raising funds to augment their capital and to finance new explorations for gas and oil. [Standard and its subsidiaries conduct these explorations in more than 100 countries.] The Hydro-Electric Power Commission of Ontario, during the same

381

week, estimated that their new capital investments for the year 1968 would cost approximately $259,000,000 and sold $75,000,000 of securities to provide part of the funds required.

All plans of this sort are made on the basis of predictions (forecasts) of future levels of business activity. Moreover, one forecast often becomes the basis of other forecasts as particular industries interpret predictions in terms of their own capacities and individual companies within industries in turn interpret the industry forecasts to suit their particular needs.

To see how such a chain of forecasts develops, let us consider the following illustration: After World War II, there was a large increase in the number of births in the United States. Consequently, it is expected that the number of people between the ages of 20 and 24 will, in the years following 1968, increase by 3 million, while those between the ages of 25 and 29 will also rise greatly thereby affecting the demand for new housing. The National Association of Home Builders expects that net household formations which in 1968 average 900,000 to 1 million per year will, by the year 1975 increase to 1.25 million per year. The Secretary of Housing and Urban Development has stated that the United States must provide 22.5 million new housing units (in addition to rehabilitation of older units) between the years 1969 to 1978 to accomodate increased population and replace dilapidated dwelling units. In anticipation of the demand for housing, the President of the United States asked Congress to take steps which would make more mortgage money available under the Federal Housing Administration and the Veterans Administration. He also asked for transfer of the Federal National Mortgage Association's secondary mortgage operations to private enterprise, hoping to increase the attractiveness of such mortgages among private investors.

Dun's Review, a Dun and Bradstreet publication, interpreted these population changes in terms of materials required for new housing. They predict that basic products such as aluminum siding, plywood, floor covering, cement, gypsum, and brick will benefit from the new construction. They also predict a wave of innovation in construction materials and techniques.

The same type of forecasting and planning is taking place in other countries as well. For example, in the growing and expanding economy of Canada, it is predicted that the population of about 20 million in 1966 will grow to over 21 million in 1970 and will increase to about 25 million in 1980. The President of Canadian Banker's Association predicted in January of 1968 that the number of bank branches would increase from 5800 to 7000 in ten years. On the basis of most population forecasts and assuming that the present ratio of branches to population is maintained, there will be 7,000 bank branches in Canada by the end of the next decade.

Population forecasts, industry forecasts, and the like are based in large part on what has happened in the past. In business and in science, as well as in everyday life, we project our experience of the past in an effort to predict what may happen in the uncertain future. *Management must, thus, spend a great deal of time in thinking about the past, in analyzing statistical information about the past, in order to gain an understanding of phenomena that can be projected into the future.*

The business world would be Utopia if it were possible to predict economic phenomena as accurately as an astronomer can predict, for example, the position of the planet Mars. Unfortunately this is far from being the case. Economic data—the raw material upon which forecasts are based—exist in such vast quantities and are so complex and unwieldy that the task of discovering sufficient generalities upon which to base forecasts is extremely difficult, indeed, up to this time largely unmanageable. Nevertheless, this does not relieve the economic or social planner of the necessity of trying to isolate and forecast certain patterns. That is, businessmen and economists must make predictions based at least in part on the regularities and patterns which they have discovered in the behavior of statistical information concerning supply, demand, prices, employment, inventories, and the like. In making such decisions, the businessman is persuaded to believe that the future follows the past with some degree of regularity, that what has happened in the past will, to a greater or lesser extent, continue to happen or will again happen in the future.

The first step in forecasting the business or economic future consists, thus, of gathering observations from the past. In this connection one usually deals with statistical data which are collected, observed, or recorded at successive intervals of time and which are generally referred to as *time series*. The term "time series" applies, for example, to annual production figures recorded for a number of years, to monthly data on the total labor force in the United States, and to daily quotations on the New York Stock Exchange. Although the term "time series" is commonly used with reference to economic data, it applies equally well if a patient's temperature is taken at regular intervals of time and recorded on a chart at the foot of his bed, if a county clerk records the number of marriage licenses issued each day, or if a government agency records the number of robberies committed each month. In this book we shall limit our discussion mostly to time series of economic data, but it should be understood that all of the techniques apply also to data belonging to any of the other natural or social sciences. Incidentally, the term "time series" is frequently abbreviated by dropping the word "time" and we thus speak of a *series* of monthly farm prices, a *series* of annual population figures, or a *series* of weekly freight car loadings.

17.2 The Behavior of Time Series

It is virtually impossible nowadays to open a newspaper, enter an office, or even go to school without seeing graphs of time series showing the behavior of stocks and bonds, reports on weekly or monthly sales, or charts of the daily attendance.   Some of these graphs look like straight lines, others look like smooth curves, but most, and above all those representing economic data, give the impression of the haphazard scrawlings of a three-year-old child.   It is for this reason that special statistical techniques have been devised to help bring some order into the irregular patterns and the seemingly erratic appearance of time series of business data.

It is customary to classify the fluctuations of a time series into *four* basic types of variation which, superimposed and acting all in concert, account for the changes in the series over a period of time and give the series its irregular appearance.   These four types of patterns, movements, or, as they are often called, *components* of a time series are:

1. *Secular trend*
2. *Seasonal variation*
3. *Cyclical variation*
4. *Irregular variation*

In traditional, or classical, time series analysis it is ordinarily assumed that there is a *multiplicative relationship* between these four components, that is, it is assumed that any particular value in a series is the *product* of factors that can be attributed to the various components. Whether or not this is reasonable is difficult to decide and we shall have more to say about this after we have discussed the general nature of the components of a time series in some detail.

17.3 Secular Trend

When we speak of the secular (or long-term) trend of a time series we ordinarily mean the smooth or regular movement of the series over a fairly long period of time.   Intuitively speaking, the trend of a time series displays the general sweep of its development, or better, it characterizes the gradual and consistent pattern of its changes.

Some series of data recorded over a given period of time show an upward trend, some series show a downward trend, while others remain more or less at a constant level.   In this country certain factors or forces have been operating to cause the trend of many important economic series to be upward.   Total industrial production has shown an upward movement, as has total personal incomes, and loans and investments of member banks of the Federal Reserve System, to mention but a few.

Another important series, population, which we have referred to before, has exhibited a strong upward movement although it has gone through periods of noticeably irregular growth.

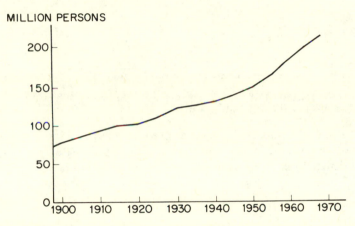

FIGURE 17.1. The Population of the United States. (Source: 1967 Statistical Abstract of the United States.)

Although there have been certain periods of slackening and increasing population, we would describe the long-term trend of this series as upward (see Figure 17.1).   New concepts of birth control, however, may have important effects on future rates of population growth.

This long-term population growth together with such other factors as technological advances and, since World War II, a marked increase in the proportion of families in the middle and upper income groups, has itself been a major force in the upward trend of various important series.   The fortunes of many different types of industries, the manufacturing of shoes for example, have closely followed population changes throughout the years.

It must not be assumed, however, that all economic series necessarily have upward trends.   Figure 17.2 shows a gradual but consistent downward trend in the number of workers employed on farms since 1950 and Figure 17.3 shows the downward trend in the percentage of the total energy produced for heat, light, and power that is supplied by coal.

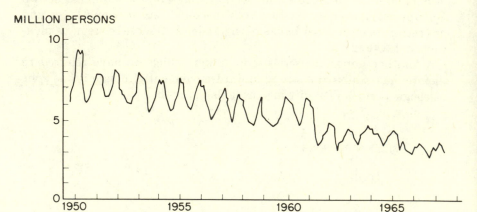

FIGURE 17.2. Farm Employment. (Source: 1967 Statistical Abstract of the United States.)

It should be apparent that there are all sorts of trends: some series increase slowly and some increase fast, others decrease at varying rates, some remain relatively constant for long periods of time, and some

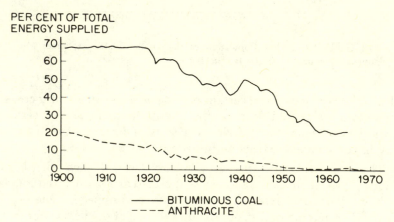

BITUMINOUS COAL
---- ANTHRACITE

FIGURE 17.3. Per Cent of Total Energy Produced for Heat, Light, and Power Supplied by Coal. (Source: Bureau of Mines and Minerals Yearbooks.)

after a period of growth or decline reverse themselves and enter a period of decline or growth. In Chapter 18 we shall study some of the standard methods that are used in the description of trends.

17.4 Seasonal Variation

The type of variation that is, perhaps, the easiest to understand is the *seasonal variation* which consists of regularly repeating patterns like those shown in Figures 17.2 and 17.4. Although the name which we have given to this type of variation implies a connection with the seasons of the year, like the variation which we might find in the monthly production of eggs or the weekly sales of department stores, it is used to indicate any kind of variation which is of a *periodic* nature and where the period is not longer than one year.

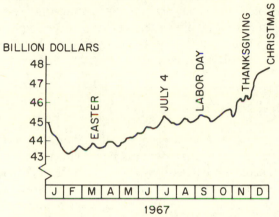

FIGURE 17.4. Currency in Circulation. (Source: Federal Reserve Board.)

Simple examples of seasonal patterns may be found also in the hourly statistics of the number of passengers traveling on New York City subways or in the monthly statistics of the price of raspberries. It is easy to see that the morning and afternoon rush hours during which persons travel to and from work present a pattern that repeats every working day. Similarly, the price of raspberries is always lowest in the months during which they are harvested and highest when the supply is scarce.

As another illustration of seasonal variation, let us consider the effect of seasonal factors on the demand for currency. One of the main objectives of the Federal Reserve Act was to provide the country with a supply of currency which would expand and contract in accordance with public needs. In a typical year (see Figure 17.4), the amount of currency in circulation changes frequently and substantially. The demand for currency varies for different days of the week, for different days of the month, and for different seasons of the year. It increases before such holidays as the Fourth of July, Labor Day, and Thanksgiving. There is a very large demand for cash before Christmas when people want money for shopping and gifts. After the holidays, the merchants, hotelkeepers, and others who have collected the money deposit it in their banks, which, in turn, deposit it with the Federal Reserve Banks.

The study and measurement of seasonal patterns constitute a very important part of the analysis of a time series. In some instances, the seasonal patterns themselves are of primary concern because little, if any, intelligent planning or scheduling (of production, inventory, personnel, advertising, and the like) can be done without a knowledge, based on adequate statistical measures, of seasonal patterns. In other cases the seasonal variation may not be of immediate concern, but it must be measured to facilitate the study of *other* types of variation (see page 454). Some of the most widely used methods for describing seasonal patterns will be discussed in Chapter 19.

17.5 Cyclical Variation

From one point of view, the so-called *business cycle* is nothing more than the variation that remains in a time series after the trend, seasonal variation, and irregular fluctuations have been eliminated. Of course, there is more to it than that, but certainly one way, perhaps the preferred way in traditional time series analysis, of arriving at a measurement of a business cycle is through such a process of elimination. More generally, we might say that *a business cycle consists of a recurrence of the up and down movements of business activity from some sort of statistical trend or "normal."* By "normal" we mean some kind of statistical average; we do not mean that there is anything very permanent or universal if we make statements such as "After a slight drop business has returned to normal" or "Carloadings are below normal for this time of the year."

Business cycles are distinguished from seasonal variations in that cycles extend over a longer period than seasonal fluctuations. Moreover, the fluctuations in a business cycle are considered to result from a different set of causes. The periods of prosperity, recession, depression, and recovery, which are sometimes viewed as the *four phases* of a complete cycle, are generated by factors other than weather, social customs, and such, which create seasonal patterns.

From the standpoint of measurement, business cycles, which show enough similarity to be identified as such, unfortunately also show so much dissimilarity as to make predictions of their future occurrence, length, and severity of little value. This is due to the fact that there exist, by virtue of our definition, not just one but many different kinds of cycles.   After a short, or perhaps long, period of prosperity there may be a crash or a panic.   The period of decline in which income, production, and employment fall may be fast or agonizingly slow.   The bottom may extend for a short or a long while and when the revival begins the recovery may be slow or fast.   Then the new prosperity may last only a couple of years or much longer.   However long the process, once we have seen a rise above, a fall below, and a return to statistical normal, there is a business cycle of the type which has been observed in all economies based on a money, rather than on a barter, system.   In Chapter 19 we shall discuss the practical problem of isolating the cyclical movements of a time series by eliminating the trend, seasonal variation, and irregular movements.

17.6 Irregular Variation

Irregular, or *erratic,* fluctuations of a time series are those variations which are either completely unpredictable or which are caused by such isolated special occurrences as good and bad news, bank failures, elections, floods, earthquakes, strikes, and wars.   Some influences which we shall class as erratic are barely noticeable, working themselves out before much of anything is felt.   Others are relatively more severe: for example, the failure of the London banking house of Baring Brothers in November, 1890, caused a sharp though short-lived financial crisis in this country.   Others, such as wars and earthquakes, have produced long-lasting and serious consequences.

The category labeled irregular variation is really intended to include all types of variation other than those accounting for the trend, seasonal, and cyclical movements.   These latter three, if they are actually at work, act in such a way as to produce certain *systematic* effects.   Irregular movements, on the other hand, are considered to be largely random, being the result of chance factors which, like those determining the fall of a coin, are wholly unpredictable taken one at a time but which, in the long run, more or less average out.

The examination of almost any time series shows that there are constant changes and this raises important questions about the nature of the particular forces that are at work.   The random forces, we presume, are constantly active, but over and above random movements it may be possible to detect forces operating in a nonrandom, that is, systematic, fashion.   For instance, there may be that persistent tendency which we called trend for a series to move in the same direction, or that persistent tendency which we called seasonal variation for certain months of the year, days of the week, or hours of the day to be above average and others to be below. If these nonrandom movements are present in a series, we would certainly like to be able to detect and identify them and to establish their patterns. Once this is done, we can make forecasts of the future, proceeding on the assumption that the same forces will continue to remain at work.

The detection of nonrandom movements in a time series is a statistical problem which we already met in Chapter 12. In order to establish the existence of a trend, for example, *we formulate the null hypothesis that there is none*. Then we might investigate the number of runs above and below the median (or use one of the other tests of randomness referred to on page 284), to test whether the observed fluctuations may be attributed to chance or whether they furnish evidence for a persistent tendency of movement in some direction. Tests of this general nature should always *precede* the actual measurement of a trend or other movement of a time series.

17.7 A Word About the Classical Approach

In Section 17.2 we mentioned that in the traditional approach to time series analysis any particular value is considered to be the product of factors attributed to trend, cyclical, seasonal, and irregular components. Now that we know something about the nature of these components, it will be well to add some further remarks about this approach.

If a time series were actually composed of the four components we have discussed and if we knew just how they are related and how they can be measured, time series analysis could take its place among the more exact sciences. We could analyze and study the components separately (assuming they are separable) by decomposing the series, and *both* our ability to predict and our understanding of economic fluctuations, which are the ultimate goals of our study, would be immeasurably improved. Unfortunately, we really do not know any of these things and, more than that, we find that there is little agreement even among experts about the validity of the various assumptions. Some economists feel that the given classification is too crude, that there are, in fact, more than four types of movements. Nothing specific is really known about how the components are related, how they combine to produce particular effects, or whether they are really separable. The effects of the various components might be additive, multiplicative, or they might be combined in any one of an indefinitely large number of other ways. Some economists think that the trend and cyclic movements are produced by the *same* set of forces, not by different ones as we indicated above, and that the two cannot really be separated. Moreover, for a given series of data there are a number of ways of measuring the individual components, and a choice of one or another is, in the final analysis, only a matter of *which seems best* and not *which is best*. As we pointed out throughout the first part of this book, the choice of statistical descriptions is always arbitrary to a very large extent.

The approach to time series analysis by way of the separation into four components is undeniably oversimplified and, regrettably, ignores the hidden interactions and interrelationships in the data and the entire "complex of individually small shifts and nuances." Admittedly, it has not been quite adequate to solve the basic problem of economic forecasting, which, in the opinion of some, is the real justification for the study of economics. The reason the traditional analysis of time series has fallen short of its goal is that it simply does not go far enough. Alternate approaches, notably the *econometric* approach, which involves a large amount of advanced mathematics, hold considerable promise for the future.

In spite of all this, the traditional approach is the place to begin one's study of time series analysis. There is still much of fundamental importance about the movements of time series that is to be learned from the principles and methods of this approach. Moreover, the traditional methods have been and continue to be widely used in practice and they have, in many instances, provided very useful results. Accordingly, we shall in the next two chapters set up a "model" (formulate a theory) which assumes that the observed values of a time series are the products of factors attributed to trend, cyclical, seasonal, and irregular sources of variation. Though this is standard in the elementary study of time series, it is important to keep in mind that there are many other possible models (assumptions or theories) which might lead to different results.

17.8 The Preliminary Adjustment of Time Series

Before beginning the actual work of computing a trend equation or a measure of seasonal variation, it is often necessary to make certain *adjustments* in the raw data. The usual adjustments which are considered, whether they are actually made or not, are adjustments for *calendar variation, price changes*, and *population changes*.

The purpose of adjusting for *calendar variation* is to eliminate certain spurious differences which are caused by peculiarities of our calendar. A production figure for the month of February may be below that of the preceding January, not because of any real drop in activity, but only because February has fewer days. This may be taken into account simply by dividing each monthly total by the number of days in the month (sometimes by the number of working days in the month), thus arriving at a daily average for each month. Comparable (adjusted) monthly data may then be obtained by multiplying each of these values by 30.4167, the average number of days in a month. (In leap years this factor is 30.5.)

An adjustment for *price changes* is necessary whenever we have a value series and are interested in quantity changes alone. Total sales volume, which fluctuates with both quantity and price changes, can very well go up when in fact the number of units sold is down, the increase in value being accounted for by higher prices. Inasmuch as price times quantity equals value, we can eliminate the effect of price changes by dividing each item in a value series by an appropriate price index. This is nothing more than the process of "deflating" a value series, which we explained earlier in Section 16.7.

Sometimes it is necessary to adjust series for *population changes* since comparisons of income, production, and consumption figures can easily be distorted by changes in the size of the population. (It can easily happen that the production of a commodity is going up whereas actually its per capita production is going down.) When it is deemed advisable to adjust data for population changes, the data are expressed on a per capita basis by dividing the original figures by the appropriate population totals.

One other matter of importance which should be mentioned in connection with the preliminary adjustment of time series is that of *comparability*. Clearly, comparability of data throughout the time period under investigation is essential to any meaningful analysis, and it should never be taken for granted. All too often it is difficult or even impossible to get strictly comparable data. Sometimes we deal with figures extending back 20, 30, 50 years or more, and over such a long period of time production, consumption, sales, or employment figures may well have been gathered by different agencies. Although the definition of an item (for example, a home furnace) may not have changed, there may have been a change in quality which renders price quotations not strictly comparable. Also, from time to time definitions are changed and this year's department store, chain store, or farm, is not necessarily the same as last year's. During World War II the government, for the purpose of fixing retail prices, developed a definition of a chain store based on volume sales; under this definition a single large store might be included and a "chain" of 12 small stores excluded. At one time, an annual sales volume of at least $100,000 was required to qualify as a department store in surveys of the federal government, but subsequent definitions dropped the sales requirement, identifying department stores instead by certain qualitative characteristics plus a minimum of 25 employees. The effect of such changes in definitions is that many recorded figures are not strictly comparable.

Sometimes there is also the danger of overlooking changes in the times (dates) to which reported figures apply. A production figure for January, 1968, may give the average for that month; some years later the corresponding production figure may give the total for the month or, perhaps, production on the 15th or last day of the month. Changes in the manner of reporting naturally make the resulting figures noncomparable.

17.9 Graphical Presentations of Time Series

After a time series has been chosen for study and all necessary adjustments, if any, have been made, the next step usually consists of plotting the series on a suitable piece of graph paper. As we pointed out earlier, graphical presentations have generally more appeal and are easier to grasp than unwieldy columns of numbers, and the presentation of a time series in the form of a graph is often the only statistical treatment that is needed to bring out relevant features or to convey specific ideas.

Although there are various ways of adding to the appeal of graphical presentations of time series, let us limit our discussion here to so-called *line charts*, in which we plot points corresponding to the annual (monthly, weekly, and so forth) values of a series at the midpoints of the time intervals which they represent and connect them with straight lines. More elaborate pictorial presentations of time series will be treated in Appendix I.

The two kinds of line charts that are most commonly used in the presentation of time series go by the names of *arithmetic* and *logarithmic line charts*. The distinction between the two lies in the kind of scales that are used. Regardless of whether we construct an arithmetic or a logarithmic line chart, it is customary to measure time along the horizontal axis, letting equal subdivisions represent successive years, months, weeks, or days. If the vertical scale along which we measure the values of a series is such that *equal intervals represent equal amounts*, the scale is referred to as an *arithmetic scale*, the paper thus constructed is referred to as *arithmetic paper*, and the graph obtained by plotting the values of the series and connecting successive points by means of straight lines is called an *arithmetic line chart*. If the vertical scale is such that *equal intervals represent equal rates of change* (see Section 17.10), the scale is referred to as a *logarithmic scale*, the paper is referred to as *semi-log paper*, and the graph obtained by plotting the values of a series and connecting successive points by means of straight lines is called a *logarithmic line chart*. In this section we shall comment briefly on the construction of arithmetic line charts, reserving logarithmic line charts for Section 17.10.

As in the case of tabular presentations (see Section 2.5), certain aspects of the construction of line charts are considered to be good practice although, strictly speaking, they are not absolutely neces-

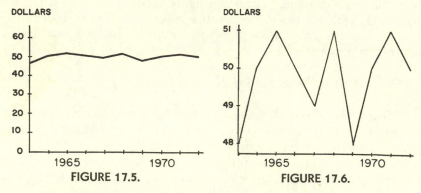

FIGURE 17.5.   FIGURE 17.6.

sary.   Regardless of whether we are drawing arithmetic or logarithmic charts, it is desirable to give each chart a *title*, preferably at the top of the diagram, and if a report or display contains several charts it is desirable also to give each chart a *number*.   Furthermore, if a chart is to tell its story without further elaboration in accompanying text, it is desirable to mention the source of the data, preferably in

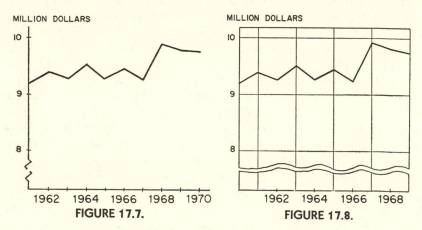

FIGURE 17.7.   FIGURE 17.8.

the bottom left-hand corner.   Other points worth mentioning are that all lettering should be horizontal; if several series are plotted on the same diagram, they should be distinguished through the use of colors or solid, dashed, and dotted lines; and if additional horizontal and vertical guide lines are used (see Figure 17.8), these lines should be so faint that they will not distract from the graph of the series.

When constructing an arithmetic chart, it is always desirable to indicate the zero of the vertical scale, but not necessarily of the horizontal scale, for the resulting graph could otherwise be quite misleading. This is apparent from Figures 17.5 and 17.6, where one diagram gives the impression that the series is stable while the other gives the impression that there are drastic changes. (See also "How to Lie with Statistics," mentioned in the Bibliography on page 109.)

The practice of including the zero of the vertical scale can involve a considerable waste of space (for instance, if all the values of a series lie between $9 million and $10 million, in which case the lower 90 per cent of the paper would be completely blank); for this reason it is often advisable to use the tricks illustrated in Figures 17.7 and 17.8. If there are no extra rulings we, so to speak, cut a piece out of the vertical scale, indicating this by means of a zig-zag line such as that of Figure 17.7. If there are additional rulings as in Figure 17.8, we make it look as if a horizontal strip has been cut out of the paper.

17.10 Logarithmic Line Charts

When plotted on one diagram with arithmetic scales, an increase from $1,000 to $2,000 looks as big as an increase from $50,000 to $51,000 or an increase from $100,000 to $101,000. On the other hand, an increase from $1,000 to $2,000 looks much smaller than an increase from $10,000 to $20,000 or an increase from $100,000 to $200,000, even though the amounts have in each case increased by 100 per cent. If we want to draw a figure in which these *percentage changes* all look the same, we can do so by using a logarithmic vertical scale. Graph paper in which the vertical scale has the rather odd looking subdivisions illustrated in Figure 17.9 is called *semi-log* or *ratio-paper*, so long as there are equally spaced subdivisions along the horizontal axis. Line graphs drawn on this kind of paper are called *logarithmic line charts* or *ratio charts*, owing to the fact that changes which are in the same *ratio* (for example, changes from $1,000 to $2,000, from $5,000 to $10,000, or from $123,000 to $246,000) all appear equally large.

Since semi-logarithmic paper has many important applications in the presentation of business data, it is unfortunate that the mere mention of logarithms seems to scare most non-mathematicians away from its use. Actually, this kind of graph paper may be used without any knowledge of logarithms as long as it is understood that, owing to the special vertical scale, changes that appear equal to the eye represent equal percentage changes, that is, equal ratios.

Semi-log paper may be bought commercially, in which case the vertical scale is usually marked off as in (a) of Figure 17.9.   Here the unusual spacing between 1, 2, 3, . . . , 9, and then again 1 is repeated three times and the paper is accordingly referred to as *three-cycle semi-log paper*.† If we actually wanted the vertical scale to begin with 1, we would have to label the remainder of the subdivisions as in (b) of Figure 17.9 and the first cycle would then represent the numbers from

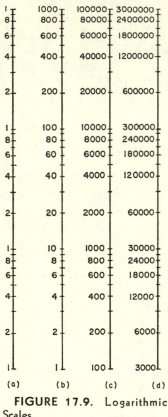

FIGURE 17.9. Logarithmic Scales.

1 to 10, the second cycle those from 10 to 100, and the third cycle those from 100 to 1,000.   *By going up one complete cycle each number is thus multiplied by 10*, and if we had wanted to start at the bottom with 100, the remainder of the subdivisions would have to be labeled as in (c) of Figure 17.9.   If, for some reason, we wanted to start with 3,000, the subdivisions of the first cycle would read 3,000, 6,000, 9,000, etc.;

---

† This kind of graph paper may be obtained in almost any college bookstore or in stores specializing in artists' supplies.   One-cycle and two-cycle paper is also available.   For a description of how to construct one's own semi-log paper, see the Bibliography on page 399.

those of the second cycle would read 30,000, 60,000, 90,000, etc.; and those of the third cycle would read 300,000, 600,000, 900,000, etc., as is shown in (d) of Figure 17.9. Alternate logarithmic scales may be obtained by multiplying all of the markings of (b) of Figure 17.9 by any positive number.

The fact that equal percentage changes look equally big on semi-log paper is apparent from Figure 17.9. For example, in Scale (b) a 100 per cent increase from 2 to 4 goes up as much as a 100 per cent increase from 4 to 8, from 10 to 20, from 40 to 80, or from 200 to 400.

In addition to the fact that semi-log paper makes equal percentage changes (ratios) look equally big, which is often of major importance

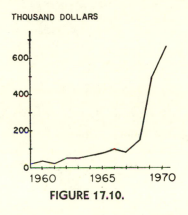

FIGURE 17.10.

FIGURE 17.11.

in the analysis of a series, it has the further advantage that it can accommodate a wide range of values without making the smaller values look too minute. To illustrate, let us suppose that a firm's profits have for a number of years fluctuated between $10,000 and $100,000 and then suddenly shot up to over half a million. If we plotted such data on arithmetic paper, we would get something like the line chart of Figure 17.10, in which most of the series is crowded into a very narrow strip at the bottom of the diagram. On the other hand, if we look at Figure 17.11, which shows the same series plotted with a logarithmic scale, it is apparent that this chart shows much more detail. This advantage of logarithmic scales is also of significance in problems in which we want to plot several series on one diagram and it so happens that the values of one series are much smaller than those of the other (for example, if the values of one series lie mostly between $500 and $1,000 while those of the other lie mostly between $80,000 and $100,000). If we used an arithmetic scale in this example, the first series would have to be crowded into a narrow strip at the bottom of the diagram covering less than 1 per cent of its area. If we used a logarithmic scale we would get a picture somewhat like that of Figure 17.12 and we would be able to present both series in considerable detail.

The various remarks made in the previous section about sound practices in the construction of arithmetic line charts apply also to logarithmic line charts with the one exception that we can never show the zero of the vertical scale. (No matter how often we may divide by 10, that is, go down one more cycle, we will never actually reach the number zero itself.)

To repeat, the use of logarithmic line charts is particularly appropriate if we wish to accentuate percentage changes (relative changes

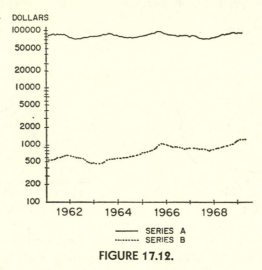

FIGURE 17.12.

or ratios) taking place in one series or if we want to compare percentage changes that are simultaneously taking place in several series. Logarithmic line charts are used, for example, to compare percentage changes in production with corresponding percentage changes in expense, or relative changes in assets with relative changes in liabilities.

An important distinction between arithmetic and logarithmic line charts is that if one series has a steeper upward slope than another series when plotted on arithmetic paper, it is actually increasing by a *numerically greater amount*. If the same thing happens when two series are plotted on semi-log paper, the series with the steeper upward slope is increasing at a *higher rate*—not necessarily by a greater amount.

EXERCISES

1. The following data represent the number of group life insurance master policies in force in the United States from 1950 to 1966 (in thousands):

56, 61, 68, 75, 81, 89, 106, 120, 134, 154, 169, 180, 193, 203, 218, 233, and 249. Draw an arithmetic line chart of this series.

2. Draw an arithmetic line chart of the series of Exercise 1 on page 404.

3. The following are the profits of a business from 1953 to 1968 in thousands of dollars: 2.1, 2.4, 1.8, 3.2, 4.7, 3.6, 9.5, 8.4, 12.0, 20.3, 37.4, 52.9, 66.3, 72.5, 98.6, and 115.4. Plot this series on semi-log paper.

4. Draw a logarithmic line chart of the series of Exercise 3 on page 422.

5. The following are the average annual yields on corporate (Aaa) bonds from 1950 to 1966: 2.6, 2.9, 3.0, 3.2, 2.9, 3.1, 3.4, 3.9, 3.8, 4.4, 4.4, 4.4, 4.3, 4.3, 4.4, 4.5, and 5.1. Draw an arithmetic line chart of these average annual yields. (Source: Moody's Investor Service)

BIBLIOGRAPHY

A discussion of the rational basis of time series analysis, together with a short history of trend analysis, is given in

Smith, J. G., and Duncan, A. J., *Elementary Statistics and Applications.* New York: McGraw-Hill, 1944, Chap. XX.

A general treatment of the nature of the components of time series may be found in most introductory texts, for example, in

Croxton, F. E., Cowden, D. J., and Klein, S., *Applied General Statistics,* 3rd ed. Englewood Cliffs, N.J.: Prentice-Hall, Inc., 1967, Chap. 11.

Mills, F. C., *Statistical Methods.* New York: Holt, Rinehart & Winston, 1956, Chap. 10.

Spurr, W. A., and Bonini, C. P., *Statistical Analysis for Business Decisions.* Homewood, Ill.: Irwin, 1967, Chap. 19.

See also

Moore, G. H. (Ed.), *Business Cycle Indicators* (National Bureau of Economic Research). Princeton: Princeton University Press, 1961.

Shiskin, J., *Signals of Recession and Recovery. An Experiment with Monthly Reporting.* New York: National Bureau of Economic Research, 1961.

The construction of semi-log graph paper is discussed in Chapter 5 of the book by Croxton, Cowden and Klein mentioned above.

CHAPTER 18

Time Series Analysis: Secular Trends

18.1 Introduction

When we try to describe the over-all movement of a time series we generally think of a smooth curve of some sort. The simplest curve to visualize, or to fit, is the straight line and although straight lines often provide a good fit, there are many occasions when

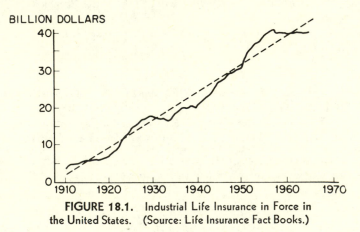

FIGURE 18.1. Industrial Life Insurance in Force in the United States. (Source: Life Insurance Fact Books.)

we have to resort to more complicated kinds of descriptions. Whereas the straight line of Figure 18.1 gives a good picture of the growth of the amount of industrial life insurance in force in the United States from 1910 to 1968, Figure 18.2 contains a series whose movements are obviously not of a linear nature. It has the shape of an elongated *S*, which is very common among series describing economic growth.

For series covering relatively short periods of time, straight lines often provide a close fit and although we may be justified in using them we cannot be sure that they really describe what we called a *secular trend*. For all we know, the seemingly linear movement may be a portion of a cycle or part of a pattern which later on "flattens out" or departs from linearity in some other fashion. Indeed, most time series of economic data tend to level off in the long run, and if we fit a straight line, we shall have to do so with the understanding that we have no guarantee that we are describing a continuing pattern. Also, to avoid mistaking parts of cycles for secular trends, it is advisable to calculate trends on the basis of data covering relatively long periods of time. The exact number of years will, of course, vary from one

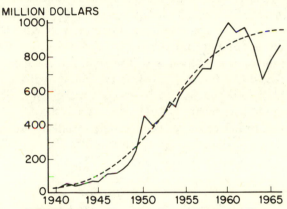

FIGURE 18.2. Logistic Curve Fitted to Factory Sales of Electron Tubes and Semi-Conductors in the United States.

study to the next, depending on the availability of data and the particular period that happens to be of interest. (Since we shall want to avoid overly lengthy computations in our illustrations, we shall sometimes refer to shorter periods than are ordinarily used in the study of secular trends.)

In Chapter 13 we saw that the problem of curve fitting may be approached with various degrees of refinement and this applies also to the description of trends. Sometimes, when a rough indication of the nature of a trend is all that is needed, we may simply draw it *freehand*. This method is very flexible in that it can be used regardless of whether the trend is a straight line or curved, but, as we pointed out before, freehand curve fitting is too subjective to be of much value if our analysis is to serve as a basis for predictions. In the remainder of this chapter we shall discuss various methods of fitting trends, linear and otherwise, and how to calculate appropriate trend equations.

18.2 Linear Trends: Semi-Averages

A simple way of fitting a straight line trend is by the *method of semi-averages*. This method has the advantage of being straightforward and easy to use, but it is crude, susceptible to distortion by the presence of one or more extreme values, and, generally speaking, it is only one step more advanced than drawing a freehand line.

To fit a line by the method of semi-averages we first divide our data into two parts, preferably with the same number of years. For instance, a series covering the period from 1953 to 1966 could be divided into one part going from 1953 to 1959 and the other going from 1960 to 1966. If a series covers an *odd* number of years we can make the two parts equal simply by omitting the middle year.

Once the data have been split into two parts, we calculate the mean of the quantities making up the series *separately* for each part, plot the two means at the midpoints of the time intervals covered by the respective parts, and finally join these two points with a straight line. This line is the trend line.

To illustrate this technique, let us consider the following data on the amount of ordinary life insurance purchased in the United States from 1953 to 1966 (Source: *Life Insurance Fact Book* 1967):

	Total Purchases of Ordinary Life Insurance (millions of dollars)
1953	23,489
1954	25,276
1955	30,827
1956	36,375
1957	45,635
1958	47,648
1959	51,678
1960	52,883
1961	55,016
1962	56,998
1963	64,267
1964	74,012
1965	83,485
1966	89,658

Dividing these annual purchases into two groups of seven years each, we find that for the 1953 to 1966 period the mean of the annual purchases of ordinary life insurance is

$$\frac{23{,}489 + 25{,}276 + 30{,}827 + 36{,}375 + 45{,}635 + 47{,}648 + 51{,}678}{7}$$

$$= \$37{,}275.4 \text{ million}$$

and that for the 1960 to 1966 period it is

$$\frac{52{,}883 + 55{,}016 + 56{,}998 + 64{,}267 + 74{,}012 + 83{,}485 + 89{,}658}{7}$$

$$= \$68{,}045.6 \text{ million}$$

Plotting these two averages at points corresponding to the years 1956 and 1963, the midpoints of the respective periods, and connecting them with a straight line, we obtain the desired trend line, which is shown in Figure 18.3 together with the original data.

For any given year the value on the line is referred to as the corresponding *trend value*. In view of our method of construction, the trend value for 1956 is \$37,275.4 million and that for 1963 is \$68,045.6 million. In order to determine the trend value for any other

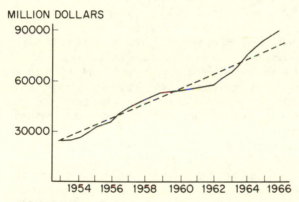

MILLION DOLLARS

FIGURE 18.3. Ordinary Life Insurance Purchased in the United States. (Source: 1967 Life Insurance Fact Book.)

year, let us first calculate the *annual trend increment*, namely, the average increase per year. Since the total increase from 1956 to 1963, a period of seven years, is $68{,}045.6 - 37{,}275.4 = 30{,}770.2$, the average increase per year is $30{,}770.2/7 = \$4{,}395.7$ million. Having found this annual trend increment, we can now determine the trend value for any year by adding or subtracting the annual trend increment an appropriate number of times to or from a trend value that is already known. For instance, the 1954 trend value is 37,275.4 minus *twice* 4,395.7 or $37{,}275.4 - 2(4{,}395.7) = \$28{,}484.0$ million. Similarly, the 1960 value is 37,275.4 plus *four times* 4,395.7 or $37{,}275.4 + 4(4{,}395.7) = \$54{,}858.2$ million. The last value could also have been obtained (except for a slight discrepancy due to rounding off) by subtracting 4,395.7 *three times* from 68,045.6.

It should also be noted that if each of the two parts into which we divide our data consisted of an *even* number of years, their means would have to be plotted half-way between two years and, to obtain the trend values of adjacent years, we would have to add or subtract *half* of the annual trend increment.

EXERCISES

1. The following are the 1953 to 1966 values of the construction index in the United States, with 1957–59 = 100: 91, 87, 97, 100, 99, 94, 106, 106, 105, 110, 116, 124, 134, and 142.   (Source: *Business Statistics 1967*)
   (a) Use the method of semi-averages to find the trend values for 1956 and 1963, plot the data on arithmetic paper, and draw a straight line through the 1956 and 1963 trend values plotted on the same paper.
   (b) Calculate the annual trend increment and use it to calculate the 1953, 1960, and 1965 trend values.
2. The following are the 1953 to 1967 figures on the total number of employees in nonagricultural establishments in the U.S. (in millions): 50.2, 49.0, 50.7, 52.4, 52.9, 51.4, 53.4, 54.2, 54.0, 55.6, 56.7, 58.3, 60.8, 64.0, and 66.0.   (Source: Bureau of Labor Statistics.)
   (a) Use the method of semi-averages to find the trend values for 1956 and 1964, omitting the year 1960 in your calculations.
   (b) Plot the original data on arithmetic paper and draw a straight line through the 1956 and 1964 trend values plotted on the same paper.
   (c) Calculate the annual trend increment and use it to calculate the 1955, 1960, and 1967 trend values.
3. The following are the 1959 to 1966 totals on the consumption of mineral energy resources and electricity produced from water power and nuclear power in the United States (in trillions of British thermal units): 43,411; 44,960; 45,705; 47,620; 49,598; 51,676; 53,962; and 56,542.   (Source: *Statistical Abstract of the United States 1967*.)   Use the method of semi-averages to calculate the trend values for the years 1959 through 1966 and plot the original data as well as the trend line.
4. Explain why it is necessary to include a relatively long period of years in an analysis of secular trend.   Use data from suitable published sources to illustrate your argument.
5. The following are the number of insured savings and loan associations in the United States (in thousands) for the years 1957 to 1967: 3.8, 3.9, 4.0, 4.1, 4.2, 4.3, 4.4, 4.5, 4.5, 4.5, 4.5.   (Source: Federal Home Loan Bank Board.)
   (a) Use the method of semi-averages to find the trend values for 1959 and 1965.
   (b) Calculate the annual trend increment and use it to calculate the trend values for 1958 and 1964.

18.3 Linear Trends: Least Squares

The most widely used method of fitting trends is the method of *least squares*, which we already studied in connection with the problem of regression. The only difference, if we can call it a difference, is that x now represents the time or date to which the measurements, the y's, refer. As before, the least squares criterion requires that the sum of the squares of the differences between the actual y values and the calculated trend values y' be a minimum. That is, in the linear equation $y' = a + bx$, which we shall consider first, a and b will have to be chosen so that $\Sigma(y - y')^2$ is a minimum.

The use of a least squares line does not eliminate all subjectivity from the analysis of a time series. Among other things, we must still decide on the number of years to use, what adjustments are to be made, and whether to fit a straight line in preference to some other type of curve. However, it does tell us where to locate a line (its height and its slope) once we have decided to use one, since, for a given series of data, there is only one line satisfying the above least squares criterion.

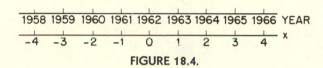

FIGURE 18.4.

As we saw in Chapter 13, the problem of fitting a least squares line is essentially that of calculating values of a and b for given data with the use of (13.2.2) and (13.2.3) or with the use of the two *normal equations* which read

$$\Sigma y = n \cdot a + b \cdot \Sigma x \tag{13.2.4}$$

and
$$\Sigma xy = a \cdot \Sigma x + b \cdot \Sigma x^2 \tag{13.2.5}$$

Since, in time series work, the x's usually refer to successive years (days, weeks, or months) we will be able to simplify these formulas considerably by changing the scale of the x's *so that their sum is equal to zero.* Taking the origin of the new scale at the middle of the series and numbering the years in both directions so that in the new scale $\Sigma x = 0$, the two normal equations, solved for a and b, reduce to

$$a = \frac{\Sigma y}{n} = \bar{y} \quad \text{and} \quad b = \frac{\Sigma xy}{\Sigma x^2} \tag{18.3.1}\star$$

To illustrate how this simplification works, we shall fit least squares trend lines to two short series of unadjusted data taken from *the* 1967 *Business Statistics.* They represent profits of manufacturers of (1) nondurable and (2) durable goods. Taking first the profits of manufacturers of nondurable goods for the nine-year period 1958–1966, we shall make the change of scale shown in Figure 18.4. There is an *odd* number of years, and in the x-scale they are numbered -4, -3, -2, -1, 0, 1, 2, 3, and 4. The middle year is 1962, and it corresponds to $x = 0$. (More generally, when there is an odd number of years we assign $x = 0$ to the middle year and number the others . . . , -5, -4, -3, -2, -1, 0, 1, 2, 3, 4, 5, . . . , so that the sum of the x's is always equal to 0.)

The sums needed to find a and b by means of (18.3.1) are given by the totals of the y, xy, and x^2 columns of the following table. The last column gives the trend values y', obtained later by substituting the x's into the trend equation.

	x	y	Profits (billions of dollars) xy	x^2	y'
1958	-4	10.0	-40.0	16	10.14
1959	-3	12.7	-38.1	9	11.01
1960	-2	12.4	-24.8	4	11.88
1961	-1	11.9	-11.9	1	12.75
1962	0	12.5	0.0	0	13.62
1963	1	13.0	13.0	1	14.49
1964	2	14.9	29.8	4	15.36
1965	3	16.5	49.5	9	16.23
1966	4	18.7	74.8	16	17.10
		122.6	52.3	60	

Substituting $n = 9$, $\Sigma y = 122.6$, $\Sigma xy = 52.3$, and $\Sigma x^2 = 60$ into the two formulas of (18.3.1), we get

$$a = \frac{122.6}{9} = 13.62$$

$$b = \frac{52.3}{60} = 0.87$$

and the equation of the trend line may be written as

$$y' = 13.62 + 0.87x$$

Having obtained this equation, we must be careful to indicate what the x's and y's mean. The value $a = 13.62$ is *not* the trend value for the first year, 1958, but for the middle year, 1962, for which we put x equal to 0. Specifying the origin of x, the units of x, and the units of y, it is advisable to add the following legend to the equation of the least squares line:

$$y' = 13.62 + 0.87x$$

(origin, 1962; x units, 1 year; y annual
profits in billions of dollars).

The trend values shown in the last column of the table were obtained by substituting the x's corresponding to the different years into the trend equation.   For instance, for 1961 we get $y' = 13.62 + 0.87(-1)$ $= 12.75$.

When a time series is given for an *even* number of years there will be no middle year and we will not be able to use the x scale shown in Figure 18.4.   Instead, there will be a *midpoint in time* between the two middle years and, letting x equal 0 at this point, we shall number the years . . . , $-5, -3, -1, 1, 3, 5, ...$ In this fashion the sum of the x's will again be zero.   To illustrate this alternate change of scale, let us fit a least squares trend line to the profits of manufacturers of durable goods for the eight-year period 1959–1966.   The middle of the series falls between the years 1962 and 1963 (that is, at

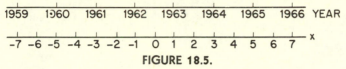

FIGURE 18.5.

the instant when 1962 becomes 1963) and the x-scale which we shall use is shown in Figure 18.5.   Arranging our calculations in a table similar to the one used in the previous example, we get

	x	y	*Profits* *(billions of dollars)* xy	x^2	y'
1959	-7	13.6	-95.2	49	10.25
1960	-5	12.0	-60.0	25	12.01
1961	-3	11.4	-34.2	9	13.77
1962	-1	14.1	-14.1	1	15.53
1963	1	15.8	15.8	1	17.29
1964	3	17.8	53.4	9	19.05
1965	5	22.2	111.0	25	20.81
1966	7	24.4	170.8	49	22.57
		131.3	147.5	168	

and substitution into (18.3.1) yields

$$a = \frac{131.3}{8} = 16.41$$

$$b = \frac{147.5}{168} = 0.88$$

The equation of the trend line thus becomes

$$y' = 16.41 + 0.88x$$

(origin 1962–1963; x units, 6 months; y,
annual profits in billions of dollars).

As before, the annual trend values are obtained by substituting the various values of x into the trend equation. Note that the origin is now *between* two years and that the x units are 6 months, not one year. The two trend lines which we have found in this section are shown in

BILLION DOLLARS

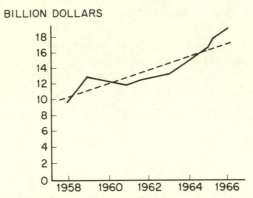

FIGURE 18.6. Profits of Manufacturers of Non-Durable Goods. (Source: 1967 Business Statistics.)

Figures 18.6 and 18.7 together with the original data. To plot them we have only to calculate, in each case, y' for two values of x and draw a line through the points representing these trend values.

BILLION DOLLARS

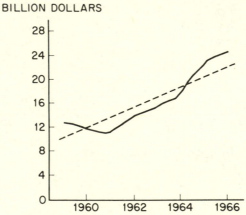

FIGURE 18.7. Profits of Manufacturers of Durable Goods. (Source: 1967 Business Statistics.)

18.4 Modified Trend Equations

In this section we shall modify trend equations by (a) changing the origin of x, that is, the time point for which $x = 0$, (b) changing the units of y, and (c) changing the units of x. These modifications are easy to perform if we remember that in a trend equation $y' = a + bx$ the constant a is the *trend value* of the year or time point for which $x = 0$ and that b is the *trend increment*. (If $x = 0, 1, 2, \ldots$, refer to successive years, b is the *annual trend increment*, namely, the amount by which trend values increase or decrease from year to year.)

Sometimes, when we want to compute trend values year by year, it is convenient to shift the origin of x so that $x = 0$ corresponds to a different year. Taking the trend equation which we obtained for the data on nondurable goods (see page 406), let us illustrate the general procedure by shifting the origin of x from 1962 to, say 1964. The original trend equation read

$$y' = 13.62 + 0.87x$$

(origin, 1962; x units, 1 year; y, annual profits in billions of dollars).

where 13.62 is the *trend value* for 1962 and 0.87 is the *annual trend increment* of the annual profits. Since changing the origin of x to 1964 will not affect the annual trend increment, the only thing we will have to do is to substitute the 1964 trend value for 13.62. Proceeding as on page 403, we obtain the 1964 trend value by adding the annual trend increment of 0.87 *twice* to the 1962 trend value of 13.62 getting $13.62 + 2(0.87) = 15.36$, and we can now write the modified trend equation as

$$y' = 15.36 + 0.87x$$

(origin, 1964; x units, 1 year; y annual profits in billions of dollars).

To give another example, let us shift the origin of the trend equation which we obtained for the data on durable goods (see page 407) to 1959. It is always desirable to have the origin at the middle of a year or, when dealing with monthly data, at the middle of a month. The equation which we obtained on page 407 read

$$y' = 16.41 + 0.88x$$

(origin, 1962–1963; x units, 6 months; y, annual profits in billions of dollars).

where 16.41 is the trend value for the time point between 1962 and 1963 while 0.88 is the *semi-annual trend increment*. Changing the origin to 1959 will again leave b unaffected and to get the new a, the trend value for 1959, we will have to subtract 0.88 *seven times* from 16.41. (The middle of 1959 is $3\frac{1}{2}$ years or *seven* 6-month periods earlier than the time point between 1962 and 1963.)   Getting $16.41 - 7(0.88) = 10.25$ the modified trend equation becomes

$$y' = 10.25 + 0.88x$$

(origin, 1959; x units, 6 months;
y, annual profits in billions of dollars).

In both of these examples y stood for *annual* profits in billions of dollars.   If we wanted to change it to average *monthly* profits in billions of dollars, we would only have to divide each y value by 12 and this division has the effect that each trend value as well as a and b are divided by 12.   Changing the trend equation obtained on page 406 for the data on nondurable goods to average monthly profits, we get $13.62/12 = 1.1350$ for a, $0.87/12 = 0.0725$ for b, and the equation now reads

$$y' = 1.1350 + 0.0725x$$

(origin, 1962; x units, 1 year;
y, average monthly profits in billions of dollars).

Having changed the y's to average monthly profits, let us modify the trend equation further by changing the x's so that they refer to successive *months*.   Since b measures the trend increment corresponding to one unit of x, it will have to be changed to the *monthly trend increment of the average monthly profits*.   (Incidentally, in the last equation 0.0725 is the *annual trend increment of the average monthly profits* and it would have been well to refer to the 0.87 of the original equation as the annual trend increment of the annual profits.)   The month to month increase (or decrease) in trend values being 1/12 of the annual trend increment, we get $0.0725/12 = 0.00604$ for the new b and, leaving a unchanged, the trend equation becomes

$$y' = 1.1350 + 0.00604x$$

(origin, 1962; x units, 1 month; y, average
monthly profits in billions of dollars).

In this equation the origin of x is the middle of 1962, i.e., the time point between June, 1962, and July, 1962.   To move it, say, to the middle of June, 1962, we have only to proceed as on page 403 and subtract a suitable multiple of the trend increment from the 1962 trend value of 1.1350.   Since the middle of June, 1962, is half a month earlier than the origin of the above equation, we find that the new a, the trend value for this date, is $1.1350 - \frac{1}{2}(0.00604)$ or approximately 1.1319. We thus get

$$y' = 1.1319 + 0.00604x$$

(origin, June, 1962 x units, 1 month;
y, average monthly profits in billions of dollars).

The reader may wish to verify that if we had wanted to shift the origin to, say, the middle of January, 1962, in the above example, we would have had to subtract $5\frac{1}{2}$ *times* the monthly trend increment of $0.00604x$ from 1.1350 to get the new a.

EXERCISES

1. Use (18.3.1) to fit a least squares line to the data of Exercise 2 on page 404. Give the equation of the line with an appropriate legend and use it to calculate the trend value for 1953 and 1967. Plot the original series together with the least squares line, obtaining the latter by drawing a straight line through the calculated 1953 and 1967 trend values.

2. Modify the trend equation obtained in Exercise 1 by shifting the origin of x to 1966.

3. Using (18.3.1), find the equation of the least squares lines for the data of Exercise 1 on page 404. Give the equation with an appropriate legend and use it to calculate the 1953 and 1966 trend values. Also, plot the original series together with the least squares line on arithmetic paper, obtaining the latter by drawing a straight line through the calculated 1953 and 1966 trend values.

4. Modify the trend equation obtained in Exercise 3 by changing the origin of x to 1953 and the x units to years.

5. The following are the shipments of steel products (in millions of net tons) of the Steel Industry for the years 1957 to 1966: 79.9, 59.9, 69.4, 71.1, 66.1, 70.6, 75.6, 84.9, 92.7, and 90.0. Use (18.3.1) to find the least squares trend line. Draw an arithmetic line chart showing the given series as well as the least squares trend line. (Source: American Iron and Steel Institute)

6. Modify the trend equation obtained in Exercise 5 by shifting the origin of x to 1961. Then modify this equation further by changing y to average monthly shipments of steel.

7. Using (18.3.1), find the equation of the least squares line for the electric power data of Exercise 3 on page 422. Use this equation to calculate the trend values for 1941 and 1966. Also, plot the original series on arithmetic paper and draw the least squares line through the 1941 and 1966 trend values.

8. Given the equation $y' = 13.62 + 0.87x$ *(origin, 1962; x units, 1 year; y, annual profits in billions of dollars),* which we obtained for the trend

of the profits of manufacturers of non-durable goods, shift the origin
to 1958 and then verify the annual trend values given in the right-hand
column of the table on page 406.

9. Shift the origin of the equation on page 407 which describes the trend
of profits of manufacturers of durable goods to the year 1959 and then
verify the trend values given in the right-hand column of the table on
page 407.

18.5 Parabolic Trends

As we pointed out earlier, trends cannot always be described
adequately by means of straight lines, and when a series departs
obviously or subtly from linearity we will have to consider fitting a
curve of some other kind. Of the many types of curves used in fitting
trends, one of the most useful is the *parabola*, whose equation is

$$y' = a + bx + cx^2 \qquad (18.5.1)$$

It is also called a *second degree polynomial equation*, owing to the fact
that *two* is the highest power to which x appears in this equation.

The problem of fitting a parabolic trend is essentially that of finding numerical values for the constants a, b, and c appearing in (18.5.1)
so that the resulting curve will provide the best possible fit. Defining
"goodness of fit" again in the sense of *least squares*, we shall have to
determine a, b, and c so that $\Sigma(y - y')^2$ is a minimum. Of course,
the y's are now calculated by means of (18.5.1).

Without going into mathematical details, let us merely state that
the method of least squares yields the following three *normal equations*
with which to calculate a, b, and c:

$$\Sigma y = n \cdot a + b \cdot \Sigma x + c \cdot \Sigma x^2 \qquad (18.5.2)$$

$$\Sigma xy = a \cdot \Sigma x + b \cdot \Sigma x^2 + c \cdot \Sigma x^3 \qquad (18.5.3)$$

$$\Sigma x^2 y = a \cdot \Sigma x^2 + b \cdot \Sigma x^3 + c \cdot \Sigma x^4 \qquad (18.5.4)$$

To simplify the use of these equations for equally spaced data, we
shall adopt the same convention as in Section 18.3, namely, that of
changing the scale of x to . . . , -3, -2, -1, 0, 1, 2, 3, . . . , or
. . . -5, -3, -1, 1, 3, 5, . . . , depending on whether we have an
even or odd number of years. Putting, as before, the 0 of this new
scale at the time point which corresponds to the *middle* of our data,
Σx and Σx^3 will both *equal* 0 and the normal equations reduce to

$$\Sigma y = n \cdot a + c \cdot \Sigma x^2 \qquad (18.5.5)\star$$

$$\Sigma xy = b \cdot \Sigma x^2 \qquad (18.5.6)\star$$

$$\Sigma x^2 y = a \cdot \Sigma x^2 + c \cdot \Sigma x^4 \qquad (18.5.7)\star$$

We can thus find b directly from (18.5.6) and determine a and c by simultaneously solving (18.5.5) and (18.5.7).

To illustrate the least squares method of fitting a parabolic trend, let us use the annual number of nonimmigrant aliens (visitors for pleasure and business, foreign government officials, students, etc.) admitted to the United States for the period 1950 to 1966. These figures, obtained from the *Annual Reports* of the Department of Justice, Immigration and Naturalization Service, are shown below in the y column of the table.

The calculation of a, b, and c by means of the simplified normal equations requires that we find n, Σy, Σxy, $\Sigma x^2 y$, Σx^2, and Σx^4. The work needed to obtain these sums is most conveniently arranged in the following kind of table:

Year	Nonimmigrants Admitted (Thousands) y	x	xy	x^2y	x^2	x^4
1950	426.8	−8	−3,414.4	27,315.2	64	4,096
1951	465.1	−7	−3,255.7	22,789.9	49	2,401
1952	516.1	−6	−3,096.6	18,579.6	36	1,296
1953	485.7	−5	−2,428.5	12,142.5	25	625
1954	566.6	−4	−2,266.4	9,065.6	16	256
1955	620.9	−3	−1,862.7	5,588.1	9	81
1956	686.3	−2	−1,372.6	2,745.2	4	16
1957	758.9	−1	− 758.9	758.9	1	1
1958	847.8	0	0.0	0.0	0	0
1959	1,024.9	1	1,024.9	1,024.9	1	1
1960	1,140.7	2	2,281.4	4,562.8	4	16
1961	1,220.3	3	3,660.9	10,982.7	9	81
1962	1,331.4	4	5,325.6	21,302.4	16	256
1963	1,507.1	5	7,535.5	37,677.5	25	625
1964	1,744.8	6	10,468.8	62,812.8	36	1,296
1965	2,076.0	7	14,532.0	101,724.0	49	2,401
1966	2,341.9	8	18,735.2	149,881.6	64	4,096
	17,761.3		45,108.5	488,953.7	408	17,544

Substituting the totals of the xy and x^2 columns into (18.5.6), we get

$$45,108.5 = b \times 408$$

or

$$b = \frac{45,108.5}{408} = 110.56$$

and substituting $n = 17$ together with the totals of the y, x^2y, x^2, and x^4 columns into (18.5.5) and (18.5.7), we get

$$17,761.3 = 17a + 408c$$
$$488,953.7 = 408a + 17,544c$$

To solve these two equations by the *method of elimination*, we could divide each term of the first by 17, each term of the second by 408, and then by subtraction obtain an equation involving only c.† Substituting this value of c into either of the two original equations would then give an equation which can be solved for a. If we actually performed these steps we would get

$$a = 850.93 \qquad \text{and} \qquad c = 8.09$$

and the equation of the parabolic trend, thus becomes

$$y' = 850.93 + 110.56x + 8.09x^2$$

(origin, 1958; x units, 1 year; y, total nonimmigrants admitted in thousands).

In this equation, $a = 850.93$ is the *trend value* for 1958, $b = 110.56$ the *slope* of the curve at $x = 0$, and $2c = 16.18$ is the rate with which the slope changes at this particular point.

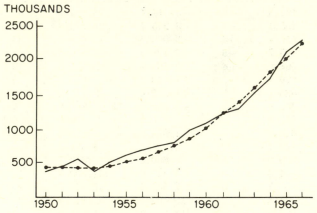

FIGURE 18.8. Nonimmigrant Aliens Admitted to the United States from 1950 to 1966. (Source: Department of Justice, Immigration and Naturalization Service; Annual Reports.)

To find the trend value for any given year we have only to substitute the appropriate x into the trend equation. For instance, to get the trend value for 1960 we have to substitute $x = 2$ and to get the trend value for 1952 we have to substitute $x = -6$. Making these substitutions we get $y' = 850.93 + 110.56(2) + 8.09(2)^2 = 1104.41$ for 1960 and $y' = 850.93 + 110.56(-6) + 8.09(-6)^2 = 478.81$ for 1952. The trend values for the years 1950 through 1966

---

† In case the reader is not familiar with solving simultaneous linear equations in two unknowns, he will find suitable treatments of this subject in college algebra texts.

are shown in Figure 18.8 together with the original data.   The curve drawn through the points representing these trend values is the parabola which we have fitted by the method of least squares.

A question that always arises in connection with parabolic trends is when, in other words, under what circumstances, such a trend should be used.   Although it is difficult *not* to be subjective, there exists a criterion that will generally be of some aid.   To illustrate it, let us consider the parabola

$$y = x^2 + 3x + 2$$

and let us calculate y for $x = 0$, 1, 2, 3, 4, 5, and 6.   The results obtained by substituting these values of x are shown below together with the *first and second differences* of the y's.   The first differences are obtained by subtracting successive values of y and the second differences are obtained by subtracting successive entries of the first-difference column.

x	y	*First Differences*	*Second Differences*
0	2		
		4	
1	6		2
		6	
2	12		2
		8	
3	20		2
		10	
4	30		2
		12	
5	42		2
		14	
6	56		

This illustrates that if a series of points (corresponding to equally spaced values of x) actually lies on a parabola, the second differences of the y's will all be the same. *Generally speaking, the fact that the second differences of a series are more or less the same may be used as an indication that a parabola will provide a good fit.*   However, even if the second differences are *not* close, it is possible that a parabola will nevertheless provide a good description of the over-all trend.

Having shown only how to fit a *second degree* polynomial equation, that is, a parabolic trend, it should be mentioned that *higher degree polynomial equations* such as $y' = a + bx + cx^2 + dx^3$, $y' = a + bx + cx^2 + dx^3 + ex^4$, etc., are also, though not very frequently, used. Since the calculations become rather involved in that case, a special method involving the use of *orthogonal polynomials* may be employed to provide some simplifications.   If the reader should ever have the occasion to fit this kind of curve, he should read the discussion of such polynomials in the bulletin referred to in the Bibliography on page 428.

18.6 Exponential Trends

The question of what type of curve to fit to a given series can often be decided by plotting it on various kinds of paper. If a series does not exhibit a linear trend when plotted on arithmetic paper, it may happen that it will "straighten out," that is, appear linear, when plotted on semi-log paper, log-log paper, or some other kind of paper with special scales.

In this section we shall discuss the problem of fitting trends to series which appear linear when plotted on *semi-log* or *ratio paper*, namely, the kind of paper which we discussed in Section 17.10. Such trends are referred to as *exponential* trends, because their equation is of the form

$$y' = ab^x \tag{18.6.1}$$

in which x appears as an *exponent*. Applying to this equation the rules of logarithms given in Appendix IV, we find that (18.6.1) can also be written as

$$\log y' = \log a + x \cdot \log b \tag{18.6.2}\star$$

namely, as a *linear* equation in x and $\log y'$.† This explains why exponential curves plot as straight lines on semi-log paper—on paper where we actually plot $\log y'$ against x—and why exponential trends are also referred to as *semi-logarithmic linear trends*. Although (18.6.1) may look complicated, the reader should have no difficulty recognizing this kind of equation from work in the mathematics of finance. The *compound interest* formula $A_n = P(1 + r)^n$, in which P is the original principal, r the interest rate per conversion period, n the number of conversion periods, and A_n the amount due, is an example of an exponential curve.

The problem of fitting an exponential trend is essentially that of determining numerical values for a and b, the two constants appearing in (18.6.1), and we shall determine these values by applying the method of least squares to (18.6.2). Since we already know how to fit a line by the method of least squares, we will be able to use the formulas of Section 18.3 with $\log a$, $\log b$, and $\log y$ substituted for a, b, and y,

---

† Writing A, B, and Y for $\log a$, $\log b$, and $\log y'$, (18.6.2) becomes $Y = A + Bx$, and this can readily be identified as the usual equation of a straight line.

provided that we have equally spaced values of x. Making these changes, formulas (18.3.1) become

$$\log a = \frac{\Sigma \log y}{n} \quad \text{and} \quad \log b = \frac{\Sigma(x \cdot \log y)}{\Sigma x^2} \qquad (18.6.3)\star$$

and we can thus find $\log a$ and $\log b$ and, hence, a and b.

To illustrate this least squares method of fitting an exponential trend, let us again refer to the data pertaining to nonimmigrant aliens admitted into the United States for the period 1950 to 1966. The necessary work proceeds just as in Section 18.3 with the only difference that $\log y$ is used instead of y. Getting the logarithms from Table IX, we obtain

	Nonimmigrants Admitted (thousands)† y	$\log y$	x	x^2	$x \log y$
1950	427	2.6304	−8	64	−21.0432
1951	465	2.6675	−7	49	−18.6725
1952	516	2.7126	−6	36	−16.2756
1953	486	2.6866	−5	25	−13.4330
1954	567	2.7536	−4	16	−11.0144
1955	621	2.7931	−3	9	− 8.3793
1956	686	2.8363	−2	4	− 5.6726
1957	759	2.8802	−1	1	− 2.8802
1958	848	2.9284	0	0	0.0
1959	1,025	3.0107	1	1	3.0107
1960	1,141	3.0573	2	4	6.1146
1961	1,220	3.0864	3	9	9.2592
1962	1,331	3.1242	4	16	12.4968
1963	1,507	3.1781	5	25	15.8905
1964	1,745	3.2417	6	36	19.4502
1965	2,076	3.3173	7	49	23.2211
1966	2,342	3.3696	8	64	26.9568
		50.2740		408	19.0291

Substituting $n = 17$ and the totals of the appropriate columns into (18.6.3), we get

$$\log a = \frac{50.2740}{17} = 2.9573$$

$$\log b = \frac{19.0291}{408} = 0.0466$$

---

† (Rounded).

and in its *logarithmic form* the trend equation may be written as

$$log\ y' = 2.9573 + 0.0466x$$

*(origin, 1958; x units, 1 year; y, annual
number of nonimmigrants admitted in thousands
of persons).*

To plot the trend which we have found on semi-log paper, we have only to calculate the trend values for two years, plot them, and join them by a straight line. Using the logarithmic form of the trend

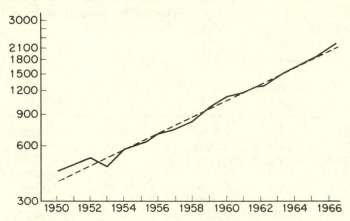

FIGURE 18.9. Nonimmigrant Aliens Admitted to the United States from 1950 to 1966. (Source: Department of Justice, Immigration and Naturalization Service; Annual Reports.)

equations, which is more convenient in this case, we thus find by substituting $x = -6$ and $x = 5$ that for 1952

$$log\ y' = 2.9573 + 0.0466(-6) = 2.6777$$
$$y' = 476.1$$

and that for 1963

$$log\ y' = 2.9573 + 0.0466(5) = 3.1903$$
$$y' = 1550$$

Plotted on semi-log paper, the original data and the exponential trend are shown in Figure 18.9. In Exercise 2 on page 421 the reader will be asked to calculate the exponential trend values for each of the years 1950 to 1966 and plot them together with the original data as well as the parabolic trend values obtained in Exercise 1 on page 421 on *arithmetic* paper. This will make it possible to compare the goodness of the fit of the two trends.

Using Table IX to look up the numbers whose logarithms are 2.9573 and 0.0466, the values which we obtained for log a and log b, we get 906.4 and 1.11. Using (18.6.1), the trend equation can then be written as $y' = 906.4(1.11)^x$ and in this form 906.4 is the trend value for 1958 while 1.11 is *1 plus the average annual rate of growth of nonimmigrants admitted into the United States over the given period.* Hence, the average annual growth is 0.11 or 11 per cent.

18.7 Other Trend Curves

Series which are not linear on either arithmetic or semi-log paper may show a linear trend when plotted on log-log paper—paper which has logarithmic scales for both x and y. If this is the case, we fit a *power function* of the form $y' = ax^b$, whose equation in logarithmic form is

$$\log y' = \log a + b \cdot \log x \qquad (18.7.1)$$

namely, a linear equation in log x and log y. Applying to it the method of least squares, we obtain the *normal equations*

$$\Sigma \log y = n \cdot \log a + b \cdot \Sigma \log x \qquad (18.7.2)^\star$$

$$\Sigma(\log x)(\log y) = (\log a) \cdot \Sigma \log x + b \cdot \Sigma(\log x)^2 \qquad (18.7.3)^\star$$

which will ultimately yield solutions for a and b. [These equations were obtained by substituting log x, log y, and log a, respectively, for x, y, and a in (13.2.4) and (13.2.5).] Equations (18.7.2) and (18.7.3) cannot readily be simplified by the method of Section 18.2 and, for that matter, the x scale of that section cannot be employed. It would lead to the logarithms of negative numbers.

The material of the previous paragraph has been included primarily for the sake of completeness. Power functions are seldom fit to business data and we shall not illustrate this kind of trend with a numerical example. In principle, we have only to obtain the required sums from our data and after substituting them into the normal equations solve for log a and b.

It may well be that for a given series none of the curves which we have studied will provide a good fit. Indeed, many series seem to go through a period of growth which is best described by a curve having the shape of an elongated S, a shape which is neither linear, parabolic, exponential, or like a power function. Some of the curves that are most commonly fit to series of this kind are briefly discussed below. (The curves which we shall introduce apply equally well to describe *upward* as well as *downward* growth, but to simplify our treatment we shall explain them with reference to series which "grow" in the usual sense, namely, series which show an upward trend.)

The two curves that are used most widely to describe growth are the *Gompertz curve* and the *Pearl-Reed or logistic curve*. The Gompertz curve, which is illustrated in Figure 18.10, serves to describe the growth of series which, while increasing, seem to approach some maximum value as a limit. Although the growth continues, it does so at a decreasing rate. Specifically, the equation of the Gompertz curve is

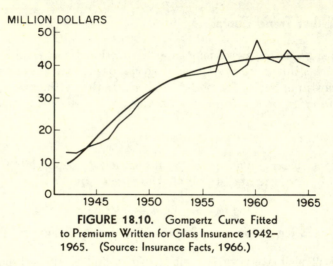

FIGURE 18.10. Gompertz Curve Fitted to Premiums Written for Glass Insurance 1942–1965. (Source: Insurance Facts, 1966.)

$y' = ka^{b^x}$ and, for purposes of curve fitting, it is usually written in the more convenient logarithmic form of

$$\log y' = \log k + b^x \cdot \log a \qquad (18.7.4)$$

The *logistic curve* has been applied widely to population data of various kinds, both human and non-human, and it has also provided a good fit to many economic series pertaining to industrial growth. Its equation is

$$y' = \frac{1}{k + ab^x} \qquad (18.7.5)$$

or, as it can also be written, $1/y' = k + ab^x$. If in the left-hand member of this equation we wrote y' instead of $1/y'$, we would get $y' = k + ab^x$, which is the equation of another curve used in fitting trends, the so-called *modified exponential*. An example of a logistic curve is shown in Figure 18.2.

Both the Gompertz curve and the logistic curve have been used extensively by the National Industrial Conference Board in its important studies on growth patterns of industries, cities, and states. Since the fitting of growth curves like the ones mentioned above involves a considerable amount of mathematical detail, we shall not go into it here.   Suitable reference to the fitting of modified exponentials, Gompertz curves, and logistic curves may be found in the Bibliography on page 428.

Although we did not exhaust all the equations that might reasonably be used, our discussion has shown that there are many ways of fitting trends.   It will be well to keep in mind that even though each series presents its individual problem, most can be handled by the methods which we have described.   Of course, what we try to do in any particular case is to select that equation or that method of measuring trend which best describes the gradual and consistent pattern of growth.   Although special kinds of graph paper, the second difference test mentioned on page 415, and other tests will aid in deciding which kind of trend is "right" for a given series, it is difficult not to be subjective, at least to some extent.   It should not be surprising, therefore, if we find that one person chooses to measure trend in one way while a different person prefers another way; in this area reasonable differences of opinion are natural and inevitable.   However, it must be clear that if one is to operate effectively in this field, a knowledge of the major types of curves and methods used is necessary, as is a considerable amount of judgment and experience.

EXERCISES

1. On page 414 we obtained $y' = 850.93 + 110.56x + 8.09x^2$ (*origin, 1958; x units, 1 year; y, total nonimmigrants admitted in thousands*) for the parabola which we fit to the annual nonimmigrants admitted to the United States for the period 1950 to 1966.

   (a) Using the above equation calculate the trend value for each year of the given period.   *Hint:* Compute $a + bx$ and cx^2 separately for each x and then add to get the desired trend values.   For the first three years we will thus get

Year	x	$a + bx$	cx^2	$y' = a + bx + cx^2$
1950	−8	−33.55	517.76	484.21
1951	−7	77.01	396.41	473.42
1952	−6	187.57	291.24	478.81

   (b) Plot these trend values together with the original data on arithmetic paper.   Draw a smooth curve through the points representing the trend values to indicate the parabolic trend.

2. For the same data as in Exercise 1 we obtained the semi-logarithmic linear trend equation $\log y' = 2.9573 + 0.0466x$ (*Origin, 1958; x units, 1 year; y, annual admissions in thousands*) on page 418.

(a) Using this equation, calculate the trend value for each year of the given period. *Hint:* Set your calculations up as follows:

Year	x	log y'	y'
1950	−8	2.5845	384.2
1951	−7	2.6311	427.7
1952	−6	2.6777	476.1

(b) Plot the trend values obtained in (a) on the chart constructed in Exercise 1 and compare the parabolic and exponential trends.

3. The following figures show the sales of electric power (in billions of kilowatt hours) for street and highway lighting in the United States for the period 1941 to 1966:

Year	Sales	Year	Sales
1941	2.1	1954	4.1
1942	2.1	1955	4.4
1943	2.1	1956	4.7
1944	2.2	1957	5.1
1945	2.2	1958	5.5
1946	2.3	1959	5.9
1947	2.4	1960	6.1
1948	2.5	1961	6.8
1949	2.7	1962	7.4
1950	3.0	1963	7.7
1951	3.3	1964	8.3
1952	3.5	1965	8.8
1953	3.8	1966	9.2

(a) Fit a parabolic trend to these data by the method of least squares.
(b) Use the equation obtained in (a) to calculate the trend values for the years 1941 through 1966. (Follow the suggested procedure of Exercise 1 above.)
(c) Plot the original data as well as the trend values on arithmetic paper. Also, draw a smooth curve through the points representing the trend values to indicate the parabolic trend.

4. Use the method of Section 18.6 to fit an exponential trend to the electric power data of Exercise 3, leaving the equation in the logarithmic form. Calculate the trend values for 1941 and 1966 and plot them together with the original data on semi-log paper. Draw a straight line through these two trend values to indicate the trend.

5. Use the procedure suggested in Exercise 2 above to calculate the trend values for the years 1941 through 1966 from the equation obtained in Exercise 4. Plot these trend values on the chart constructed in Exercise 3 and compare the parabolic and exponential trends. Also, find the average annual rate of growth in the sales of electric power for the time period considered.

6. The following are the sales of electric power in the United States for residential or domestic purposes (billions of kilowatt hours) for the period 1941 to 1966: 25.1, 26.9, 28.6, 31.3, 34.2, 38.6, 44.2, 51.0, 58.1, 70.1, 80.5, 90.5, 101.2, 113.0, 125.4, 139.0, 152.6, 164.8, 180.3, 196.4,

209.0, 226.4, 241.7, 262.0, 281.0, 306.6. (Source: *Business Statistics 1967*). Use (18.3.1) to find the least squares trend line. Fit a parabolic trend to the data. Using the trend equation, calculate the trend values for the years 1941 through 1966 and plot these trend values together with the original data on arithmetic paper. Draw a smooth curve through the points representing the trend values to indicate the parabolic trend.

7. Fit an exponential trend to the electric power data of Exercise 6 above, leaving the equation in logarithmic form. Find the trend values for 1957 and 1966 and plot them together with the original data on semi-log paper. Draw a straight line through the 1957 and 1968 trend values to indicate the trend. Also, find the average annual rate of growth in electric power for the time period considered.

18.8 The Smoothing of Time Series

In the preceding sections we devoted our attention to fitting trend curves by the use of mathematical equations. Although such techniques often produce valuable results, it is not always necessary or, in fact, wise to describe the over-all movements of time series by means of specific curves. In problems in which we are interested mainly in the general "behavior" of a series, namely, the general pattern of its growth, we may well obtain an adequate description by using what is called a *moving average*. A *moving average* is an artificially constructed time series in which each annual (monthly, daily, or hourly) figure is replaced by the average, the *mean*, of itself and values corresponding to a number of preceding and succeeding periods. For instance, in a *three-year moving average* each annual figure is replaced by the mean of itself and those of the immediately preceding and succeeding years. In a *five-year moving average* each annual figure is replaced by the mean of itself, those of the two preceding years, and those of the two succeeding years.

Moving averages need not be limited to three or five years. Their period can be any number of years (months, and so forth), and it is true in general that if we average over longer periods we will get a smoother curve. If we average over an *even* number of periods, say, 4 years or 12 months, we run into the difficulty that the moving averages will have to be plotted *between* successive years or months. (For instance, the average of figures corresponding to 1963, 1964, 1965, and 1966 will have to be plotted half-way between 1964 and 1965.) It is customary in that case to "center" these values by calculating a subsequent *2-period moving average*. Since this method belongs to the more advanced subject of *weighted moving averages*, we shall limit our discussion here to the case where the number of periods over which we average is *odd*. The calculation of a centered *12-month moving average* is illustrated in Section 19.5, where it is needed for other computations.

The central problem in the use of a moving average is that of choosing an appropriate period. Since the purpose of a moving average is to eliminate as much as possible the fluctuations of a series that tend to distract from the trend we want to describe, we must always choose its period with this objective in mind. A proper choice of the period is particularly important if a moving average is to eliminate seasonal or other cyclic patterns. *Cycles that are of uniform length and amplitude (height) can be eliminated completely by making the period of*

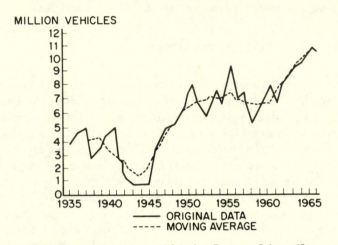

FIGURE 18.11. Motor Vehicle Factory Sales. (Source: Statistical Abstracts of the United States.)

the moving average equal to (or a multiple of) that of the cycle. It is for this reason that we shall use a centered 12-month moving average in Section 19.5 where we will need a series that is free of seasonal variations.

To illustrate the work involved in the calculation of a moving average, let us compute a five-year moving average for annual data on factory sales of automobiles and trucks in the United States from 1935 to 1966. In the table shown below, the first column contains the total annual factory sales in millions of vehicles. The next column contains the *five-year moving totals*, which for any given year consist of that year's sales *plus* those of the two preceding and the two succeeding years. The last column contains the desired five-year moving averages and they are obtained by dividing the corresponding entries of the preceding column by 5 or by multiplying them by 0.20, which is the same. In the following table all figures are in *millions of vehicles:*

	Annual Sales	Five-Year Moving Totals	Five-Year Moving Averages
1935	3.9		
1936	4.5		
1937	4.8	19.1	3.8
1938	2.3	19.7	3.9
1939	3.6	20.0	4.0
1940	4.5	16.2	3.2
1941	4.8	14.6	2.9
1942	1.0	11.7	2.3
1943	.7	7.9	1.6
1944	.7	6.2	1.2
1945	.7	10.0	2.0
1946	3.1	14.6	2.9
1947	4.8	20.2	4.0
1948	5.3	27.5	5.5
1949	6.3	31.2	6.2
1950	8.0	31.9	6.4
1951	6.8	33.9	6.8
1952	5.5	34.2	6.8
1953	7.3	35.4	7.1
1954	6.6	35.5	7.1
1955	9.2	37.2	7.4
1956	6.9	35.0	7.0
1957	7.2	35.1	7.0
1958	5.1	33.8	6.8
1959	6.7	33.6	6.7
1960	7.9	34.6	6.9
1961	6.7	38.6	7.7
1962	8.2	41.2	8.2
1963	9.1	44.4	8.9
1964	9.3	48.0	9.6
1965	11.1		
1966	10.3		

A minor disadvantage of this method is that some values are lost at each end of the series, *one* for a three-year moving average, *two* for a five-year moving average, *three* for a seven-year moving average, and so on. Generally, this is of no consequence unless the series is very short, but it becomes serious when all the trend values are needed for further computation.

The practical effect of a moving average can readily be seen from Figure 18.11, in which we have plotted the five-year moving average as well as the original data. The moving average has substantially reduced the fluctuations of the series and given it a much smoother appearance.

EXERCISES

1. Compute a three-year moving average for the factory sales of automobiles and trucks given above. Plot the original data as well as the moving average on arithmetic paper.

2. Compute a seven-year moving average for the factory sales of automobiles and trucks given on page 425. Also, plot the original series as well as the moving average on arithmetic paper.
3. Calculate a five-year moving average for the following data on the average monthly production of asphalt in the United States in millions of barrels: (Source: *Business Statistics 1967*)

Year	Production	Year	Production
1941	3.0	1954	6.2
1942	2.9	1955	6.9
1943	3.1	1956	7.6
1944	3.2	1957	7.1
1945	3.3	1958	7.4
1946	3.7	1959	8.1
1947	4.1	1960	8.2
1948	4.3	1961	8.5
1949	4.1	1962	9.1
1950	4.9	1963	9.3
1951	5.5	1964	9.6
1952	5.8	1965	10.3
1953	6.0	1966	10.8

Plot the moving average as well as the original series on arithmetic paper.
4. Calculate a three-year moving average for the data of Exercise 3 and plot it together with the original series on arithmetic paper.
5. Compute a seven-year moving average for the slaughter (federally inspected) of sheep and lambs (in millions of animals) for the years 1939 to 1966. (Source: U.S. Dept. of Agriculture, Statistical Reporting Service) 17.2, 17.4, 18.1, 21.6, 23.4, 21.9, 21.2, 19.9, 16.7, 15.3, 12.1, 11.8, 10.1, 12.7, 14.3, 14.1, 14.4, 14.2, 13.2, 12.4, 13.5, 14.0, 15.0, 14.7, 14.0, 12.9, 11.7 and 11.6.

18.9 Forecasting

In the first chapter of this book we distinguished between *descriptive* and *inductive* statistics, observing that descriptive statistics involved methods that, essentially, do not go beyond the data with which we start and that inductive statistics involved generalizations, predictions, estimations, and decisions. If this distinction is clearly understood, it will be apparent that the fitting of trends merely *describes* the movements of a series and that the work of this chapter so far belongs to the domain of descriptive statistics. Obviously, fitting a trend to describe past data is, in itself, not enough for a businessman who must go beyond the limits of past experience, beyond what has been observed, to make inferences as to what might conceivably happen in the future. As we said on page 383, a businessman must *project* past experience: he must make predictions or forecasts of future levels of business activity and he must base these forecasts, at least in part, on what has happened in the past.

Having found the equation of a trend, the obvious thing that suggests itself is to use it to *extrapolate* and, thus, forecast future events. In its technical sense, to "extrapolate" means to estimate a value which lies beyond the range of values on the basis of which an equation was originally obtained. We would be extrapolating if we substituted $x = 15$ into the equation given on page 406 and, getting $y' = 13.62 + 0.87(15) = 26.67$, predicted that in 1977 the total profits of manufacturers of nondurable goods will be \$26.67 billion.

Extrapolating from trends is a necessary, though highly speculative, procedure whose success depends on many factors. The basic question we always have to ask is whether the forces that have operated in the past will continue to operate in the future and to operate, furthermore, in the same way. In times of rapid change such as the present, the answer to this question is that in many, if not in most, instances they probably will not. This being the case, then, the crucial problem is how to assess the impact of such factors as scientific research, the development of new products, changing social patterns, and political upheavals, on the growth of economic phenomena with which we are concerned.

Taking a conservative view, let us say that a forecast based on an extrapolated curve is a starting point, a bench mark, from which to proceed to a final prediction. Of course, the more distant the future we are trying to predict, the more likely it is that changes will become operative and the more difficult it will be to evaluate their effect. In any case, extrapolations based on trend equations provide estimates which, if necessary or desirable, may be modified in accordance with one's own interpretation of the net effect of various factors.

The over-all problem of forecasting involves a number of disciplines other than statistics, and it is for this and other reasons that it cannot be treated extensively in an introductory course in statistics. To summarize our discussion, let us say that statistics, at least in its current state of development, comes somewhere between being of relatively little value and providing a complete solution of the problem of economic forecasting. If it is used intelligently—with full appreciation of its possibilities and in full recognition of its limitations—statistics can provide the businessman with invaluable aids in reaching well-informed and wise decisions.

EXERCISES

1. Use the equation of the least squares line obtained in Exercise 1 on page 411 to forecast the number of employees in nonagricultural establishments in the United States in 1975.

2. Use the equation of the least squares line obtained in Exercise 3 on page 411 to forecast the index of the volume of construction in 1974.

3. Use the equation of the least squares line obtained in Exercise 5 on page 411 to forecast the 1975 shipments of the Steel Industry.

4. A study of enrollment figures at an eastern liberal arts college yielded the following least squares trend equation: $y' = 1,600 + 200x$ (origin 1960; x units, 1 year; y, total number of students enrolled annually). The physical facilities of the college can serve a maximum of 3,600 students enrolled annually.

   (a) What is the average annual increase in students enrolled?

   (b) In what year will the college's expected enrollment reach its present physical capacity of 3,600 students?

   (c) Estimate enrollment for 1975.

5. Using the parabolic trend equation obtained on page 414, forecast the number of nonimmigrants admitted into the U.S. in 1976.

6. Using the exponential trend equation obtained on page 418, forecast the number of nonimmigrants admitted into the U.S. in 1976. Compare this estimate with the one obtained in Exercise 5 and discuss the "reasonableness" of these estimates.

BIBLIOGRAPHY

Elementary discussions of curve fitting may be found in many elementary texts including those by Mills; Croxton, Cowden, and Klein; and Spurr and Bonini, referred to in the Bibliography on page 399.

Somewhat more advanced treatments are given in

Kendall, M. G., *The Advanced Theory of Statistics*, Vol. II, 2nd ed.   London: Griffin, 1948, Chap. 29.

Rosander, A. C., *Elementary Principles of Statistics*.   New York: Van Nostrand, 1951, Chap. 32.

Whittaker, E., and Robinson, G., *The Calculus of Observations*, 4th ed.   London: Blackie, 1948, Chap. 5.

Yule, G. U., and Kendall, M. G., *Theory of Statistics*, 12th ed.   London: Griffin, 1940, Chap. 7.

For an explanation of orthogonal polynomials and their use in curve fitting, see Chapter 32 of the book by Rosander mentioned above and

Anderson, R. L., and Bancroft, T. A., *Statistical Theory in Research*.   New York: McGraw-Hill, 1952, Chap. 6.

Anderson, R. L., and Houseman, E. E., *Tables of Orthogonal Polynomial Values Extended to N = 104*.   Ames, Iowa: Iowa State College, 1942.

Two useful books on business forecasting are

Brown, R. G., *Smoothing, Forecasting, and Prediction of Discrete Time Series*.   Englewood Cliffs, N.J.: Prentice-Hall, Inc., 1962.

Forecasting Sales, Studies in Business Policy No. 106.   New York: National Industrial Conference Board, Inc., 1963.

CHAPTER 19

Time Series Analysis: Seasonal and Cyclical Variation

19.1 Seasonal Variation

In the previous chapter we discussed the measurement of trend, one of the four traditional components considered in the analysis of a time series. Although the trend may be of major concern in the statistical analysis of a series, the measurement of seasonal movements and cyclical patterns are often of equal importance and in this chapter we shall study the statistical treatment of these latter two types of variation.

As we remarked earlier, among the four components contributing to the fluctuations of a time series, seasonal variation is often the easiest to understand. The examination of almost any series of monthly data shows movements within the series which seem to occur year after year with a fair degree of regularity and the same is true also for series recorded on a weekly, daily, or hourly basis. Since the statistical problems are essentially the same, we shall limit our discussion of seasonal variation to annually repeating patterns observed in monthly data.

For a businessman concerned with the practical problem of operating his business, an understanding of seasonal patterns is absolutely essential. Otherwise, how would he know how to interpret, say, a 20 per cent drop in new orders from May to June or a 12 per cent increase from August to September. In the first case he might conclude that he is losing business and, in an attempt to reverse this situa-

429

tion, do something rash such as firing a good salesman, spending a lot of money on unnecessary advertising, or buying new equipment with the hope of reducing cost. If the 20 per cent drop is only part of an annually recurring pattern which the businessman should really have expected, the above actions might easily have dire consequences. Similarly, if the 12 per cent increase in new orders from August to September is only part of a seasonal pattern, the results might be equally disastrous if, incorrectly interpreted, they led to complacency, an unwarranted cut in the advertising budget, or other uncalled-for actions.

One need not be a statistician or an economist to *understand* seasonal variation, adjust to it, or perhaps attempt to change its pattern. The owner-operator of a small grocery store may not have any training in statistics or economics and he may not think of himself as an analyst, but the truth is that in order to operate his business intelligently he must go through a process much like that which we shall study formally in this chapter. If he doubts that he is acting as an analyst, he has only to ask himself why he orders more fresh vegetables for delivery on Friday than on Tuesday and why he employs extra help on Friday and Saturday rather than on Wednesday. He does these things because he has observed certain patterns in the past and assumes that they will continue to operate in the future. Of course, he could ignore these patterns, ordering, for example, the same amount of fresh vegetables for each day of the week, but the resulting spoilage on some days and shortage on others could easily lead to his ruin. So, he does the intelligent thing and adjusts his operations to the patterns imposed on him by seasonal variation.

In many cases there is little that can be done about changing seasonal patterns. For example, the weather in some parts of the country, a seasonal influence over which nobody has much control, forces changes in activity in construction, the sale of ice cream, the operation of drive-in theaters, and the like. In other cases, however, an understanding of seasonal patterns may serve to minimize or offset their effects. For instance, a canner of products which are delivered to him only at certain times of the year may hold his labor force together and take up the slack by adding other products to his line.

Some businesses faced with problems created by seasonal demands for their goods and services have attempted (in many cases successfully) to change the public's buying habits by means of special sales, extensive advertising, and other devices. They thus offset seasonal patterns by offering special bargains or otherwise creating demand during what would ordinarily be slack periods.

The practical benefits of studying seasonal variation are numerous and varied. It leads to more realistic planning or scheduling of production, purchasing, inventory, personnel, advertising, and such, and as we pointed out above, to more balanced operations throughout the year. Beyond this, the reader will recall that seasonal variations are of interest because they must be removed from a series before we can arrive at measurements of cyclic changes.

19.2 Some Preliminary Considerations

To obtain a statistical description of a pattern of seasonal variation we will have first to free our data from the effects of trend, cycles, and irregular variation. Once these other components have been eliminated, we can calculate, in index form, a measure of seasonal variation which is usually referred to as a *seasonal index*. *Since there are various ways of eliminating these other components, it is worth noting that there are equally many ways of defining or calculating a seasonal index.*

For monthly data, a seasonal index consists of 12 numbers, one for each month, which are given as percentages of their average; in other words, each month is represented by a figure expressing it as a percentage of the average month. For instance, if a seasonal index for January is 69, this means that January sales, orders, purchases, or whatever our data happen to be, are 69 per cent of those of the average month.

The problem of constructing seasonal indexes would be vastly simplified if a grocer sold the same percentage of the week's groceries *every* Monday, Tuesday, . . . , week in and week out, year after year, and if department store sales were always, say, 18 per cent below the monthly average in January and 40 per cent above in December. If this were the case, we could construct a seasonal index on the basis of a single week or a single year. Of course, in reality some Fridays are better than others, some Decembers are better than others, even though the grocer's sales may always be highest on Friday and department store sales may always be highest in December. This is due in part to the presence of other kinds of variation, trend, cycles, and irregular variation, and in part, perhaps, to the fact that the seasonal patterns themselves are changing.

In view of the discussion of the previous paragraph let us differentiate briefly between *specific* and *typical* seasonal patterns and also between *constant* and *changing* patterns of seasonal variation. The term "specific" implies that the pattern has been observed for *one* period, say, one year, whereas "typical" implies that the pattern is in some sense an average of the patterns observed over a number of years. *Clearly, to be of any value, a seasonal index must describe typical rather than specific patterns of variation.*

The distinction between constant and changing seasonal patterns is, as these terms would seem to imply, that in the first case the pattern remains the same year after year while in the second case it changes with time. Changing patterns have been observed in the sales of automobiles and gasoline in this country and in many other series. They may result from extraneous forces over which no one has any control or they may, in fact, be brought about by the efforts of a firm or an industry.

In this chapter we shall concentrate on the problem of measuring constant seasonal patterns—patterns whose character does not change with the passage of time. The detection and measuring of changing seasonal patterns causes different and considerably more difficult problems. A reference to a discussion of these problems is given in the Bibliography on page 457.

19.3 The Method of Simple Averages

Let us first illustrate the construction of a seasonal index by a method which, though it lacks in refinement and is subject to criticism, is quite simple and uninvolved. Although it is in many respects

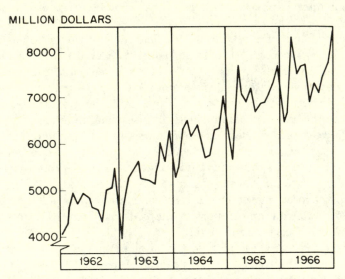

MILLION DOLLARS

FIGURE 19.1. Monthly Purchases of Regular Ordinary Life Insurance in the United States. (Source: Life Insurance Fact Books.)

less desirable than the other methods which we shall treat in this chapter, it presents a good point to begin our study of seasonal variation.

To demonstrate the so-called *method of simple averages*, let us consider the problem of describing the seasonal pattern of the monthly purchases of regular ordinary life insurance in the United States on the basis of monthly figures for the five year period 1962–1966. These figures are taken from the 1964–1967 Life Insurance Fact Books and Figure 19.1 illustrates their annually recurring pattern. This five-year period is long enough to bring out what we want to illustrate, and, in general, a five-year period may be long enough to serve even in some practical situations. In others it may be entirely too short. Of course, the longer the period, the more information we have about the typical pattern we are trying to describe.

In the following table the first five columns contain the actual monthly purchases of regular ordinary life insurance in the United States in millions of dollars. The other columns contain the calculations needed to arrive at a seasonal index.

	1962	1963	1964	1965	1966	(6)	(7)	(8)	(9)
January	4023	4230	5263	5681	6428	5125	0	5125	88.2
February	4356	4690	5551	6113	6805	5503	59	5444	93.7
March	4977	5338	6395	7560	8260	6506	118	6388	109.9
April	4772	5481	6581	7104	7567	6301	177	6124	105.4
May	4924	5664	6173	6911	7692	6273	236	6037	103.9
June	4812	5267	6431	7249	7731	6298	295	6003	103.3
July	4641	5246	6085	6697	6886	5911	354	5557	95.6
August	4635	5196	5749	6880	7396	5971	413	5558	95.7
September	4229	5147	5780	6930	7136	5844	472	5372	92.5
October	5042	6051	6366	7152	7497	6427	531	5891	101.4
November	5061	5609	6388	7377	7787	6444	590	5854	100.7
December	5526	6246	7046	7831	8473	7024	649	6375	109.7
								69728	1200.0

The first thing we do is average the respective figures for each month; in other words, we calculate the *mean* of the five January figures, the five February figures, and so on. For instance, for January we get

$$\frac{4023 + 4230 + 5263 + 5681 + 6428}{5} = \frac{25,625}{5} = 5,125$$

and this is the first entry shown in Column (6). The purpose of this is to average out, or eliminate from the data, the irregular and cyclical influences which tend to distort the seasonal pattern we seek to reveal. Actually, this is rarely, if ever, accomplished completely. The mean is a good measure to use for averaging out irregular variations which, by their nature, are random, but it generally falls short of eliminating cyclical movements. If these latter movements are to be effectively eliminated, the cycles must be of about the same duration and of uniform amplitude, and the data chosen for study must extend over a long enough period so that several whole cycles are included. In general, even though the period under study may be sufficiently long, the requirements of cycles of equal duration and uniform amplitude are seldom met. Hence, the averaging can, at best, be expected to *reduce* these distracting movements rather than *eliminate* them completely. (We might also add that, since the period under consideration in our example extends over as few as 5 years, it is really impossible to say whether the upward movement in Figure 19.1 is a trend or whether it is part of a long-range cycle. Since the purpose of this example is to illustrate the method of simple averages, let us say that it is a trend.)

The averages in Column (6) may be looked upon as an artificially constructed series which is to a greater or lesser degree free of irregular and cyclical influences. Hence, we shall attribute its movements to the other two components, seasonal variation and trend. To isolate seasonal variation, our next step will be to eliminate the trend.

The need for removing the trend is apparent when we consider that in the presence of an upward trend each month's figures are apt to be higher than those of the preceding month, even if there were no seasonal pattern. Unless we make an appropriate adjustment, one type of movement might thus easily be confused with the other. In the method of simple averages the elimination of trend is, as we shall see, accomplished by *subtraction*.

To measure the trend, we fit a least squares line to the average monthly totals for the given five years. (These figures are the means of the first five columns of the table shown above.) Leaving it to the reader to check, let us merely state that the *monthly trend increment of the average monthly purchases of ordinary life insurance* thus obtained is 59 rounded to the nearest million dollars.

Due to the effect of this trend, February figures are on the average 59 million dollars higher than January figures, March figures are on the average $2 \cdot 59 = 118$ million dollars higher than January figures, April figures are on the average $3 \cdot 59 = 177$ million dollars higher than January figures, . . . , and December figures are on the average $11 \cdot 59 = 649$ million dollars higher than January figures. These multiples of the monthly trend increment are shown in Column (7).

To eliminate trend from the averages of Column (6), we keep the January average fixed, subtract 59 from the February average, 118 from the March average, and, in general, the entry of Column (7) from the corresponding entry of Column (6). The adjusted monthly averages are shown in Column (8) and we have now succeeded, to whatever extent this method is effective, in eliminating the disturbing influences of trend, cyclical, and irregular variations from our data.

To construct a seasonal index, which, after all, is nothing but a *descriptive measure which compares each month's figures with those of the average month*, we can now write the adjusted averages of Column (8) as percentages of *their* average, namely, as percentages of the total of column (8) divided by 12. Dividing each entry of Column (8) by

69,728/12 = 5,810.7 and multiplying by 100, we finally get the values of Column (9), which constitute the desired *seasonal index*.

This index tells us, for example, that *typical* January purchases are 88.2 per cent of those of the average month, *typical* March purchases are 109.9 per cent of those of the average month, and *typical* December purchases are 109.7 per cent of those of the average month. Since all of the values of a seasonal index are given as percentages of their average, their sum must equal 1,200. If the sum does not equal exactly 1200, however, this can be attributed to rounding off, and must be adjusted (see page 439).

Having seen how trend, cyclical, and irregular variations can be eliminated from a series, the reader may have realized that there must be alternate ways of doing the same thing. As a matter of fact, there are several, each of them designed to measure seasonal variation in a different way. In the next two sections we shall take two other series of data to illustrate two alternate methods which are generally looked upon as being more refined.

One of the main criticisms raised against the method illustrated in this section is that it assumes the *wrong model* or, in other words, that it is based on the wrong theoretical assumptions. As we remarked on page 384, the theory currently held in esteem by statisticians and economists is that the effects of the four components are *multiplicative*. This means that the over-all fluctuations of a time series are the *products* of the effects of trend, cyclical, seasonal, and irregular variations. Hence, to eliminate trend, it must be divided out instead of being subtracted. How this is done will be illustrated in the next two sections.

EXERCISES

1. The following are data on the farm production of eggs in the United States in millions of cases: (Source: *Business Statistics, 1967*)

	1963	1964	1965	1966
January	14.5	15.0	15.6	15.3
February	13.5	14.6	14.3	14.0
March	15.9	16.2	16.1	16.0
April	15.8	15.9	15.8	15.8
May	16.0	16.2	16.3	16.2
June	14.9	15.1	15.3	15.3
July	14.7	15.0	15.3	15.2
August	14.3	14.6	14.8	15.0
September	13.7	14.1	14.3	14.7
October	14.3	14.7	14.9	15.5
November	14.2	14.5	14.6	15.4
December	14.8	15.3	15.2	16.2

Using the method of simple averages and a monthly trend increment of .018 million, calculate a seasonal index for the farm production of eggs in the United States. Round each figure to the nearest million after subtracting, as in Column (8) on page 433, the appropriate multiples of the trend increment.

2. Use the data on page 433 to fit a least squares trend line to the *average monthly purchases* of regular ordinary life insurance in the United States for the given five years. Verify that the monthly trend increment is 59 million and that the trend equation can be written as $y' = 6135.2 + 59x$ (*origin, January, 1964, x units, 1 month; y, average monthly purchases in millions of dollars*).

3. The following are data on the employed civilian labor force in the United States in thousands: (Source: *Business Statistics 1967*)

	1962	1963	1964	1965	1966
January	64,215	65,168	66,468	68,235	70,339
February	64,872	65,519	67,197	68,690	70,677
March	65,421	66,329	67,695	69,385	71,084
April	65,957	67,240	68,947	70,220	72,077
May	67,066	67,984	69,952	71,298	72,620
June	67,852	68,844	70,448	72,278	74,038
July	67,849	69,225	70,839	73,093	74,655
August	68,096	69,052	70,676	72,695	74,666
September	67,620	68,567	69,849	71,408	73,247
October	67,850	68,964	70,147	72,112	73,743
November	67,046	68,471	69,982	71,824	73,995
December	66,585	67,791	69,543	71,819	73,599

Using the method of simple averages and a monthly trend increment of 130.9 *thousand*, calculate a seasonal index for the employed civilian labor force in the United States.

4. The following are the number of persons patronizing commercial parking lots in a certain metropolitan area for three consecutive weeks:

	First Week	Second Week	Third Week
Monday	16,991	16,807	16,997
Tuesday	17,039	17,063	17,155
Wednesday	16,855	16,915	17,303
Thursday	16,741	16,795	16,847
Friday	17,443	17,469	17,631
Saturday	7,053	8,555	7,957
Sunday	2,237	1,929	2,081

Using the method of simple averages and a daily trend increment of 114 patrons calculate a seasonal index for the number of persons patronizing commercial parking lots each day of the week.

19.4 The Ratio-to-Trend Method

The method of calculating a seasonal index which we shall discuss in this section goes by the name of the *ratio-to-trend* or *percentage-of-*

trend method.   It is relatively simple and yet an improvement over the method of simple averages treated in the preceding section.   That method attempted to average out irregular and cyclical variation and then remove the trend by subtraction.   The present method reverses this process, eliminating first the trend by *division* and then averaging out irregular and cyclical variations.

THOUSAND DOLLARS

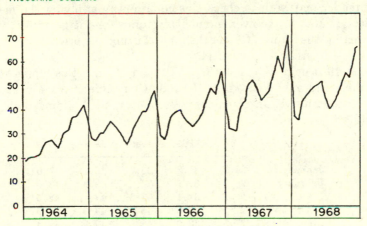

FIGURE 19.2.   Monthly Sales of the Branch Department Store.

In the example which follows we shall compute an index describing the seasonal sales pattern of a branch department store which began operations in the late 1960s in a new suburban shopping center near a large city.   The following are total monthly sales, rounded to the nearest *thousand dollars,* for the period 1964–1968.   (The source of these data is confidential.)

	1964	*1965*	*1966*	*1967*	*1968*
January	19	27	29	33	36
February	20	26	28	32	34
March	20	30	37	41	43
April	22	30	39	44	47
May	27	35	40	50	49
June	27	32	35	52	51
July	24	26	33	44	40
August	30	30	35	45	43
September	32	35	40	53	49
October	36	39	49	63	57
November	37	38	46	54	55
December	42	47	56	71	66

To measure the seasonal pattern which is apparent from Figure 19.2, we shall first fit a trend line, find its equation, and use it to calculate the trend values corresponding to the various months. Although this method is not restricted to *linear* trends, let us use the least squares trend line whose equation is

$$y' = 25.74 + 0.455x$$

(*origin, January, 1964 x units, 1 month;
y, average monthly sales in thousands of dollars*).

The reader may wish to check this equation by applying the method of Section 18.3 to the average monthly totals for the given five years, converting the x units to months, and shifting the origin to January, 1964 (see Exercise 4 on page 440).

Substituting $x = 0, 1, 2, 3, \ldots$, and 59 into this equation, we get the trend values shown in the following table: (These values are easy to obtain with a machine. We simply keep adding 0.455 and round off later on.)

	1964	1965	1966	1967	1968
January	25.7	31.2	36.7	42.1	47.6
February	26.2	31.7	37.1	42.6	48.0
March	26.7	32.1	37.6	43.0	48.5
April	27.1	32.6	38.0	43.5	48.9
May	27.6	33.0	38.5	43.9	49.4
June	28.0	33.5	38.9	44.4	49.9
July	28.5	33.9	39.4	44.9	50.3
August	28.9	34.4	39.8	45.3	50.8
September	29.4	34.8	40.3	45.8	51.2
October	29.8	35.3	40.8	46.2	51.7
November	30.3	35.8	41.2	46.7	52.1
December	30.7	36.2	41.7	47.1	52.6

The next step is to divide the original data month by month by the corresponding trend values and to multiply these *ratios* by 100. We thus arrive at the following *percentages of trend:*

	1964	1965	1966	1967	1968
January	73.9	86.5	79.0	78.4	75.6
February	76.3	82.0	75.5	75.1	70.8
March	74.9	93.5	98.4	95.3	88.7
April	81.2	92.0	102.6	101.1	96.1
May	97.8	106.1	103.9	113.9	99.2
June	96.4	95.6	90.0	117.1	102.2
July	84.2	76.7	83.8	98.0	79.5
August	103.8	87.2	87.9	99.3	84.6
September	108.8	100.6	99.3	115.7	95.7
October	120.8	110.5	120.1	136.4	110.3
November	122.1	106.1	111.7	115.6	105.6
December	136.8	129.8	134.3	150.7	125.5

These values are now free of trend and the problem that remains is to free them also of irregular and cyclical movements.† This may be done by separately averaging the figures given for the five Januaries, Februaries, and so forth, with any one of the usual measures of central location, for instance, the *median* or the *mean*.

If the data are examined month by month it is sometimes possible to ascribe a definite cause to unusually low or high values. When such outliers are found to be associated with irregular variations (extremely bad weather, an earthquake, and the like) they may be cast out and the mean of the remaining items is referred to as a *modified mean*. Since such scrutiny of the data requires considerable knowledge of prevailing conditions and is to a large extent subjective, it is often desirable to use the *median* which is generally not affected by one very high or very low value.

To complete the calculation of the seasonal index, let us therefore use the median to average the values obtained in the last table. Although this is not necessary, it is often convenient to rank each month's values as we have done in the first five columns of the following table:

	1	2	Rank 3 (Median)	4	5	Seasonal Index (6)
January	86.5	79.0	78.4	75.6	73.9	79.6
February	82.0	76.3	75.5	75.1	70.8	76.6
March	98.4	95.3	93.5	88.7	74.9	94.9
April	102.6	101.1	96.1	92.0	81.2	97.5
May	113.9	106.1	103.9	99.2	97.8	105.5
June	117.1	102.2	96.4	95.6	90.0	97.9
July	98.0	84.2	83.8	79.5	76.7	85.1
August	103.8	99.3	87.9	87.2	84.6	89.2
September	115.7	108.8	100.6	99.3	95.7	102.1
October	136.4	120.8	120.1	110.5	110.3	121.9
November	122.1	115.6	111.7	106.1	105.6	113.4
December	150.7	136.8	134.3	129.8	125.5	136.3
			1,182.2			

As we pointed out earlier, the seasonal index for each month should be expressed as a percentage of the average month and, consequently, the sum of the 12 values must equal 1,200. Since the 12 medians totaled 1,182.2, we adjust for this by multiplying each by 1,200/1,182.2 = 1.015, thus getting the final *seasonal index* shown in Column (6) above.

---

† Let us point out again that since we are using as few as 5 years it is actually impossible to differentiate between a trend and part of a long-range cycle. However, if the method were applied to a longer period it would be appropriate to refer to the line as a trend, and we shall do so in our example.

This index is to be interpreted as before. It shows that *typical* January sales are 79.6 per cent of those of the average month, *typical* August sales are 89.2 per cent of those of the average month, *typical* December sales are 136.3 per cent of those of the average month, and so on. In Exercise 5 below the reader will be asked to find an alternate seasonal index for these data by taking the means of the percentages of trend instead of their medians.

EXERCISES

1. Use the ratio-to-trend method, the data on page 433, and the trend equation of Exercise 2 on page 436, to calculate a seasonal index for the monthly purchases of regular ordinary life insurance in the United States. Use the median to average the percentages of trend obtained for each month.

2. Use the ratio-to-trend method, the data of Exercise 1 on page 435, and the trend equation $y' = 14.68 + .018x$, origin, January, 1963; x units, 1 month; y, millions of cases of eggs to calculate a seasonal index for the monthly production of eggs in the United States. Use the mean to average the four percentages of trend obtained for each month.

3. Use the ratio-to-trend method, the data of Exercise 3 on page 436, and the trend equation $y' = 65,688 + 131x$ (origin, January, 1962, x units, 1 month; y, thousands of persons), to calculate a seasonal index for the employed civilian labor force in the United States. Use the median to average the five percentages of trend obtained for each month.

4. Verify the trend equation given on page 438 by fitting a least squares line to the *average monthly sales* of the given 5 years, changing the x units, and shifting the origin.

5. Recalculate the seasonal index for the example of this section, using the mean rather than the median to average the percentages of trend obtained for each month.

6. Use the data on pages 442 and 443 to fit a least squares line to the total annual production of still wines in the United States for the years 1962 to 1966. Change the origin of this equation to January, 1962, the x units to 1 month, and y to monthly data. Then use this equation to calculate a seasonal index for production of still wines by the ratio-to-trend method. Use the median to average the percentages of trend obtained for each month.

19.5 The Ratio-to-Moving Average Method

The method we shall study in this section goes by the names of *ratio-to-moving average* or *percentages of moving average;* it is probably the most generally satisfactory way devised for measuring seasonal variation. It is relatively simple, although the arithmetic may seem involved, and it is a solid improvement over the method of simple averages, which, the reader will recall, *attempted* to average out irregular and cyclical variations and then remove the trend by subtraction. It is also superior to the ratio-to-trend method inasmuch as (a) there is no danger in mistaking seasonal variation for trend, (b) moving averages allow for more flexibility in case of non-linear trends, and (c) in the presence of short, say, 5 or 10 year, cycles, moving averages are easier to use to describe trend and cyclical variation than specific mathematical curves.

To illustrate these points, let us briefly outline how the ratio-to-moving average method works. We begin by calculating a centered 12-month moving average which eliminates most of the seasonal variation, irregular variation, and sometimes also some of the cyclical variation from our data. We thus arrive at an estimate of trend and

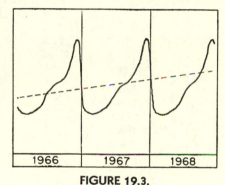

FIGURE 19.3.

cyclical variation which in subsequent steps takes the place of the trend used in the method of the previous section. In other words, once the moving average has been found, we divide our original data by the corresponding values of the moving average and, finally, average the values obtained for the different months.

How this avoids mistaking seasonal fluctuations for a trend is illustrated in Figure 19.3, in which a seasonal pattern is repeated three times. Even though there is actually no trend, the dotted least squares line shows a slight upward slope and if we used the method of Section 19.4 we would be mistaking part of the seasonal pattern for trend. The logic underlying the use of a twelve-month moving average is that it completely eliminates any patterns which regularly repeat year after year. Of course, in actual practice seasonal patterns will vary somewhat from year to year and we shall have to realize that a twelve-month moving average eliminates most, not necessarily all, of the seasonal variation. Moreover, it eliminates most of the irregular variations, and it may, thus, be looked upon as describing the trend and cyclical components of a series.

PRODUCTION OF STILL WINES IN THE UNITED STATES
(MILLIONS OF WINE GALLONS)

		(1)	(2)	(3)	(4) Centered	(5)
		Production in Millions of Wine Gallons	12-Month Moving Total	2-Period Moving Total	12-Month Moving Average (Col. 3÷24)	Percentages of 12-Month Moving Average
1962	January	3.3				
	February	2.7				
	March	2.5				
	April	2.1				
	May	2.6				
	June	1.7				
	July	1.1	189.2	378.9	15.8	7.0
	August	6.1	189.7	380.1	15.8	38.6
	September	59.7	190.4	381.2	15.9	375.5
	October	86.9	190.8	381.9	15.9	546.5
	November	14.8	191.1	382.0	15.9	93.1
	December	5.7	190.9	381.9	15.9	32.1
			191.0			
1963	January	3.8		383.0	16.0	23.8
	February	3.4	192.0	380.9	15.9	21.4
	March	2.9	188.9	363.5	15.1	19.2
	April	2.4	174.6	371.0	15.5	15.5
	May	2.4	196.4	399.5	16.6	14.5
	June	1.8	203.1	405.3	16.9	10.6
	July	2.1	202.2	403.3	16.8	12.5
	August	3.0	201.1	401.8	16.7	18.0
	September	45.4	200.7	401.9	16.7	271.9
	October	108.7	201.2	402.2	16.8	647.0
	November	21.5	201.0	401.4	16.7	128.7
	December	4.8	200.4	400.9	16.7	28.7
			200.5			
1964	January	2.7		399.7	16.7	16.2
	February	3.0	199.2	398.5	16.6	18.1
	March	3.4	199.3	412.3	17.2	19.8
	April	2.2	213.0	412.5	17.2	12.8
	May	1.8	199.5	390.8	16.3	11.0
	June	1.9	191.3	384.5	16.0	11.9
	July	.8	193.2	387.1	16.1	5.0
	August	3.1	193.9	387.8	16.2	19.1
	September	59.1	193.9	387.5	16.1	367.1
	October	95.2	193.6	389.3	16.2	587.7
	November	13.3	195.1	392.2	16.3	81.6
	December	6.7	196.5	393.6	16.4	40.9
			197.1			
1965	January	3.4		394.9	16.5	20.6
	February	3.0	197.8	396.4	16.5	18.2
	March	3.1	198.6	387.9	16.2	19.1
	April	3.7	189.3	397.0	16.5	22.4
			207.7			

PRODUCTION OF STILL WINES IN THE UNITED STATES
(MILLIONS OF WINE GALLONS)

		(1) Production in Millions of Wine Gallons	(2) 12-Month Moving Total	(3) 2-Period Moving Total	(4) Centered 12-Month Moving Average (Col. 3÷24)	(5) Percentages of 12-Month Moving Average
1965	May	3.2		438.1	18.2	17.6
	June	2.5	230.4	463.6	19.3	13.0
	July	1.5	233.2	470.4	19.6	7.7
	August	3.9	237.2	474.0	19.8	19.7
	September	49.8	236.8	473.1	19.7	252.8
	October	113.6	236.3	471.2	19.6	579.6
	November	36.0	234.9	469.6	19.6	183.7
	December	9.5	234.7	469.2	19.6	48.5
			234.5			
1966	January	7.4		469.0	19.5	37.9
	February	2.6	234.5	474.7	19.8	13.1
	March	2.6	240.2	503.5	21.0	12.4
	April	2.3	263.3	501.4	20.9	11.0
	May	3.0	238.1	458.1	19.1	15.7
	June	2.3	220.0	438.8	18.3	12.6
	July	1.5	218.8			
	August	9.6				
	September	72.9				
	October	88.4				
	November	17.9				
	December	8.3				

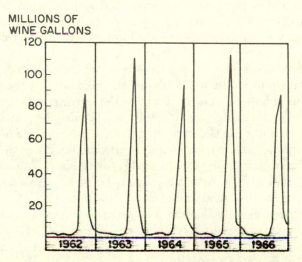

MILLIONS OF WINE GALLONS

FIGURE 19.4. Monthly Production of Still Wines in the United States (Millions of Wine Gallons). (Source: U.S. Treasury Department, I.R.S.)

In the example on pages 442–443 the ratio-to-moving average method is used to describe the seasonal pattern of production of still wines in the United States. The monthly figures shown in Column (1) below (and in Figure 19.4) are for the period 1962–1966, and they are expressed in millions of wine gallons. The source of the data is the United States Treasury Department, Internal Revenue Service.

Whereas the figures of the first column of the table on pages 442 and 443 are the actual production, Columns (2), (3), and (4) contain the calculations needed to find a centered 12-month moving average. The first figure in Column (2), 189.2, is the *sum* of the 12 month production for 1962 and it is recorded between June and July of that year. The second value, 189.7, is the sum of the 12 month production from February 1962 to January 1963, and it is recorded between July and August 1962. Progressing in this way we obtain the *12-month moving totals* of Column (2), which are recorded at the middle of the respective periods for which we added the monthly production.

The entries in Column (3) constitute 2-period moving totals of the entries of Column (2). The first is the sum of the first two entries of Column (2) or $189.2 + 189.7 = 378.9$; the second is the sum of the second and third entries of Column (2) or $189.7 + 190.4 = 380.1$, and so forth. Pairwise adding successive entries of Column (2), we thus obtain the 2-periods moving totals of Column (3). These are recorded *between* the respective entries of Column (2) and, hence, in line with the original production of Column (1).

Since each entry of Column (2) is the sum of 12 items (monthly production) and each entry of Column (3) is the sum of two entries of Column (2) or altogether 24 items, we finally obtain the centered 12-month moving average shown in Column (4) by dividing each entry of Column (3) by 24.

From here on in our work proceeds as in Section 19.4, with the moving average of Column (4) taking the place of the trend values. Dividing the original monthly data by the corresponding entries of Column (4) and multiplying by 100, we obtain the *percentages of moving average* shown in Column (5). *This step is designed to eliminate the trend and cyclical components from our data, leaving thus seasonal and irregular variations as the only kinds of variation.* (As we pointed out earlier, it might be better to say that this eliminates most, not necessarily all, of the trend and cyclical variations.)

All that remains to be done is to eliminate, so far as this is possible, the irregular variations by averaging the respective values obtained for the different months and to facilitate this we shall rearrange the entries of Column (5) in the following table.

It should be noted that there are only four values for each month since the moving average failed to provide averages for the first half of 1962 and the last half of 1966.

	1962	1963	1964	1965	1966	Median	Seasonal† Index
January		23.8	16.2	20.6	37.9	22.2	22.6
February		21.4	18.1	18.2	13.1	18.2	18.5
March		19.2	19.8	19.1	12.4	19.2	19.6
April		15.5	12.8	22.4	11.0	14.2	14.5
May		14.5	11.0	17.6	15.7	15.1	15.4
June		10.6	11.9	13.0	12.6	12.2	12.4
July	7.0	12.5	5.0	7.7		7.4	7.5
August	38.6	18.0	19.1	19.7		19.4	19.8
September	375.5	271.9	367.1	252.8		319.5	325.3
October	546.5	647.0	587.7	579.6		583.6	594.3
November	93.1	128.7	81.6	183.7		110.9	112.9
December	32.1	28.7	40.9	48.5		36.5	37.2
						1178.4	

---

† These seasonal indexes are very extreme as they result from the harvesting of the grape crop during the fall season of the year. Production of still wines represents the amount removed from fermenters, exclusive of distilling materials produced at wineries.

We are now in the same position as on page 439, having to choose some method to average the figures given for the individual months; as before, we could use the mean, median, or any of the other well-known measures of central location. An average that is sometimes used in a situation like this is a special *modified mean* which consists of the mean of the values that remain after the smallest and largest values have been cast out. Having only four values for each month, this would produce the same result as the median which, after all, for four values is the mean of the middle two. Deciding to use the median, we get the values shown above in the second column from the right and the only thing that remains to be done is to adjust these values so that their sum is 1,200. Multiplying the medians by $1,200/1178.4 = 1.0183$, we obtain the seasonal index shown in the right-hand column of the table. The interpretation of this index is the same as before: *typical* March production of wine is 19.6 per cent of those of the average month; typical October production is 594.3 per cent of those of the average month, and so on.

There are several other methods that are sometimes used in the construction of seasonal indexes. The method of *link relatives* was widely used at one time, but its disadvantages seem to outweigh its advantages and it has currently fallen into some disfavor.

Since the labor involved in computing a seasonal index by the methods which we have discussed is fairly heavy, several graphical methods requiring few calculations have been proposed. One is an all-graphical method and another a graphical shortcut to the ratio-to-moving average technique. These methods have the disadvantage that they are subjective, at least to some extent, and thus require a considerable amount of experience and judgment. We shall not treat them here, but if the reader is interested he will find suitable references on page 458. It is worth noting that graphical methods are used and actually preferred in some of the work done by the Federal Reserve Board, in whose *Bulletin* many seasonally adjusted series regularly appear.

EXERCISES

1. Express the monthly purchases of regular ordinary life insurance in the United States for the period 1962–66 (data given on page 433) in billions of dollars and round to one decimal. In other words, beginning with January 1962 write these figures as 4.0, 4.4, 5.0, etc. Then compute a seasonal index from these data by the ratio-to-moving average method, using the median to average the percentages of moving average obtained for the different months.

2. Compute a seasonal index by the ratio-to-moving average method for the department store sales data of Section 19.4. Use the mean to average the percentages of moving average obtained for the different months.

3. Compute a seasonal index by the ratio-to-moving average method for the labor force data of Exercise 3 on page 436. Use the median to average the percentages of moving average obtained for the different months.

4. The following data show retail sales (in millions of dollars) of women's apparel and accessory stores in the United States for the period 1957 to 1966: (Source: *Business Statistics, 1967*).

	1957	1958	1959	1960	1961	1962	1963	1964	1965	1966
January	310	341	359	367	347	361	371	399	427	466
February	283	278	322	324	315	312	321	365	367	428
March	347	394	455	383	457	418	421	487	433	544
April	463	420	417	526	401	496	474	449	538	604
May	422	425	461	436	434	463	452	492	496	563
June	381	358	410	406	405	407	413	454	456	532
July	338	334	356	370	358	368	383	418	440	492
August	390	373	378	394	399	414	438	451	443	524
September	398	418	428	443	438	455	439	482	496	571
October	434	457	471	467	462	468	463	530	553	596
November	448	451	451	464	483	513	499	522	602	614
December	701	744	776	749	770	801	817	895	992	979

(a) Plot this series on arithmetic paper and check whether there is a consistent seasonal pattern.

(b) Are there any adjustments that might well be made preliminary to measuring the seasonal pattern?

(c) Use the ratio-to-moving average method to compute a seasonal index for the unadjusted data, using the median to average the percentages of moving average obtained for the different months.

5. The following data are the Total Loans and Investments at Commercial Banks (in billions of dollars) in the United States for the period 1959 to 1967. (Source: Federal Reserve Bulletin, August, 1968.)

	1959	1960	1961	1962	1963	1964	1965	1966	1967
January	183.9	185.6	195.6	210.9	229.1	246.2	269.1	296.7	313.8
February	181.9	184.3	197.0	211.6	230.4	247.2	270.7	295.9	314.5
March	181.9	183.4	195.6	212.4	231.9	249.9	273.9	298.3	320.1
April	183.8	186.2	197.2	214.8	232.3	250.6	275.9	301.5	322.5
May	183.9	185.9	199.1	215.3	233.6	251.5	277.1	302.2	323.6
June	184.2	186.7	200.9	219.2	239.1	257.3	283.9	310.1	329.5
July	185.7	188.6	203.3	217.8	237.8	254.2	281.2	306.9	331.8
August	186.1	189.0	202.9	219.0	237.1	256.1	283.2	307.4	334.2
September	186.3	191.4	207.5	223.1	241.9	262.2	286.8	308.8	338.8
October	186.7	193.8	208.3	225.7	242.4	262.4	290.2	308.0	341.6
November	186.1	193.9	209.1	226.8	245.0	266.3	292.3	309.0	344.1
December	189.5	198.5	214.4	233.6	252.4	273.9	301.8	317.9	354.5

(a) Plot this series on arithmetic paper and check whether there is a consistent seasonal pattern.

(b) Use the ratio-to-moving average method to compute a seasonal index for the unadjusted data, using the median to average the percentages of moving average obtained for the different months.

19.6 Deseasonalized Data

In this and the next section we shall illustrate two important applications of seasonal indexes dealing with the removal of seasonal variations from observed series and with use of seasonal indexes in forecasting. First, let us consider the question of how a businessman can intelligently interpret month to month changes in sales, new orders, employment, inventory, or for that matter any kind of data relating to the operation of his business. As we pointed out on page 430, there could be most unfortunate consequences if he misinterpreted seasonal variations as being due to other factors. To avoid mistakes of this sort, businessmen and economists often refer to *deseasonalized data*, that is, data which, speaking very loosely, show *how things would have been or would be if there were no seasonal fluctuations.*

The process of removing seasonal variations, or *deseasonalizing*, as it is usually called, is very simple. Given a series of monthly data, we have only to *divide* each value by the seasonal index of the appropriate month and multiply by 100. The logic of this process is quite simple: if an index for March is, say, 120, this means that March sales (or whatever our data happen to be) are 120 per cent of those of the average month or, in other words, 120 per cent of what they would have been if there had been no seasonal variation. Hence, to see what March sales would have been if, again loosely speaking, there had been no seasonal variation, we *multiply* the observed sales by 100/120, or divide them by 1.20, which is the same. *To remove seasonal variations we shall thus divide by appropriate seasonal indexes written as proportions.*

To illustrate this further, let us deseasonalize some of the data for which we calculated a seasonal index in Section 19.5. The first column of the table which follows contains the 1966 production of still wines in the United States, copied from page 443, and the second column contains the seasonal index arrived at on page 445. (The values of this index are now changed to proportions and rounded to two decimals.) To deseasonalize these production figures we have only to divide each entry of the first column by the corresponding entry of the second, arriving thus at the values shown in the third column.

	Production (millions of wine gallons)	Seasonal Index	Deseasonalized Production
January	7.4	0.23	32.2
February	2.6	0.19	13.7
March	2.6	0.20	13.0
April	2.3	0.14	16.4
May	3.0	0.15	20.0
June	2.3	0.12	19.2
July	1.5	0.08	18.8
August	9.6	0.20	48.0
September	72.9	3.25	22.4
October	88.4	5.94	14.9
November	17.9	1.13	15.8
December	8.3	0.37	22.4

To understand the significance of such deseasonalized data, let us look at the above figures for August and September. The actual percentage increase in production from August to September was $(72.9 - 9.6)/9.6 = 6.59$ or 659 per cent. However, as can be seen from the index, August is traditionally low, September is traditionally high, and some kind of increase should really have been expected. If we now perform the same operation on the deseasonalized production, we find that from August to September there was a percentage change of $(22.4 - 48.0)/48.0 = -0.53$ or a decrease of 53 per cent. This shows that the increase in actual production from August to September, even though it is large, is not as large as we could have *expected* in accordance with the typical seasonal pattern. Hence, instead of celebrating the increase in production from August to September, it would be more appropriate to investigate why it was less than expected.

Deseasonalized monthly data are sometimes multiplied by 12 and then referred to as *annual rates*. The government often uses seasonally adjusted annual rates in reporting activity in various sectors of the economy. For example, in April of 1968, the Department of Commerce reported that Americans had more income in March, 1968, on an annual rate basis, than in any other prior month in history. In that month the annual rate of total personal income rose to a record $666.0 billion. As is apparent from this illustration, the use of annual rates has the advantage that it facilitates the reporting (and studying) of month to month changes in data which are ordinarily reported, and best understood, on an annual basis.

The importance of the method we have illustrated in this section cannot be overemphasized. Numerous series of business data reported in newspapers and magazines are accompanied by the legend "adjusted for seasonal variation" or "corrected for seasonal variation" and this means, precisely, that the data have been deseasonalized.

19.7 The Use of Seasonal Indexes in Forecasting

In Section 18.9 we showed how forecasts of future levels of activity may be based on extrapolated trend equations, warning the reader about the inherently speculative nature of any such procedure. Now we shall go one step further by modifying monthly forecasts based on extrapolated trend equations in accordance with seasonal patterns.

To illustrate the overall procedure, let us suppose that in 1969 the management of the department store referred to in the example of Section 19.4 asked its statistician to forecast monthly sales for 1970 and to base this forecast on the data we gave on page 437. Ignoring at first all seasonal fluctuations, he could obtain preliminary estimates by extrapolating from the least squares trend equation which read (see page 438),

$$y' = 25.74 + 0.455x$$

(origin, January, 1964; x units, 1 month;
y, average monthly sales in thousands of dollars).

Substituting $x = 72$ into this equation, he finds that the trend value for January 1970 is $25.74 + 0.455(72) = 58.50$; and after repeatedly adding 0.455 and rounding to one decimal, he obtains the trend values shown in the first column of the following table:

	Trend Values	Seasonal Index	Estimated Monthly Sales for 1970
January	58.5	0.796	46.6
February	59.0	0.766	45.2
March	59.4	0.949	56.4
April	59.9	0.975	58.4
May	60.3	1.055	63.6
June	60.8	0.979	59.5
July	61.2	0.851	52.1
August	61.7	0.892	55.0
September	62.1	1.021	63.4
October	62.6	1.219	76.3
November	63.0	1.134	71.4
December	63.5	1.363	86.6

These trend values could serve as forecasts of 1970 sales if there were no seasonal variation. However, the work of Section 19.4 (see page 439) showed that *typical* January sales of this store are 79.6 per cent of what they would be if there were no seasonal variation, *typical* February sales are 76.6 per cent of what they would be if there were no seasonal variation, . . . , and *typical* December sales are 136.3 per cent of what they would be if there were no seasonal variation. Assuming that this pattern applies also to 1970, the forecasts for that year will have to be modified by taking 79.6 per cent of the trend value obtained for January, 1970, 76.6 per cent of the trend value obtained for February, 1970, . . . , and 136.3 per cent of the trend value obtained for December, 1970. To simplify the calculation of these modified estimates, the seasonal index from page 439 has been changed from percentages to proportions and thus copied into the second column of the preceding table. The final estimated 1970 sales are then obtained by multiplying the respective entries of the first two columns and they are shown in the column on the right. Accounting, thus, for *trend* and *seasonal variation*, the statistician forecasts that in 1970 the department store's January sales will be $46.6 thousand, its February sales will be $45.2 thousand, its March sales will be $56.4 thousand, and so on.

The purpose of a forecast like the one made above is to plan the operations of a firm so as to order its daily activity in the best way possible to meet customer's needs and, more generally, to grow into the business life of the community. Of course, predictions based on trend equations and estimated seasonal patterns are not the only factors that must be considered in intelligent planning. Locally, the department store management may have to take into account new and greatly improved transportation facilities serving the store, the opening of new competing stores, the development of nearby residential areas, and many other factors.   On a wider scale, the management must consider nationwide and regional forecasts concerning overall economic activity, employment, income, consumer spending and credit, retail sales, and such.   Any or all of these factors may, and probably will, make it necessary to modify forecasts based only on the company's sales data for the period 1964–68.

It must be remembered that any projection of past experience to the uncertain future is speculative and hazardous.   Even assuming that estimated trend values arrived at by utilizing all available experience are correct, or nearly so, there is an additional problem: a seasonal index is only an *average* of monthly behavior over a period of years.   It may be quite unsatisfactory when applied to a particular year.

For example. let us consider the sale of new automobiles during 1964 and 1965.   Sales of new automobiles in October of 1964 were *contraseasonal*, which means that sales went down when they were normally expected to go up.   October sales were about one-half of what could have been expected in view of typical seasonal patterns. Automobile production was greatly restricted at that time as the result of a labor dispute.   When the strike was settled, production recovered rapidly, and sales increased dramatically.   Sales remained higher through most of the following year, 1965, than what might normally have been anticipated.   As a result, 1965 proved to be a bigger sales year than any other previous year in automotive history.   As soon as information and records like these become available, whatever parties are concerned must revise their forecasts accordingly and take the necessary steps to translate revised production or other goals into action.

Obviously, forecasting and planning, however inexact they may seem, must go on if business and government are to operate intelligently and successfully.   Improved techniques of gathering, analyzing, and interpreting data, and doing all this more rapidly, are continually appearing, and these improvements have led and will lead to increasingly better forecasting techniques.   Also, as we have indicated, no one is irrevocably committed to a forecast, to survive or perish with it once it has been made.   Intelligent forecasting and planning demands one's continual adaptation to changing conditions.

EXERCISES

1. Use the seasonal index obtained on page 433 to deseasonalize the 1962 monthly purchases of regular ordinary life insurance in the United States given on that same page.   Compare the percentage change from

May, 1962, to June, 1962, with the corresponding percentage change in the deseasonalized data.

2. Use the index obtained on page 439 to deseasonalize the 1964 to 1968 department store sales data given on page 437. Plot the given data as well as the deseasonalized data on arithmetic paper. Also express the deseasonalized 1968 data on an annual rate basis.

3. Use the seasonal index obtained in Exercise 1 on page 435 to deseasonalize the data given in that exercise. Also, plot the original series as well as the deseasonalized data on arithmetic paper.

4. Use either the seasonal index obtained in Exercise 3 on page 436 or the one obtained in Exercise 3 on page 440 to deseasonalize the labor force data given in Exercise 3 on page 436.

5. A company selling aluminum screening, had sales of $50,000 and $56,000 in April and May of 1968. The company's seasonal index for these two months stands at 105 and 140. The president of the company expressed dissatisfaction with the April sales, but the sales manager said that he was quite pleased with the $6,000 increase. What argument should the president of the company have used to reply to the sales manager? The sales manager also predicted on the basis of the May sales that the total 1968 sales were going to be $672,000, while the president of the company predicted total sales of $480,000. Criticize the sales manager's estimate and explain how the president may have arrived at his figure.

6. Use the trend equation given in Exercise 2 on page 440 and the seasonal index computed in that exercise to forecast the production of eggs in the United States for January, 1970.

7. Use the trend equation given in Exercise 3 on page 440 and the seasonal index computed in that exercise to forecast the employed civilian labor force in the United States for each month of 1970.

8. A manufacturer of sporting goods estimates its average monthly 1970 sales to be $500,000. The company's index of sales of sporting goods has been

Jan.	Feb.	Mar.	Apr.	May	Jun.	Jul.	Aug.	Sept.	Oct.	Nov.	Dec.
72	70	90	110	140	130	118	110	99	101	90	70

Ignoring the possible existence of trend, use the above information to draw up a monthly sales budget for this company for 1970.

9. Suppose that the sporting goods manufacturer referred to in the preceding exercise determines the trend in sales by means of the equation

$$y' = 3,500,000 + 500,000x$$

(origin, 1965; x units, 1 year; y, total annual sales in dollars)

(a) Use this equation, suitably modified, to calculate the trend values of monthly sales for each month of 1970.

(b) Use the trend values obtained in (a) and the seasonal index of Exercise 8 to forecast the company's monthly sales in 1970.

19.8 Cyclical Variation

So far we have worked with cyclical variation only inasmuch as we tried to eliminate it in the process of measuring seasonal variation. Now we shall investigate very briefly how cyclical variations, themselves, can be measured. The amount of time that economists and statisticians have spent on the *statistical* problem of measuring cyclical variation and the *nonstatistical* problem of trying to explain their causes has been, and will probably continue to be, enormous. So far as the problem of measuring business cycles is concerned, various methods have been proposed and some of them have met with a fair amount of success. One of the simplest of these methods, the so-called *residual method* will be discussed below. So far as the problem of explaining business cycles is concerned, numerous theories have been advanced, among them, the *changes-in-income theory*, the *underconsumption theory*, the *overinvestment theory*, the *fluctuations-in-discount theory*, the *inequality-of-foresight theory*, and others. Most of these theories recognize the relevance of various factors but they differ in the emphasis which they place on them as contributory causes of business cycles. It would not be difficult, for example, for a sophisticated economist to reconcile such seemingly diverse theories as the *monetary theory*, which attributes cycles to the expansion and contraction of bank credit, and the *nonmonetary theory*, which, in one formulation, stresses such factors as inventions and technological advances.

In the remainder of this section we shall discuss the practical problem of isolating, and thus measuring, cyclical variation by eliminating the other three components. Among the other methods that have been proposed, some isolate cycles directly without first eliminating the other components, but they are not widely used.

The method we shall discuss is called the *residual method;* it consists of eliminating in some order the other three components of a series. In some instances the process is not carried through the final stage of smoothing out irregular movements and the resulting values are referred to as *cyclical-irregulars.* If we remove irregular variations as well as seasonal variation and trend, we arrive at quantities which, in percentage form, are referred to as *cyclical relatives.*

In the residual method we first adjust our data for trend and, when dealing with monthly data, for seasonal variation. In view of the *multiplicative* theory referred to on page 384, we shall make this adjustment by *dividing* out trend and seasonal variation, arriving, thus, at *cyclical-irregulars.* This may be done in any one of the following ways:

(1) Each value of the series is divided first by the corresponding trend value and then by the corresponding value of the seasonal index.

(2) Each value of the series is divided first by the corresponding value of the seasonal index and then by the corresponding trend value.

(3) Each trend value is multiplied by the corresponding value of the seasonal index, leading, thus, to a series of trend-seasonal ($T \times S$) values called the *normal*. Then each value of the original series is divided by the corresponding value of the normal.

All these routes lead to the same set of cyclical-irregulars, which may also be called *percentages of normal*. Symbolically, we could write

$$C \times I = \frac{T \times S \times C \times I}{T \times S} \qquad (19.8.1)$$

where $T \times S \times C \times I$ stands for the values of the original series, namely, the *products* of the effects of trend, seasonal, cyclical, and irregular forces. Also, $T \times S$ stands for the *normal*, namely, the values which we would *expect* if trend and seasonal forces were the only contributing factors, and $C \times I$ stands for the cyclical-irregulars.

The choice of any one of the procedures listed above is largely a matter of convenience. For example, if a seasonal index has been computed by the ratio-to-trend method, the first step in (1) has already been completed and we will only have to divide the percentages of trend by the corresponding values of the seasonal index. If we are dealing with already-deseasonalized data, the first step in (2) has already been completed and we have only to divide the deseasonalized data by the corresponding trend values. In some situations the normal may already have been calculated for other purposes; in such cases (3) would be the most convenient. The three methods outlined above assume, of course, that we are dealing with monthly or, perhaps, weekly, daily, or hourly data. If we dealt with *yearly* data we would not have to worry at all about seasonal variation and less about irregular variations. (Irregular variations sometimes tend to average out when data cover longer periods of time.)

To illustrate the calculation of cyclical-irregulars and, subsequently, cyclical relatives, let us again refer to the department store sales of Section 19.4. Ordinarily, studies of cycles are based on data covering long periods of time, but for purposes of illustration let us limit ourselves to the department store's 1964 sales. These figures are copied from page 437 and shown below in the first column of the table. To measure trend, we shall use the least squares line of Section 19.4. Substituting $x = 0$ to $x = 11$ into $y' = 25.74 + 0.455x$ (see page 438), and rounding to one decimal, we get the trend values shown in Column (2). The seasonal index obtained earlier on page 439 is given in Column (3) and the *normal*, consisting of the products of the corresponding entries of Columns (2) and (3), is given in Column (4). To obtain the cyclical-irregulars we now have only to divide the original sales data by the corresponding values of the normal. Multiplying these quotients by 100 to express them as "percentages of normal," we finally get the cyclical-irregulars shown in Column (5).

	(1) Sales (thousands of dollars)	(2) Trend	(3) Seasonal Index	(4) Normal $T \times S$	(5) Cyclical-Irregulars $C \times I$
January	19	25.7	0.796	20.5	92.7
February	20	26.2	0.766	20.1	99.5
March	20	26.7	0.949	25.3	79.1
April	22	27.1	0.975	26.4	83.3
May	27	27.6	1.055	29.1	92.8
June	27	28.0	0.979	27.4	98.5
July	24	28.5	0.851	24.3	98.8
August	30	28.9	0.892	25.8	116.3
September	32	29.4	1.021	30.0	106.7
October	36	29.8	1.219	36.3	99.2
November	37	30.3	1.134	34.4	107.6
December	42	30.7	1.363	41.8	100.5

Cyclical-irregulars are sometimes expressed as *percentages above or below normal;* the necessary conversion is made by subtracting 100 from each entry of Column (5). We thus find that for January, 1964, the cyclical-irregular is $92.7 - 100 = -7.3$ or 7.3 per cent *below normal* and that for August, 1969 it is $116.3 - 100 = 16.3$ or 16.3 per cent *above normal.*

To isolate the cyclical component completely, insofar as this is possible, we shall now have to go one step further and eliminate irregular variations. According to the theory that the effects of the four components are multiplicative, irregular variations should be divided out, but since irregular movements are largely random and, hence, do not lend themselves to *direct* measurement, we shall instead *smooth* them out by means of a moving average.† In order to avoid using a moving average which averages out *too much* (we certainly do not want to smooth out most of the cyclical component as well) it is customary to eliminate irregular variations by means of a *weighted moving average.* In our illustration we shall use a 3-month moving

---

† As an alternative to moving averages, sine and cosine curves, giving the kind of wavy lines the reader may have seen in the reproduction of sounds on an oscillograph, have sometimes been fitted to cyclical-irregulars.

average with weights 1, 2, 1.   This means that we calculate the moving total for, say, February, 1964 by adding the January figure *once*, the February figure *twice*, and the March figure *once*, and then obtain the moving average for this month by dividing this sum by $1 + 2 + 1 = 4$.   The weights of weighted moving averages are often based on *binomial coefficients* such as 1, 2, 1, *or* 1, 4, 6, 4, 1, *or* 1, 6, 15, 20, 15, 6, 1 (see Table VI on page 507).   The advantage of using a *weighted* moving average is that we can control the extent to which a series is smoothed by choosing suitable weights.   Incidentally, the reader may wish to verify that the centered 12-month moving average which we used in Section 19.5 is actually a weighted 13-month moving average with weights 1, 2, 2, 2, 2, 2, 2, 2, 2, 2, 2, 2, 1.

The calculations needed to smooth irregular movements out of cyclical-irregulars are shown below in a continuation of the table given on page 455.   Column (5) contains the cyclical-irregulars copied from that page; Column (6), the 3-month moving totals weighted 1, 2, 1; and Column (7), the weighted 3-month moving averages which are the desired *cyclical relatives*.

	(5) Cyclical-Irregulars $C \times I$	(6) Weighted 3-month Moving Totals	(7) Cyclical Relatives C
January	92.7		
February	99.5	370.8	92.7
March	79.1	341.0	85.3
April	83.3	338.5	84.6
May	92.8	367.4	91.9
June	98.5	388.6	97.2
July	98.8	412.4	103.1
August	116.3	438.1	109.5
September	106.7	428.9	107.2
October	99.2	412.7	103.2
November	107.6	414.9	103.7
December	100.5		

The cyclical relatives of Column (7) are, in fact, *percentages of normal with irregular movements smoothed out.*

A few paragraphs above we mentioned that because of their essentially random nature irregular movements cannot be measured directly. They can be measured *indirectly*, however, by removing the three other components from a series.   In our example we could *divide* the entries of Column (5) by the corresponding entries of Column (7), or, in other words, we could isolate the irregular component by dividing $C \times I$ by C.

The purpose of this section has been to demonstrate how cyclical movements may be isolated and described.   Once they have been cal-.

culated, cyclical relatives may be subjected to further statistical treatment and various sorts of analyses. For example, one might try to separate a cyclical pattern into a number of regularly repeating cycles; in other problems it might be of interest to determine whether there exists a *correlation* between the cyclical patterns of different series. It might also be of interest to study the *lag* between the cycles of different series, for instance, those representing inventory and sales, to see whether particular series can be used as indicators, or predictors, of the movements of others. All these problems are important, but they are generally not treated in introductory texts.

EXERCISES

1. In this section we calculated the cyclical-irregulars and the cyclical relatives for the year 1964 for the department store sales data of Section 19.4. Extend these calculations through the year 1968 and plot the cyclical-irregulars on arithmetic paper. Use a 1, 2, 1 weighted moving average to smooth out the irregular movements.
2. Use the data of Exercise 1 on page 435, the trend equation given in Exercise 2 on page 440, and either of the seasonal indexes computed in these two exercises, to calculate the cyclical-irregulars and the cyclical relatives for the 1963 to 1966 farm production of eggs. Use a 1, 2, 1 weighted moving average to smooth out the irregular movements.
3. Use the data of Exercise 3 on page 436, the trend equation given in Exercise 3 on page 440, and either of the seasonal indexes computed in these two exercises, to calculate the cyclical-irregulars and the cyclical relatives for the 1962 to 1966 employed labor force in the United States. Plot the normal, obtained as an intermediate step in these calculations, together with the original data on arithmetic paper. Use a 1, 2, 1 weighted moving average to smooth out the irregular movements.

BIBLIOGRAPHY

Further discussion of the problems of measuring seasonal variations and cyclical movements may be found in various texts, for instance, in

Croxton, F. E., and Cowden, D. J., *Practical Business Statistics*, 3rd ed. Englewood Cliffs, N.J.: Prentice-Hall, Inc., 1960, Chaps. 28, 29, 30, and 34.

Mills, F. C., *Statistical Methods*, 3rd ed. New York: Holt, Rinehart & Winston, 1956, Chaps. 10, 11, and 12.

Neiswanger, W. A., *Elementary Statistical Methods*, rev. ed. New York: Macmillan, 1956, Chaps. 15, 16, and 17.

The problem of dealing with changing seasonal patterns is treated in the book by Neiswanger listed above, and the method of *link relatives* is illustrated in

Croxton, F. E., Cowden, D. J., and Klein, S., *Applied General Statistics*, 3rd ed.   Englewood Cliffs, N.J.: Prentice-Hall, Inc., 1967, p. 303.

Graphical methods used in the computation of seasonal indexes are treated in

Spurr, W. A., and Bonini, C. P., *Statistical Analysis for Business Decisions*, 3rd ed.   Homewood, Ill.: Irwin, 1967, Chap. 20.

Alternate methods used in the seasonal adjustment of data and in the measurement of cyclical variation are given in

"Adjustment for Seasonal Variation," *Federal Reserve Bulletin*, June, 1941.

Burns, A. F., and Mitchell, W. C., *Measuring Business Cycles*.   New York: National Bureau of Economic Research, 1946.
Davis, H. T., *The Analysis of Economic Time Series*.   Bloomington, Ind.: Principia Press, 1941.

APPENDICES

APPENDIX I

Pictorial Presentations

I.1 Introduction

The work of a statistician is often not completed even when his investigation has progressed to the point where the necessary samples have been obtained, the data have been analyzed, and pertinent conclusions have been drawn. Regardless of whether a study is intended for inclusion in a company's report, for newspaper publication, for public display, or advertising, the final step usually consists of making the results as appealing as possible to the reader or readers whom they are supposed to reach.

All too frequently, the results of statistical investigations are wasted, the facts which they convey do not reach their destination, simply because they are not presented in a sufficiently effective fashion. It would not be an exaggeration to say that the method of presenting facts is often more instrumental in their acceptance than the nature of the facts themselves.

Perhaps the most convincing and most appealing way in which statistical results can be presented is in pictorial form. Evidence of this can be found in the financial pages of newspapers and magazines, displays and exhibits, advertisements, and such. The number of ways in which statistical data can be displayed pictorially is almost without bounds, the only limitation being the artistic talent and imagination of the individual or agency engaged in the preparation of statistical charts and tables. In this appendix we shall illustrate briefly some of the major types of diagrams, charts, and maps used most frequently in the pictorial presentation of business data.

I.2 Bar Charts

One of the most popular kinds of charts used in the presentation of statistical data is the *bar chart*, examples of which are shown in Figures I.1, I.2, and I.3. As is apparent from these diagrams, bar

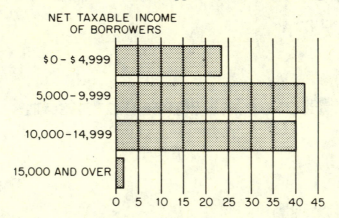

FIGURE I.1. Income of Borrowers under the Higher Education Act. Data Based on the first 387,931 loans. (Source: Business Review, Federal Reserve Bank of Boston, June, 1968.)

charts can be used to present frequency distributions (numerical and categorical) as well as time series. As their name implies, bar charts consist of bars (rectangles) which are of equal width and whose lengths are proportional to the frequencies or quantities they represent.

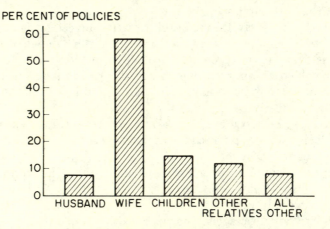

FIGURE I.2. Relationship of Beneficiary to Insured, Death Claims in the United States, October, 1966. (Source: 1967 Life Insurance Fact Book.)

So far as the construction of a bar chart is concerned, there are several rules that are considered to be sound practice and they coincide with the rules which we gave in Section 17.9. Among other things, we find that a bar chart should always have a title, if possible all

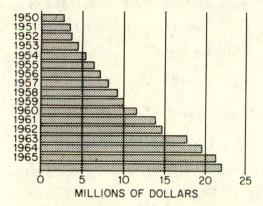

FIGURE I.3. Average Assets of Savings and Loan and Other Associations in the United States. (Source: 1967 Savings and Loan Fact Book.)

lettering should be horizontal, and there should always be mention of the source of the data on which the chart is based. Although the bars, themselves, can be horizontal or vertical, in the presentation of time series it is often preferred to measure time horizontally, thus requiring vertical bars to represent the values of a series.

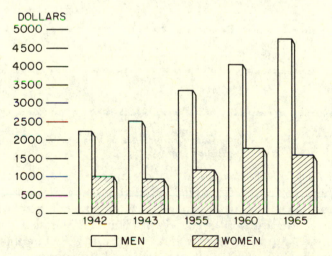

FIGURE I.4. Median Total Money Income of Men and Women. (Source: 1967 Statistical Abstract of the United States.)

There are several ways in which bar charts may be modified or refined. Figure I.4 not only shows how differently colored or shaded bars serve to represent several series (or several sets of data), but also illustrates how the eye-appeal of a bar chart may be improved by providing a three-dimensional effect. Another way of adding to the

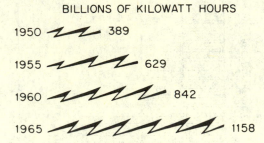

BILLIONS OF KILOWATT HOURS

FIGURE I.5. Electric Energy Production in the United States. (Source: 1967 Statistical Abstract.)

appeal of a bar chart is to replace the bars with suitably shaped objects, as illustrated in Figures I.5 and I.6.

In the preceding illustrations we have given all the bars or objects the same width, letting their height represent the quantities for which they stand. Instead, we could also draw the bars or objects as in

FIGURE I.6. Motor Vehicle Output in Millions of Vehicles 1962–66. (Source: Automobile Manufacturers Association Data.)

Figure I.7 and I.8, letting the *areas* be proportional to the quantities which they represent. This kind of chart is more difficult to draw and unless we are careful, we can easily create misleading impressions. For instance, the areas of the squares of Figure I.7 represent the value of industrial and miscellaneous bonds held by U.S. Life Insurance companies in 1950, 1955, 1960, and 1965 and the sides must be such that their *squares* are proportional to the actual amounts. Similarly,

BILLION DOLLARS

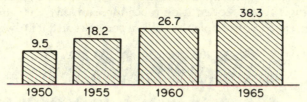

FIGURE I.7. Industrial and Miscellaneous Bonds Owned by U.S. Life Insurance Companies. (Source: 1967 Life Insurance Fact Book.)

if in a diagram like that of Figure I.8 one figure is to be twice as large as another, then each dimension must be multiplied by the *square*

THOUSAND CARS

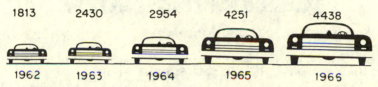

FIGURE I.8. Factory Sales of 2-Door and 4-Door "Hard Top" Passenger Cars. (Source: 1967 Statistical Abstract of the United States.)

root of 2 or, approximately, 1.41. If we multiplied each dimension by 2, the resulting figure would be *four times*, and *not twice*, as large as the first.

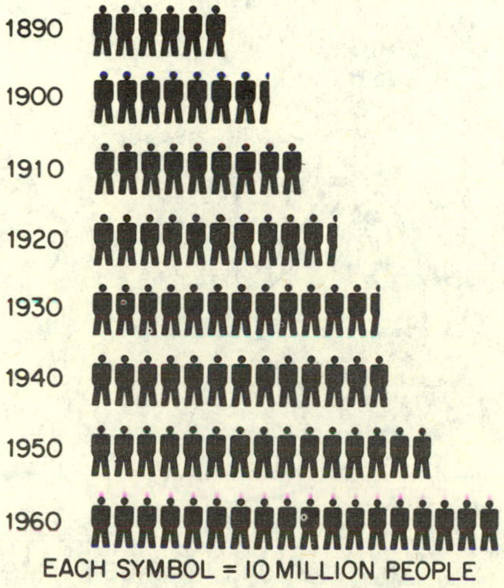

EACH SYMBOL = 10 MILLION PEOPLE

FIGURE I.9. United States Population. (Source: 1967 Statistical Abstract of the United States.)

Another variation of the idea of a bar chart is shown in Figures I.9 and I.10, where the bars are replaced by suitable numbers of small objects, each object representing a certain unit, say, 100 shares, 1,000 dollars, or, as in our example, 10 million persons and 3,000 dwelling units.

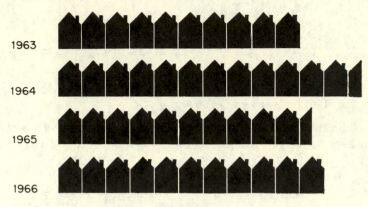

EACH SYMBOL = 3000 DWELLING UNITS

FIGURE I.10. Building Construction in Massachusetts, New Housing Units Authorized. (Source: 1967 Statistical Abstract of the United States.)

I.3 Pie Charts and Component Bar Charts

Categorical distributions, particularly percentage distributions, are frequently presented in the form of *pie charts*. Such charts consist

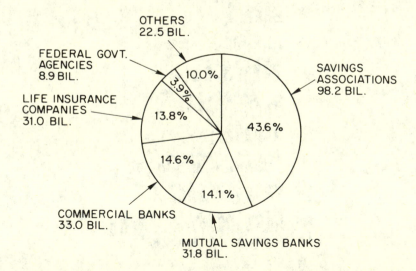

FIGURE I.11. Mortgage Loans Outstanding on One-to Four-Family Nonfarm Homes, by Type of Lender, 1966. (Source: 1967 Savings and Loan Fact Book.)

of a circle that is subdivided into sectors which, with a stretch of the imagination, look like pieces of pie proportional in size to the quantities or percentages they represent. Examples of pie charts are shown in Figure I.11 and I.12, the latter having the added feature of a three-dimensional effect.

The construction of a pie chart is very simple. Since the sum of the central angles of the sectors is 360 degrees and the entire circle represents 100 per cent, 1 per cent is represented by a central angle of

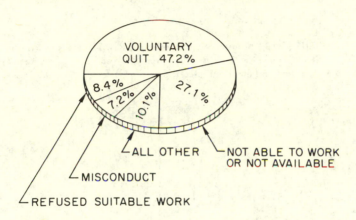

FIGURE I.12. Disqualifications by Issue in 1967 Unemployment Compensation. (Source: State of Vermont–Department of Employment Security, 1967 Annual Report.)

360/100 = 3.6 degrees. To illustrate the construction of a pie chart, let us consider the following data on the 1966 United States consumption of principal raw materials in iron and steel production:

	Consumption (millions of net tons)	Percentage Distribution	Central Angles
Iron ore	142.7	43.5	156.6°
Coking coal	85.5	26.0	93.6°
Scrap	70.1	21.4	77.0°
Limestone	29.9	9.1	32.8°
Totals	328.2	100.0%	360.0°

The percentages are obtained by dividing each of the entries of the first column by their total, namely, 328.2 million, and then multiplying by 100. The central angles are then obtained by multiplying each of the percentages by 3.6. Laying off the proper angles with a protractor, we get a pie chart like that of Figure I.13, and if we wanted to we could

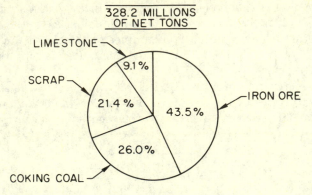

FIGURE I.13. Consumption of Principal Raw Materials in Iron and Steel Production in 1966, Millions of Net Tons. (Source: American Iron and Steel Institute.)

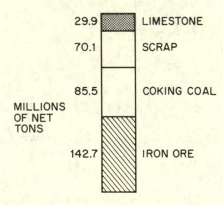

FIGURE I.14. Consumption of Principal Raw Materials in Iron and Steel Production in 1966, Millions of Net Tons. (Source: American Iron and Steel Institute.)

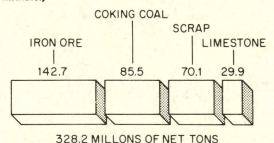

FIGURE I.15. Consumption of Principal Raw Materials in Iron and Steel Production in 1966, Millions of Net Tons. (Source: American Iron and Steel Institute.)

BILLION DOLLARS

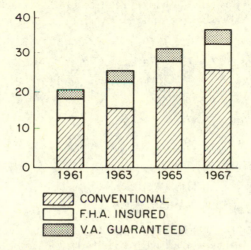

CONVENTIONAL
F.H.A. INSURED
V.A. GUARANTEED

FIGURE I.16. Residential Mortgage Loans Held by Commercial Banks. (Source: Federal Reserve Bulletin, May, 1968.)

PER CENT

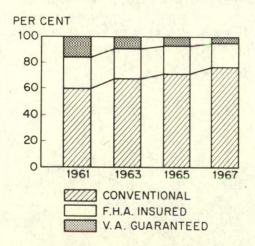

CONVENTIONAL
F.H.A. INSURED
V.A. GUARANTEED

FIGURE I.17. Residential Mortgage Loans Held by Commercial Banks. (Source: Federal Reserve Bulletin, May, 1968.)

add colors, shadings, or various other effects to increase the chart's appeal.

If in the last example we had wanted to prepare a chart showing the actual amounts allocated to the various expenditures instead of their percentages, we could also have used a pie chart or a *component bar chart* like the one shown in Figure I.14. Here the total bar is divided into "slices" that are proportional in size to the quantities

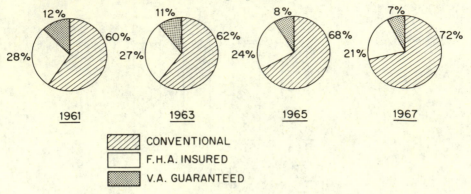

FIGURE I.18. Residential Mortgage Loans Held by Commercial Banks. (Source: Federal Reserve Bulletin, May, 1968.)

which they represent. For added effect, we might separate the slices as in Figure I.15.

Component bar charts are frequently used for comparisons in series, showing how the various components change with time (see Figure I.16). If we wanted to show how a percentage distribution changes with time, we could use a chart like that of Figure I.17 or,

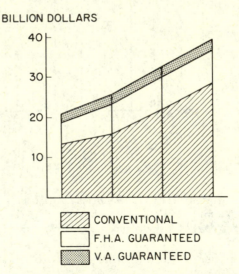

FIGURE I.19.   Residential Mortgage Loans Held by Commercial Banks.   (Source: Federal Reserve Bulletin, May, 1968.)

perhaps, a series of pie charts like that shown in Figure I.18.   Another way of illustrating how various components (as well as their total) change with time is by means of a *shaded line chart* like that of Figure I.19.   It should be noted that Figure I.16, I.17, I.18, and I.19 all present the same data pertaining to residential mortgage loans held by commercial banks, 1961 to 1967.

I.4 Statistical Maps

An important way of conveying information about geographical distributions is by means of statistical maps.   Such maps show how variables such as population density, illiteracy rates, number of branch stores, frequency of automobile accidents, incomes, farms, . . . , are distributed among continents, countries, states, counties, towns, and so forth.   Since statistical maps could be difficult to read if we simply wrote appropriate numbers (frequencies or quantities) into the corresponding regions, it is generally advisable to indicate different values by means of colors or shadings.   An example of a statistical map illustrating the 1966 life insurance in force in the United States by state is shown in Figure I.20.

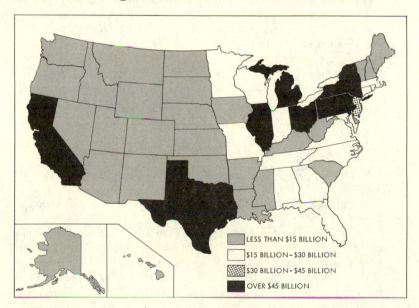

FIGURE I.20.   Life Insurance in Force in the United States by State 1966.   (Source: 1967 Life Insurance Fact Book.)

There are various ways in which statistical information can be displayed on maps.   Instead of using colors to indicate differences in automobile registration in various states, we could draw a number of small cars in the map of each state, say, one car for each 500,000 registrations. (Although this may add to the eye-appeal of a map, it will probably not contribute to its clarity.) Instead of using different shadings, we could also use dots to indicate the "density" of, say, birth registrations. When statistical maps are used in exhibits or displays, it is often possible to add to their effectiveness with the use of differently colored thumb tacks or pins.

The purpose of this appendix has been to give brief illustrations of some of the things that can be done in the pictorial presentation of statistical data. Let us repeat, statistical information is useless unless it reaches the audience for whom it is intended and this is very often facilitated by the use of pictorial techniques.

EXERCISES

1. Use the data given on page 38 to draw a bar chart showing how industrial and commercial failures were distributed among industrial groups in 1966.
2. Use the data given on page 22 to construct a bar chart showing the distribution of the orders received by the mail-order house.
3. Draw a bar chart of the series given in Exercises 2 on page 56.
4. Use the data given on page 370 to construct a pie chart of the December, 1966, Relative Importance Percentages of the *Consumer Price Index*.
5. The following is the percentage distribution of the size of Private and Public Nonfarm Housing Units in the U.S. in 1967 (prelim.):

Size	Percentage
One-Family	63.3
Two-Family	3.7
Three-and Four-Family	2.8
Five or more	30.2

   Construct a pie-chart of this distribution. (Source: *1968 Savings and Loan Fact Book*.)
6. Foreign Liquid Assets in the U.S. (as percentage of total) By holder on December 31, 1967, consisted of the following: Foreign commercial banks, 33.6 per cent; Nonbanking concerns and individuals, 14.0 per cent; Foreign central banks and governments, 47.2 per cent; and International and Regional organizations, 5.2 per cent. Draw a pie chart of this distribution. (Source: U.S. Dept. of the Treasury and the Federal Reserve Bank of New York.)
7. The following table shows the distribution of securities of business and industry owned by U.S. Life Companies *in millions of dollars:* (Source: *1967 Life Insurance Fact Book*).

	Industrial and Misc. Bonds	Public Utility Bonds	Railroad Bonds	Stocks
1920	49	125	1,775	75
1925	97	687	2,238	81
1930	367	1,631	2,931	519
1935	575	2,114	2,625	583
1940	1,542	4,273	2,830	605
1945	1,900	5,212	2,948	999
1950	9,526	10,587	3,187	2,103
1955	18,179	13,968	3,912	3,633
1960	26,728	16,719	3,774	4,981
1965	38,338	17,046	3,314	9,126

Draw a component bar chart like that of Figure I.16.

8. Use the data of Exercise 7 to draw a shaded line chart like that of Figure I.19.

9. Convert the data of Exercise 7 into percentages, in other words, divide each year's figures by their total and multiply by 100; and draw a percentage component bar chart like that of Figure I.17.

BIBLIOGRAPHY

The subject of graphical (pictorial) presentation of statistical data is treated in detail in

Croxton, F. E., Cowden, D. J., and Klein, S., *Applied General Statistics*, 3rd ed. Englewood Cliffs, N.J.: Prentice-Hall, Inc., 1967, Chaps. 4, 5, 6.

Smart, L. E., and Arnold, S., *Practical Rules for Graphic Presentation of Business Statistics*. Columbus, Ohio: Bureau of Business Research, Ohio State University, 1951.

The problem of charting, particularly *what not to do*, is treated in an amusing and informative fashion in

Huff, D., and Geiss, I., *How to Lie with Statistics*. New York: W. W. Norton, 1954.

Reichmann, W. J., *Use and Abuse of Statistics*. London: Methuen & Co., Ltd., 1961.

APPENDIX II

Quality Control

II.1 Introduction

Although there is a tendency to think of quality control as a recent development, there is really nothing new about the idea of making a quality product characterized by a high degree of uniformity. For centuries, highly skilled artisans have striven to make products distinctive through superior quality, and once a standard of quality was achieved, to eliminate insofar as possible all variability between products that were nominally alike.

However, the idea that statistics might be instrumental in controlling the quality of manufactured products is new, going back no farther than the 1920's, and the widespread use of what is now called *statistical quality control* (or *quality control*, or *Q.C.*, for short) is even more recent than that. Although the new methods were readily accepted in Great Britain, it was not until World War II that statistical quality control methods were used on any scale in the United States. As a result of the war, it became necessary for many different industries to devote themselves to an all-out production effort, requiring the production of tremendous amounts of war material made to more exacting specifications than ever before and, in many cases, made by new methods, with substitute materials, poorly trained help, on machines designed for other purposes; it was this exigency which led to the wide acceptance of statistical quality control in this country.

Despite the claims of a few enthusiasts, quality control did not win the war single-handedly, but it undoubtedly made a substantial contribution. Like any other statistical tool, there are certain things that Q.C. can do and some that it cannot. Obviously, the problem of designing machines capable of making products to very exact spec-

474

ifications is an engineering problem and not a statistical one. But like other statistical tools properly used, quality control proved to be extremely helpful in a wide variety of situations, and it was only natural that its success during the war should be followed by its continued and expanded use in the post-war period. At the present time, statistical quality control is used to some extent in virtually every kind of industry in existence and in virtually every country having an industry within its borders. Precisely what economies are being achieved by these methods through reductions in time and manpower requirements, waste, scrap, rework, and inspection cost, is impossible to determine, but they are undoubtedly large. Statistical quality control, which was once viewed by some with distrust and even alarm, has passed its qualifying tests with distinction and there is no apparent limit to its use in the years ahead.

Many problems arising in the manufacture of a product are amenable to statistical treatment. (Some of them have already been treated in earlier parts of this book.) In a broad sense, what is sometimes called *industrial statistics* embraces all the statistical techniques which can be used anywhere in the solution of these problems. Thus, quality control is merely one branch of industrial statistics and is not, as is sometimes thought, synonymous with industrial statistics.

When we speak of (statistical) quality control, we are referring to two specific statistical techniques—the control chart and acceptance sampling—used as aids in assuring that desired quality standards are being met as economically as possible. Used technically, as in this discussion, the *quality* of a product refers to some property of the product such as the outside diameter of a ball bearing, the breaking strength of yarn, the drained weight of a No. $2\frac{1}{2}$ can of fruit salad, or the potency of a drug product.

It may surprise some persons to learn that two apparently identical parts made under carefully controlled conditions, from the same batch of raw material, and only seconds apart by the same machine, can nevertheless be quite different in many respects. This is due to the fact that it is seldom, if ever, possible to duplicate the exact conditions existing at a given time, no matter how great an effort is made. Thus, any manufacturing process, however good, is characterized by a certain amount of variability which is of the same *chance* or *random* nature as the variation we might find between repeated rolls of a pair of dice. Chance variation is an inherent and inevitable part of any process and there is no way in which it can be completely eliminated. When the variability present in a production process is confined to chance variation, the process is said to be in a *state of statistical control*.

Statistical control is usually achieved in a process by finding and eliminating trouble of the sort causing another kind of variation called *assignable variation.*   Under this heading we include variations in a process due to poorly trained operators, substandard or poor quality raw materials, faulty machine settings, broken or worn parts, and the like.   Inasmuch as manufacturing processes are rarely free from troubles of this sort for any length of time, it is important to have some systematic method of detecting serious deviations from a state of statistical control when, or if possible before, they occur.   It is to this end that *control charts*, the subject matter of the next few sections, are principally used.

II.2 The Control Chart

A control chart is a simple chart characterized essentially by three horizontal lines—a *central line* to indicate the desired standard, or

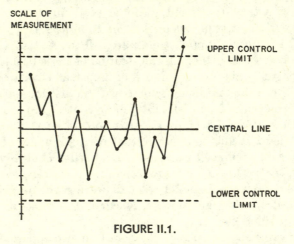

FIGURE II.1.

level, of the process, and an *upper and a lower control limit* (see Figure II.1).   By plotting results obtained from samples taken periodically at frequent intervals (for example, each hour, half-day, or day), it is possible to check by means of such a chart whether the variation between the samples may be attributed to chance or whether trouble of the sort indicated above has entered the process.   The upper and lower control limits serve as the decision criteria.   When a sample point falls beyond them, one looks for trouble, that is, for sources of assignable variation; otherwise, the process is left alone.   The detection of "lack of control" is a very important use of control charts, but it is not the only one.   If a process is known to be in control, it is possible to indicate by means of a control chart the process capability, namely, the average level at which the process is capable of operating, and also the amount of chance variation that is inherent in the process.

In short, control charts constitute powerful tools which provide management with logical bases for many important actions.

Since the word "quality" is used in a general way to refer to any characteristic, or property, of a product, let us distinguish between those characteristics which are *measurable* and those which are merely *noted*, or *observed*. If the relevant quality characteristic is the length of a bolt, the initial angle of a torsion spring, the bore of a ball bearing, the seam strength of a shirt, the weight of a bag of coffee, . . . , this characteristic can in some way be assigned a numerical value. Control charts designed to control such measurable characteristics are called *control charts for variables*.

In contrast, there are properties of products that are not measurable, referring to differences in kind rather than differences in degree. Bolts do or do not have heads, phonograph records do or do not have labels, shirts do or do not have all the necessary buttons, and, more generally, items are classified as non-defective or defective, satisfactory or unsatisfactory. Charts designed to control such quality characteristics as these are called *control charts for attributes;* they will be treated briefly in Section II.5. Control charts for variables will be discussed in Sections II.3 and II.4.

If the reader understands the principles underlying statistical estimation and tests of hypotheses (see Chapters 9 and 10), the logic of a control chart, its construction, and the reason why it works, will be fairly obvious. A control chart simply provides a convenient way of repeatedly testing hypotheses relative to the quality of a manufactured product. The purpose of these tests is to decide on the basis of relatively few observations whether a desired standard of quality is actually being met. The hypothesis being tested is that the process is "in control." If the hypothesis is accepted and the process is allowed to continue without modification, it is *presumed* that the process is operating satisfactorily or, in other words, that whatever variability there is can be attributed to chance. If the hypothesis is rejected, it is *presumed* that there is trouble of some sort, that the process is "out of control," and a search is made to detect the source (or sources) of assignable nonrandom variations.

As always in the testing of statistical hypotheses, there is the possibility of committing errors and this accounts for the use of the word "presume" in the preceding paragraph. If we *reject* the hypothesis that the process is in control when it should actually be accepted, we will be committing a Type I error (see page 218). The consequences of such a Type I error are that we look for nonexistent trouble, and such a search can easily lead to an increase in cost owing to loss in time and production. If we *accept* the hypothesis that the process is in control when actually it is not, we will be committing a Type II error.   The consequences of such a Type II error are that we fail to look for existing trouble and thus may manufacture and ship products which fail to meet required standards.

Naturally, manufacturers would like to avoid these errors completely and always follow the right course of action, but, as we have seen, this is generally impossible.   In practice, there is no choice but to seek an economic balance between the possible risks (losses) associated with the two kinds of errors, and in this country it has become more or less accepted practice to use *3-sigma control limits*, that is, charts on which the upper and lower control limits are drawn three standard deviations (standard errors) above and below the central line.   When dealing with sampling distributions that can be approximated closely with normal curves, the probability of committing a Type I error will, thus, be 0.003.   (The reader can easily check for himself that 99.7 per cent of the area under the normal curve lies between $z = -3$ and $z = 3$.)   The use of 3-sigma control limits does not provide any guarantee, or for that matter any information, about the probabilities of committing Type II errors.   As we saw in Chapter 10, questions about Type II errors can only be decided by studying operating characteristic curves (see page 220).   Nevertheless, the use of 3-sigma control limits can be justified on the grounds of long experience and satisfactory performance in practice, and it is recommended that they be used unless there are very good reasons why other control limits should be preferred.

II.3 $\bar{X}$ and R Charts†

To illustrate the construction of control charts for variables, let us assume that a manufacturer of electric motors, has started producing a new type of armature shaft and wishes to set up a control procedure for the *diameter* of the armature shafts, specifications for which are $1.000'' \pm 0.005''$.   Their objective is twofold: the company not only wants to know whether the production process designed for the new armature shaft is under control and in accordance with specifications, but also wants to use control charts in the future to keep a continual check on the process.

In the beginning the company plans to set up control procedures for *both* the average and the variability of the diameters.   Later

---

† In this and in subsequent sections we shall use the symbolism that is generally found in the literature on quality control.   It is important to note that this symbolism differs somewhat from that which has become standard in many other branches of statistics and which we used in other parts of this book.

on it may prove possible to dispense with one or the other. Generally speaking, a continual check on the quality of a product requires that we take samples at frequent intevals of time, calculate certain statistics, and decide on the basis of the values thus obtained whether or not the process may be presumed to be under control. In quality control it is customary to refer to such samples as "subgroups," or as "rational subgroups," if they are such that the variation *within* a sample can be due only to chance while the variation *between* samples can be due either to chance or to assignable causes of variation. Time, order of production, or source of production, are common bases used for forming rational subgroups, and frequent small subgroups of 4 or 5 are usually better and more informative than infrequent larger groups. It is generally recommended that at least 25 subgroups be used in the construction of trial control limits.

To check on average quality, it is customary in quality control to use the *mean*, $\bar{X}$, of the observations within each subgroup, and to check on variability, it is customary to use either the *standard deviation*, σ, or the *range*, R. As we pointed out in Chapter 4, the standard deviation is in many respects a more desirable measure of variation than the range, but, and this is important in quality control, the range is much easier to find. Since we have only to subtract the smallest sample value from the largest, the range is often used in problems of quality control where it is important to obtain results quickly and with a minimum of arithmetic.

To return now to the problem of the manufacturer of electric motors, let us suppose that samples of size 5 have been taken from each hour's production for 30 hours and that the diameters of the armature shafts are measured to the nearest thousandth of an inch. To simplify our notation, we shall "code" these measurements by writing each as so many thousandths of an inch above or below 1 inch. We shall thus write 2 instead of 1.002", -1 instead of 0.999", 0 instead of 1.000", 0 ± 5. The table which follows contains the (coded) measurements obtained in 30 samples (subgroups), and since we shall construct control charts based on the mean and the range, the values of these statistics are shown below the respective samples:

Subgroup:	1	2	3	4	5	6	7	8	9	10	11	12	13	14	15
	-1	0	1	1	0	-2	2	1	-1	1	2	1	0	2	1
	-1	1	1	-2	3	-2	1	-2	2	-1	-1	-3	0	-1	-1
	0	-3	2	-1	-2	1	-1	-1	1	0	-2	-2	-1	-1	2
	-2	2	-1	0	-2	1	0	-1	1	0	-1	1	0	0	0
	1	1	0	-1	1	2	0	0	2	-1	-1	0	1	-2	-2
$\bar{X}$	$-.6$	.2	.6	$-.6$	0	0	.4	$-.6$	1.0	$-.2$	$-.6$	$-.6$	0	$-.4$	0
R	3	5	3	3	5	4	3	3	3	2	4	4	2	4	4

Subgroup:	16	17	18	19	20	21	22	23	24	25	26	27	28	29	30
	2	1	0	-2	-3	1	-2	2	1	3	2	1	1	2	0
	-1	1	0	1	3	1	1	2	0	-1	0	-2	0	0	0
	0	0	1	2	1	0	3	-1	1	0	0	-2	0	3	2
	0	0	-2	0	2	-1	1	1	2	1	1	1	-2	1	1
	1	1	-1	1	0	2	-3	0	-2	-1	1	1	-1	-2	-1
$\bar{X}$	.4	.6	-.4	.4	.6	.6	0	.8	.4	.4	.8	-.2	-.4	.8	.4
R	3	1	3	4	6	3	6	3	4	4	2	3	3	5	3

To estimate the true average of the (coded) diameters of the armature shafts produced by the new process, we shall use the *mean* of the thirty $\bar{X}$'s which, in the symbolism of quality control, is written as $\bar{\bar{X}}$. In general, if we have k subgroups whose means are $\bar{X}_1$, $\bar{X}_2, \ldots$, and $\bar{X}_k$, the over-all mean, *which will serve as the central line of the control chart for the mean,* is given by the formula

$$\bar{\bar{X}} = \frac{\bar{X}_1 + \bar{X}_2 + \cdots + \bar{X}_k}{k} \qquad (\text{II.3.1})\star$$

Substituting the means of the 30 samples given above, we get $\bar{\bar{X}} = 3.8/30 = 0.13$.

The *average range* that can be expected for samples (of the given size) from this process is estimated by means of the formula

$$\bar{R} = \frac{R_1 + R_2 + \cdots + R_k}{k} \qquad (\text{II.3.2})\star$$

where $R_1, R_2, \ldots$, are the ranges of the individual subgroups. In our example, this average range, *which will serve as the central line of the control chart for the range,* is $\bar{R} = 105/30 = 3.50$.

To construct upper and lower control limits for $\bar{X}$ and R, namely, criteria for testing hypotheses about the mean and variation of the population from which these samples were obtained, we shall, as always (see Chapter 10), have to investigate the sampling distributions of the given statistics. Leaving all theoretical considerations about these sampling distributions aside, let us merely state that the required 3-sigma control limits for the *mean* are usually calculated by means of the formula

$$\bar{\bar{X}} \pm A_2\bar{R} \qquad (\text{II.3.3})\star$$

where A_2 is a constant which depends on the size of the sample (subgroup) and which can be looked up in Table VII on page 508. Similarly, the 3-sigma control limits for the *range* are usually calculated by means of the formulas

$$D_4\bar{R} \quad \text{and} \quad D_3\bar{R} \qquad (\text{II.3.4})\star$$

where D_4 and D_3 are also given in Table VII.

Since $A_2 = 0.577$, $D_4 = 2.115$, and $D_3 = 0$, for samples of size 5, we find that in our example the upper and lower control limits for $\bar{X}$ are $0.13 + 0.577(3.5) = 2.15$ and $0.13 - 0.577(3.5) = -1.89$, while the upper and lower control limits for the range are $2.115(3.5) = 7.40$ and $0(3.5) = 0$. The resulting control charts together with the means and ranges of the original 30 samples are shown in Figure II.2.

The fact that on both charts all sample points fall within the 3-sigma control limits can be interpreted as implying that the process is in a state of statistical control or, in other words, that the only kind of variation present is chance variation. Had the charts indicated

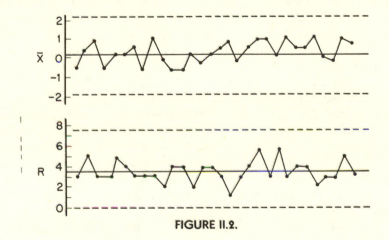

FIGURE II.2.

that at one time or another the process was *not in control*, it would have been necessary to eliminate the out-of-control data, recalculate $\bar{\bar{X}}$ and $\bar{R}$ on the basis of the remaining samples, and establish new control limits.

As we pointed out earlier, the purpose of constructing control charts is not only to check whether there is any assignable variation, but also to see whether the process will meet specifications. To answer this kind of question, we shall have to estimate the true mean and standard deviation of the diameters of armature shafts produced by the process, and in the notation of quality control these quantities are written as $\bar{X}'$ and σ'. (In the notation used otherwise in this book we would refer to these population values as μ and σ.) Although there are other, and perhaps better, ways in which this can be done, it is customary in quality control problems of this kind to estimate $\bar{X}'$ by means of $\bar{\bar{X}}$ and σ' by means of $\bar{R}/d_2$, where d_2 is another constant which can be looked up in Table VII. Since $d_2 = 2.326$ for samples of size 5 and $\bar{\bar{X}}$ and $\bar{R}$ equalled 0.13 and 3.5, we shall thus estimate the true mean of the (coded) diameters of the armature shafts as 0.13 and their standard deviation as 3.5/2.326 = 1.50. Assuming that the diameters of the armature shafts produced by the given process are normally distributed with the above mean and standard deviation, we can say that 99.7 per cent of the armature shafts will have (coded) diameters lying *between their mean plus and minus three standard deviations* and, hence, between 0.13 − 3(1.50) = −4.37 and 0.13 + 3(1.50) = 4.63. Since the specifications in terms of the coded data read 0 ± 5, it appears that almost all of the product will meet specifications so long as the process will continue operating at the same level and with the same amount of variability.

To keep a continual check on the process, we simply extend the control lines of Figure II.2 and plot, as before, the means and ranges obtained from repeated (hourly or daily) samples of 5 observations. If at any time in the future a point falls outside the control limits, appropriate actions will have to be taken to search for possible sources of assignable variation. Of course, it is possible to obtain a point outside the control limits by chance, but the probability of this happening is less than 0.003. It should also be remembered that the control limits which we established are really only *estimates* and after more information will have been obtained it may well become desirable to recalculate and, if necessary, adjust them. Furthermore, it could happen that changes will take place in the process itself later on, and such changes, of course, would make it necessary to calculate new control limits.

Control charts are not only watched for points falling outside the control limits; they are also scrutinized for unusual patterns suggesting trouble. For instance, a run of 7 successive points on the same side of the central line of an $\bar{X}$ chart is looked upon as roughly equivalent to a point outside the control limits, and it is usually interpreted to mean that the process average has shifted or that the process will soon go out of control. In this way, impending trouble can often be detected and prevented at considerable savings.

II.4 $\bar{X}$ and σ Charts

Although sample ranges are very easy to compute and, for this reason, used in many problems of quality control, a standard deviation will generally provide more information about the variability of a set of data. For samples of size 10 or less there is relatively little to gain by using the standard deviation instead of the range and for $n = 2$ there is no gain at all, but when n is greater than 10 it is recommended that the standard deviation be used instead of the range. In quality control, sample standard deviations are defined by means of the formula

$$\sigma = \sqrt{\frac{\sum\limits_{i=1}^{n} (X_i - \bar{X})^2}{n}} \qquad \text{(II.4.1)}^\star$$

where the X's are the individual observations. The use of σ in connection with *sample* standard deviations is frowned upon in some branches of statistics, but it has become more or less established in quality control work.

If we have k samples (subgroups) whose standard deviations are $\sigma_1, \sigma_2, \ldots,$ and σ_k, we write their mean as

$$\bar{\sigma} = \frac{\sigma_1 + \sigma_2 + \cdots + \sigma_k}{k}. \qquad \text{(II.4.2)}^\star$$

In terms of $\overline{\overline{X}}$ and $\bar{\sigma}$, the control limits of the $\bar{X}$-chart become

$$\overline{\overline{X}} \pm A_1\bar{\sigma} \qquad \text{(II.4.3)}^\star$$

while the central line is again $\overline{\overline{X}}$. The constant A_1 can also be looked up in Table VII. To control *variability*, we now use a σ-chart whose central line is $\bar{\sigma}$ and whose upper and lower control limits are

$$B_4\bar{\sigma} \quad \text{and} \quad B_3\bar{\sigma} \qquad \text{(II.4.4)}^\star$$

where B_4 and B_3 are given in Table VII. It will be left as an exercise for the reader to construct such $\bar{X}$ and σ-charts for the data of the preceding section.

The methods which we have illustrated so far are designed to establish control limits and central lines when $\bar{X}'$ and σ', the true mean and standard deviation, are unknown. If we wanted to construct control charts for processes for which these quantities are known or specified, we could write the central line and control limits for $\bar{X}$ as $\bar{X}'$ and $\bar{X}' \pm A\sigma'$ and those for σ as $c_2\sigma'$, $B_2\sigma'$, and $B_1\sigma'$. All these constants are given in Table VII.

EXERCISES

1. Construct $\bar{X}$ and σ charts for the coded diameters of the armature shafts given on page 479. The mean of the standard deviations of the 30 subgroups, $\bar{\sigma}$, is 1.30. Plot the means of the 30 subgroups on the $\bar{X}$-chart and comment on the state of control.

2. In order to determine whether or not a process producing bronze castings is in control, 20 subgroups of size 6 are taken. The quality characteristic of interest is the weight of the castings and it is found that $\overline{\overline{X}}$ is 3.126 grams and $\bar{R} = 0.009$ grams.

(a) Estimate σ', the true standard deviation of the weights of the castings.

(b) Assuming that the process is in control, find upper and lower control limits for the subgroup means.

(c) Assuming that the process is in control, find upper and lower control limits for the subgroup ranges.

(d) Using (a), within what limits would you expect 99.7 per cent of all individual measurements to fall?

3. It is proposed to establish control over a machine which turns out some 5,000 nails per hour. Subgroups of 10 nails are taken from each hour's production, and lengths of the nails are measured. The following are the means and ranges (in inches) obtained in 20 subgroups:

Subgroup:	1	2	3	4	5	6	7
$\bar{X}$	1.524	1.520	1.488	1.521	1.505	1.510	1.495
R	.039	.028	.035	.033	.041	.025	.030
Subgroup:	8	9	10	11	12	13	14
$\bar{X}$	1.491	1.491	1.482	1.475	1.478	1.522	1.531
R	.037	.028	.043	.032	.027	.041	.038
Subgroup:	15	16	17	18	19	20	
$\bar{X}$	1.531	1.502	1.490	1.465	1.529	1.444	
R	.028	.040	.054	.060	.020	.029	

Calculate control limits for $\bar{X}$ and R, plot the given data, and determine whether or not the process may be considered to be in control.

4. Tests of the breaking strength of a certain type of worsted yarn yielded the following (coded) data. The values given below are computed from subgroups of size 10 taken from the production of 30 consecutive periods:

Subgroup:	1	2	3	4	5	6	7	8	9	10
$\bar{X}$	9.12	8.57	9.43	7.81	8.96	9.32	8.34	7.81	9.23	10.36
σ	1.19	7.00	1.64	1.13	1.29	2.19	.79	1.14	1.20	1.10
Subgroup:	11	12	13	14	15	16	17	18	19	20
$\bar{X}$	10.52	7.76	8.20	8.97	8.54	9.06	8.56	7.73	9.31	8.56
σ	1.78	.91	.93	1.30	.89	1.02	1.45	.80	1.62	.71
Subgroup:	21	22	23	24	25	26	27	28	29	30
$\bar{X}$	9.53	7.49	9.07	7.85	8.14	8.81	8.43	9.12	7.95	8.10
σ	2.30	1.01	.78	1.12	.84	1.00	1.12	.84	.65	1.07

Construct $\bar{X}$ and σ charts, plot the given data, and determine to what extent the process is under statistical control. If there are out-of-control samples, eliminate them from the data and recalculate the control values.

II.5 Control Charts for Attributes

To illustrate the construction of control charts for attributes, let us suppose that the manufacturer of electric motors also produces small battery-powered electric motors used in toy boats, cars and plastic model kits.   The specifications for these toy motors prescribe a certain testing procedure according to which each motor can be classified as satisfactory or unsatisfactory (defective).   To set up a control chart for the proportion of defective motors, called the *fraction defective* samples of size 100 are tested from each half-day's production until 30 such groups are at hand.   Although it is not always necessary, or possible, that all the samples be of the same size, it is certainly desirable.   Also, to establish control charts for attributes it is usually recommended that the size of the samples be 50 or more.   The following are the results obtained in the 30 samples:

Lot Number	Sample Size	Number of Defectives	Fraction Defective
1	100	5	.05
2	100	2	.02
3	100	4	.04
4	100	6	.06
5	100	3	.03
6	100	2	.02
7	100	0	.00
8	100	9	.09
9	100	15	.15
10	100	3	.03
11	100	2	.02
12	100	2	.02
13	100	0	.00
14	100	6	.06
15	100	3	.03
16	100	4	.04
17	100	2	.02
18	100	0	.00
19	100	5	.05
20	100	4	.04
21	100	2	.02
22	100	1	.01
23	100	1	.01
24	100	2	.02
25	100	1	.01
26	100	0	.00
27	100	1	.01
28	100	2	.02
29	100	2	.02
30	100	1	.01
	3,000	90	

To estimate the true proportion of defectives, we combine all of the samples, getting 90 defectives among 3,000 motors tested and an estimate of $90/3,000 = 0.03$. The symbol used for this estimate, *which provides the central line of the control chart for the fraction defective,* is $\bar{p}$. To construct 3-sigma control limits, we use the fact that when n, the sample size, is large, the sampling distribution of a proportion can be approximated closely with a normal curve having the mean p' and the standard deviation $\sqrt{p'(1 - p')/n}$, where p' is the true proportion of defectives (see Section 9.4). Substituting $\bar{p}$ for p', the upper and lower control limits become

$$\bar{p} \pm 3 \sqrt{\frac{\bar{p}(1 - \bar{p})}{n}} \qquad \text{(II.5.1)}\star$$

and in our example we get $0.03 + 3 \sqrt{(0.03)(0.97)/100} = 0.081$ and $0.03 - 3 \sqrt{(0.03)(0.97)/100} = -0.021$. For the latter value we shall substitute 0 since there can obviously not be a negative proportion of

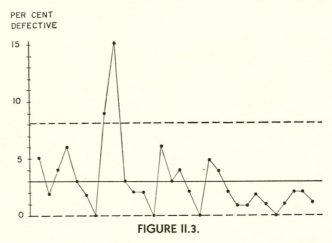

FIGURE II.3.

defectives. Converting these control limits as well as the central line into *percentages,* we get the control chart shown in Figure II.3. As can be seen from this diagram, the points representing samples from Lots 8 and 9 are *above* the upper control limits and, hence, indications of a possible lack of control. For the sake of argument, let us suppose that an investigation revealed an assignable cause of variation present at these times—namely, a breakdown which the operator failed to record. For maintaining control in the future, a new control chart will have to be constructed, in the calculation of whose central line and control limits the two out-of-control samples will have to be eliminated. It will be left as an exercise for the reader to show that the remaining 28 samples yield a central line of 2.4 per cent, an upper control limit of 6.9 per cent, and a lower control limit of 0. It should also be observed that the remaining 28 points all fall within these new limits.

The interpretation of a control chart for fraction (or percentage) defectives is the same as that of the control charts of the preceding sections. Any sample point falling outside the control limits is evidence of a possible lack of control inasmuch as the probability of getting such a value by chance is less than 0.003. It should be noted that in our illustration there seems to be a downward trend in the proportion of defectives and this suggests that it would probably be wise to recalculate the control limits after more data has been obtained.

To simplify the work of the inspector who plots the necessary points on control charts, we could modify the chart of Figure II.3 so that he can directly plot the *number*, rather than the fraction or percentage of defectives. Multiplying the central line as well as the control limits by the sample size n, the central line of the control chart for the *number of defectives* becomes $n\bar{p}$ while the control limits become

$$n\bar{p} \pm 3 \sqrt{n\bar{p}(1 - \bar{p})} \qquad (II.5.2)\star$$

In (II.5.1) and (II.5.2) we *estimate* p', the true proportion of defectives, with $\bar{p}$. If p' were known or specified, we could construct appropriate control limits and central lines with the same formulas as before, but with p' substituted for $\bar{p}$.

There are some quality control problems concerned with counting numbers of defectives which cannot be handled with the methods described above, because n, the sample size, is essentially unspecified though presumably very large. This happens, for example, if we count the number of imperfections in a piece of cloth, the number of blemishes in a sheet of paper, or the number of air bubbles in a piece of glass. Visualizing such a piece of cloth (sheet of paper, or piece of glass) as subdivided into many small squares, each being perfect or having an imperfection, we are faced with a situation in which the sampling distribution of the total number of imperfections can be approximated closely with a *Poisson* distribution (see page 165). Letting c stand for the number of defects counted in one unit of cloth (paper, glass, or whatever), and $\bar{c}$ for the mean of the defects counted in several (usually 25 or more) such units of cloth, the central line of the control chart for c is $\bar{c}$ and the 3-sigma control limits are

$$\bar{c} \pm 3 \sqrt{\bar{c}} \qquad (II.5.3)\star$$

These formulas are based on a normal curve approximation to the Poisson distribution.

To illustrate this last technique, let us suppose that 25 pieces (units) of cloth contained, respectively, 2, 4, 3, 6, 5, 4, 8, 1, 4, 2, 3, 5, 6, 4, 2, 2, 1, 7, 5, 3, 2, 1, 5, 6, and 4 imperfections. Getting $\bar{c} = \frac{95}{25} = 3.80$, we find that the upper and lower control limits for the number of imperfections per unit of cloth become $3.80 + 3\sqrt{3.80} = 9.65$ and $3.80 - 3\sqrt{3.80} = -2.05$. The latter value will again be replaced by 0. The resulting control chart containing the points representing the given data is shown in Figure II.4. Since none of the points fall outside the control limits, it may be presumed that the process is in a state of statistical control. If the given level of quality, on the

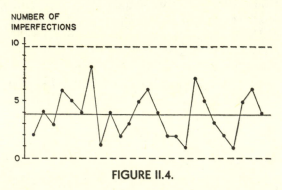

FIGURE II.4.

average 3.8 imperfections per unit of cloth, is satisfactory, the control chart which we have constructed may be used in the future to keep a continual check on the process.

EXERCISES

1. Recalculate the control limits for the battery powered electric motors data on page 485 after the two out-of-control samples are eliminated.

2. A manufacturer of transistors found the following number of defectives in 25 subgroups of 50 transistors:

   3, 5, 4, 2, 3, 2, 7, 0, 2, 4, 2, 3, 4,
   1, 2, 4, 8, 2, 4, 2, 6, 4, 3, 1, 4.

   Construct a control chart for the *fraction defective*, plot the sample data on this chart, and comment on the state of control.

3. A manufacturer of charcoal briquets of the type used in barbecue grills desires to establish control of his production process at a level p' 0.05. Using this value of p', construct a control chart for the number of defectives in subgroups of size 200. Given that 20 consecutive samples of this size contained 10, 16, 20, 3, 4, 12, 12, 14, 12, 14, 11, 13, 9, 8, 9, 8, 7, 8, 10, and 5 defectives, plot these sample values on the chart and comment on the possibility of establishing control at the desired level.

4. For the last three months of 1968, a production department completed

500 units a day. The average number of defective parts was 30, though on some days there were only 15 defectives and on others there were as many as 46. Is this evidence of a possible lack of control? Base your argument on 3-sigma control limits.

5. The following are the number of defects noted in the final inspection of 30 bolts of woolen cloth: 0, 3, 1, 4, 2, 2, 1, 3, 5, 0, 2, 0, 0, 1, 2, 4, 3, 0, 0, 0, 1, 2, 4, 5, 0, 9, 4, 10, 0, and 3. Construct a control chart for c, the number of defects, plot the given data, and comment on the state of control.

II.6 Acceptance Sampling

Control chart techniques are by no means suited to all types of problems that arise in maintaining the quality of manufactured products. Suppose, for example, that a manufacturer has contracted to buy parts of a certain standard of quality and that these parts are shipped to him in large lots. For simplicity, let us confine our discussion of so-called *acceptance sampling* to *lot-by-lot sampling inspection* and let us suppose that the quality characteristic in which we are interested is an attribute, rather than a variable. Individual items in a lot are thus non-defective or defective depending on whether they possess a certain characteristic. The buyer of these parts is, of course, interested in making sure that when he accepts and pays for a lot, the product meets quality standards agreed upon. He would like to accept all good lots, reject all bad ones, using some inspection scheme to decide which lots are good and which ones are bad.† Naturally, the buyer is also interested in making this decision as economically as possible within the limitations imposed by his desire for accuracy.

In some cases it may be possible to inspect each item in a lot and to reject those which do not conform to specifications. This type of 100 per cent inspection is called "screening," and although it may be appropriate in some instances, it is often impractical or impossible. If the testing is destructive, 100 per cent inspection can obviously not be used. Moreover, screening is generally wasteful of both time and money, and, because neither men nor machines are infallible, it cannot even guarantee that all defective items will be eliminated.

---

† When we say that a lot is "good," this is not meant to imply that it cannot contain any defectives at all. Some defectives in submitted lots are generally inevitable, and the buyer, recognizing this, sets a practical limit on the proportion of defectives he is willing to accept at the price he is paying for the product.

As its name implies, *lot-by-lot sampling inspection*, which is now widely used in industry, involves taking samples and deciding on the basis of the items they contain whether to accept or reject entire lots. Since generalizations from samples to lots (populations) cannot be expected to be infallible, there is the possibility that in *lot-by-lot sampling inspection* a good lot will be rejected or a bad lot accepted.

It stands to reason that both the producer and the consumer, the seller and the buyer, will want to know the risks to which they are exposed by an inspection plan before agreeing to it as an acceptance criterion. They will want to know the probability of rejecting a good lot, the probability of accepting a bad one, or, in other words, the probabilities of committing Type I and Type II errors. These are the probabilities which in Exercise 3 on page 221 were referred to as the *producer's and consumer's risks*.

If the producer and consumer specify these probabilities as well as the proportion of defectives above which a lot is considered to be *bad* and the proportion of defectives below which a lot is considered to be *good*, it is possible to construct inspection plans which will meet these requirements. For instance, if we wanted the probability of rejecting a lot with 2 per cent or fewer defectives to be less than or equal to 0.05 and the probability of accepting a lot with 6 per cent or more defectives to be also less than or equal to 0.05, we could refer to a suitable table and come up with the following inspection plan:† *take a sample of 235 items from each lot and accept it if the number of defectives is 8 or less, reject it if the number of defectives is 9 or more.* (It is assumed here that the lots are so large that for all practical purposes they may be looked upon as infinite populations.)

Numerous tables are available for the construction of acceptance sampling plans based on such things as the AQL (*acceptable quality level*), the worst quality at which the consumer is willing to accept a lot with a specified high probability, the LTPD (*lot tolerance percent defective*), the quality above which there is only a specified small probability that a lot will be accepted, the AOQL (*average outgoing quality limit*), the maximum quality that can be expected in the long run when all rejected lots are subjected to 100 per cent inspection, with all defectives removed and replaced with good items, and other criteria.

We shall not go into the problem of acceptance sampling in any further detail except to remind the reader of the possibility of affecting a reduction in cost by using *double or multiple sampling plans* of the sort outlined in Section 12.5. It is also feasible at times to use an

---

† The table used here is Table 13.17 in Bowker, A. H., and Lieberman, G. J., *Handbook of Industrial Statistics.* Englewood Cliffs, N.J.: Prentice-Hall, Inc., 1955

item-by-item sequential plan, deciding whether to accept, reject, or continue sampling after each item is inspected. All these plans are nowadays used quite widely in the inspection of the quality of manufactured products and also, with considerable success, in such non-manufacturing problems as auditing and controlling clerical accuracy.

BIBLIOGRAPHY

The various topics which we have touched upon in this Appendix are treated in more detail in

Bowker, A. H., and Lieberman, G. J., *Handbook of Industrial Statistics.* Englewood Cliffs, N.J.: Prentice-Hall, Inc., 1955.

Burr, I. W., *Engineering Statistics and Quality Control.* New York: Mc-Graw-Hill, 1953.

Cowden, D. J., *Statistical Methods in Quality Control.* Englewood Cliffs, N.J.: Prentice-Hall, Inc., 1957.

Grant, E. L., *Statistical Quality Control,* 3rd ed. New York: McGraw-Hill, 1964.

Calculations with Rounded Numbers

III.1 Rounded Numbers

In statistics one generally deals with rounded numbers at some stages of one's work, so it will be well to review briefly some of the questions that are involved. First, there is the mechanics of rounding quantities, say, to the nearest dollar, the nearest ounce, the nearest thousand employees, the nearest tenth of a pound; and then there is the problem of deciding in each case how far to round off.

If we wanted to round $125.37, which lies between $125 and $126, *to the nearest dollar*, we would simply substitute for it whichever of the two values is closest, namely, $125. Similarly, if we wanted to round the number of vessels in the U.S. Merchant Marine in 1965 to the nearest thousand, we would write 46 thousand instead of 45,579, and if we wanted to round the 1966 population of New York to the nearest million, we would write 18 million instead of 18,258,000. In the first case we rounded up since 45,579 is closer to 46 thousand than to 45, and in the second case we rounded down since 18,258,000 is closer to 18 million than it is to 19. *The general rule to follow when rounding a number lying between two chosen units (dollars, cents, inches, 1,000 cars, etc.) is to substitute the nearest unit.*

The only time that the above rule can lead to difficulties is when a number lies exactly half-way between two units. This would happen, for example, if we wanted to round $12.50 to the nearest dollar, 5.75 inches to the nearest tenth of an inch, 28.5 cents to the nearest cent, or 0.015 seconds to the nearest hundredth of a second. *The rule to follow here is always to round off in such a way that the last digit remaining on the right is even, that is, 0, 2, 4, 6, or 8.* Accordingly, we shall round the above quantities to $12, 5.8 inches, 28 cents, and 0.02

seconds. If we follow this rule we will sometimes round up, sometimes down, and in the long run the error due to rounding should more or less average out. If we consistently rounded up or down, we would be introducing what is called a *systematic error* or a *systematic bias*.

Although the above rule is obeyed in most scientific work, it is often ignored in practice. For instance, a grocer who sells 2 cans of frozen orange juice for 37 cents does not round $\frac{37}{2} = 18.5$ to 18; he sells single cans for 19 cents. Similarly, a bank which pays 1 per cent interest on savings accounts every 6 months will credit an account of \$15.50 with 15 cents, rounding 15.5 to 15 instead of 16. These examples show that methods used to round numbers can be dictated by practical considerations.

The problem of how far to round a given number can actually be decided only on an individual basis. There is always the danger of *not rounding enough* and the equally important, and sometimes more serious, danger of *rounding too much*. The trouble with not rounding enough is that a number can thus give an impression of spurious accuracy. For instance, an insurance official would probably be more impressed if a company reported the average cost of certain claims as \$246.4285714 than if it reported this figure as \$246.43. If this average were actually based on as few as 7 claims, the extra digits could easily create the unwarranted impression that the average represents much more than it does.

The danger of rounding too much is best explained by means of an example. Let us suppose, for instance, that the accounting department of a firm finds that the over-all manufacturing cost of an item is 82.67 cents and that it reports this figure as 83 cents. If this value were used to estimate the manufacturing cost of, say, 1,000,000 of the given items, the result would be \$830,000, and the error due to rounding would be $830,000 - 826,700 = \$3,300$. To avoid this kind of error, it would have been better to report the cost as being \$82.67 per 100 items. *The rule to follow in situations of this kind is never to round too much if the quantities will subsequently be multiplied by large numbers.*

One question that is particularly important in statistical work is how much to round the results of various kinds of computations with numbers that are, themselves, rounded off. Since this will depend largely on the nature of the computations, we shall treat this question separately in the following section.

III.2 Calculations with Rounded Numbers

When adding or subtracting rounded numbers we must always remember that a chain is only as strong as its weakest link. For example, if we were asked to add 25.8743, 3.45, 1005.2, and 16.8, which are all rounded to the indicated numbers of decimals, we cannot give the result as 1051.3234. After all, 1005.2 stands for a number between 1005.15 and 1005.25, 16.8 stands for a number between 16.75 and 16.85, . . . , and we will therefore have to round the sum to 1051.3, or perhaps to 1051, since we cannot be sure even of the .3. *A relatively safe rule to follow in the addition and subtraction of rounded numbers is never to carry a digit to the right of the last digit carried in any one of the numbers which we add or subtract.* (Sometimes, when adding a great many numbers, it is permissible to carry extra digits in view of the fact that the rounding error can be expected to average out.)

When multiplying or dividing rounded numbers we will have to watch what is called the *number of significant digits.* By this we mean the number of digits which remain after we have discarded all zeros to the left of the first non-zero digit, and possibly also some of the zeros to the right of the last non-zero digit. For example, 34.5, rounded to the nearest tenth, has 3 significant digits; 0.014, rounded to the nearest thousandth, has 2 significant digits; 476.81, rounded to the nearest hundredth, has 5 significant digits; and 0.000072, rounded to the nearest millionth, has 2 significant digits. If we rounded 13,038 to the nearest thousand, we could write 13 *thousand* or 13,000, and there would be only 2 significant digits. If we rounded this number to the nearest hundred, we could write 130 hundreds, 13.0 thousands, or 13,000, and there would be 3 significant digits. As can be seen from this example, if there are zeros on the right we will have to be careful to investigate which ones are significant and which ones are not. There exists a notation which avoids this difficulty, but it is seldom used in business applications.

The rule to be followed in the multiplication and division of rounded numbers is very simple: the result must not have more significant digits than any one of the numbers which we multiply or divide. For instance, the product of 7.6 and 3.8445, both of which are rounded, should be given as 29 and not as 29.21820. There are 2 significant digits in 7.6, 5 in 3.8445, so there should be only 2 significant digits in the product. Of course, when multiplying or dividing a rounded number by an *exact* number, the above rule does not apply. If we divide a rounded number, say 24.56, by 2, we can write the result as 12.28.

By following the above rules and using some judgment, the reader should not only avoid making mistakes but save a good deal of time and effort by not carrying uncalled-for digits.

The Use of Logarithm and Square Root Tables

IV.1 The Use of Logarithm Tables

Logarithms usually serve to simplify calculations and they are used in this book to determine geometric means, to fit certain kinds of trends, and in some other instances. The advantage of using logarithms lies in the fact that instead of multiplying or dividing numbers we add or subtract their logarithms, and instead of going through the tedious process of calculating a root we multiply a logarithm by a constant.

To illustrate how two numbers are multiplied with the use of logarithms, let us consider the rather trivial example of multiplying 100 and 1,000. Writing $100 = 10^2$ and $1,000 = 10^3$, the product of these numbers becomes $100 \cdot 1000 = 10^2 \cdot 10^3 = 10^{2+3} = 10^5 = 100,000$. If a number N equals 10 raised to some power, we refer to this power as the *common logarithm of N* and write it as log N. Thus, the common logarithm of 100 is 2 since $100 = 10^2$, the common logarithm of 1,000 is 3 since $1,000 = 10^3$, and the common logarithm of 100,000 is $2 + 3 = 5$ since $100,000 = 10^5$. This illustrates how the logarithm of a product equals the sum of the logarithms of the individual factors.

The logarithms of 100 and 1,000 were easy to find since these numbers are integral powers of 10, but most other logarithms will have to be looked up in an appropriate table. To explain the use of Table VIII, *a four place logarithm table*, let us discuss separately the integral part of a logarithm, called its *characteristic*, and the fractional part, called its *mantissa*. Thus, if the common logarithm of some number is 2.3010, the characteristic is 2 and the mantissa is .3010.

Whereas a mantissa can always be looked up in a table, for example, in Table VIII, the characteristic of a logarithm must be determined by following any one of a number of rules. For example, if we followed the rule that if one number lies between two other numbers, its logarithm lies between the logarithms of these numbers, we could argue that log 683 must have the characteristic 2 since 683 lies between 100 and 1,000.   Most beginners will find it easier to use the following rules:

(a) *If a positive number is greater than 1, the characteristic of its logarithm equals the number of places the decimal point has to be moved to the left so that it comes after the first non-zero digit.*

(b) *If a positive number is less than 1, the characteristic of its logarithm is minus the number of places the decimal point has to be moved to the right so that it comes after the first non-zero digit.*

We thus find that the characteristic of log 67.2 is 1, the characteristic of log 0.00139 is -3, the characteristic of log 3,470 is 3, and the characteristic of log 0.572 is -1.

To find the mantissa of the logarithm of any positive number in Table VIII, we must first round the number to 3 significant digits unless, of course, the number already has 3 or less. (If more accuracy is desired, the reader will have to use another table or interpolate, a process which we shall not go into here.)   Then, the first two digits determine the *row* of Table VIII in which we will find the mantissa and the third digit determines the *column*.   For example, to find the mantissa of log 683, we go down the left-hand column of Table VIII until we come to 68 and then we go over to the column marked 3. We, thus, find that the desired mantissa is .8344.   Since the characteristic is 2, we finally get log 683 = 2.8344.   Similarly, the reader can check for himself that log 67.2 = 1.8274, log 3,470 = 3.5403, and log 1.29 = 0.1106.   When the characteristic is *negative*, it is customary to add and subtract 10, writing, for example, log 0.00139 as 7.1430 $-$ 10 and log 0.572 = 9.7574 $-$ 10.   This has the advantage that the fractional part is always kept positive, which is necessary if we want to call it the mantissa.

As we pointed out above, logarithms serve to simplify multiplications, divisions, and the extractions of roots. If M and N are two positive numbers, three rules used for this purpose are

$$\log M \cdot N = \log M + \log N \qquad\qquad \text{(IV.1.1)}$$

$$\log \frac{M}{N} = \log M - \log N \qquad\qquad \text{(IV.1.2)}$$

$$\log M^k = k \cdot \log M \qquad\qquad \text{(IV.1.3)}$$

where k is a constant. To find roots, for example, $\sqrt{51}$ or $\sqrt[3]{17}$, we write these quantities as $51^{\frac{1}{2}}$ and $17^{\frac{1}{3}}$ and then apply (IV.1.3).

To find the number that corresponds to a given logarithm, we simply perform all of the above work in reverse. Given, for example, that the logarithm of a number is 1.6191, we first look for the number (series of digits) whose logarithm has the mantissa .6191. According to Table VIII this is 416, and the only thing that remains to be done is to put the decimal point in the right place. Since the characteristic is 1, we begin after the first non-zero digit on the left, in this case after the 4, and move the decimal point 1 place *to the right*. Hence, the number whose logarithm equals 1.6191 is 41.6. Similarly, to find the number whose logarithm is 8.5065 − 10, we use Table VIII to get 321 corresponding to a mantissa of .5065, and then, beginning after the 3, move the decimal point 2 places *to the left*. Hence, the number whose logarithm is 8.5065 − 10 is 0.0321. The reader may wish to check for himself that the number whose logarithm equals 0.9759 is 9.46, the number whose logarithm equals 9.7076 − 10 is 0.510, and the number whose logarithm equals 3.5024 is 3,180, all rounded to 3 significant digits. To find the number which corresponds to a logarithm whose mantissa is not given exactly in Table VIII, we simply round to the nearest entry that can be found in this table.

IV.2 The Use of Square Root Tables

Although square root tables are relatively easy to use, most beginners seem to have some difficulty in choosing the right column and placing the decimal point correctly in the answer. Table IX, in addition to containing the *squares* of the numbers from 1.00 to 9.99 spaced at intervals of 0.01, gives the square roots of these numbers rounded to 6 decimals. To find the square root of any positive number rounded to 3 significant digits, we have only to use the following rule in deciding whether to take the entry of the $\sqrt{n}$ or the $\sqrt{10n}$ column:

> *Move the decimal point an even number of places to the right or to the left until a number greater than or equal to 1 but less than 100 is reached. If the resulting number is less than 10 go to the $\sqrt{n}$ column, if it is 10 or more go to the $\sqrt{10n}$ column.*

Thus, to find the square roots to 12,500, 374, and 0.0514 we go to the $\sqrt{n}$ column since the decimal point has to be moved, respectively, 4 places to the left, 2 places to the left, and 2 places to the right, to give 1.25, 3.74, and 5.14. Similarly, to find the square roots of 1,060, 0.182, and 0.0000352 we go to the $\sqrt{10n}$ column since the decimal point has to be moved, respectively, 2 places to the left, 2 places to the right, and 6 places to the right, to give 10.6, 18.2, and 35.2.

Having found the entry in the appropriate column of Table IX, the only thing that remains to be done is to put the decimal point in the right position. Here it will help to use the following rule:

> *Having moved the decimal point an even number of places to the left or right to get a number greater than or equal to 1 but less than 100, the decimal point of the entry of the appropriate column is moved half as many places in the opposite direction.*

For example, to determine the square root of 12,500 we first note that the decimal point has to be moved *4 places to the left* to give 1.25. We thus take the entry of the $\sqrt{n}$ column corresponding to 1.25, move the decimal point *2 places to the right*, and get $\sqrt{12,500} = 111.8034$. Similarly, to determine the square root of 0.0000352, we note that the decimal point has to be moved *6 places to the right* to give 35.2. We thus take the entry of the $\sqrt{10n}$ column corresponding to 3.52, move the decimal point *3 places to the left*, and get $\sqrt{0.0000352} = 0.005932959$. In actual practice, if a number whose square root we want to find is rounded, the square root will have to be rounded to as many significant digits as the original number.

Statistical Tables

I. Normal Curve Areas
II. Values of t
III. Values of χ^2
IV. Values of F
V. 95 Per Cent Confidence Intervals for Proportions
VI. Binomial Coefficients
VII. Control Chart Constants
VIII. Logarithms
IX. Squares and Square Roots
X. Random Numbers

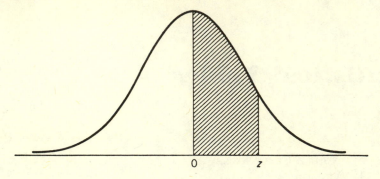

The entries in Table I are the probabilities that a random variable having the standard normal distribution assumes a value between 0 and z; they are given by the area under the curve shaded in the figure shown above.

TABLE I

The Standard Normal Distribution

z	.00	.01	.02	.03	.04	.05	.06	.07	.08	.09
0.0	.0000	.0040	.0080	.0120	.0160	.0199	.0239	.0279	.0319	.0359
0.1	.0398	.0438	.0478	.0517	.0557	.0596	.0636	.0675	.0714	.0753
0.2	.0793	.0832	.0871	.0910	.0948	.0987	.1026	.1064	.1103	.1141
0.3	.1179	.1217	.1255	.1293	.1331	.1368	.1406	.1443	.1480	.1517
0.4	.1554	.1591	.1628	.1664	.1700	.1736	.1772	.1808	.1844	.1879
0.5	.1915	.1950	.1985	.2019	.2054	.2088	.2123	.2157	.2190	.2224
0.6	.2257	.2291	.2324	.2357	.2389	.2422	.2454	.2486	.2517	.2549
0.7	.2580	.2611	.2642	.2673	.2704	.2734	.2764	.2794	.2823	.2852
0.8	.2881	.2910	.2939	.2967	.2995	.3023	.3051	.3078	.3106	.3133
0.9	.3159	.3186	.3212	.3238	.3264	.3289	.3315	.3340	.3365	.3389
1.0	.3413	.3438	.3461	.3485	.3508	.3531	.3554	.3577	.3599	.3621
1.1	.3643	.3665	.3686	.3708	.3729	.3749	.3770	.3790	.3810	.3830
1.2	.3849	.3869	.3888	.3907	.3925	.3944	.3962	.3980	.3997	.4015
1.3	.4032	.4049	.4066	.4082	.4099	.4115	.4131	.4147	.4162	.4177
1.4	.4192	.4207	.4222	.4236	.4251	.4265	.4279	.4292	.4306	.4319
1.5	.4332	.4345	.4357	.4370	.4382	.4394	.4406	.4418	.4429	.4441
1.6	.4452	.4463	.4474	.4484	.4495	.4505	.4515	.4525	.4535	.4545
1.7	.4554	.4564	.4573	.4582	.4591	.4599	.4608	.4616	.4625	.4633
1.8	.4641	.4649	.4656	.4664	.4671	.4678	.4686	.4693	.4699	.4706
1.9	.4713	.4719	.4726	.4732	.4738	.4744	.4750	.4756	.4761	.4767
2.0	.4772	.4778	.4783	.4788	.4793	.4798	.4803	.4808	.4812	.4817
2.1	.4821	.4826	.4830	.4834	.4838	.4842	.4846	.4850	.4854	.4857
2.2	.4861	.4864	.4868	.4871	.4875	.4878	.4881	.4884	.4887	.4890
2.3	.4893	.4896	.4898	.4901	.4904	.4906	.4909	.4911	.4913	.4916
2.4	.4918	.4920	.4922	.4925	.4927	.4929	.4931	.4932	.4934	.4936
2.5	.4938	.4940	.4941	.4943	.4945	.4946	.4948	.4949	.4951	.4952
2.6	.4953	.4955	.4956	.4957	.4959	.4960	.4961	.4962	.4963	.4964
2.7	.4965	.4966	.4967	.4968	.4969	.4970	.4971	.4972	.4973	.4974
2.8	.4974	.4975	.4976	.4977	.4977	.4978	.4979	.4979	.4980	.4981
2.9	.4981	.4982	.4982	.4983	.4984	.4984	.4985	.4985	.4986	.4986
3.0	.4987	.4987	.4987	.4988	.4988	.4989	.4989	.4989	.4990	.4990

STATISTICAL TABLES

TABLE II

Values of t†

d.f.	80% $t_{.100}$	90% $t_{.050}$	95% $t_{.025}$	98% $t_{.010}$	99% $t_{.005}$	d.f.
1	3.078	6.314	12.706	31.821	63.657	1
2	1.886	2.920	4.303	6.965	9.925	2
3	1.638	2.353	3.182	4.541	5.841	3
4	1.533	2.132	2.776	3.747	4.604	4
5	1.476	2.015	2.571	3.365	4.032	5
6	1.440	1.943	2.447	3.143	3.707	6
7	1.415	1.895	2.365	2.998	3.499	7
8	1.397	1.860	2.306	2.896	3.355	8
9	1.383	1.833	2.262	2.821	3.250	9
10	1.372	1.812	2.228	2.764	3.169	10
11	1.363	1.796	2.201	2.718	3.106	11
12	1.356	1.782	2.179	2.681	3.055	12
13	1.350	1.771	2.160	2.650	3.012	13
14	1.345	1.761	2.145	2.624	2.977	14
15	1.341	1.753	2.131	2.602	2.947	15
16	1.337	1.746	2.120	2.583	2.921	16
17	1.333	1.740	2.110	2.567	2.898	17
18	1.330	1.734	2.101	2.552	2.878	18
19	1.328	1.729	2.093	2.539	2.861	19
20	1.325	1.725	2.086	2.528	2.845	20
21	1.323	1.721	2.080	2.518	2.831	21
22	1.321	1.717	2.074	2.508	2.819	22
23	1.319	1.714	2.069	2.500	2.807	23
24	1.318	1.711	2.064	2.492	2.797	24
25	1.316	1.708	2.060	2.485	2.787	25
26	1.315	1.706	2.056	2.479	2.779	26
27	1.314	1.703	2.052	2.473	2.771	27
28	1.313	1.701	2.048	2.467	2.763	28
29	1.311	1.699	2.045	2.462	2.756	29
inf.	1.282	1.645	1.960	2.326	2.576	inf.

† This table is abridged from Table IV of R. A. Fisher, *Statistical Methods for Research Workers*, published by Oliver and Boyd, Ltd., Edinburgh, by permission of the author and publishers.

TABLE III
The Chi-square Distribution (Values of χ_α^2)*

d.f.	$\chi^2_{.995}$	$\chi^2_{.99}$	$\chi^2_{.975}$	$\chi^2_{.95}$	$\chi^2_{.05}$	$\chi^2_{.025}$	$\chi^2_{.01}$	$\chi^2_{.005}$	d.f.
1	.0000393	.000157	.000982	.00393	3.841	5.024	6.635	7.879	1
2	.0100	.0201	.0506	.103	5.991	7.378	9.210	10.597	2
3	.0717	.115	.216	.352	7.815	9.348	11.345	12.838	3
4	.207	.297	.484	.711	9.488	11.143	13.277	14.860	4
5	.412	.554	.831	1.145	11.070	12.832	15.086	16.750	5
6	.676	.872	1.237	1.635	12.592	14.449	16.812	18.548	6
7	.989	1.239	1.690	2.167	14.067	16.013	18.475	20.278	7
8	1.344	1.646	2.180	2.733	15.507	17.535	20.090	21.955	8
9	1.735	2.088	2.700	3.325	16.919	19.023	21.666	23.589	9
10	2.156	2.558	3.247	3.940	18.307	20.483	23.209	25.188	10
11	2.603	3.053	3.816	4.575	19.675	21.920	24.725	26.757	11
12	3.074	3.571	4.404	5.226	21.026	23.337	26.217	28.300	12
13	3.565	4.107	5.009	5.892	22.362	24.736	27.688	29.819	13
14	4.075	4.660	5.629	6.571	23.685	26.119	29.141	31.319	14
15	4.601	5.229	6.262	7.261	24.996	27.488	30.578	32.801	15
16	5.142	5.812	6.908	7.962	26.296	28.845	32.000	34.267	16
17	5.697	6.408	7.564	8.672	27.587	30.191	33.409	35.718	17
18	6.265	7.015	8.231	9.390	28.869	31.526	34.805	37.156	18
19	6.844	7.633	8.907	10.117	30.144	32.852	36.191	38.582	19
20	7.434	8.260	9.591	10.851	31.410	34.170	37.566	39.997	20
21	8.034	8.897	10.283	11.591	32.671	35.479	38.932	41.401	21
22	8.643	9.542	10.982	12.338	33.924	36.781	40.289	42.796	22
23	9.260	10.196	11.689	13.091	35.172	38.076	41.638	44.181	23
24	9.886	10.856	12.401	13.848	36.415	39.364	42.980	45.558	24
25	10.520	11.524	13.120	14.611	37.652	40.646	44.314	46.928	25
26	11.160	12.198	13.844	15.379	38.885	41.923	45.642	48.290	26
27	11.808	12.879	14.573	16.151	40.113	43.194	46.963	49.645	27
28	12.461	13.565	15.308	16.928	41.337	44.461	48.278	50.993	28
29	13.121	14.256	16.047	17.708	42.557	45.722	49.588	52.336	29
30	13.787	14.953	16.791	18.493	43.773	46.979	50.892	53.672	30

* This table is based on Table 8 of *Biometrika Tables for Statisticians*, *Volume I*, 3rd ed., Cambridge: University Press, 1966, by permission of the *Biometrika* trustees.

TABLE IVa: Values of $F_{.05}$ †

Degrees of freedom for numerator

Den. df	1	2	3	4	5	6	7	8	9	10	12	15	20	24	30	40	60	120	∞
1	161	200	216	225	230	234	237	239	241	242	244	246	248	249	250	251	252	253	254
2	18.5	19.0	19.2	19.2	19.3	19.3	19.4	19.4	19.4	19.4	19.4	19.4	19.4	19.5	19.5	19.5	19.5	19.5	19.5
3	10.1	9.55	9.28	9.12	9.01	8.94	8.89	8.85	8.81	8.79	8.74	8.70	8.66	8.64	8.62	8.59	8.57	8.55	8.53
4	7.71	6.94	6.59	6.39	6.26	6.16	6.09	6.04	6.00	5.96	5.91	5.86	5.80	5.77	5.75	5.72	5.69	5.66	5.63
5	6.61	5.79	5.41	5.19	5.05	4.95	4.88	4.82	4.77	4.74	4.68	4.62	4.56	4.53	4.50	4.46	4.43	4.40	4.37
6	5.99	5.14	4.76	4.53	4.39	4.28	4.21	4.15	4.10	4.06	4.00	3.94	3.87	3.84	3.81	3.77	3.74	3.70	3.67
7	5.59	4.74	4.35	4.12	3.97	3.87	3.79	3.73	3.68	3.64	3.57	3.51	3.44	3.41	3.38	3.34	3.30	3.27	3.23
8	5.32	4.46	4.07	3.84	3.69	3.58	3.50	3.44	3.39	3.35	3.28	3.22	3.15	3.12	3.08	3.04	3.01	2.97	2.93
9	5.12	4.26	3.86	3.63	3.48	3.37	3.29	3.23	3.18	3.14	3.07	3.01	2.94	2.90	2.86	2.83	2.79	2.75	2.71
10	4.96	4.10	3.71	3.48	3.33	3.22	3.14	3.07	3.02	2.98	2.91	2.85	2.77	2.74	2.70	2.66	2.62	2.58	2.54
11	4.84	3.98	3.59	3.36	3.20	3.09	3.01	2.95	2.90	2.85	2.79	2.72	2.65	2.61	2.57	2.53	2.49	2.45	2.40
12	4.75	3.89	3.49	3.26	3.11	3.00	2.91	2.85	2.80	2.75	2.69	2.62	2.54	2.51	2.47	2.43	2.38	2.34	2.30
13	4.67	3.81	3.41	3.18	3.03	2.92	2.83	2.77	2.71	2.67	2.60	2.53	2.46	2.42	2.38	2.34	2.30	2.25	2.21
14	4.60	3.74	3.34	3.11	2.96	2.85	2.76	2.70	2.65	2.60	2.53	2.46	2.39	2.35	2.31	2.27	2.22	2.18	2.13
15	4.54	3.68	3.29	3.06	2.90	2.79	2.71	2.64	2.59	2.54	2.48	2.40	2.33	2.29	2.25	2.20	2.16	2.11	2.07
16	4.49	3.63	3.24	3.01	2.85	2.74	2.66	2.59	2.54	2.49	2.42	2.35	2.28	2.24	2.19	2.15	2.11	2.06	2.01
17	4.45	3.59	3.20	2.96	2.81	2.70	2.61	2.55	2.49	2.45	2.38	2.31	2.23	2.19	2.15	2.10	2.06	2.01	1.96
18	4.41	3.55	3.16	2.93	2.77	2.66	2.58	2.51	2.46	2.41	2.34	2.27	2.19	2.15	2.11	2.06	2.02	1.97	1.92
19	4.38	3.52	3.13	2.90	2.74	2.63	2.54	2.48	2.42	2.38	2.31	2.23	2.16	2.11	2.07	2.03	1.98	1.93	1.88
20	4.35	3.49	3.10	2.87	2.71	2.60	2.51	2.45	2.39	2.35	2.28	2.20	2.12	2.08	2.04	1.99	1.95	1.90	1.84
21	4.32	3.47	3.07	2.84	2.68	2.57	2.49	2.42	2.37	2.32	2.25	2.18	2.10	2.05	2.01	1.96	1.92	1.87	1.81
22	4.30	3.44	3.05	2.82	2.66	2.55	2.46	2.40	2.34	2.30	2.23	2.15	2.07	2.03	1.98	1.94	1.89	1.84	1.78
23	4.28	3.42	3.03	2.80	2.64	2.53	2.44	2.37	2.32	2.27	2.20	2.13	2.05	2.01	1.96	1.91	1.86	1.81	1.76
24	4.26	3.40	3.01	2.78	2.62	2.51	2.42	2.36	2.30	2.25	2.18	2.11	2.03	1.98	1.94	1.89	1.84	1.79	1.73
25	4.24	3.39	2.99	2.76	2.60	2.49	2.40	2.34	2.28	2.24	2.16	2.09	2.01	1.96	1.92	1.87	1.82	1.77	1.71
30	4.17	3.32	2.92	2.69	2.53	2.42	2.33	2.27	2.21	2.16	2.09	2.01	1.93	1.89	1.84	1.79	1.74	1.68	1.62
40	4.08	3.23	2.84	2.61	2.45	2.34	2.25	2.18	2.12	2.08	2.00	1.92	1.84	1.79	1.74	1.69	1.64	1.58	1.51
60	4.00	3.15	2.76	2.53	2.37	2.25	2.17	2.10	2.04	1.99	1.92	1.84	1.75	1.70	1.65	1.59	1.53	1.47	1.39
120	3.92	3.07	2.68	2.45	2.29	2.18	2.09	2.02	1.96	1.91	1.83	1.75	1.66	1.61	1.55	1.50	1.43	1.35	1.25
∞	3.84	3.00	2.60	2.37	2.21	2.10	2.01	1.94	1.88	1.83	1.75	1.67	1.57	1.52	1.46	1.39	1.32	1.22	1.00

Degrees of freedom for denominator

† This table is reproduced from M. Merrington and C. M. Thompson, "Tables of percentage points of the inverted beta (F) distribution," *Biometrika*, Vol. 33 (1943), by permission of the *Biometrika* trustees.

TABLE IVb: Values of $F_{.01}$†

Degrees of freedom for numerator

	1	2	3	4	5	6	7	8	9	10	12	15	20	24	30	40	60	120	∞
1	4,052	5,000	5,403	5,625	5,764	5,859	5,928	5,982	6,023	6,056	6,106	6,157	6,209	6,235	6,261	6,287	6,313	6,339	6,366
2	98.5	99.0	99.2	99.2	99.3	99.3	99.4	99.4	99.4	99.4	99.4	99.4	99.4	99.5	99.5	99.5	99.5	99.5	99.5
3	34.1	30.8	29.5	28.7	28.2	27.9	27.7	27.5	27.3	27.2	27.1	26.9	26.7	26.6	26.5	26.4	26.3	26.2	26.1
4	21.2	18.0	16.7	16.0	15.5	15.2	15.0	14.8	14.7	14.5	14.4	14.2	14.0	13.9	13.8	13.7	13.7	13.6	13.5
5	16.3	13.3	12.1	11.4	11.0	10.7	10.5	10.3	10.2	10.1	9.89	9.72	9.55	9.47	9.38	9.29	9.20	9.11	9.02
6	13.7	10.9	9.78	9.15	8.75	8.47	8.26	8.10	7.98	7.87	7.72	7.56	7.40	7.31	7.23	7.14	7.06	6.97	6.88
7	12.2	9.55	8.45	7.85	7.46	7.19	6.99	6.84	6.72	6.62	6.47	6.31	6.16	6.07	5.99	5.91	5.82	5.74	5.65
8	11.3	8.65	7.59	7.01	6.63	6.37	6.18	6.03	5.91	5.81	5.67	5.52	5.36	5.28	5.20	5.12	5.03	4.95	4.86
9	10.6	8.02	6.99	6.42	6.06	5.80	5.61	5.47	5.35	5.26	5.11	4.96	4.81	4.73	4.65	4.57	4.48	4.40	4.31
10	10.0	7.56	6.55	5.99	5.64	5.39	5.20	5.06	4.94	4.85	4.71	4.56	4.41	4.33	4.25	4.17	4.08	4.00	3.91
11	9.65	7.21	6.22	5.67	5.32	5.07	4.89	4.74	4.63	4.54	4.40	4.25	4.10	4.02	3.94	3.86	3.78	3.69	3.60
12	9.33	6.93	5.95	5.41	5.06	4.82	4.64	4.50	4.39	4.30	4.16	4.01	3.86	3.78	3.70	3.62	3.54	3.45	3.36
13	9.07	6.70	5.74	5.21	4.86	4.62	4.44	4.30	4.19	4.10	3.96	3.82	3.66	3.59	3.51	3.43	3.34	3.25	3.17
14	8.86	6.51	5.56	5.04	4.70	4.46	4.28	4.14	4.03	3.94	3.80	3.66	3.51	3.43	3.35	3.27	3.18	3.09	3.00
15	8.68	6.36	5.42	4.89	4.56	4.32	4.14	4.00	3.89	3.80	3.67	3.52	3.37	3.29	3.21	3.13	3.05	2.96	2.87
16	8.53	6.23	5.29	4.77	4.44	4.20	4.03	3.89	3.78	3.69	3.55	3.41	3.26	3.18	3.10	3.02	2.93	2.84	2.75
17	8.40	6.11	5.19	4.67	4.34	4.10	3.93	3.79	3.68	3.59	3.46	3.31	3.16	3.08	3.00	2.92	2.83	2.75	2.65
18	8.29	6.01	5.09	4.58	4.25	4.01	3.84	3.71	3.60	3.51	3.37	3.23	3.08	3.00	2.92	2.84	2.75	2.66	2.57
19	8.19	5.93	5.01	4.50	4.17	3.94	3.77	3.63	3.52	3.43	3.30	3.15	3.00	2.92	2.84	2.76	2.67	2.58	2.49
20	8.10	5.85	4.94	4.43	4.10	3.87	3.70	3.56	3.46	3.37	3.23	3.09	2.94	2.86	2.78	2.69	2.61	2.52	2.42
21	8.02	5.78	4.87	4.37	4.04	3.81	3.64	3.51	3.40	3.31	3.17	3.03	2.88	2.80	2.72	2.64	2.55	2.46	2.36
22	7.95	5.72	4.82	4.31	3.99	3.76	3.59	3.45	3.35	3.26	3.12	2.98	2.83	2.75	2.67	2.58	2.50	2.40	2.31
23	7.88	5.66	4.76	4.26	3.94	3.71	3.54	3.41	3.30	3.21	3.07	2.93	2.78	2.70	2.62	2.54	2.45	2.35	2.26
24	7.82	5.61	4.72	4.22	3.90	3.67	3.50	3.36	3.26	3.17	3.03	2.89	2.74	2.66	2.58	2.49	2.40	2.31	2.21
25	7.77	5.57	4.68	4.18	3.85	3.63	3.46	3.32	3.22	3.13	2.99	2.85	2.70	2.62	2.53	2.45	2.36	2.27	2.17
30	7.56	5.39	4.51	4.02	3.70	3.47	3.30	3.17	3.07	2.98	2.84	2.70	2.55	2.47	2.39	2.30	2.21	2.11	2.01
40	7.31	5.18	4.31	3.83	3.51	3.29	3.12	2.99	2.89	2.80	2.66	2.52	2.37	2.29	2.20	2.11	2.02	1.92	1.80
60	7.08	4.98	4.13	3.65	3.34	3.12	2.95	2.82	2.72	2.63	2.50	2.35	2.20	2.12	2.03	1.94	1.84	1.73	1.60
120	6.85	4.79	3.95	3.48	3.17	2.96	2.79	2.66	2.56	2.47	2.34	2.19	2.03	1.95	1.86	1.76	1.66	1.53	1.38
∞	6.63	4.61	3.78	3.32	3.02	2.80	2.64	2.51	2.41	2.32	2.18	2.04	1.88	1.79	1.70	1.59	1.47	1.32	1.00

Degrees of freedom for denominator

† This table is reproduced from M. Merrington and C. M. Thompson, "Tables of percentage points of the inverted beta (F) distribution," *Biometrika*, Vol. 33 (1943), by permission of the *Biometrika* trustees.

TABLE V

0.95 Confidence Intervals for Proportions*

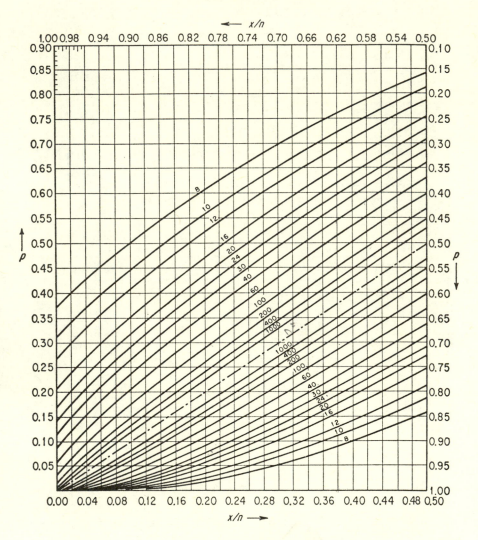

TABLE VI

Binomial Coefficients

n	$\binom{n}{0}$	$\binom{n}{1}$	$\binom{n}{2}$	$\binom{n}{3}$	$\binom{n}{4}$	$\binom{n}{5}$	$\binom{n}{6}$	$\binom{n}{7}$	$\binom{n}{8}$	$\binom{n}{9}$	$\binom{n}{10}$
0	1										
1	1	1									
2	1	2	1								
3	1	3	3	1							
4	1	4	6	4	1						
5	1	5	10	10	5	1					
6	1	6	15	20	15	6	1				
7	1	7	21	35	35	21	7	1			
8	1	8	28	56	70	56	28	8	1		
9	1	9	36	84	126	126	84	36	9	1	
10	1	10	45	120	210	252	210	120	45	10	1
11	1	11	55	165	330	462	462	330	165	55	11
12	1	12	66	220	495	792	924	792	495	220	66
13	1	13	78	286	715	1287	1716	1716	1287	715	286
14	1	14	91	364	1001	2002	3003	3432	3003	2002	1001
15	1	15	105	455	1365	3003	5005	6435	6435	5005	3003
16	1	16	120	560	1820	4368	8008	11440	12870	11440	8008
17	1	17	136	680	2380	6188	12376	19448	24310	24310	19448
18	1	18	153	816	3060	8568	18564	31824	43758	48620	43758
19	1	19	171	969	3876	11628	27132	50388	75582	92378	92378
20	1	20	190	1140	4845	15504	38760	77520	125970	167960	184756

TABLE VII

Control Chart Constants†

Number of observations in sample, n	Chart for averages			Chart for standard deviations					Chart for ranges				
	Factors for control limits			Factor for central line	Factors for control limits				Factor for central line	Factors for control limits			
	A	A_1	A_2	c_2	B_1	B_2	B_3	B_4	d_2	D_1	D_2	D_3	D_4
2	2.121	3.760	1 880	0.5642	0	1.843	0	3.267	1.128	0	3.686	0	3.267
3	1.732	2.394	1.023	0.7236	0	1.858	0	2.568	1.693	0	4.358	0	2.575
4	1.500	1.880	0.729	0.7979	0	1.808	0	2.266	2.059	0	4.698	0	2.282
5	1.342	1.596	0.577	0 8407	0	1.756	0	2.089	2.326	0	4.918	0	2.115
6	1 225	1.410	0.483	0.8686	0 026	1.711	0.030	1.970	2.534	0	5.078	0	2.004
7	1 134	1.277	0.419	0.8882	0.105	1 672	0.118	1.882	2.704	0.205	5.203	0.076	1.924
8	1.061	1.175	0.373	0 9027	0.167	1.638	0.185	1.815	2.847	0.387	5.307	0.136	1.864
9	1.000	1.094	0.337	0.9139	0.219	1.609	0.239	1.761	2.970	0.546	5.394	0.184	1.816
10	0.949	1.028	0.308	0.9227	0.262	1 584	0.284	1.716	3.078	0.687	5.469	0.223	1.777
11	0.905	0.973	0.285	0.9300	0.299	1.561	0.321	1.679	3.173	0.812	5.534	0.256	1.744
12	0.866	0.925	0.266	0.9359	0.331	1.541	0.354	1.646	3.258	0.924	5.592	0.284	1.716
13	0.832	0.884	0.249	0.9410	0.359	1.523	0.382	1.618	3.336	1.026	5.646	0.308	1.692
14	0.802	0.848	0.235	0.9453	0.384	1 507	0.406	1.594	3.407	1.121	5.693	0.329	1 671
15	0 775	0.816	0 223	0 9490	0 406	1.492	0.428	1.572	3.472	1.207	5.737	0.348	1.652

Statistic	Standards Given		Analysis of Past Data	
	Central line	Limits	Central line	Limits
$\bar{X}$	$\bar{X}'$	$\bar{X}' \pm A\sigma'$	$\bar{\bar{X}}$	$\bar{\bar{X}} \pm A_1\bar{\sigma}$ or $\bar{\bar{X}} \pm A_2\bar{R}$
σ	$c_2\sigma'$	$B_1\sigma'$, $B_2\sigma'$	$\bar{\sigma}$	$B_3\bar{\sigma}$, $B_4\bar{\sigma}$
R	$d_2\sigma'$	$D_1\sigma'$, $D_2\sigma'$	$\bar{R}$	$D_3\bar{R}$, $D_4\bar{R}$

† This table is reproduced, by permission, from the *ASTM Manual on Quality Control of Materials*, American Society for Testing Materials, Philadelphia, Pa., 1951.

TABLE VIII
Logarithms

N	0	1	2	3	4	5	6	7	8	9
10	0000	0043	0086	0128	0170	0212	0253	0294	0334	0374
11	0414	0453	0492	0531	0569	0607	0645	0682	0719	0755
12	0792	0828	0864	0899	0934	0969	1004	1038	1072	1106
13	1139	1173	1206	1239	1271	1303	1335	1367	1399	1430
14	1461	1492	1523	1553	1584	1614	1644	1673	1703	1732
15	1761	1790	1818	1847	1875	1903	1931	1959	1987	2014
16	2041	2068	2095	2122	2148	2175	2201	2227	2253	2279
17	2304	2330	2355	2380	2405	2430	2455	2480	2504	2529
18	2553	2577	2601	2625	2648	2672	2695	2718	2742	2765
19	2788	2810	2833	2856	2878	2900	2923	2945	2967	2989
20	3010	3032	3054	3075	3096	3118	3139	3160	3181	3201
21	3222	3243	3263	3284	3304	3324	3345	3365	3385	3404
22	3424	3444	3464	3483	3502	3522	3541	3560	3579	3598
23	3617	3636	3655	3674	3692	3711	3729	3747	3766	3784
24	3802	3820	3838	3856	3874	3892	3909	3927	3945	3962
25	3979	3997	4014	4031	4048	4065	4082	4099	4116	4133
26	4150	4166	4183	4200	4216	4232	4249	4265	4281	4298
27	4314	4330	4346	4362	4378	4393	4409	4425	4440	4456
28	4472	4487	4502	4518	4533	4548	4564	4579	4594	4609
29	4624	4639	4654	4669	4683	4698	4713	4728	4742	4757
30	4771	4786	4800	4814	4829	4843	4857	4871	4886	4900
31	4914	4928	4942	4955	4969	4983	4997	5011	5024	5038
32	5051	5065	5079	5092	5105	5119	5132	5145	5159	5172
33	5185	5198	5211	5224	5237	5250	5263	5276	5289	5302
34	5315	5328	5340	5353	5366	5378	5391	5403	5416	5428
35	5441	5453	5465	5478	5490	5502	5514	5527	5539	5551
36	5563	5575	5587	5599	5611	5623	5635	5647	5658	5670
37	5682	5694	5705	5717	5729	5740	5752	5763	5775	5786
38	5798	5809	5821	5832	5843	5855	5866	5877	5888	5899
39	5911	5922	5933	5944	5955	5966	5977	5988	5999	6010
40	6021	6031	6042	6053	6064	6075	6085	6096	6107	6117
41	6128	6138	6149	6160	6170	6180	6191	6201	6212	6222
42	6232	6243	6253	6263	6274	6284	6294	6304	6314	6325
43	6335	6345	6355	6365	6375	6385	6395	6405	6415	6425
44	6435	6444	6454	6464	6474	6484	6493	6503	6513	6522
45	6532	6542	6551	6561	6571	6580	6590	6599	6609	6618
46	6628	6637	6646	6656	6665	6675	6684	6693	6702	6712
47	6721	6730	6739	6749	6758	6767	6776	6785	6794	6803
48	6812	6821	6830	6839	6848	6857	6866	6875	6884	6893
49	6902	6911	6920	6928	6937	6946	6955	6964	6972	6981
50	6990	6998	7007	7016	7024	7033	7042	7050	7059	7067
51	7076	7084	7093	7101	7110	7118	7126	7135	7143	7152
52	7160	7168	7177	7185	7193	7202	7210	7218	7226	7235
53	7243	7251	7259	7267	7275	7284	7292	7300	7308	7316
54	7324	7332	7340	7348	7356	7364	7372	7380	7388	7396

TABLE VIII

Logarithms (Continued)

N	0	1	2	3	4	5	6	7	8	9
55	7404	7412	7419	7427	7435	7443	7451	7459	7466	7474
56	7482	7490	7497	7505	7513	7520	7528	7536	7543	7551
57	7559	7566	7574	7582	7589	7597	7604	7612	7619	7627
58	7634	7642	7649	7657	7664	7672	7679	7686	7694	7701
59	7709	7716	7723	7731	7738	7745	7752	7760	7767	7774
60	7782	7789	7796	7803	7810	7818	7825	7832	7839	7846
61	7853	7860	7868	7875	7882	7889	7896	7903	7910	7917
62	7924	7931	7938	7945	7952	7959	7966	7973	7980	7987
63	7993	8000	8007	8014	8021	8028	8035	8041	8048	8055
64	8062	8069	8075	8082	8089	8096	8102	8109	8116	8122
65	8129	8136	8142	8149	8156	8162	8169	8176	8182	8189
66	8195	8202	8209	8215	8222	8228	8235	8241	8248	8254
67	8261	8267	8274	8280	8287	8293	8299	8306	8312	8319
68	8325	8331	8338	8344	8351	8357	8363	8370	8376	8382
69	8388	8395	8401	8407	8414	8420	8426	8432	8439	8445
70	8451	8457	8463	8470	8476	8482	8488	8494	8500	8506
71	8513	8519	8525	8531	8537	8543	8549	8555	8561	8567
72	8573	8579	8585	8591	8597	8603	8609	8615	8621	8627
73	8633	8639	8645	8651	8657	8663	8669	8675	8681	8686
74	8692	8698	8704	8710	8716	8722	8727	8733	8739	8745
75	8751	8756	8762	8768	8774	8779	8785	8791	8797	8802
76	8808	8814	8820	8825	8831	8837	8842	8848	8854	8859
77	8865	8871	8876	8882	8887	8893	8899	8904	8910	8915
78	8921	8927	8932	8938	8943	8949	8954	8960	8965	8971
79	8976	8982	8987	8993	8998	9004	9009	9015	9020	9025
80	9031	9036	9042	9047	9053	9058	9063	9069	9074	9079
81	9085	9090	9096	9101	9106	9112	9117	9122	9128	9133
82	9138	9143	9149	9154	9159	9165	9170	9175	9180	9186
83	9191	9196	9201	9206	9212	9217	9222	9227	9232	9238
84	9243	9248	9253	9258	9263	9269	9274	9279	9284	9289
85	9294	9299	9304	9309	9315	9320	9325	9330	9335	9340
86	9345	9350	9355	9360	9365	9370	9375	9380	9385	9390
87	9395	9400	9405	9410	9415	9420	9425	9430	9435	9440
88	9445	9450	9455	9460	9465	9469	9474	9479	9484	9489
89	9494	9499	9504	9509	9513	9518	9523	9528	9533	9538
90	9542	9547	9552	9557	9562	9566	9571	9576	9581	9586
91	9590	9595	9600	9605	9609	9614	9619	9624	9628	9633
92	9638	9643	9647	9652	9657	9661	9666	9671	9675	9680
93	9685	9689	9694	9699	9703	9708	9713	9717	9722	9727
94	9731	9736	9741	9745	9750	9754	9759	9763	9768	9773
95	9777	9782	9786	9791	9795	9800	9805	9809	9814	9818
96	9823	9827	9832	9836	9841	9845	9850	9854	9859	9863
97	9868	9872	9877	9881	9886	9890	9894	9899	9903	9908
98	9912	9917	9921	9926	9930	9934	9939	9943	9948	9952
99	9956	9961	9965	9969	9974	9978	9983	9987	9991	9996

TABLE IX

Squares and Square Roots

n	n^2	$\sqrt{n}$	$\sqrt{10n}$	n	n^2	$\sqrt{n}$	$\sqrt{10n}$
1.00	1.0000	1.000000	3.162278	1.50	2.2500	1.224745	3.872983
1.01	1.0201	1.004988	3.178050	1.51	2.2801	1.228821	3.885872
1.02	1.0404	1.009950	3.193744	1.52	2.3104	1.232883	3.898718
1.03	1.0609	1.014889	3.209361	1.53	2.3409	1.236932	3.911521
1.04	1.0816	1.019804	3.224903	1.54	2.3716	1.240967	3.924283
1.05	1.1025	1.024695	3.240370	1.55	2.4025	1.244990	3.937004
1.06	1.1236	1.029563	3.255764	1.56	2.4336	1.249000	3.949684
1.07	1.1449	1.034408	3.271085	1.57	2.4649	1.252996	3.962323
1.08	1.1664	1.039230	3.286335	1.58	2.4964	1.256981	3.974921
1.09	1.1881	1.044031	3.301515	1.59	2.5281	1.260952	3.987480
1.10	1.2100	1.048809	3.316625	1.60	2.5600	1.264911	4.000000
1.11	1.2321	1.053565	3.331666	1.61	2.5921	1.268858	4.012481
1.12	1.2544	1.058301	3.346640	1.62	2.6244	1.272792	4.024922
1.13	1.2769	1.063015	3.361547	1.63	2.6569	1.276715	4.037326
1.14	1.2996	1.067708	3.376389	1.64	2.6896	1.280625	4.049691
1.15	1.3225	1.072381	3.391165	1.65	2.7225	1.284523	4.062019
1.16	1.3456	1.077033	3.405877	1.66	2.7556	1.288410	4.074310
1.17	1.3689	1.081665	3.420526	1.67	2.7889	1.292285	4.086563
1.18	1.3924	1.086278	3.435113	1.68	2.8224	1.296148	4.098780
1.19	1.4161	1.090871	3.449638	1.69	2.8561	1.300000	4.110961
1.20	1.4400	1.095445	3.464102	1.70	2.8900	1.303840	4.123106
1.21	1.4641	1.100000	3.478505	1.71	2.9241	1.307670	4.135215
1.22	1.4884	1.104536	3.492850	1.72	2.9584	1.311488	4.147288
1.23	1.5129	1.109054	3.507136	1.73	2.9929	1.315295	4.159327
1.24	1.5376	1.113553	3.521363	1.74	3.0276	1.319091	4.171331
1.25	1.5625	1.118034	3.535534	1.75	3.0625	1.322876	4.183300
1.26	1.5876	1.122497	3.549648	1.76	3.0976	1.326650	4.195235
1.27	1.6129	1.126943	3.563706	1.77	3.1329	1.330413	4.207137
1.28	1.6384	1.131371	3.577709	1.78	3.1684	1.334166	4.219005
1.29	1.6641	1.135782	3.591657	1.79	3.2041	1.337909	4.230839
1.30	1.6900	1.140175	3.605551	1.80	3.2400	1.341641	4.242641
1.31	1.7161	1.144552	3.619392	1.81	3.2761	1.345362	4.254409
1.32	1.7424	1.148913	3.633180	1.82	3.3124	1.349074	4.266146
1.33	1.7689	1.153256	3.646917	1.83	3.3489	1.352775	4.277850
1.34	1.7956	1.157584	3.660601	1.84	3.3856	1.356466	4.289522
1.35	1.8255	1.161895	3.674235	1.85	3.4225	1.360147	4.301163
1.36	1.8496	1.166190	3.687818	1.86	3.4596	1.363818	4.312772
1.37	1.8769	1.170470	3.701351	1.87	3.4969	1.367479	4.324350
1.38	1.9044	1.174734	3.714835	1.88	3.5344	1.371131	4.335897
1.39	1.9321	1.178983	3.728270	1.89	3.5721	1.374773	4.347413
1.40	1.9600	1.183216	3.741657	1.90	3.6100	1.378405	4.358899
1.41	1.9881	1.187434	3.754997	1.91	3.6481	1.382027	4.370355
1.42	2.0164	1.191638	3.768289	1.92	3.6864	1.385641	4.381780
1.43	2.0449	1.195826	3.781534	1.93	3.7249	1.389244	4.393177
1.44	2.0736	1.200000	3.794733	1.94	3.7636	1.392839	4.404543
1.45	2.1025	1.204159	3.807887	1.95	3.8025	1.396424	4.415880
1.46	2.1316	1.208305	3.820995	1.96	3.8416	1.400000	4.427189
1.47	2.1609	1.212436	3.834058	1.97	3.8809	1.403567	4.438468
1.48	2.1904	1.216553	3.847077	1.98	3.9204	1.407125	4.449719
1.49	2.2201	1.220656	3.860052	1.99	3.9601	1.410674	4.460942

TABLE IX

Squares and Square Roots (Continued)

n	n^2	$\sqrt{n}$	$\sqrt{10n}$	n	n^2	$\sqrt{n}$	$\sqrt{10n}$
2.00	4.0000	1.414214	4.472136	2.50	6.2500	1.581139	5.000000
2.01	4.0401	1.417745	4.483302	2.51	6.3001	1.584298	5.009990
2.02	4.0804	1.421267	4.494441	2.52	6.3504	1.587451	5.019960
2.03	4.1209	1.424781	4.505552	2.53	6.4009	1.590597	5.029911
2.04	4.1616	1.428286	4.516636	2.54	6.4516	1.593738	5.039841
2.05	4.2025	1.431782	4.527693	2.55	6.5025	1.596872	5.049752
2.06	4.2436	1.435270	4.538722	2.56	6.5536	1.600000	5.059644
2.07	4.2849	1.438749	4.549725	2.57	6.6049	1.603122	5.069517
2.08	4.3264	1.442221	4.560702	2.58	6.6564	1.606238	5.079370
2.09	4.3681	1.445683	4.571652	2.59	6.7081	1.609348	5.089204
2.10	4.4100	1.449138	4.582576	2.60	6.7600	1.612452	5.099020
2.11	4.4521	1.452584	4.593474	2.61	6.8121	1.615549	5.108816
2.12	4.4944	1.456022	4.604346	2.62	6.8644	1.618641	5.118594
2.13	4.5369	1.459452	4.615192	2.63	6.9169	1.621727	5.128353
2.14	4.5796	1.462874	4.626013	2.64	6.9696	1.624808	5.138093
2.15	4.6225	1.466288	4.636809	2.65	7.0225	1.627882	5.147815
2.16	4.6656	1.469694	4.647580	2.66	7.0756	1.630951	5.157519
2.17	4.7089	1.473092	4.658326	2.67	7.1289	1.634013	5.167204
2.18	4.7524	1.476482	4.669047	2.68	7.1824	1.637071	5.176872
2.19	4.7961	1.479865	4.679744	2.69	7.2361	1.640122	5.186521
2.20	4.8400	1.483240	4.690416	2.70	7.2900	1.643168	5.196152
2.21	4.8841	1.486607	4.701064	2.71	7.3441	1.646208	5.205766
2.22	4.9284	1.489966	4.711688	2.72	7.3984	1.649242	5.215362
2.23	4.9729	1.493318	4.722288	2.73	7.4529	1.652271	5.224940
2.24	5.0176	1.496663	4.732864	2.74	7.5076	1.655295	5.234501
2.25	5.0625	1.500000	4.743416	2.75	7.5625	1.658312	5.244044
2.26	5.1076	1.503330	4.753946	2.76	7.6176	1.661325	5.253570
2.27	5.1529	1.506652	4.764452	2.77	7.6729	1.664332	5.263079
2.28	5.1984	1.509967	4.774935	2.78	7.7284	1.667333	5.272571
2.29	5.2441	1.513275	4.785394	2.79	7.7841	1.670329	5.282045
2.30	5.2900	1.516575	4.795832	2.80	7.8400	1.673320	5.291503
2.31	5.3361	1.519868	4.806246	2.81	7.8961	1.676305	5.300943
2.32	5.3824	1.523155	4.816638	2.82	7.9524	1.679286	5.310367
2.33	5.4289	1.526434	4.827007	2.83	8.0089	1.682260	5.319774
2.34	5.4756	1.529706	4.837355	2.84	8.0656	1.685230	5.329165
2.35	5.5225	1.532971	4.847680	2.85	8.1225	1.688194	5.338539
2.36	5.5696	1.536229	4.857983	2.86	8.1796	1.691153	5.347897
2.37	5.6169	1.539480	4.868265	2.87	8.2369	1.694107	5.357238
2.38	5.6644	1.542725	4.878524	2.88	8.2944	1.697056	5.366563
2.39	5.7121	1.545962	4.888763	2.89	8.3521	1.700000	5.375872
2.40	5.7600	1.549193	4.898979	2.90	8.4100	1.702939	5.385165
2.41	5.8081	1.552417	4.909175	2.91	8.4681	1.705872	5.394442
2.42	5.8564	1.555635	4.919350	2.92	8.5264	1.708801	5.403702
2.43	5.9049	1.558846	4.929503	2.93	8.5849	1.711724	5.412947
2.44	5.9536	1.562050	4.939636	2.94	8.6436	1.714643	5.422177
2.45	6.0025	1.565248	4.949747	2.95	8.7025	1.717556	5.431390
2.46	6.0516	1.568439	4.959839	2.96	8.7616	1.720465	5.440588
2.47	6.1009	1.571623	4.969909	2.97	8.8209	1.723369	5.449771
2.48	6.1504	1.574802	4.979960	2.98	8.8804	1.726268	5.458938
2.49	6.2001	1.577973	4.989990	2.99	8.9401	1.729162	5.468089

TABLE IX

Squares and Square Roots (Continued)

n	n^2	$\sqrt{n}$	$\sqrt{10n}$	n	n^2	$\sqrt{n}$	$\sqrt{10n}$
3.00	9.0000	1.732051	5.477226	3.50	12.2500	1.870829	5.916080
3.01	9.0601	1.734935	5.486347	3.51	12.3201	1.873499	5.924525
3.02	9.1204	1.737815	5.495453	3.52	12.3904	1.876166	5.932959
3.03	9.1809	1.740690	5.504544	3.53	12.4609	1.878829	5.941380
3.04	9.2416	1.743560	5.513620	3.54	12.5316	1.881489	5.949790
3.05	9.3025	1.746425	5.522681	3.55	12.6025	1.884144	5.958188
3.06	9.3636	1.749286	5.531727	3.56	12.6736	1.886796	5.966574
3.07	9.4249	1.752142	5.540758	3.57	12.7449	1.889444	5.974948
3.08	9.4864	1.754993	5.549775	3.58	12.8164	1.892089	5.983310
3.09	9.5481	1.757840	5.558777	3.59	12.8881	1.894730	5.991661
3.10	9.6100	1.760682	5.567764	3.60	12.9600	1.897367	6.000000
3.11	9.6721	1.763519	5.576737	3.61	13.0321	1.900000	6.008328
3.12	9.7344	1.766352	5.585696	3.62	13.1044	1.902630	6.016644
3.13	9.7969	1.769181	5.594640	3.63	13.1769	1.905256	6.024948
3.14	9.8596	1.772005	5.603570	3.64	13.2496	1.907878	6.033241
3.15	9.9225	1.774824	5.612486	3.65	13.3225	1.910497	6.041523
3.16	9.9856	1.777639	5.621388	3.66	13.3956	1.913113	6.049793
3.17	10.0489	1.780449	5.630275	3.67	13.4689	1.915724	6.058052
3.18	10.1124	1.783255	5.639149	3.68	13.5424	1.918333	6.066300
3.19	10.1761	1.786057	5.648008	3.69	13.6161	1.920937	6.074537
3.20	10.2400	1.788854	5.656854	3.70	13.6900	1.923538	6.082763
3.21	10.3041	1.791647	5.665686	3.71	13.7641	1.926136	6.090977
3.22	10.3684	1.794436	5.674504	3.72	13.8384	1.928730	6.099180
3.23	10.4329	1.797220	5.683309	3.73	13.9129	1.931321	6.107373
3.24	10.4976	1.800000	5.692100	3.74	13.9876	1.933908	6.115554
3.25	10.5625	1.802776	5.700877	3.75	14.0625	1.936492	6.123724
3.26	10.6276	1.805547	5.709641	3.76	14.1376	1.939072	6.131884
3.27	10.6929	1.808314	5.718391	3.77	14.2129	1.941649	6.140033
3.28	10.7584	1.811077	5.727128	3.78	14.2884	1.944222	6.148170
3.29	10.8241	1.813836	5.735852	3.79	14.3641	1.946792	6.156298
3.30	10.8900	1.816590	5.744563	3.80	14.4400	1.949359	6.164414
3.31	10.9561	1.819341	5.753260	3.81	14.5161	1.951922	6.172520
3.32	11.0224	1.822087	5.761944	3.82	14.5924	1.954483	6.180615
3.33	11.0889	1.824829	5.770615	3.83	14.6689	1.957039	6.188699
3.34	11.1556	1.827567	5.779273	3.84	14.7456	1.959592	6.196773
3.35	11.2225	1.830301	5.787918	3.85	14.8225	1.962142	6.204837
3.36	11.2896	1.833030	5.796551	3.86	14.8996	1.964688	6.212890
3.37	11.3569	1.835756	5.805170	3.87	14.9769	1.967232	6.220932
3.38	11.4244	1.838478	5.813777	3.88	15.0544	1.969772	6.228965
3.39	11.4921	1.841195	5.822371	3.89	15.1321	1.972308	6.236986
3.40	11.5600	1.843909	5.830952	3.90	15.2100	1.974842	6.244998
3.41	11.6281	1.846619	5.839521	3.91	15.2881	1.977372	6.252999
3.42	11.6964	1.849324	5.848077	3.92	15.3664	1.979899	5.260990
3.43	11.7649	1.852026	5.856620	3.93	15.4449	1.982423	6.268971
3.44	11.8336	1.854724	5.865151	3.94	15.5236	1.984943	6.276942
3.45	11.9025	1.857418	5.873670	3.95	15.6025	1.987461	6.284903
3.46	11.9716	1.860108	5.882176	3.96	15.6816	1.989975	6.292853
3.47	12.0409	1.862794	5.890671	3.97	15.7609	1.992486	6.300794
3.48	12.1104	1.865476	5.899152	3.98	15.8404	1.994994	6.308724
3.49	12.1801	1.868154	5.907622	3.99	15.9201	1.997498	6.316645

TABLE IX

Squares and Square Roots (Continued)

n	n^2	$\sqrt{n}$	$\sqrt{10n}$	n	n^2	$\sqrt{n}$	$\sqrt{10n}$
4.00	16.0000	2.000000	6.324555	4.50	20.2500	2.121320	6.708204
4.01	16.0801	2.002498	6.332456	4.51	20.3401	2.123676	6.715653
4.02	16.1604	2.004994	6.340347	4.52	20.4304	2.126029	6.723095
4.03	16.2409	2.007486	6.348228	4.53	20.5209	2.128380	6.730527
4.04	16.3216	2.009975	6.356099	4.54	20.6116	2.130728	6.737952
4.05	16.4025	2.012461	6.363961	4.55	20.7025	2.133073	6.745369
4.06	16.4836	2.014944	6.371813	4.56	20.7936	2.135416	6.752777
4.07	16.5649	2.017424	6.379655	4.57	20.8849	2.137756	6.760178
4.08	16.6464	2.019901	6.387488	4.58	20.9764	2.140093	6.767570
4.09	16.7281	2.022375	6.395311	4.59	21.0681	2.142429	6.774954
4.10	16.8100	2.024846	6.403124	4.60	21.1600	2.144761	6.782330
4.11	16.8921	20.27313	6.410928	4.61	21.2521	2.147091	6.789698
4.12	16.9744	2.029778	6.418723	4.62	21.3444	2.149419	6.797058
4.13	17.0569	2.032240	6.426508	4.63	21.4369	2.151743	6.804410
4.14	17.1396	2.034699	6.434283	4.64	21.5296	2.154066	6.811755
4.15	17.2225	2.037155	6.442049	4.65	21.6225	2.156386	6.819091
4.16	17.3056	2.039608	6.449806	4.66	21.7156	2.158703	6.826419
4.17	17.3889	2.042058	6.457554	4.67	21.8089	2.161018	6.833740
4.18	17.4724	2.044505	6.465292	4.68	21.9024	2.163331	6.841053
4.19	17.5561	2.046949	6.473021	4.69	21.9961	2.165641	6.848357
4.20	17.6400	2.049390	6.480741	4.70	22.0900	2.167948	6.855655
4.21	17.7241	2.051828	6.488451	4.71	22.1841	2.170253	6.862944
4.22	17.8084	2.054264	6.496153	4.72	22.2784	2.172556	6.870226
4.23	17.8929	2.056696	6.503845	4.73	22.3729	2.174856	6.877500
4.24	17.9776	2.059126	6.511528	4.74	22.4676	2.177154	6.884766
4.25	18.0625	2.061553	6.519202	4.75	22.5625	2.179449	6.892024
4.26	18.1476	2.063977	6.526868	4.76	22.6576	2.181742	6.899275
4.27	18.2329	2.066398	6.534524	4.77	22.7529	2.184033	6.906519
4.28	18.3184	2.068816	6.542171	4.78	22.8484	2.186321	6.913754
4.29	18.4041	2.071232	6.549809	4.79	22.9441	2.188607	6.920983
4.30	18.4900	2.073644	6.557439	4.80	23.0400	2.190890	6.928203
4.31	18.5761	2.076054	6.565059	4.81	23.1361	2.193171	6.935416
4.32	18.6624	2.078461	6.572671	4.82	23.2324	2.195450	6.942622
4.33	18.7489	2.080865	6.580274	4.83	23.3289	2.197726	6.949820
4.34	18.8356	2.083267	6.587868	4.84	23.4256	2.200000	6.957011
4.35	18.9225	2.085665	6.595453	4.85	23.5225	2.202272	6.964194
4.36	19.0096	2.088061	6.603030	4.86	23.6196	2.204541	6.971370
4.37	19.0969	2.090454	6.610598	4.87	23.7169	2.206808	6.978539
4.38	19.1844	2.092845	6.618157	4.88	23.8144	2.209072	6.985700
4.39	19.2721	2.095233	6.625708	4.89	23.9121	2.211334	6.992853
4.40	19.3600	2.097618	6.633250	4.90	24.0100	2.213594	7.000000
4.41	19.4481	2.100000	6.640783	4.91	24.1081	2.215852	7.007139
4.42	19.5364	2.102380	6.648308	4.92	24.2064	2.218107	7.014271
4.43	19.6249	2.104757	6.655825	4.93	24.3049	2.220360	7.021396
4.44	19.7136	2.107131	6.663332	4.94	24.4036	2.222611	7.028513
4.45	19.8025	2.109502	6.670832	4.95	24.5025	2.224860	7.035624
4.46	19.8916	2.111871	6.678323	4.96	24.6016	2.227106	7.042727
4.47	19.9809	2.114237	6.685806	4.97	24.7009	2.229350	7.049823
4.48	20.0704	2.116601	6.693280	4.98	24.8004	2.231591	7.056912
4.49	20.1601	2.118962	6.700746	4.99	24.9001	2.233831	7.063993

TABLE IX

Squares and Square Roots (Continued)

n	n^2	$\sqrt{n}$	$\sqrt{10n}$	n	n^2	$\sqrt{n}$	$\sqrt{10n}$
5.00	25.0000	2.236068	7.071068	5.50	30.2500	2.345208	7.416198
5.01	25.1001	2.238303	7.078135	5.51	30.3601	2.347339	7.422937
5.02	25.2004	2.240536	7.085196	5.52	30.4704	2.349468	7.429670
5.03	25.3009	2.242766	7.092249	5.53	30.5809	2.351595	7.436397
5.04	25.4016	2.244994	7.099296	5.54	30.6916	2.353720	7.443118
5.05	25.5025	2.247221	7.106335	5.55	30.8025	2.355844	7.449832
5 06	25.6036	2.249444	7.113368	5.56	30.9136	2.357965	7.456541
5.07	25.7049	2.251666	7.120393	5.57	31.0249	2.360085	7.463243
5.08	25.8064	2.253886	7.127412	5.58	31.1364	2.362202	7.469940
5.09	25.9081	2.256103	7.134424	5.59	31.2481	2.364318	7.476630
5.10	26.0100	2.258318	7.141428	5.60	31.3600	2.366432	7.483315
5.11	26.1121	2.260531	7.148426	5.61	31.4721	2.368544	7.489993
5.12	26.2144	2.262742	7.155418	5.62	31.5844	2.370654	7.496666
5.13	26.3169	2.264950	7.162402	5.63	31.6969	2.372762	7.503333
5.14	26.4196	2.267157	7.169379	5.64	31.8096	2.374868	7.509993
5.15	26.5225	2.269361	7.176350	5.65	31.9225	2.376973	7.516648
5.16	26.6256	2.271563	7.183314	5.66	32.0356	2.379075	7.523297
5.17	26.7289	2.273763	7.190271	5.67	32.1489	2.381176	7.529940
5.18	26.8324	2.275961	7.197222	5.68	32.2624	2.383275	7.536577
5.19	26.9361	2.278157	7.204165	5.69	32.3761	2.385372	7.543209
5.20	27.0400	2.280351	7.211103	5.70	32.4900	2.387467	7.549834
5.21	27.1441	2.282542	7.218033	5.71	32.6041	2.389561	7.556454
5.22	27.2484	2.284732	7.224957	5.72	32.7184	2.391652	7.563068
5.23	27.3529	2.286919	7.231874	5.73	32.8329	2.393742	7.569676
5.24	27.4576	2.289105	7.238784	5.74	32.9476	2.395830	7.576279
5.25	27.5625	2.291288	7.245688	5.75	33.0625	2.397916	7.582875
5.26	27.6676	2.293469	7.252586	5.76	33.1776	2.400000	7.589466
5.27	27.7729	2.295648	7.259477	5.77	33.2929	2.402082	7.596052
5.28	27.8784	2.297825	7.266361	5.78	33.4084	2.404163	7.602631
5.29	27.9841	2.300000	7.273239	5.79	33.5241	2.406242	7.609205
5.30	28.0900	2.302173	7.280110	5.80	33.6400	2.408319	7.615773
5.31	28.1961	2.304344	7.286975	5.81	33.7561	2.410394	7.622336
5.32	28.3024	2.306513	7.293833	5.82	33.8724	2.412468	7.628892
5.33	28.4089	2.308679	7.300685	5.83	33.9889	2.414539	7.635444
5.34	28.5156	2.310844	7.307530	5.84	34.1056	2.416609	7.641989
5.35	28.6225	2.313007	7.314369	5.85	34.2225	2.418677	7.648529
5.36	28.7296	2.315167	7.321202	5.86	34.3396	2.420744	7.655064
5.37	28.8369	2.317326	7.328028	5.87	34.4569	2.422808	7.661593
5.38	28.9444	2.319483	7.334848	5.88	34.5744	2.424871	7.668116
5.39	29.0521	2.321637	7.341662	5.89	34.6921	2.426932	7.674634
5.40	29.1600	2.323790	7.348469	5.90	34.8100	2.428992	7.681146
5.41	29.2681	2.325941	7.355270	5.91	34.9281	2.431049	7.687652
5.42	29.3764	2.328089	7.362065	5.92	35.0464	2.433105	7.694154
5.43	29.4849	2.330236	7.368853	5.93	35.1649	2.435159	7.700649
5.44	29.5936	2.332381	7.357636	5.94	35.2836	2.437212	7.707140
5.45	29.7025	2.334524	7.382412	5.95	35.4025	2.439262	7.713624
5.46	29.8116	2.336664	7.389181	5.96	35.5216	2.441311	7.720104
5.47	29.9209	2.338803	7.395945	5.97	35.6409	2.443358	7.726578
5.48	30.0304	2.340940	7.402702	5.98	35.7604	2.445404	7.733046
5.49	30.1401	2.343075	7.409453	5.99	35.8801	2.447448	7.739509

TABLE IX

Squares and Square Roots (Continued)

n	n^2	$\sqrt{n}$	$\sqrt{10n}$	n	n^2	$\sqrt{n}$	$\sqrt{10n}$
6.00	36.0000	2.449490	7.745967	6.50	42.2500	2.549510	8.062258
6.01	36.1201	2.451530	7.752419	6.51	42.3801	2.551470	8.068457
6.02	36.2404	2.453569	7.758866	6.52	42.5104	2.553429	8.074652
6.03	36.3609	2.455606	7.765307	6.53	42.6409	2.555386	8.080842
6.04	36.4816	2.457641	7.771744	6.54	42.7716	2.557342	8.087027
6.05	36.6025	2.459675	7.778175	6.55	42.9025	2.559297	8.093207
6.06	36.7236	2.461707	7.784600	6.56	43.0336	2.561250	8.099383
6.07	36.8449	2.463737	7.791020	6.57	43.1649	2.563201	8.105554
6.08	36.9664	2.465766	7.797435	6.58	43.2964	2.565151	8.111720
6.09	37.0881	2.467793	7.803845	6.59	43.4281	2.567100	8.117881
6.10	37.2100	2.469818	7.810250	6.60	43.5600	2.569047	8.124038
6.11	37.3321	2.471841	7.816649	6.61	43.6921	2.570992	8.130191
6.12	37.4544	2.473863	7.823043	6.62	43.8244	2.572936	8.136338
6.13	37.5769	2.475884	7.829432	6.63	43.9569	2.574879	8.142481
6.14	37.6996	2.477902	7.835815	6.64	44.0896	2.576820	8.148620
6.15	37.8225	2.479919	7.842194	6.65	44.2225	2.578759	8.154753
6.16	37.9456	2.481935	7.848567	6.66	44.3556	2.580698	8.160882
6.17	38.0689	2.483948	7.854935	6.67	44.4889	2.582634	8.167007
6.18	38.1924	2.485961	7.861298	6.68	44.6224	2.584570	8.173127
6.19	38.3161	2.487971	7.867655	6.69	44.7561	2.586503	8.179242
6.20	38.4400	2.489980	7.874008	6.70	44.8900	2.588436	8.185353
6.21	38.5641	2.491987	7.880355	6.71	45.0241	2.590367	8.191459
6.22	38.6884	2.493993	7.886698	6.72	45.1584	2.592296	8.197561
6.23	38.8129	2.495997	7.893035	6.73	45.2929	2.594224	8.203658
6.24	38.9376	2.497999	7.899367	6.74	45.4276	2.596151	8.209750
6.25	39.0625	2.500000	7.905694	6.75	45.5625	2.598076	8.215838
6.26	39.1876	2.501999	7.912016	6.76	45.6976	2.600000	8.221922
6.27	39.3129	2.503997	7.918333	6.77	45.8329	2.601922	8.228001
6.28	39.4384	2.505993	7.924645	6.78	45.9684	2.603843	8.234076
6.29	39.5641	2.507987	7.930952	6.79	46.1041	2.605763	8.240146
6.30	39.6900	2.509980	7.937254	6.80	46.2400	2.607681	8.246211
6.31	39.8161	2.511971	7.943551	6.81	46.3761	2.609598	8.242272
6.32	39.9424	2.513961	7.949843	6.82	46.5124	2.611513	8.258329
6.33	40.0689	2.515949	7.956130	6.83	46.6489	2.613427	8.264381
6.34	40.1956	2.517936	7.962412	6.84	46.7856	2.615339	8.270429
6.35	40.3225	2.519921	7.968689	6.85	46.9225	2.617250	8.276473
6.36	40.4496	2.521904	7.974961	6.86	47.0596	2.619160	8.282512
6.37	40.5769	2.523886	7.981228	6.87	47.1969	2.621068	8.288546
6.38	40.7044	2.525866	7.987490	6.88	47.3344	2.622975	8.294577
6.39	40.8321	2.527845	7.993748	6.89	47.4721	2.624881	8.300602
6.40	40.9600	2.529822	8.000000	6.90	47.6100	2.626785	8.306624
6.41	41.0881	2.531798	8.006248	6.91	47.7481	2.628688	8.312641
6.42	41.2164	2.533772	8.012490	6.92	47.8864	2.630589	8.318654
6.43	41.3449	2.535744	8.018728	6.93	48.0249	2.632489	8.324662
6.44	41.4736	2.537716	8.024961	6.94	48.1636	2.634388	8.330666
6.45	41.6025	2.539685	8.031189	6.95	48.3025	2.636285	8.336666
6.46	41.7316	2.541653	8.037413	6.96	48.4416	2.638181	8.342661
6.47	41.8609	2.543619	8.043631	6.97	48.5809	2.640076	8.348653
6.48	41.9904	2.545584	8.049845	6.98	48.7204	2.641969	8.354639
6.49	42.1201	2.547548	8.056054	6.99	48.8601	2.643861	8.360622

TABLE IX

Squares and Square Roots (Continued)

n	n^2	$\sqrt{n}$	$\sqrt{10n}$	n	n^2	$\sqrt{n}$	$\sqrt{10n}$
7.00	49.0000	2.645751	8.366600	7.50	56.2500	2.738613	8.660254
7.01	49.1401	2.647640	8.372574	7.51	56.4001	2.740438	8.660026
7.02	49.2804	2.649528	8.378544	7.52	56.5504	2.742262	8.671793
7.03	49.4209	2.651415	8.384510	7.53	56.7009	2.744085	8.677557
7.04	49.5616	2.653300	8.390471	7.54	56.8516	2.745906	8.683317
7.05	49.7025	2.655184	8.396428	7.55	57.0025	2.747726	8.689074
7.06	49.8436	2.657066	8.402381	7.56	57.1536	2.749545	8.694826
7.07	49.9849	2.658947	8.408329	7.57	57.3049	2.751363	8.700575
7.08	50.1264	2.660827	8.414274	7.58	57.4564	2.753180	8.706320
7.09	50.2681	2.662705	8.420214	7.59	57.6081	2.754995	8.712061
7.10	50.4100	2.664583	8.426150	7.60	57.7600	2.756810	8.717798
7.11	50.5521	2.666458	8.432082	7.61	57.9121	2.758623	8.723531
7.12	50.6944	2.668333	8.438009	7.62	58.0644	2.760435	8.729261
7.13	50.8369	2.670206	8.443933	7.63	58.2169	2.762245	8.734987
7.14	50.9796	2.672078	8.449852	7.64	58.3696	2.764055	8.740709
7.15	51.1225	2.673948	8.455767	7.65	58.5225	2.765863	8.746428
7.16	51.2656	2.675818	8.461678	7.66	58.6756	2.767671	8.752143
7.17	51.4089	2.677686	8.467585	7.67	58.8289	2.769476	8.757854
7.18	51.5524	2.679552	8.473488	7.68	58.9824	2.771281	8.763561
7.19	51.6961	2.681418	8.479387	7.69	59.1361	2.773085	8.769265
7.20	51.8400	2.683282	8.485281	7.70	59.2900	2.774887	8.774964
7.21	51.9841	2.685144	8.491172	7.71	59.4441	2.776689	8.780661
7.22	52.1284	2.687006	8.497058	7.72	59.5984	2.778489	8.786353
7.23	52.2729	2.688866	8.502941	7.73	59.7529	2.780288	8.792042
7.24	52.4176	2.690725	8.508819	7.74	59.9076	2.782086	8.797727
7.25	52.5625	2.692582	8.514693	7.75	60.0625	2.783882	8.803408
7.26	52.7076	2.694439	8.520563	7.76	60.2176	2.785678	8.809086
7.27	52.8529	2.696294	8.526429	7.77	60.3729	2.787472	8.814760
7.28	52.9984	2.698148	8.532292	7.78	60.5284	2.789265	8.820431
7.29	53.1441	2.700000	8.538150	7.79	60.6841	2.791057	8.826098
7.30	53.2900	2.701851	8.544004	7.80	60.8400	2.792848	8.831761
7.31	53.4361	2.703701	8.549854	7.81	60.9961	2.794638	8.837420
7.32	53.5824	2.705550	8.555700	7.82	61.1524	2.796426	8.843076
7.33	53.7289	2.707397	8.561542	7.83	61.3089	2.798214	8.848729
7.34	53.8756	2.709243	8.567380	7.84	61.4656	2.800000	8.854377
7.35	54.0225	2.711088	8.573214	7.85	61.6225	2.801785	8.860023
7.36	54.1696	2.712932	8.579044	7.86	61.7796	2.803569	8.865664
7.37	54.3169	2.714774	8.584870	7.87	61.9369	2.805352	8.871302
7.38	54.4644	2.716616	8.590693	7.88	62.0944	2.807134	8.876936
7.39	54.6121	2.718455	8.596511	7.89	62.2521	2.808914	8.882567
7.40	54.7600	2.720294	8.602325	7.90	62.4100	2.810694	8.888194
7.41	54.9081	2.722132	8.608136	7.91	62.5681	2.812472	8.893818
7.42	55.0564	2.723968	8.613942	7.92	62.7264	2.814249	8.899438
7.43	55.2049	2.725803	8.619745	7.93	62.8849	2.816026	8.905055
7.44	55.3536	2.727636	8.625543	7.94	63.0436	2.817801	8.910668
7.45	55.5025	2.729469	8.631338	7.95	63.2025	2.819574	8.916277
7.46	55.6516	2.731300	8.637129	7.96	63.3616	2.821347	8.921883
7.47	55.8009	2.733130	8.642916	7.97	63.5209	2.823119	8.927486
7.48	55.9504	2.734959	8.648699	7.98	63.6804	2.824889	8.933085
7.49	56.1001	2.736786	8.654479	7.99	63.8401	2.826659	8.938680

TABLE IX

Squares and Square Roots (Continued)

n	n^2	$\sqrt{n}$	$\sqrt{10n}$	n	n^2	$\sqrt{n}$	$\sqrt{10n}$
8.00	64.0000	2.828427	8.944272	8.50	72.2500	2.915476	9.219544
8.01	64.1601	2.830194	8.949860	8.51	72.4201	2.917190	9.224966
8.02	64.3204	2.831960	8.955445	8.52	72.5904	2.918904	9.230385
8.03	64.4809	2.833725	8.961027	8.53	72.7609	2.920616	9.235800
8.04	64.6416	2.835489	8.966605	8.54	72.9316	2.922328	9.241212
8.05	64.8025	2.837252	8.972179	8.55	73.1025	2.924038	9.246621
8.06	64.9636	2.839014	8.977750	8.56	73.2736	2.925748	9.252027
8.07	65.1249	2.840775	8.983318	8.57	73.4449	2.927456	9.257429
8.08	65.2864	2.842534	8.988882	8.58	73.6164	2.929164	9.262829
8.09	65.4481	2.844293	8.994443	8.59	73.7881	2.930870	9.268225
8.10	65.6100	2.846050	9.000000	8.60	73.9600	2.932576	9.273618
8.11	65.7721	2.847806	9.005554	8.61	74.1321	2.934280	9.279009
8.12	65.9344	2.849561	9.011104	8.62	74.3044	2.935984	9.284396
8.13	66.0969	2.851315	9.016651	8.63	74.4769	2.937686	9.289779
8.14	66.2596	2.853069	9.022195	8.64	74.6496	2.939388	9.295160
8.15	66.4225	2.854820	9.027735	8.65	74.8225	2.941088	9.300538
8.16	66.5856	2.856571	9.033272	8.66	74.9956	2.942788	9.305912
8.17	66.7489	2.858321	9.038805	8.67	75.1689	2.944486	9.311283
8.18	66.9124	2.860070	9.044335	8.68	75.3424	2.946184	9.316652
8.19	67.0761	2.861818	9.049862	8.69	75.5161	2.947881	9.322017
8.20	67.2400	2.863564	9.055385	8.70	75.6900	2.949576	9.327379
8.21	67.4041	2.865310	9.060905	8.71	75.8641	2.951271	9.332738
8.22	67.5684	2.867054	9.066422	8.72	76.0384	2.952965	9.338094
8.23	67.7329	2.868798	9.071935	8.73	76.2129	2.954657	9.343447
8.24	67.8976	2.870540	9.077445	8.74	76.3876	2.956349	9.348797
8.25	68.0625	2.872281	9.082951	8.75	76.5625	2.958040	9.354143
8.26	68.2276	2.874022	9.088454	8.76	76.7376	2.959730	9.359487
8.27	68.3929	2.875761	9.093954	8.77	76.9129	2.961419	9.364828
8.28	68.5584	2.877499	9.099451	8.78	77.0884	2.963106	9.370165
8.29	68.7241	2.879236	9.104944	8.79	77.2641	2.964793	9.375500
8.30	68.8900	2.880972	9.110434	8.80	77.4400	2.966479	9.380832
8.31	69.0561	2.882707	9.115920	8.81	77.6161	2.968164	9.386160
8.32	69.2224	2.884441	9.121403	8.82	77.7924	2.969848	9.391486
8.33	69.3889	2.886174	9.126883	8.83	77.9689	2.971532	9.396808
8.34	69.5556	2.887906	9.132360	8.84	78.1456	2.973214	9.402127
8.35	69.7225	2.889637	9.137833	8.85	78.3225	2.974895	9.407444
8.36	69.8896	2.891366	9.143304	8.86	78.4996	2.976575	9.412757
8.37	7.00569	2.893095	9.148770	8.87	78.6769	2.978255	9.418068
8.38	70.2244	2.894823	9.154234	8.88	78.8544	2.979933	9.423375
8.39	70.3921	2.896550	9.159694	8.89	79.0321	2.981610	9.428680
8.40	70.5600	2.898275	9.165151	8.90	79.2100	2.983287	9.433981
8.41	70.7281	2.900000	9.170605	8.91	79.3881	2.984962	9.439280
8.42	70.8964	2.901724	9.176056	8.92	79.5664	2.986637	9.444575
8.43	71.0649	2.903446	9.181503	8.93	79.7449	2.988311	9.449868
8.44	71.2336	2.905168	9.186947	8.94	79.9236	2.989983	9.455157
8.45	71.4025	2.906888	9.192388	8.95	80.1025	2.991655	9.460444
8.46	71.5716	2.908608	9.197826	8.96	80.2816	2.993326	9.465728
8.47	71.7409	2.910326	9.203260	8.97	80.4609	2.994996	9.471008
8.48	71.9104	2.912044	9.208692	8.98	80.6404	2.996665	9.476286
8.49	72.0801	2.913760	9.214120	8.99	80.8201	2.998333	9.481561

TABLE IX

Squares and Square Roots (Continued)

n	n^2	$\sqrt{n}$	$\sqrt{10n}$	n	n^2	$\sqrt{n}$	$\sqrt{10n}$
9.00	81.0000	3.000000	9.486833	9.50	90.2500	3.082207	9.746794
9.01	81.1801	3.001666	9.492102	9.51	90.4401	3.083829	9.751923
9.02	81.3604	3.003331	9.497368	9.52	90.6304	3.085450	9.757049
9.03	81.5409	3.004996	9.502631	9.53	90.8209	3.087070	9.762172
9.04	81.7216	3.006659	9.507891	9.54	91.0116	3.088689	9.767292
9.05	81.9025	3.008322	9.513149	9.55	91.2025	3.090307	9.772410
9.06	82.0836	3.009983	9.518403	9.56	91.3936	3.091925	9.777525
9.07	82.2649	3.011644	9.523655	9.57	91.5849	3.093542	9.782638
9.08	82.4464	3.013304	9.528903	9.58	91.7764	3.095158	9.787747
9.09	82.6281	3.014963	9.534149	9.59	91.9681	3.096773	9.792855
9.10	82.8100	3.016621	9.539392	9.60	92.1600	3.098387	9.797959
9.11	82.9921	3.018278	9.544632	9.61	92.3521	3.100000	9.803061
9.12	83.1744	3.019934	9.549869	9.62	92.5444	3.101612	9.808160
9.13	83.3569	3.021589	9.555103	9.63	92.7369	3.103224	9.813256
9.14	83.5396	3.023243	9.560335	9.64	92.9296	3.104835	9.818350
9.15	83.7225	3.024897	9.565563	9.65	93.1225	3.106445	9.823441
9.16	83.9056	3.026549	9.570789	9.66	93.3156	3.108054	9.828530
9.17	84.0889	3.028201	9.576012	9.67	93.5089	3.109662	9.833616
9.18	84.2724	3.029851	9.581232	9.68	93.7024	3.111270	9.838699
9.19	84.4561	3.031501	9.586449	9.69	93.8961	3.112876	9.843780
9.20	84.6400	3.033150	9.591663	9.70	94.0900	3.114482	9.848858
9.21	84.8241	3.034798	9.596874	9.71	94.2841	3.116087	9.853933
9.22	85.0084	3.036445	9.602083	9.72	94.4784	3.117691	9.859006
9.23	85.1929	3.038092	9.607289	9.73	94.6729	3.119295	9.864076
9.24	85.3776	3.039737	9.612492	9.74	94.8676	3.120897	9.869144
9.25	85.5625	3.041381	9.617692	9.75	95.0625	3.122499	9.874209
9.26	85.7476	3.043025	9.622889	9.76	95.2576	3.124100	9.879271
9.27	85.9329	3.044667	9.628084	9.77	95.4529	3.125700	9.884331
9.28	86.1184	3.046309	9.633276	9.78	95.6484	3.127299	9.889388
9.29	86.3041	3.047950	9.638465	9.79	95.8441	3.128898	9.894443
9.30	86.4900	3.049590	9.643651	9.80	96.0400	3.130495	9.899495
9.31	86.6761	3.051229	9.648834	9.81	96.2361	3.132092	9.904544
9.32	86.8624	3.052868	9.654015	9.82	96.4324	3.133688	9.909591
9.33	87.0489	3.054505	9.659193	9.83	96.6289	3.135283	9.914636
9.34	87.2356	3.056141	9.664368	9.84	96.8256	3.136877	9.919677
9.35	87.4225	3.057777	9.669540	9.85	97.0225	3.138471	9.924717
9.36	87.6096	3.059412	9.674709	9.86	97.2196	3.140064	9.929753
9.37	87.7969	3.061046	9.679876	9.87	97.4169	3.141656	9.934787
9.38	87.9844	3.062679	9.685040	9.88	97.6144	3.143247	9.939819
9.39	88.1721	3.064311	9.690201	9.89	97.8121	3.144837	9.944848
9.40	88.3600	3.065942	9.695360	9.90	98.0100	3.146427	9.949874
9.41	88.5481	3.067572	9.700515	9.91	98.2081	3.148015	9.954898
9.42	88.7364	3.069202	9.705668	9.92	98.4064	3.149603	9.959920
9.43	88.9249	3.070831	9.710819	9.93	98.6049	3.151190	9.964939
9.44	89.1136	3.072458	9.715966	9.94	98.8036	3.152777	9.969955
9.45	89.3025	3.074085	9.721111	9.95	99.0025	3.154362	9.974969
9.46	89.4916	3.075711	9.726253	9.96	99.2016	3.155947	9.979980
9.47	89.6809	3.077337	9.731393	9.97	99.4009	3.157531	9.984989
9.48	89.8704	3.078961	9.736529	9.98	99.6004	3.159114	9.989995
9.49	90.0601	3.080584	9.741663	9.99	99.8001	3.160696	9.994999

TABLE X

Random Numbers

04433	80674	24520	18222	10610	05794	37515
60298	47829	72648	37414	75755	04717	29899
67884	59651	67533	68123	17730	95862	08034
89512	32155	51906	61662	64130	16688	37275
32653	01895	12506	88535	36553	23757	34209
95913	15405	13772	76638	48423	25018	99041
55864	21694	13122	44115	01601	50541	00147
35334	49810	91601	40617	72876	33967	73830
57729	32196	76487	11622	96297	24160	09903
86648	13697	63677	70119	94739	25875	38829
30574	47609	07967	32422	76791	39725	53711
81307	43694	83580	79974	45929	85113	72268
02410	54905	79007	54939	21410	86980	91772
18969	75274	52233	62319	08598	09066	95288
87863	82384	66860	62297	80198	19347	73234
68397	71708	15438	62311	72844	60203	46412
28529	54447	58729	10854	99058	18260	38765
44285	06372	15867	70418	57012	72122	36634
86299	83430	33571	23309	57040	29285	67870
84842	68668	90894	61658	15001	94055	36308
56970	83609	52098	04184	54967	72938	56834
83125	71257	60490	44369	66130	72936	69848
55503	52423	02464	26141	68779	66388	75242
47019	76273	33203	29608	54553	25971	69573
84828	32592	79526	29554	84580	37859	28504
68921	08141	79227	05748	51276	57143	31926
36458	96045	30424	98420	72925	40729	22337
95752	59445	36847	87729	81679	59126	59437
26768	47323	58454	56958	20575	76746	49878
42613	37056	43636	58085	06766	60227	96414
95457	30566	65482	25596	02678	54592	63607
95276	17894	63564	95958	39750	64379	46059
66954	52324	64776	92345	95110	59448	77249
17457	18481	14113	62462	02798	54977	48349
03704	36872	83214	59337	01695	60666	97410
21538	86497	33210	60337	27976	70661	08250
57178	67619	98310	70348	11317	71623	55510
31048	97558	94953	55866	96283	46620	52087
69799	55380	16498	80733	96422	58078	99643
90595	61867	59231	17772	67831	33317	00520
33570	04981	98939	78784	09977	29398	93896
15340	93460	57477	13898	48431	72936	78160
64079	42483	36512	56186	99098	48850	72527
63491	05546	67118	62063	74958	20946	28147
92003	63868	41034	28260	79708	00770	88643
52360	46658	66511	04172	73085	11795	52594
74622	12142	68355	65635	21828	39539	18988
04157	50079	61343	64315	70836	82857	35335
86003	60070	66241	32836	27573	11479	94114
41268	80187	20351	09636	84668	42486	71303

* Based on parts of *Table of 105,000 Random Decimal Digits*, Interstate Commerce Commission, Bureau of Transport Economics and Statistics, Washington, D.C.

TABLE X

Random Numbers (Continued)

48611	62866	33963	14045	79451	04934	45576
78812	03509	78673	73181	29973	18664	04555
19472	63971	37271	31445	49019	49405	46925
51266	11569	08697	91120	64156	40365	74297
55806	96275	26130	47949	14877	69594	83041
77527	81360	18180	97421	55541	90275	18213
77680	58788	33016	61173	93049	04694	43534
15404	96554	88265	34537	38526	67924	40474
14045	22917	60718	66487	46346	30949	03173
68376	43918	77653	04127	69930	43283	35766
93385	13421	67957	20384	58731	53396	59723
09858	52104	32014	53115	03727	98624	84616
93307	34116	49516	42148	57740	31198	70336
04794	01534	92058	03157	91758	80611	45357
86265	49096	97021	92582	61422	75890	86442
65943	79232	45702	67055	39024	57383	44424
90038	94209	04055	27393	61517	23002	96560
97283	95943	78363	36498	40662	94188	18202
21913	72958	75637	99936	58715	07943	23748
41161	37341	81838	19389	80336	46346	91895
23777	98392	31417	98547	92058	02277	50315
59973	08144	61070	73094	27059	69181	55623
82690	74099	77885	23813	10054	11900	44653
83854	24715	48866	65745	31131	47636	45137
61980	34997	41825	11623	07320	15003	56774
99915	45821	97702	87125	44488	77613	56823
48293	86847	43186	42951	37804	85129	28993
33225	31280	41232	34750	91097	60752	69783
06846	32828	24425	30249	78801	26977	92074
32671	45587	79620	84831	38156	74211	82752
82096	21913	75544	55228	89796	05694	91552
51666	10433	10945	55306	78562	89630	41230
54044	67942	24145	42294	27427	84875	37022
66738	60184	75679	38120	17640	36242	99357
55064	17427	89180	74018	44865	53197	74810
69599	60264	84549	78007	88450	06488	72274
64756	87759	92354	78694	63638	80939	98644
80817	74533	68407	55862	32476	19326	95558
39847	96884	84657	33697	39578	90197	80532
90401	41700	95510	61166	33757	23279	85523
78227	90110	81378	96659	37008	04050	04228
87240	52716	87697	79433	16336	52862	69149
08486	10951	26832	39763	02485	71688	90936
39338	32169	03713	93510	61244	73774	01245
21188	01850	69689	49426	49128	14660	14143
13287	82531	04388	64693	11934	35051	68576
53609	04001	19648	14053	49623	10840	31915
87900	36194	31567	53506	34304	39910	79630
81641	00496	36058	75899	46620	70024	88753
19512	50277	71508	20116	79520	06269	74173

TABLE X

Random Numbers (Continued)

24418	23508	91507	76455	54941	72711	39406
57404	73678	08272	62941	02349	71389	45605
77644	98489	86268	73652	98210	44546	27174
68366	65614	01443	07607	11826	91326	29664
64472	72294	95432	53555	96810	17100	35066
88205	37913	98633	81009	81060	33449	68055
98455	78685	71250	10329	56135	80647	51404
48977	36794	56054	59243	57361	65304	93258
93077	72941	92779	23581	24548	56415	61927
84533	26564	91583	83411	66504	02036	02922
11338	12903	14514	27585	45068	05520	56321
23853	68500	92274	87026	99717	01542	72990
94096	74920	25822	98026	05394	61840	83089
83160	82362	09350	98536	38155	42661	02363
97425	47335	69709	01386	74319	04318	99387
83951	11954	24317	20345	18134	90062	10761
93085	35203	05740	03206	92012	42710	34650
33762	83193	58045	89880	78101	44392	53767
49665	85397	85137	30496	23469	42846	94810
37541	82627	80051	72521	35342	56119	97190
22145	85304	35348	82854	55846	18076	12415
27153	08662	61078	52433	22184	33998	87436
00301	49425	66682	25442	83668	66236	79655
43815	43272	73778	63469	50083	70696	13558
14689	86482	74157	46012	97765	27552	49617
16680	55936	82453	19532	49988	13176	94219
86938	60429	01137	86168	78257	86249	46134
33944	29219	73161	46061	30946	22210	79302
16045	67736	18608	18198	19468	76358	69203
37044	52523	25627	63107	30806	80857	84383
61471	45322	35340	35132	42163	69332	98851
47422	21296	16785	66393	39249	51463	95963
24133	39719	14484	58613	88717	29280	77360
67253	67064	10748	16006	16767	57345	42285
62382	76941	01635	35829	77516	98468	51686
98011	16503	09201	03523	87192	66483	55649
37366	24386	20654	85117	74078	64120	04643
73587	83993	54176	05221	94119	20108	78101
33583	68291	50547	96085	62180	27453	18567
02878	33223	39199	49536	56199	05993	71201
91498	41673	17195	33175	04994	09879	70337
91127	19815	30219	55591	21725	43827	78862
12997	55013	18662	81724	24305	37661	18956
96098	13651	15393	69995	14762	69734	89150
97627	17837	10472	18983	28387	99781	52977
40064	47981	31484	76603	54088	91095	00010
16239	68743	71374	55863	22672	91609	51514
58354	24913	20435	30965	17453	65623	93058
52567	65085	60220	84641	18273	49604	47418
06236	29052	91392	07551	83532	68130	56970

TABLE X

Random Numbers (Continued)

94620	27963	96478	21559	19246	88097	44926
60947	60775	73181	43264	56895	04232	59604
27499	53523	63110	57106	20865	91683	80688
01603	23156	89223	43429	95353	44662	59433
00815	01552	06392	31437	70385	45863	75971
83844	90942	74857	52419	68723	47830	63010
06626	10042	93629	37609	57215	08409	81906
56760	63348	24949	11859	29793	37457	59377
64416	29934	00755	09418	14230	62887	92683
63569	17906	38076	32135	19096	96970	75917
22693	35089	72994	04252	23791	60249	83010
43413	59744	01275	71326	91382	45114	20245
09224	78530	50566	49965	04851	18280	14039
67625	34683	03142	74733	63558	09665	22610
86874	12549	98699	54952	91579	26023	81076
54548	49505	62515	63903	13193	33905	66936
73236	66167	49728	03581	40699	10396	81827
15220	66319	13543	14071	59148	95154	72852
16151	08029	36954	03891	38313	34016	18671
43635	84249	88984	80993	55431	90793	62603
30193	42776	85611	57635	51362	79907	77364
37430	45246	11400	20986	43996	73122	88474
88312	93047	12088	86937	70794	01041	74867
98995	58159	04700	90443	13168	31553	67891
51734	20849	70198	67906	00880	82899	66065
88698	41755	56216	66852	17748	04963	54859
51865	09836	73966	65711	41699	11732	17173
40300	08852	27528	84648	79589	95295	72895
02760	28625	70476	76410	32988	10194	94917
78450	26245	91763	73117	33047	03577	62599
50252	56911	62693	73817	98693	18728	94741
07929	66728	47761	81472	44806	15592	71357
09030	39605	87507	85446	51257	89555	75520
56670	88445	85799	76200	21795	38894	58070
48140	13583	94911	13318	64741	64336	95103
36764	86132	12463	28385	94242	32063	45233
14351	71381	28133	68269	65145	28152	39087
81276	00835	63835	87174	42446	08882	27067
55524	86088	00069	59254	24654	77371	26409
78852	65889	32719	13758	23937	90740	16866
11861	69032	51915	23510	32050	52052	24004
67699	01009	07050	73324	06732	27510	33761
50064	39500	17450	18030	63124	48061	59412
93126	17700	94400	76075	08317	27324	72723
01657	92602	41043	05686	15650	29970	95877
13800	76690	75133	60456	28491	03845	11507
98135	42870	48578	29036	69876	86563	61729
08313	99203	00990	13595	77457	79969	11339
90974	83965	62732	85161	54330	22406	86253
33273	61993	88407	69399	17301	70975	99129

Answers to Odd Exercises†

Page 15

3. (a) $x_1 f_1 + x_2 f_2 + x_3 f_3 + x_4 f_4 + x_5 f_5$.
   (b) $k(x_1 + x_2 + x_3 + x_4 + x_5 + x_6)$.
   (c) $x_1^2 + x_2^2 + x_3^2 + x_4^2$.
   (d) $y_2 + z_2 + y_3 + z_3 + y_4 + z_4 + y_5 + z_5$
   (e) $z_1 + z_2 + z_3 + z_4 + z_5 + z_6 - 6c$.
   (f) $x_1^2 y_1 + x_2^2 y_2 + x_3^2 y_3 + x_4^2 y_4 + x_5^2 y_5$

5. (a) 15 (b) 1 (c) 158 (d) 138 (e) 23 (f) 8738.
7. Yes.

Page 28

1. (a) \$−0.005, 9.995, 19.995, 29.995, 39.995, 49.995.
   (b) \$4.995, 14.995, 24.995, 34.995, 44.995.
   (c) \$10.00.
3. (a) 3.95, 4.95, 5.95, 6.95, 7.95, 8.95, 9.95 inches.
   (b) 4.45, 5.45, 6.45, 7.45, 8.45, 9.45 inches.
   (c) 1.0 inch.
5. For example, 100–199, 200–299, 300–399, . . . etc., bushels.
7. The class frequencies are 2, 43, 46, 15, 7, and 1. The cumulative "less than" frequencies are 2, 45, 91, 106, 113, and 114.
9. For example, 8.0–8.9, 9.0–9.9, 10.0–10.9, . . . etc., per cent. For example, "less than" 8.0, "less than" 9.0, "less than" 10.0, . . . etc., per cent.
11. For example, 45,000–45,999, 46,000–46,999, 47,000–47,999, . . . etc. For example, "less than" 45,000, "less than" 46,000, "less than" 47,000, . . . etc.
13. (a) The class frequencies are 14, 48, 28, 14, 4, and 2.
    (b) The class frequencies are 6, 46, 33, 17, 6, and 2.

Page 50

1. 10.2 per cent.	3. \$158.7 million.	5. 225 pounds.
7. \$24,247.	9. 11.5 per cent.	11. \$47,775 mil.
13. \$68.76.	15. 225 pounds.	

---

† It should be noted that in some instances the reader's answers will differ from the ones given here due to rounding off at various stages of the calculations.

Page 56

1. 180.43 thousands.   5. (a) 160.0 physicians.
                            (b) 161 (rounded from 161.1) physicians.

Page 65

1. 52 m.p.h.   3. 114.5.   5. 24.   7. $135.9.
11. Firm A uses the means, Firm B uses the medians to make the comparison.
13. 51 and 55.
15. (a) 118–157, 137.5; (b) 110–149, 129.5.
17. Blue.

Page 71

1. (a) 10; (b) 14; (c) 3.   3. 26.62.   5. 8.
7. $312; yes.   9. $1.53.
11. 0.307.

Page 75

1. 87.8 thousands and 234.7 thousands.
7. $88.76 and $103.04.
11. 138 (rounded from 138.2) physicians and 168 (rounded from 168.1) physicians.

Page 84

1. 52 minutes.   3. 90 pounds.   5. 1.83.
7. $40.2 millions.

Page 90

1. $s = 3.9$.   3. $s = 5.15$.   5. $\sigma = 5.6$ years.
7. $\sigma = 29$ pounds.

Page 94

1. $7.65.   3. $s = 42$ (rounded from 41.9) physicians.
5. Mr. Holmes' performance relative to the law class was slightly better than his performance relative to the marketing class. The respective values of z are 1.80 and 1.67.

Page 97

1. 5.6 per cent.   3. 26.2 per cent.   5. 17.8 per cent.

Page 108

1. 0.05.   3. 0.6.   5. 0.1 and 2.3.

Page 118

1. (a) $P(B|A)$; (b) $P(A|B)$; (c) $P(A$ and $B)$.
3. (a), (c), and (f) are mutually exclusive; (b), (d), and (e) not.
5. 0.77.   7. 1/36.   9. 0.32.
11. 2/3.

Page 122

1. 2/13.
3. 6/13.
5. 2/12.
7. 8/18.
9. 35/36.
11. 1/16.
13. (a) 1/4; (b) 25/102.
15. 2/169.

Page 125

1. $0.10.
3. $0.02.
5. $555.56.

Page 130

1. The expected frequencies for 0, 1, 2, 3, and 4 heads are 15, 60, 90, 60, and 15, respectively.
3. The probabilities for 0, 1, 2, 3, 4, and 5 heads are 1/32, 5/32, 10/32, 10/32, 5/32, and 1/32, while the corresponding expected frequencies are 10, 50, 100, 100, 50, and 10.

Page 137

1. 495/4096.
3. 540/4096.
5. 56/256.
7. 1,966,080/9,765,625.
9. The probabilities that 0, 1, 2, 3, or 4 insurance salesmen will make a sale are 0.2401, 0.4116, 0.2646, 0.0756, and 0.0081.
11. The probabilities are 4096/15,625, 6144/15,625, 3840/15,625, 1280/15,625, 240/15,625, 24/15,625, and 1/15,625.

Page 141

1. $\mu = 1.5$ and $\sigma = 0.87$.
3. (a) $\mu = 50$ and $\sigma = 5$; (b) $\mu = 4$ and $\sigma = 1.732$; (c) $\mu = 40$ and $\sigma = 6$; (d) $\mu = 45$ and $\sigma = 6$.
5. $\mu = 1$ and $\sigma = 0.913$.

Page 152

1. (a) 1.0; (b) 2.5; (c) -1.5; (d) -2.25.
3. (a) 0.01; (b) 1.19; (c) 2.00; (d) -2.50; (e) ± 2.58.

Page 155

1. (a) .62 per cent; (b) 67 per cent; (c) 71.16 inches.
3. (a) 21,040 persons; (b) 46.5 per cent; (c) 13.4 per cent.

Page 160

1. (a) 0.205; (b) 0.203.
3. 0.217.
5. 0.1335.
7. 0.96.

Page 167

1. 4.98 per cent.
3. 0.15.

Page 183

3. 0.73 and 0.68.
7. It is one-fifth as large as before.
11. 0.81.

Page 194

1. (a) 47,216–48,784 pounds; (b) We can assert with a probability of 0.95 that our error is less than 784 pounds.
3. (a) $26.82–$29.18; (b) We can assert with a probability of 0.98 that our error is less than $1.065.
5. 0.89. 7. At least 139. 9. 43 measurements.

Page 199

1. (a) 13.47–15.53 oz.; (b) 228.74–271.86 gals.; (c) $1345.69–$1654.31.
3. (a) 13.10–15.90 oz.; (b) 219.63–280.97 gals.; (c) $1244.06–$1755.94.
5. We can assert with a probability of 0.98 that our error is less than 254.27 miles.
7. We can assert with a probability of 0.99 that our error is less than $13.91.
9. They can assert with a probability of 0.95 that our error is less than $6.39.

Page 205†

1. (a) 0.29–0.42; (b) We can assert with a probability of 0.95 that our error is less than 0.06.
3. (a) 0.35–0.45; (b) 0.34–0.46.
5. 0.78–0.80.
7. At least 385. 9. At least 849. 11. At least 1,025.

Page 209

1. 3120–4255.

Page 211

1. 108.6–111.4.
3. 0.117–0.243.

Page 216

1. (a) 0.005; (b) 0.079. 3. (a) 0.019; (b) 0.024.

Page 221

1. The marketing expert will be committing a Type I error if he predicts that less than 1 million units of the new toy will be sold and at least 1 million units are sold; he will be committing a Type II error if he predicts that at least 1 million units of the new toy will be sold and fewer than 1 million units are sold.
3. Here a Type I error is to the disadvantage of the producer who has a good lot rejected; a Type II error is to the disadvantage of the consumer who accepts a bad shipment.

Page 231

1. $z = 2.48$, reject. 3. $z = -2.80$, reject.
5. $z = -1.70$, cannot reject. 7. $z = -2.26$, difference is significant.
9. $z = 1.8$, difference is not significant.

† In Exercises 7 through 11 of this section the final answers are always rounded up.

Page 241

1. $z = -2.74$, reject.   3. $z = -3.10$, reject.
5. $t = 3.00$, reject.
7. $z = -4.24$, difference is significant.
9. $z = 10.24$, difference is significant.
11. $t = -1.09$, difference is not significant.

Page 252

1. $\chi^2 = 3.98$, cannot reject null hypothesis.
3. $\chi^2 = 20.95$, reject null hypothesis.
5. $\chi^2 = 0.31$, cannot reject. $z^2 = (0.57)^2 = 0.32$ and the difference may be attributed to rounding.

Page 256

1. $\chi^2 = 33.80$, reject null hypothesis.
3. $\chi^2 = 20.71$, reject null hypothesis.

Page 260

1. $\chi^2 = 0.719$, excellent fit.   3. $\chi^2 = 19.39$, fit is not good.
7. $\chi^2 = 5.74$, very good fit.

Page 267

1. $F = 5.7$, there are significant differences.
3. $F = 2.51$, the differences are not significant.

Page 276

1. $z = -3.71$, significantly nonrandom.
3. $z = -3.16$, significantly nonrandom.
5. $z = -2.76$, arrangement is nonrandom.

Page 278

1. $z = -2.97$, significantly nonrandom.
3. $z = .097$, random.

Page 294

1. (a) $y' = 0.151 + 0.332x$; (c) 1.15 million tons.
3. (a) $y' = \$3183.29 + 201.94x$; (b) \$5606.57.
5. (a) $y' = \$2207 + (-136)x$; \$1799.
7. (a) $y' = 8.9 + 2.7x$; (b) $x' = 0.484 + 0.148x$; (c) $x = -3.296 + 0.370y'$.

Page 303

1. \$27.4 thousand.
3. $x = -12.98 + .0044y$.
5. \$797–\$1803.

Page 313

1. 0.97.   3. 0.972.
5. One can always draw a straight line through any two distinct points.

7. 0.87.
9. (a) A positive correlation; (b) a negative correlation; (c) no correlation; (d) a negative correlation; (e) a positive correlation.

Page 319

1. (a) Significant; (b) significant; (c) not significant; (d) significant; (e) significant; (f) not significant.
3. Significant.

Page 329

1. 0.92.   3. 0.96.   5. 0.68.

Page 335

1. $\chi^2 = 24.29$, the correlation is significant, $C = 0.23$.
3. $\chi^2 = 45.4$, the correlation is significant, $C = 0.28$.

Page 355

1. (a) 92.5, 91.8, 95.0, and 94.7; (b) 102.3; (c) 93.8, and 95.6; (d) 102.5.
3. (a) 80.3; (b) 93.0.

Page 361

1. (a) 124.8, 137.8, 159.4; (b) 157.9; (c) 158.7; (d) 158.6.
3. (a) 205.6, 233.6; (b) 202.9, 230.4; (c) 231.2; (d) 232.4; (e) 231.8.

Page 365

1. 161.9.   3. 233.7.   5. 98.1.

Page 371

1. 94.3   5. 111.8.

Page 375

1. 96.3, 98.2, 100.4, 100.0, 102.7, 106.2, 107.0, 108.1, 110.3, and 111.1.
3. 89.2, 94.7, 93.6, 89.9, 92.9, 93.4, 95.1, 100.0, 106.2, 105.8.
5. 89.3, 88.9, 88.8, 88.6, 88.3, 87.9, 87.6, 87.3, 87.3, 87.2, 87.2, 87.1, 87.0, and 86.7.

Page 378

1. (16.3.3) and (16.3.4) do not satisfy the time reversal test; the others do.

Page 404

1. (a) The 1956 and 1963 trend values are 96.3 and 119.6; (b) The annual trend increment is 3.33 and the 1953, 1960, and 1965 trend values are 86.3, 109.6, and 126.3.
3. The trend values are 42,604, 44,484, 46,364, 48,244, 50,124, 52,004, 53,884, and 55,764.

Page 411

1. $y' = 55.31 + 1.05x$, (Origin, 1960; x units, 1 year; y, million employees); The 1953 and 1967 trend values are 47.96 and 62.66.

3. $y' = 107.93 + 1.78x$, (Origin, 1959–60; x units, 6 months; y, per cent); The 1953 and 1966 trend values are 84.79 and 131.07.

5. $y' = 76.02 + 1.26x$, (Origin, 1961–62; x units, 6 months; y, annual shipments in millions of net tons.)

7. $y' = 4.55 + .147x$, (Origin, 1953–54; x units, 6 months; y, annual data in billions of kilowatt hours.); The 1941 and 1966 trend values are 0.87 and 8.23.

9. $y' = 10.25 + 0.88x$, (Origin, 1959; x units, 6 months; y, annual profits in billions of dollars.)

Page 421

1. (a) The trend values are: 484.21, 473.42, 478.81, 478.81, 500.38, 538.13, 592.06, 662.17, 748.46, 850.93, 969.58, 1104.41, 1255.42, 1422.61, 1605.98, 1805.53, 2021.26, and 2253.17.

3. (a) $y' = 3.900 + .147x + .003x^2$, (Origin, 1953–54, x units, 6 months; y, annual sales in billions of kilowatt hours.)
   (b) 2.1, 2.1, 2.1, 2.2, 2.3, 2.4, 2.5, 2.6, 2.8, 3.0, 3.2, 3.5, 3.8, 4.0, 4.4, 4.7, 5.1, 5.5, 5.9, 6.3, 6.8, 7.3, 7.8, 8.3, 8.9, 9.4.

Page 425

1. The 1936 to 1965 values of the moving average are: 4.4, 3.9, 3.6, 3.5, 4.3, 3.4, 2.2, 0.8, 0.7, 1.5, 2.9, 4.4, 5.5, 6.5, 7.0, 6.8, 6.5, 6.5, 7.7, 7.6, 7.8, 6.4, 6.3, 6.6, 7.1, 7.6, 8.0, 8.9, 9.8, and 10.2.

3. The 1943 to 1964 values of the moving average are: 3.1, 3.2, 3.5, 3.7, 3.9, 4.2, 4.6, 4.9, 5.3, 5.7, 6.1, 6.5, 6.8, 7.0, 7.4, 7.7, 7.9, 8.3, 8.6, 8.9, 9.4, and 9.8.

5. The 1942 to 1963 values of the moving average are: 20.1, 20.5, 20.4, 20.0, 18.6, 17.0, 15.3, 14.1, 13.3, 12.9, 12.8, 13.1, 13.3, 13.6, 13.7, 13.7, 13.8, 13.9, 13.8, 13.8, 13.7, and 13.4.

Page 427

1. 71.06 million.   3. 111.30 million net tons.
5. 5,872.06 thousand.

Page 435

1. 100.8, 94.0, 106.8, 104.8, 107.4, 100.8, 100.0, 97.3, 94.0, 96.0, 96.7, and 101.4.
3. 97.2, 97.7, 98.4, 99.5, 100.6, 101.8, 102.1, 101.9, 100.4, 100.8, 100.2, 99.4.

Page 440

1. 88.9, 93.4, 110.4, 104.5, 103.2, 102.8, 97.5, 96.4, 92.3, 101.5, 99.9, and 109.2.
3. 97.0, 97.5, 98.3, 99.7, 100.6, 101.9, 102.2, 102.1, 100.4, 100.9, 100.1, and 99.3.
5. 79.2, 76.4, 90.8, 95.2, 104.9, 100.9, 85.0, 93.2, 104.7, 120.4, 113.0, and 136.3.

Page 446

1. 87.7, 93.1, 110.2, 105.3, 103.3, 103.9, 96.7, 95.9, 92.3, 102.2, 101.5, and 107.9.

3. 97.0, 97.7, 98.4, 99.5, 100.6, 101.8, 102.1, 101.8, 100.5, 101.4, 100.0 and 99.2

5. 99.8, 99.4, 99.6, 99.6, 99.2, 100.6, 99.6, 99.2, 100.2, 100.4, 100.2 and 102.2.

Page 451

1. The deseasonalized 1962 data are: 4561, 4649, 4529, 4528, 4739, 4658, 4855, 4843, 4572, 4972, 5026 and 5037; there are percentage decreases of 2.3 and 1.7 per cent, respectively.

3. *1963:* 144.1, 143.8, 149.4, 150.5, 148.8, 148.0, 148.0, 147.1, 146.2, 146.5, 146.7, 146.1; *1964:* 149.1, 155.5, 152.3, 151.4, 150.7, 150.0, 151.0, 150.2, 150.5, 150.6, 149.8, 151.0; *1965:* 155.1, 152.3, 151.3, 150.5, 151.6, 151.9, 154.1, 152.3, 152.6, 152.7, 150.8, 150.0; *1966:* 152.1, 149.1, 150.4, 150.5, 150.7, 151.9, 153.1, 154.3, 156.9, 158.8, 159.1, 159.9.

5. When deseasonalized, sales declined from \$47,619 to \$40,000. Sales Manager did not take seasonal into consideration. President multiplied deseasonalized sales of \$40,000 by 12 months.

Page 452

7. 75,916, 76,435, 77,191, 78,421, 79,261, 80,418, 80,789, 80,844, 79,629, 80,158, 79,654, 79,147.

9. (a) 5.77, 5.81, 5.85, 5.90, 5.94, 5.98, 6.02, 6.06, 6.10, 6.15, 6.19, 6.23 million dollars.
(b) 4.15, 4.07, 5.26, 6.49, 8.32, 7.77, 7.10, 6.67, 6.04, 6.21, 5.57, 4.36 million dollars.

Page 483

1. The central line for $\overline{X}$ is 0.13; the control limits are 2.20 and -1.94. The central line for σ is 1.30; the control limits are 2.72 and 0.

3. The central line for $\overline{X}$ is 1.4997; the control limits are 1.4888 and 1.5106. The central line for R is 0.0354; the control limits are 0.0079 and 0.0629.

Page 488

1. The central line is 0.0236; the control limits are 0.0692 and 0.

3. The central line is 10; the control limits are 19.24 and 0.76.

5. The central line is 2.37; the control limits are 6.99 and 0.

Index

A

Absolute value, 82
Acceptable quality level (AQL), 490
Acceptance sampling, 489–490
Alpha, α, probability of Type I error, 219
Alpha, α, regression coefficient, 298, 299
Alpha-four, α_4, 107
Alpha-three, α_3, 105–106
Analysis of variance, 268–269, 314
Annual rate, 449
Annual trend increment, 403, 409
Area sample, 281
Area under distribution curve, 143–144
Arithmetic line chart, 393–395
Arithmetic mean (*see* Mean)
Arithmetic paper, 393
Assignable variation, 476
Average (*see* Mean)
Average deviation, 81–84
  short-cut formulas, 99
Average outgoing quality limit (AOQL), 490

B

Bar chart, 462
  component, 470
Base year, 345
Beta, β, probability of Type II error, 219
Beta, β, regression coefficient, 298, 299
Bias:
  Laspeyres' index, 359
  Paasche's index, 359
  sampling, 8, 271
Binary comparison, 338
Binomial coefficient, 135, 507
  and weighted moving average, 456

533

Binomial distribution, 131–137
  formula, 134
  mean, 138
  normal curve approximation, 146, 156–160, 200, 213, 226
  standard deviation, 140
Business cycle, 388, 453

C

Carli, G. R., 354
Categorical distribution, 19, 37–39
Causation and correlation, 316–317, 331
Central limit theorem, 179
Central line, control chart, 476
Central tendencies, 45
Chance variation, 78
  and assignable variation, 475–476
Change of scale, 53–54, 92, 106
Chain index numbers, 365–371
Characteristic, 495
Chi-square distribution, 248–249
  analysis of 2 by k table, 249
  analysis of r by k table, 255, 334
  goodness of fit, 257–260
  table, 503
Chi-square statistic, χ^2, 248, 258
Class:
  boundary, 23
  frequency, 22
  interval, 21
  limit, 23
  mark, 24
  open, 25
Clopper, C. J., 506
Cluster sample, 281
Coding, 54
Coefficient of correlation (*see* Correlation coefficient)
Coefficient of quartile variation, 97

Coefficient of rank correlation, 327
Coefficient of variation, 96
Comparability of data, 342–344
Comparison, binary and in series, 338
Component bar chart, 470
Conditional distribution, 297
Conditional probability, 117
Confidence coefficient, 190
Confidence interval:
  and point estimate, 187–188
  interpretation, 191
  mean, 190–191
    small sample, 197
  proportion, 202–204
  regression coefficient, 299, 306
  standard deviation, 208
Confidence limits (see Confidence interval)
Consumer Price Index, 339–340, 345, 367–369, 378–379
  calculation, 367
  relative importance weights, 370
Consumer's risk, 221, 490
Contingency coefficient, 335
Contingency table, 334
Continuous distribution, 144
  mean and standard deviation, 145
Continuous variable, 142
Control chart, 476
  attributes, 477, 484–488
  constants, 508
  fraction defective, 486
  mean, 480–481, 483
  number defective, 487
  percentage defective, 486
  range, 480–481
  specified standards, 483
  standard deviation, 483
  variables, 477, 478–483
Control limits, 478
Correlation:
  and causation, 316–317, 331
  curvilinear, 316
  linear, 309, 316
  multiple, 330
  negative, 311
  partial, 331–332
  positive, 311
  qualitative data, 333–335
  rank, 327–329
  spurious, 317, 336
Correlation coefficient, 309
  abuse of, 316

Correlation coefficient (cont.):
  computation, 310
  grouped data, 323
  interpretation, 315
  multiple, 330
  partial, 332
  significance test, 318
  standard error, 318
Correlation table, 323
Cross-stratification, 280
Cumulative distribution, 26–27
Curve fitting:
  exponential function, 416–418
  line, 288–293, 401–408
  parabola, 412–416
  power function, 419
Cyclical irregulars, 453–455
Cyclical relatives, 453, 456
Cyclical variation, 388–389

D

Data:
  deseasonalized, 447–449
  external, 6
  internal, 6
  primary, 6
  published, 8–12
  raw, 20
  secondary, 6
  sources of, 6–12
Deciles, 74
Decision theory, 125
Deflating, 374
Degrees of freedom, 98
  chi-square distribution, 249, 255, **259**
  F-distribution, 265–266
  standard error of estimate, 301
  t-distribution, 196, 236, 241
de Moivre, A., 145
Density function, 144
Descriptive statistics, 3
Deseasonalized data, 447–449
Destructive sampling, 173
Deviations from mean, 81
Difference between means, standard error, 238
Differences, first and second, 415
Discrete variable, 142
Distribution (see also Frequency, Theoretical, and Sampling distributions):

Distribution (*cont.*):
  binomial, 131–137
  chi-square, 248–249, 255, 259
  conditional, 297
  continuous, 142–145
  F, 265–266
  normal, 145–152
  Poisson, 165–167, 487
  probability, 129
  sampling, 175–176
  Student-*t*, 196, 236, 240, 302
  theoretical, 127
Double sampling, 283
Drobisch index, 359
Dun & Bradstreet, Index of Wholesale
  Food Prices, 344, 352

E

Edgeworth, F. Y., 354, 360
Equation:
  exponential curve, 416
  Gompertz curve, 420
  linear, 287, 330
  logistic curve, 420
  modified exponential, 420
  parabola, 412
  power function, 419
Errors, Type I and Type II, 218
Estimate, point and interval, 187–188
Events:
  independent and dependent, 116
  mutually exclusive, 114–115
Exhaustive sampling, 172
Expectation, mathematical, 123–125
Expected frequencies:
  normal curve, 162–164, 258
  r by *k* table, 254
  theoretical distribution, 127–130
  2 by *k* table, 246
Experimental sampling distribution,
  175–183
  means, 175, 178, 179
  medians, 181
Exponential trend, 416–419
External data, 6
Extrapolation, 426–427

F

Factor reversal test, 377
F-distribution, 265–266, 504, 505

Finite population, 170
Fisher, I., 359
Fisher, R. A., 502, 503
Fixed-weight aggregative index, 360
Forecasting, 381–383, 426–427, 449–451
Fraction defective, 485
Frequencies, expected (*see* Expected
  frequencies)
Frequency distribution, 17–42
  bell-shaped, 100, 102
  categorical, 19, 37–39
  class boundaries, 23
  class frequencies, 22
  class intervals, 21
  class limits, 23
  class marks, 24
  cumulative, 26–27
  graphical presentation, 31–37, 462–472
  J-shaped, 101–102
  leptokurtic, 107
  number of classes, 20
  numerical, 18
  open classes, 25
  peakedness, 107
  percentage, 27
  platykurtic, 107
  qualitative and quantitative, 19
  skewness, 104–106
  symmetry, 101
  tabular presentation, 40
  two-way, 320–321
  unequal classes, 33
  U-shaped, 101–102
Frequency polygon, 35

G

Gauss, C., 146
General purpose index, 344
Geometric mean, 66–68, 353
Given year, 345
Gompertz curve, 420
Goodness of fit:
  binomial distribution, 260
  least squares line, 307–309
  normal curve, 258
Gosset, W. S., 196
Graphical presentations:
  frequency distributions, 31–36
  pictorial, 461–472
  time series, 393–398
Grouping error, 92, 99

H

Harmonic mean, 68
Histogram, 31–34
  three-dimensional, 322
Hypothesis testing (*see* Tests of
  hypotheses)

I

Ideal index, 359
Independent events, 116
Independent samples, 238
Index:
  Consumer Price, 339–340, 345, 367–
    369, 378–379
  Drobisch, 359
  Edgeworth, 360
  fixed-weight aggregative, 360
  general purpose, 344
  Ideal, 359
  Industrial Production, 364
  Laspeyres', 357
  Paasche's, 357
  Spot Market Prices, 345, 355
  weighted aggregative, 357.
  weighted mean of price relatives, 362–
    363
  Wholesale Food Prices, 344, 352
  Wholesale Price, 344, 355, 360
Index Numbers:
  arithmetic mean of price relatives,
    353
  chain, 365–371
  factor reversal test, 377
  geometric mean of price relatives, 353
  history, 354
  properties, 376–378
  sampling of items, 345
  shifting of base, 372–373
  simple aggregative, 351
  symbolism, 348, 366
  time reversal test, 376
  units test, 351
  unweighted, 350–355
  use in deflating, 373–375
  weighted, 356–365
Index of seasonal variation (*see* Seasonal
  index)
Inductive statistics, 3–5
Industrial Production, Index of, 364
Infinite population, 170

Internal data, 6
Interquartile range, 95
Interval estimate, 188
Irregular variation, 389

J

Jevons, W. S., 354
J-shaped distribution, 101–102
Judgment sample, 270

K

Kurtosis, 107

L

Laplace, P., 145
Laspeyres' index, 357
  bias, 359
Least squares, method of, 290–291
Legendre, A., 290
Leptokurtic distribution, 107
Level of significance, 224
Limits of prediction, 299–303
  limitations of, 303
Linear curve fitting, 288–293, 401–408
  normal equations, 293
Linear equations:
  several unknowns, 304
  two unknowns, 287
Linear regression, 296–299
  multiple, 304
Linear trend:
  least squares, 404–408
    modifications, 408–411
  method of semi-averages, 402–404
Link relatives, 445, 457
Logarithmic line chart, 393, 395–398
Logistic curve, 420
Lot-by-lot sampling inspection, 490
Lot tolerance percent defective
  (LTPD), 490
Lower control limit, 476

M

Mantissa, 495
Mathematical expectation, 123–125

Mean:
  arithmetic, 46–56
  assumptions for grouped data, 52
  binomial distribution, 138
  combined data, 49
  confidence interval for 190–191, 197–198
  continuous distribution, 145
  distribution of differences between means, 229
  geometric, 66–68, 353
  grouped data, 51–56
  harmonic, 68
  modified, 439
  point estimate of, 192–193
  probable error, 208
  properties, 47, 59–60, 76
  relation to median and mode, 59, 104–105
  sample and population, 46
  sampling distribution of, 176–180
  short-cut formula, 55
  small sample confidence interval, 197
  standard error, 177
  theoretical sampling distribution of, 176–178
  weighted, 69–70
Mean deviation, 81–84
  short-cut formulas, 99
Median:
  assumption for grouped data, 61
  definitions, 57, 60–61
  properties, 58–59, 76
  relation to mean and mode, 59, 104–105
  sampling distribution of, 180–182
  standard error, 182
Merrington, M., 504, 505
Method of least squares, 290–291
Method of semi-averages, 402–404
Method of simple averages, 432–435
Mitchell, W. C., 355
Modal class, 65
Mode, 63
  properties, 64
  relation to mean and median, 104–105
Modified exponential curve, 420
Modified mean, 439, 445
Money wages and real wages, 374–375
Monthly trend increment, 410
Moving average, 423–425
  weighted, 455–456
Mu, μ, population mean, 47

Multiple correlation, 330–331
  coefficient, 330
Multiple linear regression, 304
Multiple sampling, 284
Mutually exclusive events, 114

N

Negative correlation, 31
Neyman, J., 501
Normal, of time series, 454
Normal curve, 145–152
  and binomial distribution, 146, 156–160
  applications, 152–155
  areas under, 147–152
  fitting, 160–164
  formula, 147
  goodness of fit, 258–259
  paper, 160
  standard form, 148
  table, 501
  use of table, 147–152
Normal equations:
  linear regression, 293
  parabolic trend, 412
  power function, 419
Null hypothesis, 222
Numbers, random, 170–172
  table, 171
Numerical distribution, 18

O

Ogive, 35
One-sided alternative, 227
One-tail test, 227, 235
Open class, 25
Operating characteristic curve, 220
Optimum allocation, stratified sampling, 280
Orthogonal polynomials, 415

P

Paasche's index, 357
  bias, 359
Parabolic trend, 412–416
  second difference test, 415
Parameter, 47

Parity Index, 339
Partial correlation, 331–332
  coefficient, 332
Peakedness, 107
Pearl-Reed curve, 420
Pearsonian coefficient of skewness, 104–105
Pearson product-moment coefficient of correlation, 309
Percentage distribution, 27
Percentage-of-moving average method, seasonal index, 440–446
Percentage-of-trend method, seasonal index, 436–440
Percentiles, 74
Pie chart, 466
Platykurtic distribution, 107
Point estimate, 187
Poisson distribution, 165–167, 487
Population, 43
  and sample, 43–45
  finite, 170
  infinite, 170
  mean, 47, 138
  standard deviation, 85–86, 139–140
Positive correlation, 311
Power function, 220, 419
Prediction, limits of, 299–303
Price relative, 352
Primary data, 6
Probability:
  and area under curve, 143–145
  and relative frequency, 110
  conditional, 117
  density function, 144
  distribution, 129
  general rule of addition, 118
  graph paper, 160
  meaning of, 110–113
  special rule for equiprobable events, 120
  special rule of addition, 115
  special rule of multiplication, 116
    extension, 133
Probability sample, 270
Probable error, 208
Producer's risk, 221, 490
Proportions:
  confidence intervals for, 202, 203
  differences between, 229, 245–249
  mean of sampling distribution, 201
  point estimation, 204

Proportions (cont.):
  probable error, 208
  standard error, 201
Proportional stratified sampling, 280

Q

Qualitative distribution, 19
Quantitative distribution, 19
Quantity index, 341
Quantity relative, 341
Quartile deviation, 95
Quartile variation, coefficient of, 97
Quartiles, 72
Quota sampling, 281

R

r (see Correlation coefficient)
Random numbers, 170–172
  table, 171
Random sample, 169–170, 270–272
  simple, 279
Randomness, tests of, 272–278, 284
Range, 80–81, 479–480
  interquartile, 95
Rank correlation, 327–329
  coefficient of, 327
  significance test, 329
  ties in rank, 328
Rate of growth, 419
Ratio paper, 395
Ratio-to-moving average method, seasonal index, 440–446
Ratio-to-trend method, seasonal index, 436–440
Rational subgroup, 479
Raw data, 20
Real wages and money wages, 374–375
Regression:
  coefficients, 292, 298
    calculation for equally spaced data, 405–407
  curve, 298
  line of y on x, 298
  linear, 298
  multiple linear, 304, 330
Relative frequency, limit of, 110
Relative importance weights, Consumer Price Index, 370
Relative variation, 95–97

Residual method, time series analysis, 453
Root-mean-square deviation, 85
Runs:
  above and below the median, 277–278
  and cyclical patterns, 278
  and differences between means, 277
  and trends, 278
  definition, 273
  distribution of total number, 274

S

Sample:
  and population, 43–44
  area, 281
  bias, 8, 271
  cluster, 281
  design, 279
  judgment, 270
  probability, 270
  proportional stratified, 280
  quota, 281
  random, 169–170, 270–272
    simple, 279
  size, 170
    and estimation of means, 193–194
    and estimation of proportions, 204
  stratified, 279–280
  systematic, 282
Sampling:
  acceptance, 489–490
  bias in, 271
  destructive, 173
  double, 283
  exhaustive, 172
  from small populations, 209
  multiple, 284
  random, 169–170
    simple, 279
  sequential, 284
  single, 283
Sampling distribution:
  correlation coefficient, 318
  differences between means, 238, 240
  differences between proportions, 229
  experimental, 175
  means, 176–180
  medians, 180–182
  theoretical, 176, 177, 182
Screening, 489

Seasonal Index:
  definition, 431
  method of simple averages, 432–435
  ratio-to-moving average method, 440–446
  ratio-to-trend method, 436–440
  use in deseasonalizing, 447–449
  use in forecasting, 449–451
Seasonal variation (see also Seasonal index), 387–388, 429–431
  specific and typical, 431
Secondary data, 6
Secular trend, 384–386
Semi-averages, method of, 402–404
Semi-log paper, 395–396
Semi-logarithmic linear trend, 416
Sequential sampling, 284
Sheppard's correction, 93, 99
Siegel, I. H., 379
Sigma, σ, population standard deviation, 85
Sigma, Σ, summation sign, 13
Significance test (see also Tests of hypotheses), 223
Significant digits, 494
Simple aggregative index, 351
Simple random sample, 279
Single sampling, 283
Skewness, 104–106
  alpha-three, 106
  Pearsonian coefficient, 104–105
  positive and negative, 105
Smoothing of time series, 423–425
Specific seasonal pattern, 431
Spot Market Prices, Index of, 345, 355
Spurious correlation, 317, 336
Standard deviation, 84–95
  and chance variation, 93
  binomial distribution, 140
  confidence interval, 208
  grouped data, 91–94
  population, 85–86
  probable error, 209
  sample, 85
  short-cut formulas, 88, 91, 92
  standard error, 207
  theoretical distributions, 139–140, 145
Standard error, 207–209
  correlation coefficient, 318
  difference between means, 238
  difference between proportions, 229–230

Standard error (*cont.*):
  mean, 177
  median, 182
  number of runs, 274
  proportion, 201
  standard deviation, 207
Standard error of estimate, 300
Standard form, normal curve, 148
Standard units, 94, 148
Statistic, 3, 47
Statistical map, 471
Statistics, descriptive and inductive,
    3–5
Stratified sampling, 279–280
  optimum allocation, 280
  proportional, 280
Student-*t* distribution, 196, 236, 240,
    302, 502
Subscripts, 13
Summation, 13–15
Systematic sample, 282

T

t-distribution, 196, 236, 240, 302, 502
Tests of hypotheses:
  correlation coefficient, 318
  correlation of qualitative data, 334
  differences between means, 239, 241
  differences between proportions, 231,
    249
  general procedure, 224–225
  goodness of fit, 257–260
  means, 234, 235, 236, 237, 264–265
  proportions, 226, 228, 249, 255
  randomness, 274
  rank correlation, 329
  r by *k* table, 255
Tests of significance (*see also* Tests of
    hypotheses), 223
Theoretical distribution (*see also* Bi-
    nomial, *F*, Normal, Poisson, *t*,
    and Chi-square distributions),
    127–129
Thompson, C. M., 504, 505
Ties in rank, 328
Time reversal test, 376
Time series (*see also* Time series analy-
    sis), 383
  components, 384
  smoothing, 423–425

Time series analysis:
  cyclical variation, 388–389, 453–457
  exponential trends, 416–419
  forecasting from trends, 426–427
  forecasting from seasonal variations,
    449–457
  graphical presentations, 393–398,
    461–470
  linear trends, 402–411
  multiplicative theory, 384, 453–454
  parabolic trend, 412–416
  preliminary adjustments, 391–393
  residual method, 453
  seasonal variation, 387–388, 429–453
  secular trend, 384–386, 400–422
Trend:
  exponential, 416
  linear, least squares, 404, 408
    modifications of equations, 408–411
  linear, method of semi-averages, 402–
    404
  parabolic, 412–416
  secular, 384–386
  semi-logarithmic linear, 416
  significance test, 390
Trend increment:
  annual, 403, 409
  monthly, 410
Two-sided alternative, 227
Two-tail test, 227
Type I and Type II errors, 218
Typical seasonal pattern, 431

U

u, number of runs, 274
Units test, 351
Universe, 44
Upper control limit, 476

V

Variable, continuous and discrete, 142
Variance, 87
Variation:
  chance, 78–80
  chance and assignable, 475–476
  cyclical, 388–389
  irregular, 389
  measures of, 77–99
  relative, 95–97

Variation (*cont.*):
  seasonal, 387–388
Value index, 357, 377

W

Weighted aggregative index, 357
Weighted mean, 69–70
Weighted mean of price relatives, 362–363

Weighted moving average, 455–456
Wholesale Food Prices, Index of, 344, 352
Wholesale Price Index, 344, 355, 360

Z

z-scale, 148
z-transformation, 337